Sustainable Agriculture and the Environment in the HUMID TROPICS

Committee on Sustainable Agriculture and the
Environment in the Humid Tropics

Board on Agriculture
and
Board on Science and Technology
for International Development

National Research Council

NATIONAL ACADEMY PRESS
Washington, D.C. 1993

NATIONAL ACADEMY PRESS • 2101 Constitution Avenue • Washington, DC 20418

NOTICE: The project that is the subject of this report was approved by the Governing Board of the National Research Council, whose members are drawn from the councils of the National Academy of Sciences, the National Academy of Engineering, and the Institute of Medicine. The members of the committee responsible for the report were chosen for their special competences and with regard for appropriate balance.

This report has been reviewed by a group other than the authors according to procedures approved by a Report Review Committee consisting of members of the National Academy of Sciences, the National Academy of Engineering, and the Institute of Medicine.

This report has been prepared with funds provided by the Office of Agriculture, Bureau for Research and Development, U.S. Agency for International Development, under Amendment No. 2 of Cooperative Agreement No. DPE-5545-A-00-8068-02. Partial funding was also provided by the Office of Policy Analysis of the U.S. Environmental Protection Agency through this cooperative agreement. The U.S. Agency for International Development reserves a royalty-free and nonexclusive and irrevocable right to reproduce, publish, or otherwise use and to authorize to use the work for government purposes.

Cover illustration by Michael David Brown © 1987.

Library of Congress Cataloging-in-Publication Data

National Research Council (U.S.). Committee on Sustainable
  Agriculture and the Environment in the Humid Tropics.
   Sustainable agriculture and the environment in the humid tropics /
  Committee on Sustainable Agriculture and the Environment in the
  Humid Tropics, Board on Agriculture and Board on Science and
  Technology for International Development, National Research Council.
     p.   cm.
  Includes bibliographical references and index.
  ISBN 0-309-04749-8
  1. Agricultural systems—Tropics. 2. Sustainable agriculture—
Tropics. 3. Land use, Rural—Tropics. 4. Agricultural ecology—
Tropics. I. Title.
S481.N38   1992   92-36869
333.76'15'0913—dc20   CIP

Any opinions, findings, conclusions, or recommendations expressed in this publication are those of the author(s) and do not necessarily reflect the view of the organizations or agencies that provided support for this project.

Printed in the United States of America

v

The National Academy of Sciences is a private, nonprofit, self-perpetuating society of distinguished scholars engaged in scientific and engineering research, dedicated to the furtherance of science and technology and to their use for the general welfare. Upon the authority of the charter granted to it by the Congress in 1863, the Academy has a mandate that requires it to advise the federal government on scientific and technical matters. Dr. Frank Press is president of the National Academy of Sciences.

The National Academy of Engineering was established in 1964, under the charter of the National Academy of Sciences, as a parallel organization of outstanding engineers. It is autonomous in its administration and in the selection of its members, sharing with the National Academy of Sciences the responsibility for advising the federal government. The National Academy of Engineering also sponsors engineering programs aimed at meeting national needs, encourages education and research, and recognizes the superior achievements of engineers. Dr. Robert M. White is president of the National Academy of Engineering.

The Institute of Medicine was established in 1970 by the National Academy of Sciences to secure the services of eminent members of appropriate professions in the examination of policy matters pertaining to the health of the public. The Institute acts under the responsibility given to the National Academy of Sciences by its congressional charter to be an adviser to the federal government and, upon its own initiative, to identify issues of medical care, research, and education. Dr. Kenneth I. Shine is president of the Institute of Medicine.

The National Research Council was organized by the National Academy of Sciences in 1916 to associate the broad community of science and technology with the Academy's purposes of furthering knowledge and advising the federal government. Functioning in accordance with general policies determined by the Academy, the Council has become the principal operating agency of both the National Academy of Sciences and the National Academy of Engineering in providing services to the government, the public, and the scientific and engineering communities. The Council is administered jointly by both Academies and the Institute of Medicine. Dr. Frank Press and Dr. Robert M. White are chairman and vice-chairman, respectively, of the National Research Council.

Preface

The increasingly adverse effects of human activities on the earth's land, water, atmospheric, and biotic resources have clearly demonstrated that a new attitude of stewardship and sustainable management is required if our global resources are to be conserved and remain productive. Nowhere is this need more urgent than in the world's humid tropics. Its populations, many subsisting at or below the poverty level, will continue to rely on the resource base to meet their needs. That base must be stabilized while becoming increasingly productive. Thoughtful and prompt actions, especially positive policy changes, are required to break the current pattern of unplanned deforestation in the humid tropics, to reverse environmental degradation caused by improper or mismanaged crop and animal production systems, and to revitalize abandoned lands.

At the request of the U.S. Agency for International Development (USAID), the National Research Council's Board on Agriculture and the Board on Science and Technology for International Development convened the 15-member Committee on Sustainable Agriculture and the Environment in the Humid Tropics. The U.S. Environmental Protection Agency also provided support, emphasizing its interest in the global environmental implications of the problem.

The study responds to the recognized need for sustainable land use systems that (1) maintain the long-term biological and ecological integrity of natural resources, (2) provide economic returns at the farm level, (3) contribute to quality of life of rural populations, and

(4) integrate into national economic development strategies. In particular, the committee was asked to identify and analyze key problems of agricultural practices that contribute to environmental degradation and result in declining agricultural production in humid tropic environments.

The committee began its work in March 1990. It sought to understand the overarching environmental, social, and policy contexts of land conversion and deforestation—and the promise of sustainable land uses—by integrating the views of experts in the broad areas of agriculture, ecology, and social sciences. Its work focused on the range of land use systems appropriate to the forest boundary, an area where agriculture and forestry merge in a continuum of production types involving trees, agricultural crops, and animals. The committee addressed intensive, high-input agriculture only as it relates to common environmental problems. The committee undertook supplemental analyses of tropical forest land use policies and the effects of tropical land use on global climate change. We sought a wide range of scientific data, specialized information, and expert views to address our broad charge.

A critical component of the humid tropics equation that was not within the scope of the study is human population. The committee acknowledges population dynamics as a major factor in achieving sustainable land use and development in the humid tropics; the land use systems it describes fit a broad range of population densities. We stress the importance of population issues, particularly in this region of the world, but an analysis of population densities, pressures, and trends was not part of our study, nor does the composition of the committee reflect the demographic expertise necessary to address population issues.

This report, *Sustainable Agriculture and the Environment in the Humid Tropics,* will contribute to the elusive "solution" to tropical deforestation through its outline of a variety of approaches to tropical land use and conservation. Each land use option would take advantage of the opportunities inherent in physical resource patterns, labor, market availability, and social setting, and each would contribute to the common goal of sustainability in the humid tropics.

The land use options scheme in Chapter 2 and its accompanying table for evaluating land use attributes can be used as a guide in decision making. The presentation makes the information usable by in-country decision makers, from the local level on up, as well as by governmental and nongovernmental agencies. We believe the information in this report will be helpful to researchers, planners, and policymakers in industrialized countries and in developing countries.

Part One is the committee's deliberative report. It emphasizes the restoration of degraded land, the importance of general economic growth as an alternative to forest exploitation, and the need for comprehensive management of forest and agricultural resources. The underlying premise of the committee's work is that under conditions of economic and social pressure, what is not managed today is at risk of being lost tomorrow.

Within Part One, the Executive Summary discusses the findings of the committee and presents key recommendations. Chapter 1 describes the humid tropics, the consequences of forest conversion and deforestation, environmental factors affecting agriculture, and the fostering of sustainable land use in the humid tropics. Chapter 2 discusses major land use options that local, regional, and national managers might choose in making decisions to achieve food production goals, maintain or increase local income levels, and protect the natural resource base. Chapter 3 discusses technical research needs and presents recommendations on land use options. Chapter 4 presents policy imperatives to promote sustainability. The Appendix to Part One presents a discussion of emissions of greenhouse gases associated with land use change.

To enhance its understanding, the committee commissioned a series of country profiles to gather information on land use and forest conversion in different countries, to evaluate general causes and consequences within specific contexts, to identify sustainable land use alternatives, and to compare policy implications. Seven country profiles are presented in Part Two. Authors review agricultural practices and environmental issues in Brazil, Côte d'Ivoire, Indonesia, Malaysia, Mexico, the Philippines, and Zaire.

The committee's intent in this report is to make a positive statement about the potential benefits of sustainable agriculture in the humid tropics, rather than to condemn the forces that have contributed to the current situation. It is an attempt to promote the restoration and rehabilitation of already deforested lands, to increase their productivity, and to explore the policy changes required to take the next steps toward sustainability. Guidelines for future research and policy, whether for conserving natural ecosystems or for encouraging sustainable agroecosystems, must be designed with a global perspective and within the context of each country's environment, history, and culture.

The committee underscores the fact that sustainable agriculture in any given country will consist of many diverse production systems, each fitting specific environmental, social, and market niches. Some alternatives require higher inputs, labor, or capital—depending on their makeup, resource base, and environment—but each must become more sustainable. Conversely, each system can contribute toward the sustainability of the agricultural system in general by helping to meet the varied and changing needs facing countries in the humid tropics. To maintain a diversity of approaches while making real progress toward common goals is the challenge that confronts all who are concerned with the future of the lands and people of the humid tropics.

RICHARD R. HARWOOD, *Chair*
Committee on Sustainable
Agriculture and the Environment
in the Humid Tropics

Acknowledgments

The disciplines, multidisciplinary experiences, expertise, and countries of the world that are represented by the many individuals who have generously contributed to this report constitute a very long list. Because of the efforts of the many who shared ideas and offered background knowledge, the committee was able to expand its views of issues relating to sustainable agriculture and the environment in the humid tropics and benefit from a variety of perspectives.

Among the many individuals whose work was of special significance to this report are the authors of the appended paper, the country profiles, and their collaborators. The descriptive data and analyses presented in the seven country profiles, contained in Part Two of the report, provided much of the foundation for the committee's work. In addition to the authors and their collaborators, the committee acknowledges the contributions of Cyril B. Brown, Purdue University; Avtar Kaul, Winrock International; Daniel Nepstad, Woods Hole Research Center; and Christopher Uhl, Pennsylvania State University. (Both Nepstad and Uhl are associated with the Center for Agroforestry Research of the Eastern Amazon, Belém, Brazil.) Michael Hayes provided valuable editorial assistance in preparing the country profiles for publication.

To broaden its information resources, the committee convened two regional meetings on agricultural and environmental practices and policies in the humid tropics. The first meeting was held at the

Faculty of Agronomy, University of Costa Rica, in San Jose. The second was held in Bangkok, Thailand, under the auspices of the Asian Regional Office of the National Research Council.

During the course of its deliberations, the committee sought the counsel and advice of independent scholars and individuals representing a range of organizations. Among those who gave generously of their experience were Robert O. Blake, Committee on Agricultural Sustainability for Developing Countries; Erick Fernandes, Thurman Grove, and Cheryl Palm, North Carolina State University; Douglas Lathwell, Cornell University; Charles H. Murray, Food and Agriculture Organization of the United Nations; Stephen L. Rawlins, U.S. Department of Agriculture; R. D. H. Rowe, World Bank; Roger A. Sedjo, Resources for the Future; and John S. Spears, Consultative Group on International Agricultural Research. The assistance of Andrea Kaus and Veronique M. Rorive, University of California at Riverside, was also helpful to the committee.

Research assistance was provided by three student interns, who were sponsored by the Midwest Universities Consortium for International Activities, Inc. The committee extends special thanks to Joi Brooks, University of Illinois at Urbana, and Jil Reifschneider and Kristine Agard, University of Wisconsin.

The committee is grateful to Curt Meine and Barbara Rice, whose skill and teamwork transformed imperfect and incomplete draft materials into a comprehensive report. We are particularly grateful to Jay Davenport, whose insights and support were invaluable to the committee throughout the course of the study.

And the committee especially recognizes the efforts of Pedro Sanchez, who served as committee chairman until assuming responsibilities as director general of the International Center for Research in Agroforestry, Nairobi, Kenya.

SUSTAINABLE AGRICULTURE AND THE ENVIRONMENT IN THE HUMID TROPICS

Rain forests are rapidly being cleared in the humid tropics to keep pace with food demands, economic needs, and population growth. At the same time, important natural resources are being eroded or lost due to unsound practices. This book provides critically needed direction for developing strategies that both mitigate land degradation, deforestation, and biological resource losses and help the economic status of tropical countries. It includes a practical discussion of 12 major land use options for boosting food production and enhancing local economies while protecting the natural resource base, recommendations for developing technologies needed for sustainable agriculture, and a strategy for changing policies that erode natural resources and biodiversity.
ISBN 0-309-04749-8; 1993, 720 pages, 6 x 9, index, hardbound, $49.95

FOREST TREES

News reports concerning the decline of the world's forests are becoming sadly familiar. As forests disappear, so do their genetic resources. This book assesses the status of the world's tree genetic resources and management efforts, and presents strategies for meeting future needs.
ISBN 0-309-04390-5; 1990, 196 pages, 6 x 9, index, hardbound, $24.95

THE U.S. NATIONAL PLANT GERMPLASM SYSTEM

In the United States, the critical task of preserving our plant genetic resources is the responsibility of the National Plant Germplasm System (NPGS). NPGS undergoes a thorough analysis in this book, which offers wide-ranging recommendations for better meeting U.S. needs and taking the lead in international conservation efforts.
ISBN 0-309-04390-5; 1990, 196 pages, 6 x 9, index, hardbound, $19.95

Use the form on the reverse of this card to order your copies today.

SUSTAINABLE AGRICULTURE AND THE ENVIRONMENT IN THE HUMID TROPICS

Use this card to order additional copies of SUSTAINABLE AGRICULTURE AND THE ENVIRONMENT IN THE HUMID TROPICS and the book described on the reverse. All orders must be prepaid. Please add $4.00 for shipping and handling for the first copy ordered and $0.50 for each additional copy. If you live in CA, MD, MO, TX, VA or Canada, add applicable sales tax or GST. Prices apply only in the United States, Canada, and Mexico and are subject to change without notice.

___ I am enclosing a U.S. check or money order.

___ Please charge my VISA/MasterCard/American Express account.

Number: _____

Expiration date: _____

Signature: _____

Quantity Discounts:
5-24 copies   15%
25-499 copies   25%

To be eligible for a discount, all copies must be shipped and billed to one address.

Return this card with your payment to NATIONAL ACADEMY PRESS, 2101 Constitution Avenue, NW, Box 285, Washington, DC 20055.

PLEASE SEND ME:

Qty.	Code	Title	Price
___	HUMTRO	Sustainable Agriculture and the Environment in the Humid Tropics	$49.95
___	FORESC	Forest Trees	$24.95
___	GERM	The U.S. National Plant Germplasm System	$19.95
		SUSC	

Please print.
Name _____

Address _____

City _____ State _____ Zip Code _____

To order by phone using VISA/MasterCard/American Express, call toll-free 1-800-624-6242 or call 202-334-3313 in the Washington metropolitan area.

Contents

PART TWO: COUNTRY PROFILES

Executive Summary

Agriculture and forestry are major human activities on the global landscape. Increasingly, data show that many widely employed agricultural and forestry practices are having significant adverse effects on local and regional soil conditions, water quality, biological diversity, climatic patterns, and long-term biological and agricultural productivity. These local and regional adverse effects are now being felt on a global scale, and have become matters of international concern. These issues are especially acute in the world's humid tropic regions.

Timing is critical. Land transformation in northern Europe, for example, from a natural state to its present-day highly intensive agriculture and land use, occurred over thousands of years. Changes in the humid tropics are occurring at a more rapid rate. Shifts in economics and population, internal and external to the region, have ultimately yielded radical changes to the landscape, with mixed results. Widespread, inappropriate use of fragile landscapes is also causing significant reduction in production potential. Within one generation, in some cases, areas will be degraded beyond economically feasible restoration.

Agricultural production practices in tropical regions are frequently unsustainable because the capacity of land to support crop production is rapidly exhausted. This fundamental problem is exacerbated by the pressures arising from poverty and the demand for food. Principal factors undermining crop production capacity include soil erosion,

1

loss of soil nutrients, water management problems, and pest outbreaks, as well as socioeconomic environments that frequently limit the use of alternative solutions for more sustainable agricultural development. Faced with declining yields, farmers in many areas of the humid tropics typically seek new forestlands to clear for crop production. Unsustainable logging practices and the conversion of environmentally fragile lands to crop production and cattle ranching pose difficulties in achieving long-term economic development and food production goals, and often contribute to environmental degradation.

This report focuses on the world's humid tropics. It examines the potential of improved agricultural and land use systems to provide lasting benefits for these regions and to alleviate adverse environmental effects at local and global levels. In assessing agricultural sustainability, development, and resource management in the humid tropics, the committee recognized the need for sustainable land use systems that

- Maintain the long-term biological and ecological integrity of natural resources,
- Provide economic returns to individual farmers and farm-related industries,
- Contribute to the quality of life of rural populations, and
- Strengthen the economic development strategies of countries in the humid tropics.

The committee also identified constraints to adopting sustainable land use systems.

A key factor in attaining improved resource management, which can lead to agricultural sustainability and development, is population. Population issues—and the accompanying and overwhelming incidence of poverty—are critical in many regions of the world, and certainly in the humid tropics. However, it was not within the scope of this study to specifically analyze or draw conclusions about data on population densities, pressure, or trends. In this report, the committee does, however, evaluate land use options not only from a biophysical basis, but also from social and economic bases.

FINDINGS

The committee's assessment confirms that land degradation and deforestation are severe in many areas. But, more important, the committee has found that farmers are employing a wide range of alternative strategies, albeit in limited areas, for confronting land use problems and for moving toward sustainability. In spite of obstacles, innovative farmers, foresters, researchers, and land managers continue to develop and refine land use practices, many of which, if broadly implemented, will ultimately benefit agricultural production, the economy, and the environment. With appropriate changes in policies, research, and information and extension networks, the committee believes the rate of progress in developing and adopting sustainable land use systems could be accelerated.

Based on its study, the committee arrived at three major findings.

1. *Throughout the humid tropics, degraded lands can be found that have the potential to be restored.* The country profiles included in this report cite examples of successful restoration, although in many cases, a scientific understanding and documentation of the process is incomplete. The committee notes, however, that as researchers move into complex, interrelated issues involving land use in the humid tropics, some standard scientific practices such as replications, retesting over large areas, and statistical analysis will be difficult if not impossible. Experience and observation over time, however, will validate the restoration methods that lead to the more sustainable land uses. The application of restoration methods can be accelerated along with the scientific analysis of their effectiveness.

2. *A continuum of land use systems exists ranging from those that entail minimal disturbance of natural resources to those that involve substantial clearing of forests.* Many of the successful systems involve integrative approaches to farming and forestry that are characterized by a high level of environmental stability, increased productivity, and social and economic improvements, while only modestly reducing biodiversity. A wide variety of sustainable land use methods are available and can be adapted to the specific needs, limitations, resource bases, and economic conditions of different land sites. Farmers, foresters, and land managers will need to receive information and technical assistance in developing new management skills to select and employ sustainable land use systems.

3. *Some locales of the humid tropics are successfully shifting from economic growth that is based largely on forest harvest to a more diversified economy involving substantial nonfarm employment.* Economic gains from further harvest of forestlands are increasingly marginal. Development of new markets for the products of the local farmer is often essential if necessary incentives for diversification are to exist. Market development can be an effective means of encouraging sustainable, diversified land use. Successful diversification can offer increased

employment as well as stimulate both investment in transportation, storage, and processing and expansion of marketing and trade opportunities. If diversification is to be attained, however, a management systems approach is required for the research necessary to fuel and continue development. The result can be general economic growth that is less dependent on forest conversion.

The three findings—the potential to restore degraded lands, the range of appropriate land uses, and the capacity for general economic growth—have brought the committee to conclude that more effective management of forests and other lands will be required to resolve natural resource and economic issues in the humid tropics.

LANDSCAPE MANAGEMENT: A GLOBAL REQUIREMENT

Superficially, the underlying cause for the transformation and degradation of the landscape in the humid tropics may appear to be excessive forest conversion, but in reality there are many underlying causes that are interrelated and cumulative in their effects. The committee strongly believes, however, that optimal and balanced management of the entire landscape is integral to resolving problems related to forest conversion, agricultural production, and land use options in all countries of the humid tropics and in all their unique local situations.

The committee envisions that a comprehensive development scheme could

• Provide an enabling environment for institution building, credit and financing, and improved marketing of products;
• Increase incentives and opportunities for sustainable agricultural practices; and
• Strengthen research, development, and dissemination.

This report is based on the committee's conclusion that it will be necessary, within the next generation, to achieve effective management of all land resources for sustained use. These land resources include the pristine forest, which should be protected in perpetuity, to lands transformed into plantations or small landholdings. Management will include decision-making at every step: by the farmer or landholder, by the village or community, and by regional and national agencies. Failure to implement sustainable resource management systems will mean the loss of much of the remaining tropical forests and wetlands, the endemic plant and animal species, and the values they represent.

Agricultural lands and forested lands are often viewed as managed ecosystems. But now, with the increasing rate of change in human activity across the face of the land, the earth itself must be viewed as a managed ecosystem.

Timing is critical. What is not managed is at risk of being lost.

THE HUMID TROPICS

Technically, the humid tropics is a bioclimatic region of the world characterized by consistently high temperatures, abundant precipitation, and high relative humidity. Gradients of temperature, rainfall, soils, and slope of the land contribute to variations in vegetation. Tropical lowland vegetation constitutes about 80 percent of the vegetation in the humid tropics. Although a variety of distinct plant associations and forest formations exist in the region, the forests of the humid tropics are often referred to as tropical rain forests. Collectively, however, lowland, premontane, and montane forest formations that include moist, wet, and rain forests can be generally referred to as humid tropic or tropical moist forests.

Humid tropic conditions are found over nearly 50 percent of the tropical land mass and 20 percent of the earth's total land surface—an area of about 3 billion ha. This total is distributed among three principal regions. Tropical Central and South America contain about 45 percent of the world's humid tropics, Africa about 30 percent, and Asia about 25 percent. As many as 62 countries are located partly or entirely within the humid tropics.

Forest Conversion

Forest conversion is defined as the alteration of forest cover and forest conditions through human intervention. Deforestation is a conversion extreme that reduces crown cover to less than 10 percent. Available data suggest that the annual rate of deforestation in the (primarily humid) tropics increased from 9.2 million ha per year in the late 1970s to an average of 16.8 million ha per year in the 1980s. Deforestation currently affects about 1.2 percent of the total tropical forest area annually. Forest degradation—changes in forest structure and function of sufficient magnitude to have long-term negative effects on the forest's productive potential—also affects a large area.

CAUSES OF FOREST CONVERSION

The leading direct causes of forest loss and degradation include large-scale commercial logging and timber extraction, the advance-

Convoluted rows of oil palms stretch along the border of a tropical rain forest in Malaysia. As a result of farming projects sponsored by the Malaysian government, thousands of hectares of rain forest have been converted to farmlands. The government's drive to reduce landlessness and unemployment began in the 1950s. Credit: James P. Blair © 1983 National Geographic Society.

ment of agricultural frontiers and subsequent use of land by subsistence farmers, conversion of forests to perennial tree plantations and other cash crops, conversion to commercial livestock production, land speculation, the cutting and gathering of wood for fuel and charcoal, and large-scale colonization and resettlement projects. The demand for land by shifting cultivators, small-scale farmers, and landless migrants accounts for a significant portion of forest conversion in some regions. Most of the farmers in the humid tropics, however, are acting in response to a socioeconomic environment that offers few alternatives.

CONSEQUENCES OF FOREST CONVERSION

Forest conversion, especially deforestation, can have far-ranging environmental, economic, and social effects. Environmental consequences can include the disruption of natural hydrological processes, soil erosion and degradation, nutrient depletion, loss of biological diversity, increased susceptibility to fires, and changes in local distribution and amount of rainfall.

The social consequences of unsustainable conversion practices may include the decline of indigenous cultural groups and the loss of knowledge of local resources and resource management practices; dislocation of small communities of farmers and forest dwellers as forestlands are appropriated for more profitable land uses; and continued poverty and rural migration as farmers abandon lands degraded through soil-depleting agricultural practices. The economic consequences include the loss of production potential as soil is degraded; the loss of biological resources, such as foods or pharmaceuticals, from primary forests; the destabilization of watersheds, with the attendant downstream effects of flooding and siltation; and, at the global level, the long-term impacts of deforestation on global climate change.

Agriculture in the Humid Tropics

The efficiency of tropical agriculture is determined by a combination of environmental factors (including climate, soil, and biological conditions) and social, cultural, and economic factors. Agricultural systems and techniques that have evolved over time to meet the special environmental conditions of the humid tropics include the paddy rice systems of Southeast Asia; terrace, mound, and drained field systems; raised bed systems, such as the *chinampas* of Mexico and Central America; and a variety of agroforestry, shifting cultivation, home garden, and natural forest systems. Although diverse in their adaptations, these systems often share many traits, such as high retention of essential nutrients, maintenance of vegetative cover, high diversity of crops and crop varieties, complex spatial and temporal cropping patterns, and the integration of domestic and wild animals into the system.

Shifting cultivation is a common agricultural approach in the tropics. Traditionally, it incorporates practices that maintain or conserve the natural resource base, including a natural restoration or fallow cycle. Today, however, the hallmarks of unstable shifting cultivation, or slash-and-burn agriculture, are shortened fallow periods that lead to fertility decline, weed infestation, disruption of forest regeneration, and excessive soil erosion.

Monocultural systems have been successfully introduced over large areas of the humid tropics, and include production of coffee, tea, bananas, citrus fruits, palm oil, rubber, sugarcane, and other commodities produced primarily for export. Plantations and other monocultural systems provide employment and earn foreign exchange.

Adopting an Integrated Approach to Land Use

The committee has focused its analysis on the relationship between forest conversion and agriculture, and on how the problems of both might be better addressed through developing and implementing more sustainable land use systems. Improved land use in the humid tropics requires an approach that recognizes the characteristic cultural and biological diversity of these lands, incorporates ecological processes, and involves local communities at all stages of the development process.

Fundamental scientific, social, and economic questions—and certainly the more applied problems—are multifaceted. Steady progress toward sustainability and the resolution of problems in the humid tropics requires that several scientific disciplines be integrated and managed to ensure collaboration and synergy.

SUSTAINABLE LAND USE OPTIONS

No single type of land use can simultaneously meet all the requirements for sustainability or fit the diverse socioeconomic and ecological conditions. In this report, the committee describes 12 overlapping categories within the complete range of sustainable land use options. The committee also presents a scheme, for comparing the attributes of each of the 12 categories (see Chapter 3), that can be used as a tool for management and decision making in evaluating land use options for a specific area. The attributes are grouped as biophysical, economic, and social benefits. With proper management, these land use options have the potential to stabilize forest buffer zone areas, reclaim cleared lands, restore degraded and abandoned lands, improve small farm productivity, and provide rural employment. They are described below:

• *Intensive cropping systems* are concentrated on lands with adequate water, naturally fertile soils, low to modest slope, and other environmental characteristics conducive to high agricultural productivity. The best agricultural lands in most parts of the humid tropics have been cleared and converted to high-productivity agriculture. High-productivity technologies, if improperly applied, can lead to resource degradation through, for example, nutrient loading from fertilizers, water contamination from pesticides and herbicides, and waterlogging and salinization of land. Food needs require that these systems remain productive and possibly expand in area, but that they be stabilized through biological pest management, nutrient containment, and improved water management.

- *Shifting cultivation systems* are traditional and remain in widespread use throughout the humid tropics. Temporary forest clearings are planted for a few years with annual or short-term perennial crops, and then allowed to remain fallow for a longer period than they were cropped. Migration has brought intensified shifting cultivation to newly cleared lands, where it is often inappropriate. In these areas, however, shifting cultivation can be stabilized by adopting local cropping practices and varieties, observing sufficient fallow periods, maintaining continuous ground cover, diversifying cropping systems, and introducing fertility-restoring plants and mulches into natural fallows.

- *Agropastoral systems* combine crop and animal production, allowing for enhanced agroecosystem productivity and stability through efficient nutrient management, integrated management of soil and water resources, and a wider variety of both crop and livestock products. Agropastoral systems may provide relatively high levels of income and employment in resource-poor areas.

- *Cattle ranching* on a large scale has been identified as a leading contributor to deforestation and environmental degradation in the humid tropics, primarily in Latin America and some Asian countries. However, cattle ranching operations can be made more sustainable by reclaiming degraded pastures in deforested lands through the use of improved forages, fertilization, weed control, and appropriate mechanization, and by integrating pasture-based production systems with agroforestry and annual crop systems. Medium- to small-scale ranching systems have proved economical, but require changes in land tenure and ownership incentives.

- *Agroforestry systems* include a range of options in which woody and herbaceous perennials are grown on land that also supports agricultural crops, animals, or both. Under ideal conditions, these systems offer multiple agronomic, environmental, and socioeconomic benefits for resource-poor small-scale farmers, including enhanced nutrient cycling, fixing of atmospheric nitrogen through the use of perennial legumes, efficient allocation of water and light, conservation of soils, natural suppression of weeds, and diversification of farm products. Agroforestry systems require market access for widespread use.

• *Mixed tree systems* are common throughout the humid tropics. In contrast to modern plantations, in which one tree species is grown to yield a single commercial product, mixed tree systems employ a variety of useful species, planted together, to yield different products (including fruits, forage, fiber, and medicines). These systems also protect soil and water resources, provide pest control, serve as habitat for game and other animal species, and offer opportunities for small-scale reforestation efforts that are economically productive and environmentally sound.

• *Perennial tree crop plantations* are part of a broad category of plantation agriculture that includes short rotation crops (such as sugarcane and pineapple) as well as tree crops. Large areas of primary forest have been converted to tree crop plantations. Despite social and environmental problems inherent in these systems, modifications to enhance their sustainability could allow plantation crops to play a role in converting deforested or degraded land to more ecologically and economically sustainable use.

• *Plantation forestry systems* in the tropics cover about 11 million ha of land. Most have been established only in the past 30 years, usually in deforested or degraded lands, primarily for fuelwood, pulpwood, and lumber production, and for environmental protection. Increasingly, however, attention is focusing on the ability of plantations to accumulate biomass, sequester atmospheric carbon, and rehabilitate damaged lands. Because these systems offer flexibility in design and purpose, they provide a potentially important tool for land managers in the humid tropics.

• *Regenerating and secondary forests* have followed forest conversion and land abandonment in many areas of the humid tropics. Regenerating forests can be viewed as a type of land use in that they provide valuable goods and services to society, while preparing degraded lands for conversion to more intensive agricultural uses or alternative purposes. The regeneration process protects soils from erosion, restores the capacity of the land to retain rainfall, sequesters atmospheric carbon, and allows biological diversity to increase. This process can be guided and accelerated through fire protection, supplemental planting, and other management methods. Regenerating forests will, if other options are not implemented, mature into secondary forests, providing many ecological and economic benefits and preparing the way for the restoration of primary forest. Properly managed secondary forests, by supplying a variety of products, increasing site fertility, and restoring biological diversity, can be critical for attaining the goals of sustainability.

- *Natural forest management systems* show promise for ameliorating the effects of destructive logging practices. The ecological characteristics, biological diversity, and structural complexity of moist tropical forest ecosystems make them more vulnerable than temperate forests to the impacts of conventional intensive forest management techniques. Management techniques (for example, selective cutting procedures) that are more appropriate to tropical systems may provide sustainable alternatives to destructive logging and other more intensive land uses.

- *Modified forests* are often difficult, if not impossible, to distinguish from pristine primary forests. In these areas, indigenous people have subtly altered the native plant and animal community, but without significantly affecting the rate of primary productivity, the efficiency of nutrient cycling, or other ecosystem functions. Modified forests should be considered a viable land use that allows indigenous peoples and local communities to sustain their ways of life while protecting large areas of forestland.

- *Forest reserves* have been established through a variety of protection mechanisms, including biological and extractive reserves, wildlife preserves, national parks, national forests, refuges, private land trusts, crown lands, and sanctuaries. Reserves allow for the protection of ecosystem functions, environmental services, cultural values, and biological diversity, and provide important opportunities for research, education, recreation, and tourism.

The continuum of options from intensive cropping systems to forest reserves constitutes a spectrum of potential land uses. They meet different goals and involve varying degrees of forest conversion, management skill, and investment. Each confers a mix of biophysical, economic, and social benefits. Consequently, trade-offs are involved in choosing among them. Agroforestry systems, for example, require fewer purchased inputs (although initial soil fertility treatments may be required on degraded lands), but they generally do not generate the high levels of employment or income on a per unit area basis that intensive crop or animal agriculture does. They are, however, adapted to less fertile soils. Perennial tree plantations, such as for oil palm or rubber, require considerable chemical inputs and labor to maintain productivity, but generate more employment and income on a per unit area basis than do agroforestry systems. Sustainability, in this context, largely entails meeting unique needs, minimizing negative effects, and offering a range of opportunities for land areas that vary in size from the local farmer's field to the surrounding landscape to the country as a whole.

In the Amazon River Basin in Brazil, tropical rain forest is burned to prepare the land for cattle pastures and other agricultural uses. Credit: James P. Blair © 1983 National Geographic Society.

RECOMMENDATIONS

Progress toward sustainability in the humid tropics depends not only on the availability of improved techniques of land use, but on the creation of a more favorable environment for their development, dissemination, and implementation. For this to happen, substantial changes will need to take place in the national and international institutions that determine the character of public policy. The committee's recommendations fall into the categories of technical research needs and policy strategies.

Technical Research Needs

The committee has found that publicly supported development efforts are confined to a range of land use choices that is too narrow. In this report, the committee identifies sustainable land use options suitable for a broad range of conditions in the humid tropics. That so many instances of diverse production systems were found is not surprising; that they appear to have such broad applicability across the humid tropics is of great development interest.

Recommendations on technical research needs are based on the success of land uses that are chronicled in the country profiles (see Part Two, this volume) and on the potential that exists in many locales throughout the humid tropics.

DOCUMENTATION OF LAND USE SYSTEMS

To be readily usable by development planners, land use systems should be defined according to their environmental, social, and economic attributes, and described in detail. The place and role for each system, which will depend on the level of national or local development, should be identified along with conditions required for their implementation and evolution.

In Chapter 3, the committee provides a scheme for comparing the biophysical, social, and economic attributes of land use systems. Biophysical attributes are grouped as nutrient cycling capacity, soil and water conservation capacity, stability toward pests and diseases, biodiversity level, and carbon storage. Social attributes are grouped as health and nutritional benefits, cultural and communal viability, and political acceptability. Economic attributes are grouped as level of external inputs necessary to maintain optimal production, employment per land unit, and income generated.

In all attribute categories, intensive cropping, agroforestry, agropastoral systems, mixed tree plantations, and, to some extent, modified forests offer significant benefits. For many low resource areas, the newly researched and demonstrated technologies for mixed cropping systems show considerable promise. In general, changes in social and economic attributes will be gradual.

INDIGENOUS KNOWLEDGE

The vast body of indigenous knowledge on land use systems must be recorded and made available for use in national development planning.

Traditional systems and indigenous knowledge will not yield panaceas for land use problems in the humid tropics. However, traditional ways of making a living, refined over many generations by intelligent land users, provide insights into managing tropical forests, soils, waters, crops, animals, and pests. Research can assess the benefits of aspects of traditional systems: their structure, genetic diversity, species composition, and function as agroecosystems, as well as their social and economic characteristics and potential for wider application. The research process can have additional benefits by fostering collaborative relationships between researchers and indigenous people, and providing the groundwork for successful local development projects. Sustainable systems will often combine traditional practices and structure with more modern, scientifically derived technologies.

MONITORING

Resources should be available for linking national monitoring agencies with global satellite-based data sources so these agencies can refine, update, and verify their data bases for tracking land use changes and effects.

Monitoring systems and methodologies must be improved to trace land use changes and their effects. Only within the past 2 decades in the United States has satellite-generated information made it possible to estimate the magnitude of soil loss and its effect on productivity. In most countries of the humid tropics, only rudimentary data are available on soil loss, groundwater contamination, salinization, sedimentation rates, levels of biological diversity, and greenhouse gas emissions. Modern-day international data bases employing satellite-generated information should be more effectively linked with national monitoring systems.

Policy Strategies

The goal of the committee's policy-related recommendations is to meet human needs without further undermining the long-term integrity of tropical soils, waters, plants, and animals. Sustainable agriculture will not automatically slow forest conversion, or deforestation, in the humid tropics. However, the combination of forest management and the use of sustainable land use options will provide a framework that each country can use to fit its capabilities, natural resources, and stage of economic and technological development.

POLICY REVIEWS

Policy reviews under way at local, national, and international levels must be broadened to consider the negative effects that policies have had on sustainable land use.

Many international and bilateral development agencies have reassessed their forest policies in response to escalating rates of deforestation. Few, however, focus on the need for agricultural sustainability. At national and regional levels, policy reviews should respond to the specific biophysical, social, and economic circumstances that affect land use patterns within countries and regions. At the international level, the review process will vary from institution to institution, depending on its size and objectives and the range of its activities.

In general, policy reviews should involve multidisciplinary teams; evaluate externalized costs of policies that encourage large-scale land clearing; assign value to the forests in standard economic terms; integrate forest and agriculture sectors; and integrate infrastructure, land use, and development policies.

GLOBAL EQUITY

The adoption of sustainable agriculture and land use practices in the humid tropics should be encouraged through the equitable distribution of costs on a global scale.

Industrialized countries have a responsibility to assume some proportionate share of the costs related to the adoption of sustainable land use practices. They must use their financial and institutional resources to encourage the conservation of natural resources and the development of human resources in developing countries. Global distribution of costs can be directed through technical assistance, research, and institution building; financing; and international trade reforms. In other words, if industrialized countries want developing countries to preserve their resources for global benefit, financial and other assistance must be transferred to developing countries specifically to protect global common resources. Assistance could be provided for in situ protection of genetic resources, enhancement of the capacity to sequester carbon, and new markets for high-value products of the humid tropics.

Supporting Sustainable Agriculture

Changes in policies that contribute to forest conversion, deforestation, and natural resource degradation in the humid tropics alone will not encourage the adoption of sustainable agricultural systems. The committee makes the following recommendations for efforts to support sustainable agriculture.

CREATION OF AN ENABLING ENVIRONMENT

National governments in the humid tropics should promote policies that provide an enabling environment for developing land use systems that simultaneously address social and economic pressures and environmental concerns.

Based on studies of successful experience in moving toward sustainable agricultural practices, the committee concluded that essential components of an enabling environment include assurance of resource access through land titling or other tenure-related instruments, access to credit, investment in infrastructure, local community empowerment in the decision-making process, and social stability and security.

More than any other factor, the status of land tenure determines the destiny of land and forest resources in the humid tropics. Land tenure arrangements that provide long-term access to land resources are the prerequisite to efficient land use decision making and to the implementation of sustainable land use systems. Formalization of property rights is important in many countries.

INCENTIVES

National governments in the humid tropics and international aid agencies should develop and provide incentives to encourage long-term investment in increasing the production potential of degraded lands, for settling and restoring abandoned lands, and for creating market opportunities for the variety of products available through sustainable land use.

To attain the most efficient use of limited funds, it will be necessary to determine where natural regeneration of degraded lands is proceeding without major investment, and alternatively, where regeneration and economic development will require a financial boost. As regeneration and economic development proceeds, the mix of land use inputs is likely to change and so too will the mix of appropriate incentives. For example, labor-intensive agroforestry systems that might be suitable in low-wage countries may be less financially viable in high-wage countries.

In the case of abandoned lands, securing tenure is a critical step in rehabilitation, but special concessions may be necessary to attract farmers to these areas. Depending on local tenure arrangements, villages and communities, rather than individuals, might more appropriately be the recipient of subsidies, tax concessions, and other incentives where, for example, the stabilization of entire watersheds is critical.

PARTNERSHIPS

New partnerships must be formed among farmers, the private sector, nongovernmental organizations, and public institutions to address the broad needs for research and development and the needs for knowledge transfer of the more complex, integrated land use systems.

The international community has given substantial support for research to increase the productivity of major crops such as rice and maize, and for research on tropical soils, livestock, chemical methods of pest control, and human nutrition. Additional support will be necessary in the areas of small-landholder agroforestry systems, tree crops, improved fallow and pasture management, low input cropping, corridor systems, methods of integrated pest management, and other agricultural systems and technologies appropriate to higher risk lands.

National and international development agencies should foster the productive involvement of local nongovernmental organizations (NGOs) as intermediaries between themselves, national government agencies, universities, and local communities in support of the methods and goals of sustainable land use. In particular, NGOs can assume a prominent role in training and education at the community level, in partnership with (or in the absence of) official extension services. Local NGOs are likely to be more effective than external organizations in shaping environmentally and socially acceptable land use policies based on local needs and priorities.

CONCLUSION

The boundary around what was once pristine, unmanaged forests has blurred. Lands on either side of the so-called boundary can be used and managed in innovative and, eventually, sustainable ways along a continuum of land use choices. The committee has documented some of the most promising options.

The gains sought through the further conversion of forests in the humid tropics are becoming increasingly marginal. When the full environmental, social, and economic costs are considered—even if they cannot be precisely quantified—the nations of the humid tropics stand to gain little from the further depletion of forests and land resources. Likewise, nations beyond the humid tropics will reap few benefits by contributing to the forces behind accelerated forest conversion and deforestation.

Decisions will continue to be made, necessarily in the absence of complete data. But the committee strongly believes that the continuum of land use options presented in this report and the accompanying evaluation of attributes can provide a foundation for decision making and the management of all lands—the key to sustainability in the humid tropics.

PART ONE

Part One

1

Agriculture and the Environment in the Humid Tropics

The wide belt of land and water that lies between the tropics of Cancer and Capricorn is home to half of the world's people and some of its most diverse and productive ecosystems. Citizens and governments within and beyond the tropics are increasingly aware of this region's unique properties, problems, and potential. As scientific understanding of tropical ecosystems has expanded, appreciation of their biological diversity and the vital role they play in the functioning of the earth's biophysical systems has risen. The fate of tropical rain forests, in particular, has come to signify growing scientific and public interest in the impact of human activities on the global environment.

At the same time, the people and nations of the tropics face a difficult future. Most of the world's developing countries are in the tropics, where agriculture is important to rural and national economies. About 60 percent of the people in these countries are rural residents, and a large proportion of these are small-scale farmers and herders with limited incomes (Population Reference Bureau, 1991). The need to stimulate economic growth, reduce poverty, and increase agricultural production to feed a rapidly growing population is placing more pressures on the natural resource base in developing countries (see Part Two, this volume). The deterioration of natural resources, in turn, impedes efforts to improve living conditions. This dilemma, however, has stimulated a growing commitment to sustainable development among tropical and nontropical countries alike, with special concern for the world's humid tropics.

This report focuses on the humid tropics, a biogeographical area within the tropical zone that contains most of its population and biologically rich natural resources. The problems associated with unstable shifting cultivation and tropical monocultures, together with the need to improve productivity on degraded and resource-poor lands, have prompted farmers, researchers, and agricultural development officials to search for more sustainable agricultural and land use systems suitable for the humid tropics. This chapter describes the agricultural resources of the humid tropics, outlines the processes of forest conversion that have affected wide areas, and examines the potential of improved agricultural practices to prevent continued resource degradation. It stresses the need for a more integrated approach to research, policy, and development activities in managing resources on a more sustainable basis.

The definition of agricultural sustainability varies by individual, discipline, profession, and area of concern. Common characteristics include the following: long-term maintenance of natural resources and agricultural productivity; minimal adverse environmental impacts; adequate economic returns to farmers; optimal production with purchased inputs used only to supplement natural processes that are carefully managed; satisfaction of human needs for food, nutrition, and shelter; and provision for the social needs of health, welfare, and social equity of farm families and communities. All definitions embrace environmental, economic, and social goals in their efforts to clarify and interpret the meaning of sustainability. In addition, they suggest that farmers and farm systems must be able to respond effectively to environmental and economic stresses and opportunities. In the humid tropics, priority must be given to soil protection and the efficient recycling of nutrients (including those derived from external sources); to implementation of mixed forest and crop systems; and to secondary forest management that incorporates forest fallow practices (Ewel, 1986; Hart, 1980).

THE HUMID TROPICS

The humid tropics are defined by bioclimates that are characterized by consistently high temperatures; abundant, at times seasonal, precipitation; and high relative humidity (Lugo and Brown, 1991). Annual precipitation exceeds or equals the potential return of moisture to the atmosphere through evaporation. Total annual rainfall amounts usually range from 1,500 mm to 2,500 mm, but levels of

6,000 mm or more are not uncommon. In general, seasons in the humid tropics are determined by variations in rainfall, not temperature. Most areas experience no more than 4 months with less than 200 mm of precipitation per year.

About 60 countries, with a total population of 2 billion, are located partly or entirely within the humid tropics (Table 1-1). About 45 percent of the world's humid tropics are found in the Americas (essentially Latin America), 30 percent in Africa, and 25 percent in Asia. Small portions of the humid tropics can be found in other areas such as Hawaii and portions of the northeastern coast of Australia.

The typical vegetation for the humid tropics consists of moist, wet, and rain forests in the lowlands and in the hill and montane uplands. Estimates of their extent vary. The most current effort to provide reliable and globally consistent information on tropical forest cover, deforestation, and degradation is by the Forest Resources Assessment 1990 Project of the Food and Agriculture Organization (FAO) of the United Nations, using remote sensing imagery and national survey data as part of its methodology (Forest Resources Assessment 1990 Project, 1992). It defines forests as ecological systems with a minimum of 10 percent crown cover of trees (minimum height 5 m) and/or bamboos, generally associated with wild flora, fauna, and natural soil conditions, and not subject to agricultural practices.

The project estimates that forests cover 1.46 billion ha, or 48 percent of the land area (3.02 billion ha) in the tropical rain forest, moist deciduous forest, and hill and montane forest zones. These forests constitute 30 percent of the land area within the tropical region (4.82 billion ha) and 86 percent of the total tropical forest area (1.7 billion ha). Although they cover only 10 percent of the land area of the world (15 billion ha), they contain one-third of the world's plant matter. Nearly two-thirds of the world's humid forests are found in Latin America, with the remainder split between Africa and Asia.

The soils of the humid tropics are highly variable. Table 1-2 shows the geographical distribution of soil orders and major suborders based on the soil classification system developed in the United States. Oxisols and Ultisols are the most abundant soils in the humid tropics, together covering almost two-thirds of the region. Oxisols, found mostly in tropical Africa and South America, are deep, generally well-drained red or yellowish soils, with excellent granular structure and little contrast between horizon layers. As a result of extreme weathering and resultant chemical processes, however, Oxisols are acidic, low in phosphorus, nitrogen, and other nutrients, and limited in their ability to store nutrients, but have relatively high soil organic matter content. Ultisols are the most abundant soils of tropical Asia,

TABLE 1-1 Population Data for Selected Countries with Tropical Moist Forests

Region or Country	Population Estimate, Mid-1991 (millions)	Urban Population (%)	Rate of Natural Increase (annual %)	Number of Years to Double Population	Population Projection to 2025 (millions)	1989 Per Capita GNP ($)
Middle South Asia						
Bangladesh	116.6	14	2.4	28	226.4	180
India	859.2	27	2.0	34	1,365.5	350
Sri Lanka	17.4	22	1.5	47	24.0	430
Continental Southeast Asia						
Brunei	0.3	59	2.5	27	0.5	14,120
Cambodia	7.1	11	2.2	32	12.9	—
Laos	4.1	16	2.2	32	7.4	170
Myanmar	42.1	24	1.9	36	72.2	—
Thailand	58.8	18	1.3	53	78.1	1,170
Vietnam	67.6	20	2.3	31	107.8	—
Insular Southeast Asia						
Indonesia	181.4	31	1.7	41	237.9	490
Malaysia	18.3	35	2.5	28	34.7	2,130
Papua New Guinea	3.9	19	2.3	31	7.6	900
Philippines	62.3	42	2.6	27	100.7	700
Subtotal	1,439.1	26[a]	2.1[a]	34[a]	2,275.7	—
Middle America						
Belize	0.2	50	3.3	21	0.5	1,600
Costa Rica	3.1	45	2.4	28	5.6	1,790
Dominican Republic	7.3	58	2.3	30	11.4	790
El Salvador	5.4	43	2.8	25	9.4	1,040
Guatemala	9.5	39	3.0	23	21.7	920
Haiti	6.3	28	2.9	24	12.3	400

Honduras	5.3	43	3.1	23	11.5	900
Mexico	85.7	71	2.3	30	143.3	1,990
Nicaragua	3.9	57	3.4	21	8.2	830
Panama	2.5	52	2.1	34	3.9	1,780
Puerto Rico (U.S.)	3.3	72	1.1	62	4.2	6,010
Trinidad and Tobago	1.3	64	1.6	44	1.8	3,160
Tropical South America						
Bolivia	7.5	50	2.6	27	14.3	600
Brazil	153.3	75	1.9	36	245.8	2,550
Colombia	33.6	68	2.0	35	54.2	1,190
Ecuador	10.8	55	2.4	29	19.2	1,040
French Guiana	0.1	81	2.2	31	0.2	—
Guyana	0.8	35	1.8	39	1.2	310
Peru	22.0	69	2.3	30	37.4	1,090
Suriname	0.4	48	2.0	35	0.7	3,020
Venezuela	20.1	83	2.3	30	35.4	2,450
Subtotal	382.4	56[a]	2.4[a]	31[a]	642.2	—
West Africa						
Côte d'Ivoire	12.5	39	3.5	20	39.3	790
Ghana	15.5	32	3.2	22	35.4	380
Guinea	7.5	22	2.6	27	16.0	430
Guinea-Bissau	1.0	27	2.0	35	1.9	180
Liberia	2.7	44	3.2	22	7.4	450
Nigeria	122.5	16	2.8	25	305.4	250
Sierra Leone	4.3	30	2.7	26	10.0	200
Togo	3.8	22	3.7	19	11.3	390
Central Africa						
Cameroon	11.4	42	2.6	26	26.1	1,010
Central African Republic	3.0	43	2.6	27	6.6	390
Congo	2.3	41	3.0	23	5.5	930
Equatorial Guinea	0.4	60	2.6	26	0.9	430

continued

TABLE 1-1 *Continued*

Region or Country	Population Estimate, Mid-1991 (millions)	Urban Population (%)	Rate of Natural Increase (annual %)	Number of Years to Double Population	Population Projection to 2025 (millions)	1989 Per Capita GNP ($)
Central Africa—*continued*						
Gabon	1.2	41	2.3	31	2.9	2,770
São Tomé and Principe	0.1	38	2.5	28	0.3	360
Zaire	37.8	40	3.1	22	101.1	260
Eastern Africa						
Burundi	5.8	5	3.2	21	15.5	220
Kenya	25.2	22	3.8	18	63.2	380
Madagascar	12.4	23	3.2	22	34.0	230
Mauritius	1.1	41	1.4	51	1.4	1,950
Mozambique	16.1	23	2.7	26	35.4	80
Rwanda	7.5	7	3.4	20	22.9	310
Tanzania	26.9	20	3.7	19	78.9	120
Uganda	18.7	10	3.5	20	55.0	250
Subtotal	339.7	30[a]	2.9[a]	25[a]	876.4	—

NOTES: A dash denotes information that was not available; GNP, gross national product.

[a]Average.

SOURCE: Population Reference Bureau. 1991. World Population Data Sheet 1991. Washington, D.C.: Population Reference Bureau.

TABLE 1-2 Geographical Distribution of Soils of the Humid
Tropics (in Millions of Hectares)[a]

Soil Order or Suborder	Humid Tropics Total	Humid Tropic America[b]	Humid Tropic Africa[c]	Humid Tropic Asia[d]
Oxisols	525	332	179	14
Ultisols	413	213	69	131
Inceptisols				
Aquepts	120	42	55	23
Andepts	12	2	1	9
Tropepts	94	17	19	58
Subtotal	226	61	75	90
Entisols				
Fluvents	50	6	10	34
Psamments	90	6	67	17
Lithic	72	19	14	39
Subtotal	212	31	91	90
Alfisols	53	18	20	15
Histosols	27	—	4	23
Spodosols	19	10	3	6
Mollisols	7	—	—	7
Vertisols	5	1	2	2
Aridisols[e]	2	—	1	1
Total	1,489	666	444	379

[a]Based on dominant soil in maps (scale of 1:5 million) of the Food and Agriculture Organization (FAO) of the United Nations.
[b]From Sanchez and Cochrane (1980) plus recent adjustments.
[c]From the FAO (1975) and Dudal (1980).
[d]From the FAO (1977, 1978). Includes 46 million ha of the humid tropics of Australia and Pacific Islands.
[e]Saline soils only (Salorthids).

SOURCE: National Research Council. 1982. Ecological Aspects of Development in the Humid Tropics. Washington, D.C.: National Academy of Sciences.

and are also found in Central America, the Amazon Basin, and humid coastal Brazil. Ultisols are usually deep, well-drained red or yellowish soils, somewhat higher in weatherable minerals than Oxisols but also acidic and low in nutrients.

Inceptisols and Entisols account for most of the remaining soils of the humid tropics (about 16 percent and 14 percent, respectively). These are younger soils, more limited in distribution, and range from highly fertile soils of alluvial and volcanic origin to very acidic and nutrient-poor sands.

TABLE 1-3   Summary of Forested Bioclimates in the Tropical Zone

Bioclimate	Mean Annual Biotemperature	Mean Annual Precipitation (mm)	Other Precipitation Characteristics
Lowland	>24°C		
Moist forest		1,500–4,000	No more than 4 months with <200 mm
Wet forest		4,000–8,000	No more than 2 months with <200 mm
Rain forest		>8,000	No months with <200 mm
Premontane	18°–24°C		
Moist forest		1,000–2,000	2–4 months with <100 mm
Wet forest		2,000–4,000	No more than 2 months with <100 mm
Rain forest		>4,000	No months with <100 mm
Lower montane	12°–18°C		
Moist forest		1,000–2,000	2–4 months with <100 mm
Wet forest		2,000–4,000	No more than 2 months with <100 mm
Rain forest		>4,000	No months with <100 mm
Montane	6°–12°C		
Moist forest		500–1,000	2–4 months with <50 mm
Wet forest		1,000–2,000	No more than 2 months with <50 mm
Rain forest		>2,000	No months with <50 mm

NOTE: At any given latitude, the treeline lies at a mean annual biotemperature of 6°C.

Although many humid tropic soils are acidic and low in reserves of essential nutrients, the constant warm temperatures, plentiful rainfall, and even allocation of sunlight throughout the year permit abundant plant growth. Broadleaf evergreen forests are the dominant form of vegetation. The generally infertile soils are able to support these biologically diverse, high-biomass forests because they have fast rates of nutrient cycling and have reached maturity without frequent disturbances.

While the forests of the humid tropics are often referred to generically as tropical rain forests, they in fact include a variety of distinct plant associations. Holdridge's (1967) System for the Classification of World Life Zones provides the basis for differentiating forest formations over broad gradients of temperature and rainfall (see Table 1-3). Tropical lowland forests are the most abundant, constituting some 80 percent of humid tropic vegetation. Lowland areas are also significant from the standpoint of human economic activity, environmental impacts, development potential, and scientific interest. Although tropical premontane forest formations comprise only about 10 percent of humid tropic vegetation, they are disproportionately modified by human activity, especially toward the drier end of the gradient, because of their suitability for plantation culture and crop agriculture. The remainder of the humid tropic forests consists of relatively uncommon lower montane and montane formations. Collectively, lowland, premontane, and montane forest formations can be referred to as humid tropic or tropical moist forests.

The small nonforest component of humid tropic vegetation includes aquatic and wetland flora and treeless plant communities that exist above timberline on the highest mountaintops. At the latitudinal and climatic limits of the humid tropics, the tropical moist forests grade into more seasonal (monsoonal), semievergreen types and eventually into savannah ecosystems. The term "closed tropical forests" is sometimes used to distinguish the unbroken forests of the humid tropics from drier, more open tropical forest types.

FOREST CHARACTERISTICS AND BENEFITS

The forests of the humid tropics provide multiple goods, values, and environmental services. At the global scale, tropical moist forests, through photosynthesis, evapotranspiration, decomposition, succession, and other natural processes, play a significant role in the functioning of the atmosphere and biosphere. At local and regional scales, the ecological processes and biological diversity of forests provide the foundations for stable human communities and opportunities for sustainable development. The special characteristics of tropical moist forests, and the direct and indirect benefits they afford, are described in numerous publications (for example, Myers, 1984; National Research Council, 1982; Office of Technology Assessment, 1984; Wilson and Peter, 1988) and summarized below. These characteristics underscore the need to begin with an understanding of ecosystem components and processes in the humid tropics in moving toward more sustainable land uses.

Although the environmental characteristics and benefits described pertain fundamentally to primary tropical moist forests, they are also provided to varying degrees by secondary forests, regenerating forests, managed forests, forest plantations, and agroforestry systems. These distinctions become important in weighing the impacts of different types of forest conversion and formulating sustainable agricultural systems suited to humid tropic conditions.

Local and Global Climatic Interactions

Local and global climatic patterns are influenced by the interaction of tropical moist forests and the atmosphere. At the continental scale, forests are thought to influence convection currents, wind and precipitation patterns, and rainfall regimes because of their ability to reflect solar heat back into space and to receive and release large volumes of water (Houghton et al., 1990; Salati and Vose, 1984). It is estimated, for example, that as much as half the atmospheric moisture in the Amazon basin originates in local forests by transpiration (Salati et al., 1983).

At the global level, tropical moist forests play an important role in large-scale biogeophysical cycles (especially those of carbon, water, nitrogen, and other elements) that are critical in determining atmospheric conditions. Particularly important is the function of the forests in the carbon cycle. The total biomass accumulations in mature tropical moist forests are the highest in the tropics and among the highest of any terrestrial ecosystem (Brown and Lugo, 1982). In primary forests, carbon exists in essentially a steady state—the amount of carbon accumulated is about equal to the amount released, although there may be a small net accumulation (Lugo and Brown, In press). Secondary and recovering forests act as important carbon sinks (Brown et al., 1992). Carbon stored within forest biomass and soils is prevented from reaching the atmosphere in the form of carbon dioxide or methane, both of which contribute to global warming.

Biological Diversity

The unusually high concentration of species in tropical moist forests is widely recognized, and the accelerated loss of that diversity—especially of plant species—has drawn much attention in recent years (Ehrlich and Wilson, 1991; Myers, 1984; Raven, 1988; Wilson and Peter, 1988). Although tropical moist forests cover about 7 percent of the earth's land surface, they are believed to harbor more than half of the world's plant and animal species. Estimates of the total number of species in tropical moist forests range between 2 million and 20 million (Ehrlich and Wilson, 1991). The majority of these species have yet to be described, much less studied. Basic taxonomic work in tropical moist forests remains a high research priority (National Research Council, 1992).

Beyond the high levels of diversity of wild species found in the forests themselves, the humid tropics are also important centers of germplasm diversity for rice, beans, cassava, cocoa, banana, sugarcane, citrus fruits, and other economically important crops. These

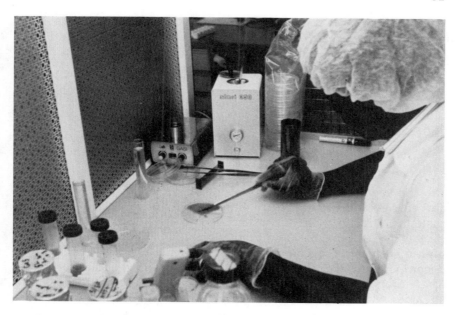

Germplasm collected from the tropics is used in crop improvement research in laboratories around the world. Friable callus of cassava, an important root crop in the tropics, is chopped for suspension in an Austrian laboratory. Credit: Food and Agriculture Organization of the United Nations.

germplasm resources include wild relatives of domesticated plants as well as highly localized crop varieties and landraces developed over centuries by farmers. To boost productivity, provide resistance against pests and other environmental stresses, and improve overall quality, plant breeders have already incorporated genetic material from these wild and domesticated strains into breeding lines of rice, cocoa, sugar, and other major crops.

Products and Commodities

The high degree of biological diversity within tropical moist forests is reflected not only in germplasm resources, but also in the array of established and potential products and commodities they contain. Tropical forests are sources not only of widely exploited timber and plantation products, but also of foods (including animal protein), spices, medicines, resins, oils, gums, pest control agents, fuels, fibers, and forages for forest dwellers and small-scale farmers. Many of the products used for subsistence purposes at the local level hold promise for broader economic use within a sustainable development framework.   In addition to known forest products and food germplasm resources, many plants and animals of the humid tropics contain genetic material and chemical compounds useful in developing new pharmaceuticals and other products.   Others are likely to have agronomic and environmental applications (for example, as multipurpose tree species and biocontrol agents) within sustainable agroecosystems.

Nutrient Cycling

The vegetation within tropical moist forests thrives by retaining and efficiently recycling scarce but essential nutrients within the ecosystem.   Root growth is concentrated in the topsoil.   When litter (leaves, twigs, branches, and whole trees) falls to the forest floor, the high-quality litter decomposes rapidly, while the low-quality litter decomposes slowly.   Plant nutrients are mineralized and adsorbed by forest roots.   Adsorption by deep roots minimizes nutrient loss into streams.   Most of the nutrients are efficiently recycled, with nutrient additions through rain, dust, and biological nitrogen fixation in balance with losses through leaching, denitrification, and volatilization.   However, in steep areas with relatively young soils, there can be significant nutrient losses from pristine rain forest.   These losses provide nutrients to streams and rivers that support large fish populations.   The closed nutrient cycle between the tropical rain forest and the soil operates only if there is no net harvest of biomass from the system.   In agriculture, the biomass removal through harvest is large.

Protection of Soils

Forest cover protects the topsoil of humid tropic ecosystems from the erosive effects of rainfall.   In forested areas, the lack of exposed ground and the interception of rainfall by multiple layers of vegetation minimizes soil loss.   The dense mat of interwoven roots in the topmost soil layers allows rainfall to be absorbed and released while lower soil horizons are protected.   These features are especially important for lands that are steeply sloped and for lands with shallow soils (Sanchez, 1991).

Stabilization of Hydrological Systems

Forests stabilize watersheds by regulating the rates at which rainfall is absorbed and released.   Intact forest cover allows rainfall to reach the ground, percolate through soils, and flow into streams at a gradual rate. Because soil loss through erosion is low, sedimentation and deposition rates downstream are also minimized. As a result, flood and drought cycles are moderated within the watershed as a whole. This is especially important in areas where irrigated agriculture is concentrated in fertile alluvial valleys downstream from forests.

Water Availability and Quality

The quantity and quality of water delivered to cities and rural villages depend on conditions within the entire hydrological system, and thus in part on upstream forests. In areas where urban population growth is rapid, people depend on surface waters, reservoirs, or groundwater stocks for cleaning, cooking, and drinking water. Cholera, typhoid, and other water-related diseases and parasites are significant public health concerns in the humid tropics. Forests, by providing steady flows of good quality water, are a line of defense against the spread of these maladies, followed by sewage facilities, water treatment plants, and public health programs, many of which are lacking in developing countries (Latin American and Caribbean Commission on Development and Environment, 1990; World Bank, 1992).

Mitigation of Storm Impacts

Forest cover provides protection against the impacts of intense tropical storms, known regionally as cyclones, hurricanes, or typhoons. While forests cannot prevent the loss of life and property that storms inflict, they can mitigate some of their effects, particularly storm surges in coastal zones and mud slides on sloping lands.

CONVERSION OF HUMID TROPIC FORESTS

Forest conversion is the alteration of forest cover and forest conditions through human intervention, ranging from marginal modification to fundamental transformation. At one extreme, forests that have been slightly modified (through, for example, selective extraction, traditional shifting cultivation, or gradual substitution of perennial species) maintain most of their cover, with little long-term impact on ecosystem components, processes, and regeneration rates. Deforestation—changes in land use that reduce forest cover to less than 10 percent—represents the opposite extreme. Between these extremes, conversion happens to varying degrees, entailing changes in forest structure, species diversity, biomass, successional processes,

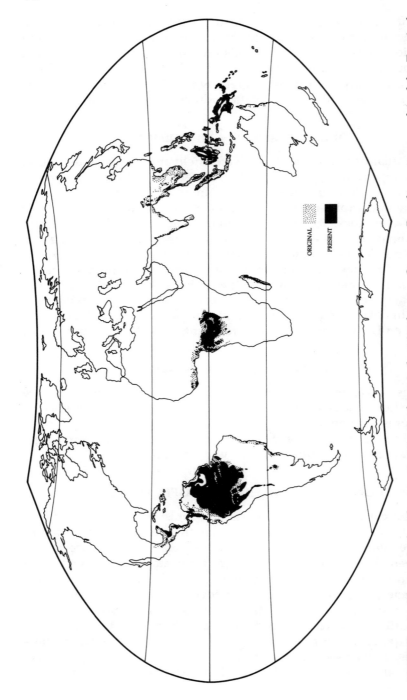

FIGURE 1-1 The original and present extents of tropical moist forests. Source: Based on maps produced for Tropical Rainforests: A Disappearing Treasure, Smithsonian Institution Travelling Exhibition Service, 1988. Courtesy of the Office of Environmental Awareness, Smithsonian Institution, Washington, D.C. © 1988 by Smithsonian Institution.

and ecosystem dynamics. Land or forest degradation occurs when these changes are of sufficient magnitude to have a long-term negative effect on productive potential. Forest transformation occurs when the original forest is eliminated and replaced with permanent agriculture, plantations, pasturelands, and urban or industrial developments.

Estimates of the original and current humid tropic forests are difficult to present, especially concerning forest type. The original extent of tropical rain forests (apparently excluding tropical moist deciduous forests) has been estimated to total 1.5 billion ha, with 600 million ha having been cleared and converted over the past several centuries (Ehrlich and Wilson, 1991; Food and Agriculture Organization and United Nations Environment Program, 1981). The current extent of tropical rain forests and tropical moist deciduous forests has been estimated to be 1.5 billion ha, with 1 billion ha considered to be intact or primary forests in which human activity has had little impact (World Bank, 1991). Apparently Africa has lost the greatest proportion of its original tropical moist forests (about 52 percent), followed by Asia (42 percent) and Latin America (37 percent) (Lean et al., 1990). Figure 1-1 illustrates the original and present extent of tropical rain forests historically and at present.

During the past two decades, the rate of conversion in the humid tropics has accelerated (Table 1-4), although comparisons of data collected over several decades are unreliable due to differences in data gathering methodologies and definitions of area, type of forest, and deforestation. However, the accuracy of more recent information on the rate, extent, and nature of forest conversion is improving.

Forest resources appraisals are part of the mandate of the FAO. The last worldwide assessment was carried out with 1980 as the reference year (Lanly, 1982). An assessment with 1990 as the reference year was launched in 1989 to provide reliable and globally consistent information on tropical forest cover and trends of deforestation and forest degradation. Deforestation refers to change of land use or depletion of crown cover to less than 10 percent. Forest degradation is defined as change within the forest that negatively affects the stand or site and, in particular, lowers its regenerative capacity.

The first interim report of the Forest Resources Assessment 1990 Project (1990) contained preliminary area estimates at the regional level for 62 countries lying mostly in the humid tropic zone. Comparison with the 1980 assessment is possible for 52 countries covered by both assessments; definitions of forest and deforestation are basically the same. The estimated deforestation rate for the period 1976

TABLE 1-4 Provisional Estimates of Forest Cover and Deforestation for 62 Countries in the Humid Tropics

Continent	Number of Countries Studied	Area in Thousands of Hectares				Rate of Change 1981–1990 (percent/year)
		Total Land	Forest 1980	Forest 1990	Annual Deforestation 1981–1990	
Africa	15	609,800	289,700	241,800	4,800	-1.7
Latin America	32	1,263,600	825,900	753,000	7,300	-0.9
Asia	15	891,100	334,500	287,500	4,700	-1.4
Total	62	2,764,500	1,450,100	1,282,300	16,800	-1.2

NOTE: Countries include almost all of the moist tropical forest zone, along with some dry areas. Figures are indicative, and should not be taken as regional averages. Forests are defined as ecological systems with a minimum of 10 percent crown cover of trees and bamboos, generally associated with wild flora, fauna, and natural soil conditions, and not subject to agricultural practices. Deforestation refers to change of land use or depletion of crown cover to less than 10 percent.

SOURCE: Forest Resources Assessment 1990 Project. 1990. Interim report on Forest Resources Assessment 1990 Project. Item 7 of the Provisional Agenda presented at the Tenth Session of the Committee on Forestry of the Food and Agriculture Organization of the United Nations, Rome, Italy, September 24–28, 1990.

to 1980 is 9.2 million ha per year, and it is 16.8 million ha per year for the period 1981 to 1990, an annual rate increase of 83 percent.

The project cautions that this significant difference can be attributed to an actual increase in the deforestation rate, an underestimation of the rate in the 1980 assessment, or an overestimation of the rate in the 1990 assessment. It is known that the 1980 assessment underestimated the rate of deforestation in some large Asian countries. Regardless of the relative contribution of these components, deforestation has accelerated in the humid tropics as a whole. The final results of the project will be based on uniform remote sensing observations of tropical forests specifically made for the project.

Preliminary indications concerning forest degradation indicate that the loss of biomass in the tropical forest is occurring at a significantly higher rate than the loss of area due to deforestation (Forest Resources Assessment 1990 Project, 1991). The project offers two explanations: (1) deforestation is occurring disproportionately on forest-land with higher biomass levels; and (2) remaining forests are being degraded through the removal of biomass. The analysis points to the need for improved land use planning to conserve forest resources. FAO scientists believe the crisis can be corrected. They point to the experience of industrialized countries, where widespread deforestation is being reversed, although at a slow rate. Between 1980 and 1985, forest resources in the developed world increased by 5 percent, from 2 billion ha (4.94 billion acres) to 2.1 billion ha (5.187 billion acres).

Deforestation Rates Within Regions of the Humid Tropics

Although the rate of deforestation rose substantially through the 1980s, the impact has varied from country to country and from region to region (Table 1-4). The rate was highest in Africa (1.7 percent), followed by Asia (1.4 percent) and Latin America (0.9 percent). The areal extent of deforestation, however, was highest in Latin America (7.3 million ha), followed by Africa (4.8 million ha) and Asia (4.7 million ha) (Forest Resources Assessment 1990 Project, 1990). At the country level, deforestation statistics should be interpreted in the context of the total area of original and remaining forest cover. Table 1-5 lists 20 of the principal countries with threatened forests in the humid tropics. In Costa Rica, Côte d'Ivoire, and Nigeria, closed forests were lost at rates exceeding 4 percent per year during the 1980s (World Resources Institute, 1990a). The deforestation rate in Brazil in the 1980s was lower, about 2 percent per year, but the area of forest affected was far greater—about 8 million ha annually (World Resources

TABLE 1-5   Countries with Threatened Closed Forests
(Thousands of Hectares)

Country	Closed Forest Area	Annual Deforestation Rate
Latin America and the Caribbean		
Bolivia	44,010	87
Brazil	375,480	8,000[a]
Colombia	46,400	820
Ecuador	14,250	340
Mexico	46,250	595
Peru	69,680	270
Venezuela	31,870	125
Sub-Saharan Africa		
Cameroon	16,500	100
Central African Republic	3,590	5
Congo	21,340	22
Côte d'Ivoire	4,458	290
Gabon	20,500	15
Madagascar	10,300	150
Zaire	105,750	182
Asia and the Pacific		
India	36,540	1,500
Indonesia	113,895	900
Malaysia	20,996	255
Myanmar	31,941	677
Papua New Guinea	34,230	22
Philippines	9,510	143
Total	1,057,490	14,498

NOTE: A closed forest has a stand density greater than 20 percent of the area and tree crowns approach general contact with one another.

[a]More recent estimates suggest that the rate of deforestation may have declined to 2 million ha per year.

SOURCE: World Bank. 1991. The Forest Sector: A World Bank Policy Paper. Washington, D.C.: World Bank. Reprinted, with permission, from the World Bank. © 1991 International Bank for Reconstruction.

Institute, 1990a). (This rate, which includes open forests outside the Amazon Basin, appears to have fallen in recent years.)

Data on the subsequent fate of converted forestlands are likewise inadequate. Some deforested lands degrade to such a degree that they support little biological recovery or economic activity. Grainger (1988) estimates that as many as 1 billion ha of degraded land may have accumulated in tropical countries, of which 750 million are suitable for reforestation. Only rarely, however, are cleared lands completely barren or abandoned. Large areas are converted to subsistence cultivation, rice production, permanent plantations, and pastures. The spatial extent of each of these, especially on a global basis, is poorly quantified.

Natural regeneration and managed reforestation may return forest cover to some lands that have been cleared. However, reliable information on the extent of secondary forests in the tropics is not available. In a number of areas, secondary forests may not reach advanced stages of restoration due to the activities of subsistence farmers and the impacts of fires, soil degradation and nutrient depletion, inadequate tree regeneration, and invasion by grasses and shrubs.

Causes of Forest Conversion

People do not make the enormous investments in capital, time, and energy that forest conversion can entail without valid social,

Lumber workers transport dipterocarp logs, which command high prices on the international market, out of the tropical rain forest on the island of Borneo, Indonesia. If these tall trees are not harvested carefully, significant damage can be done to the surrounding forest. Credit: James P. Blair © 1983 National Geographic Society.

economic, and political reasons. Analysis reveals a variety of direct and indirect causes, usually acting in combination, behind the increased rates of forest conversion in the humid tropics (Hecht and Cockburn, 1989; Myers, 1984; Office of Technology Assessment, 1984; Repetto and Gillis, 1988). The leading direct causes of forest loss and degradation include large-scale commercial logging and timber extraction, the advancement of agricultural frontiers and subsequent use of land by subsistence farmers, conversion of forests to perennial tree plantations and other cash crops, conversion to commercial livestock production, land speculation, the cutting and gathering of wood

Population Issues in the Tropics

Population growth is one of many factors contributing to resource degradation in the humid tropics. It does not occur independent of other socioeconomic factors. High fertility rates are closely associated with underdevelopment and poverty. However, population growth statistics offer some insight into the level and intensity of land development pressures to meet more immediate food and income needs.

Population growth increases the demand for goods and services and the need for employment and livelihoods, exerting additional pressure on natural resources. Countries with higher population growth rates have experienced faster conversion of land to agricultural uses and greater demands for wood for fuel and building materials. Few government programs help low-income people improve their earning potential or their quality of life. Most development policies have helped the medium- and large-scale agricultural units to capitalize, modernize, and sell their products, and not necessarily in a manner that enhances sustainability and protects natural resources. Because they lack resources and technology, land-hungry farmers often abandon traditional land uses in favor of agricultural practices that produce more food or income in the short term but may involve long-term social, economic, and environmental costs. Sustainable land use cannot be achieved as long as high rates of poverty and population growth continue.

Although demographic and socioeconomic statistics for the humid tropics as a distinct region do not exist, available information does illustrate the population situation in the humid tropics. About 60 countries, representing 90 percent of the world's developing countries, lie within or border on the humid tropics. During the past 4 decades, the population of developing countries, excluding the People's Republic of China, increased by 1.5 billion (Population Reference Bureau, 1988). During the same period about 350 million people were added to the population in developed countries.

for fuel and charcoal, and large-scale colonization and resettlement projects.

In many areas of the humid tropics, agricultural expansion is one of the most important direct causes of forest conversion. For example, shifting cultivation practices in Africa account for 70 percent of the clearing of closed-canopy forests (Brown and Thomas, 1990). In general, shifting cultivators fall into two broad categories: local or native farmers, who tend to be resource conserving and use sustainable traditional agricultural practices, and more recent farmers, who have migrated to frontier lands to make a living and tend to be less

---

In much of Africa and Latin America throughout the 1980s per capita income declined, although it grew in Asia and in industrialized countries (World Health Organization, 1990). Average per capita income in industrialized countries is about 50 times that of the least developed countries, and the annual increase alone in the richer countries is about as large as the whole per capita income in the poorest countries ($300).

It took about 130 years (from around 1800 to 1927) for the world to increase its population from 1 billion to 2 billion. Only 33 years (1927–1960) were necessary for the third billion, 14 years (1960–1974) for the fourth, and 13 years (1974–1987) for the fifth (World Health Organization, 1990). The world's population is expected to increase by 1 billion each decade well into the twenty-first century. Most of this growth will occur in developing countries. Their population (excluding China) is expected to increase from a total of 3 billion today to about 5.6 billion by the year 2035 (Population Reference Bureau, 1991). The percentage of the world's population living in developing countries will increase from 55 percent to 65 percent.

Leaders of developing countries in the humid tropics are also confronted by financial circumstances that have contributed to poverty. In the early 1980s, international assistance provided developing countries with a surplus of some $40 million. A decade later, developing countries had accumulated a total debt burden in excess of $1.3 trillion (Lean et al., 1990), partly as the result of inflation, global recession, increasing interest rates, poor returns on development investments, and trade imbalances. The costs of servicing these debts now outpace the amount of aid. As a result, spending to reduce poverty and help the poor is cut, and continued poverty contributes to population growth rates. Some of the highest debt loads (both absolute and relative to gross national product) have been incurred by Brazil, Mexico, and the Philippines.

knowledgeable about local environments and sustainable practices. Estimates of the number of farmers engaged in the clearing of forest-lands in the humid tropics (including both primary and secondary forests) each year have ranged from 300 million to 500 million (Andriesse and Schelhaas, 1987; Denevan, 1982; Myers, 1989). Assessments of the area of forestland affected are similarly divergent, ranging from 7 million to 20 million ha each year (Gradwohl and Greenberg, 1988; Lanly, 1982; National Academy of Sciences, 1980).

Agricultural expansion, as well as the other immediate causes of forest conversion and degradation, is driven by a network of forces operating at national and international levels. In general, develop-ment efforts have been unable to relieve these forces and in some cases have aggravated them. Widespread poverty, the unequal distri-bution of income, flawed food distribution policies, and high-popula-tion density and growth rates act as exacerbating factors throughout the humid tropics (Ehui, Part Two, this volume; Kartasubrata, Part Two, this volume; Gómez-Pompa et al., Part Two, this volume). High fiscal deficits, underemployment, and other symptoms of economic stress lead many countries to encourage the conversion of forests through favorable tax policies, forest concessions, rents, credits, and other financial incentives, which often lead to enhanced disparities of income distribution (Serrão and Homma, Part Two, this volume).

Infrastructure development policies have opened forestlands through road building, mining operations, dam construction, and other large-scale projects, while agricultural development has devoted inadequate resources to the needs of farmers and local communities in areas with low-quality soil and water resources (Serrão and Homma, Part Two, this volume). Many of these projects have been funded by bilateral and multilateral assistance agencies. In settling these newly opened lands, farmers are seldom provided with the means or the knowledge to secure sustainable livelihoods. Rural development ef-forts that might give small-scale farmers greater security are hin-dered by inequitable land tenure arrangements and a lack of access to scientific knowledge, improved technologies, and credit facilities.

Forestry, agriculture, and environmental ministries in many countries are insufficiently integrated and often unable to enforce existing con-servation policies, while officials lack opportunities for further edu-cation or professional training (Ngandu and Kolison, Part Two, this volume). Agronomic strategies proposed by research agencies and extension services at times have suggested inappropriate technolo-gies that left farmers in debt (Gómez-Pompa et al., Part Two, this volume). In some countries political corruption, warfare, and na-

About 20,000 prospectors and laborers work tiny claims at a makeshift gold mine at Serra Pelada (Naked Mountain) in Brazil's Amazon rain forest. The gold is sold to the Brazilian government, which is counting on the region's mineral wealth, including iron ore, bauxite, and manganese, to offset its foreign debt. However, this type of land use may destroy both the extraction site and downstream watershed areas through runoff of soil and contaminants. Credit: James P. Blair © 1983 National Geographic Society.

tional security concerns have also contributed to ineffective resource management (Garrity et al., Part Two, this volume; Rush, 1991).

Other causal factors are international in scope. Over the past 20 years, many humid tropic countries have incurred large foreign debts, even as the global economic climate has made it more difficult to service these debts. To meet debt obligations, a number of tropical countries have tried to increase their export earnings through rapid extraction of forest resources and conversion of forestlands. International commodity prices and trade policies have also contributed to forest conversion by failing to reflect social and economic costs and by rewarding land uses that provide higher short-term economic returns.

The relationship between people and land resources in the many countries of the humid tropics vary widely as a function of their cultures, rates of population growth, economic circumstances, and environmental conditions. As a result, the degree to which different causal factors contribute to forest conversion varies from country to country and even within countries. Furthermore, the influence of these factors relative to one another changes over time. For example, in Côte d'Ivoire the expansion of the agricultural frontier has been the leading direct cause of forest conversion and is primarily responsible for a two-thirds reduction in the area of forest between 1965 and 1985. The deforested area is often in sloping uplands with marginal soils that cannot support intensive permanent cropping (Ehui, Part Two, this volume). In the Philippines, a combination of intensified commercial logging, agricultural expansion, increased use of fuelwood and other wood products, and a lack of alternative means of livelihood has greatly accelerated the rate of forest conversion since World War II (Garrity et al., Part Two, this volume). In Brazil, the formerly extensive Atlantic coast forest has been reduced to remnants through conversion to agricultural use over the centuries. Large-scale conversion of forestlands to cattle pastures and the opening of access roads was the leading cause of deforestation in the Amazon Basin (Serrão and Homma, Part Two, this volume). The removal of incentives to clear forestlands appears to have slowed the conversion to cattle ranching, but the migration of people to establish small-scale farms in forest areas has increased.

Historical Patterns of Forest Conversion

Subsistence farmers and forest dwellers have modified forestlands in the humid tropics for hundreds and even thousands of years (Gómez-Pompa, 1987a; Gómez-Pompa and Kaus, 1992). The scale of these modifications, however, was generally small, and the rate at which they occurred allowed time for forests to adapt and regenerate. As a result, their effects on the total area of forest cover and on nutrient cycling, watershed stability, biological diversity, and other ecosystem characteristics were limited.

Although forest conversion has expanded steadily over the past five centuries, the three continental expanses of humid tropic forest remained largely intact prior to the late nineteenth century (Tucker, 1990). Extraction of woods, spices, nuts, and other commercial products, although widespread, seldom exceeded the forests' productive capacities. The expansion of sugarcane, coffee, cacao, and other plantation systems was confined primarily to lowlands and adjacent uplands

BUSINESS REPLY MAIL

First Class   Permit No. 10207   Washington, D.C.

Postage will be paid by addressee

ISSUES IN
SCIENCE AND TECHNOLOGY

National Academy of Sciences
2101 Constitution Avenue, N.W.
Washington, D.C. 20077-5576

You are cordially
invited to send for a
free copy of
ISSUES
N SCIENCE AND TECHNOLOGY

ISSUES, a journal of ideas and opinions, is published quarterly by the National Academy of Sciences, National Academy of Engineering, and Institute of Medicine.

YES. Please send me my FREE copy of ISSUES and enter my trial subscription for one year (four quarterly issues including the first free issue) at the special Individual Rate of $36. If for some reason I choose not to continue my subscription, I will return your invoice marked "cancel" and owe nothing.

*Institution Rate: U.S. $65; Outside U.S. $75 payable in U.S. $
Send no money now; we will be happy to bill you. (U.S. ONLY)*

Name _____

Address _____

City _____

State _____

ZIP _____

along rivers and coastlines. Rates of population growth in the humid tropics were generally low, and although ownership and control over prime agricultural land became increasingly concentrated in many areas, small-scale farmers migrated to intact forestlands on a relatively limited basis. Deforestation on the scale that has occurred more recently was technically and economically infeasible.

During the twentieth century, and especially in the past 5 decades, the rate of forest conversion has accelerated in response to economic pressures, population growth, technological developments, and programs and incentives to open lands for development (Hecht and Cockburn, 1989; Repetto and Gillis, 1988). Many of the physical constraints, such as the lack of roads and machinery for timber extraction, that had previously limited the intensity and extent of forest conversion have been overcome. At the same time, global markets for timber and other tropical products have expanded (Kartasubrata, Part Two, this volume; Ngandu and Kolison, Part Two, this volume; Serrão and Homma, Part Two, this volume). These factors have combined to encourage resource-poor countries to clear forests for timber and to convert forestlands to cash crops, plantations, pastures, and other uses of higher but shorter-term economic value (World Bank, 1992).

A classic example of deforestation brought about by population pressures and demand for agricultural land is that of the islands of Java and Bali in Indonesia (Kartasubrata, Part Two, this volume). In Côte d'Ivoire, which has one of the highest population growth rates in the world, population pressures combined with unstable shifting cultivation and logging have been a principal cause of deforestation. Part of the country's agricultural growth has been achieved at the expense of the natural resource base (Ehui, Part Two, this volume).

Forest conversion has followed diverse pathways in the humid tropics, but a general pattern can be discerned. The clearing of forests usually occurred first in areas where the soils and climatic conditions were most favorable for agriculture and for densely populated settlements and where transportation was not a major problem—islands and coastal zones, river basins, lowlands, and the more fertile uplands (Tosi, 1980; Tosi and Voertman, 1964). It then expanded to both wetter and drier life zones, initially affecting easily accessible forestlands. Less accessible lands are now being deforested, including areas unfavorable for human habitation and agriculture, such as steep slopes, mangrove swamps, and flood plains (Green and Sussman, 1990; Harrison, 1991; Kangas, 1990; Sader and Joyce, 1988; Smiet, 1990).

Consequences of Forest Conversion

The effects of forest conversion on the long-term stability and productivity of land resources depend on the characteristics of the original forest, the nature of the conversion that occurs, the methods used in the process of conversion, the social and economic context of conversion, and the subsequent use and management of the land. At one extreme—complete deforestation of primary forest on marginal soils and subsequent abandonment to weed cover—virtually all of the environmental values and services as well as the long-term social and economic benefits provided by the forest are lost. Selective extraction, small-scale sustainable forest management, and other conservative land uses can maintain most of the advantages of primary forests, although biological diversity is likely to decrease to varying degrees.

It is difficult at present to determine with precision the magnitude of these interrelated environmental, social, and economic impacts. Most areas of the humid tropics lack reliable baseline data on ecosystem composition and function, and little systematic long-term ecological (or agroecological) research has been undertaken in the region. Watershed-level research that combines information on forestry, agriculture, and land use is scarce, as are integrative studies of the social and economic consequences of forest conversion. The need for further research on these questions should not, however, delay efforts to forestall expected negative impacts. Because of the nature of land use problems in the humid tropics, many of the negative effects may not be felt until they are irreversible.

ENVIRONMENTAL CONSEQUENCES

The environmental consequences of forest conversion involve the degree to which ecosystem functions are disrupted, forest biomass and composition altered, and forest cover lost. If conversion entails large-scale loss (hundreds of square kilometers) of forest cover on steep lands and the subsequent adoption of inappropriate land uses, natural hydrological processes can be substantially altered, increasing the discharge of water into streams and the amplitude of flood and drought cycles within the watershed. Under these circumstances, rivers, reservoirs, and canals receive increased sediment loads, with negative effects on irrigated agriculture, fishing, hydroelectric power generation, and water quality. Exposed soils, particularly following mechanical clearing, are subject to erosion, compaction, and crusting until a new vegetative cover or canopy is established (Lal, 1987; Sanchez, 1991).

A scientist measures 50.8 cm (20 in) of silt deposited in 1 year on a riverbank in the Amazon River Basin. Credit: James P. Blair © 1983 National Geographic Society.

Large-scale conversion of primary tropical forests is a leading factor in the worldwide loss of biological diversity (Ehrlich and Wilson, 1991; Raven, 1988; Wilson, 1988). Due to the high levels of species diversity, the limited distribution of most of these species, and the specialized relationships and reproductive strategies within tropical forest ecosystems, forest clearing and fragmentation result in high levels of species loss. Because current scientific knowledge can provide only rough estimates of total species diversity within tropical moist forests, the rate at which species are being lost cannot be accurately determined. Even conservative estimates, however, suggest that tropical deforestation results in a loss of at least 4,000 species per year (Ehrlich and Wilson, 1991; Wilson, 1988).

Forest conversion in the humid tropics also has climatic consequences (Bunyard, 1985; Intergovernmental Panel on Climate Change, 1990a,b). Changes in regional hydrological cycles may affect the distribution and amount of rainfall, impairing agricultural productivity and water availability. The risk of fire rises as forest cover diminishes due to hotter and drier microclimatic conditions (Crutzen and Andreae, 1990). At the global scale, forest conversion affects atmospheric concentrations of carbon dioxide, methane, nitrous oxide, and

---

Climate Change and Land Use

Emissions of trace gases as a result of human activities could change the atmosphere's radiative properties enough to alter the earth's climate. Greenhouse gases, including water vapor, carbon dioxide, methane, nitrous oxide, chlorofluorocarbons, and ozone, insulate the earth, letting sunlight through to the earth's surface while trapping outgoing radiation. Atmospheric concentrations of all of these gases are rising due to human industrial and agricultural processes. Atmospheric models indicate that, at the rate these gases are accumulating, the global mean temperature will increase by between 0.2°C and 0.5°C per decade over the next century (Houghton et al., 1990). This increase could have widespread effects on global sea level, seawater temperatures, rainfall distribution, seasonal weather patterns, plant and animal populations, agricultural production, and human settlement and economic systems.

Carbon dioxide is believed to be responsible for about half of the total global warming potential. If current trends continue, carbon dioxide is expected to account for 55 percent of global warming over the next century, or four times more than methane, the second most important heat-trapping gas (Houghton et al., 1990). According to recent estimates, 75 percent of total carbon dioxide emissions from human activities occur as a result of the combustion of fossil fuels, mostly in nontropical countries (Intergovernmental Panel on Climate Change, 1990a). Land use changes are responsible for most of the remainder.

The most significant of these land use changes are occurring in the humid tropics (Dale et al., Appendix, this volume). As forest conversion occurs, carbon stored in vegetation and soils is released as carbon dioxide through the burning and decomposition of biomass and the oxidation of soil organic matter. Agricultural activities that follow forest conversion—including paddy rice culture, cattle raising, and the use of nitrogen fertilizers—are sources of methane and nitrous oxide.

other greenhouse gases. Assessments of the effects of tropical deforestation on greenhouse gas levels vary. Dale et al. (Appendix, this
volume) estimate that tropical deforestation is responsible for about
25 percent of the total radiative effect of greenhouse gases emitted as
a result of human activities.

SOCIAL CONSEQUENCES

The social consequences of forest conversion, like the environmental consequences, vary according to its extent and type. In areas

---

On a global basis, the conversion of tropical forests and the expansion
of crop- and pasturelands on former forestlands account for about 20 to
25 percent of carbon dioxide emissions and 25 percent of the total radiative effect of greenhouse gas emissions (Dale et al., Appendix, this
volume; Houghton, 1990a).

In terms of potential impact on climate change, the most important
feature of land use in the humid tropics is the net release of carbon that
occurs as a result of forest conversion. The carbon release represents
the difference between the pre- and postconversion levels of carbon
stocks. This figure can range widely, depending on the nature of the
original forest, the degree and rate of conversion, and the subsequent
land use. Permanent agriculture based on annual crops, for example,
reduces by more than 90 percent the amount of carbon stored in the
original vegetation, while the loss from selective logging can be as
small as 10 percent (Dale et al., Appendix, this volume). (Tropical
vegetation and soils can also naturally release greenhouse gases, such
as nitrous oxide and methane.) As secondary forests regrow, or are
replaced by forest fallows, plantations, agroforestry systems, or other
agricultural land uses, carbon is sequestered again within the biomass
and soil (Wisniewski and Lugo, 1992).

These differential releases and accumulations become important in
weighing the land use options described in Chapter 2. Some activities,
such as logging, might allow a virgin forest landscape to actually accumulate and store more carbon than it would if it was left as virgin
forest, where the storage and release of carbon are in balance. In logging, the sawn boards are not destroyed but used for long periods of
time. Hence, carbon remains stored in the harvested wood and, mean-
while, carbon continually accumulates through vegetation growth in
the open spaces left after cutting. If the forest is not treated carefully,
or the sawn wood is not put to wise long-term uses, even logged forests could act as sources, instead of collectors of carbon.

where indigenous cultural groups have maintained ways of life that depend on the forest, the loss of forests disrupts traditional social systems and threatens communal land claims (Lynch, 1990). As these groups are dislocated or acculturated, their knowledge of forest resources and methods of resource management are lost. Deforestation activities have also brought new diseases to tribal peoples, especially in areas where previous contacts with outsiders had been infrequent.

Forest conversion has consequences for both the forest frontier and the cities in tropical countries. Often the ownership and use of the best lands by those who possess the resources and the technology to exploit them relegate the very poor to land of inferior quality (Latin American and Caribbean Commission on Development and Environment, 1990). Over large areas of cleared forestland, nonsustainable land uses have degraded soil and water resources and failed to raise living standards for small-scale farmers. Deforested lands that are subjected to soil-depleting production practices must be abandoned after only a few years, forcing many large- and small-scale farmers to move to newly cleared forestlands (Sanchez, 1991). Economic, demographic, and political pressures have increased the level of migration to forest frontier areas. At the same time, the degradation of natural resources has contributed to the migration of millions of people into cities in search of livelihoods. Population pressures, in turn, diminish the capacity of cities to contribute to sustainable development through efficient production of nonagricultural goods and services (Lugo, 1991).

This cycle of nonsustainability can be addressed, in part, by providing employment alternatives and better managing the degraded and abandoned lands outside the urban core. The loss of soil fertility, shortages of essential natural resources such as water, and the reduced productivity of damaged natural systems reduce job and subsistence opportunities and constitute a clear cause of poverty. The need for sustainable production methods for cleared lands is paramount to rural social well-being. In many parts of the humid tropics, however, the expanses of degraded land between the cities and the remaining forests continue to grow.

ECONOMIC CONSEQUENCES

The conversion of forests involves costs at the local, regional, and global levels that are hard to quantify and that are not reflected in markets (Norgaard, 1989; Randall, 1988; Repetto and Gillis, 1988). These include, for example, the loss of proven or potential biological resources, such as foods or pharmaceuticals, from primary forests; the destabilization of watersheds, with the attendant downstream effects of flooding and siltation; and, at the global level, the long-term impacts of deforestation on global climate change. At the same time, market prices inadequately reflect the benefits secured through the adoption of sustainable land uses (Repetto and Gillis, 1988).

Resource depletion has often been justified as the only way for nations in the humid tropics, faced with growing populations, large foreign debts, nascent industrial capacity, and an often undereducated rural populace, to develop. Especially in recent decades, a number of tropical countries have depleted forest resources in the effort to solve social, political, and economic problems in their societies, and to reduce large and growing international debt burdens (Ehui, Part Two, this volume; Serrão and Homma, Part Two, this volume; Vincent and Hadi, Part Two, this volume). These countries, however, have often found themselves coming under even tighter fiscal constraints as a result. In other cases, the link between deforestation and the need for foreign exchange to service external debt is tenuous; deforestation is more accurately associated with in-country uses of wood (Ngandu and Kolison, Part Two, this volume). Nevertheless, forest conversion may provide only short-term economic benefits, while undermining long-term productivity and social well-being through depletion of soil, water, atmospheric, and biotic resources and reduction of resource development options available to future generations (Ehui, Part Two, this volume; Norgaard, 1992).

SUSTAINABLE AGRICULTURE IN THE HUMID TROPICS

The challenges facing farmers in the humid tropics, and the connections between agricultural expansion, deforestation, land degradation, and rural poverty, have long been recognized. Development policies, however, have tended to overlook the large proportion of small farms on resource-poor land. In broad terms, national and international policies have emphasized urban development and large-scale infrastructure projects over rural development needs. The resources available for agricultural development were applied to the best lands, where economic returns were highest. Most agricultural research and development programs, in turn, focused on the refinement of input-intensive production systems suited to resource-rich areas. The practical difficulties facing the resource-poor farmer have thus been neglected, despite the multiple socioeconomic and environmental benefits that solutions would offer. At the same time, efforts to curb deforestation have usually approached the problem only from the perspective of forest management or environmental protection.

Sustainable agriculture can provide opportunities to address productivity and environmental goals simultaneously. By adopting alternative land use practices that can reduce the need to abandon established farmland and that can restore degraded land to economic and biological productivity, farmers can meet their food needs and make an adequate living without contributing to the further depletion of forests and other natural resources.

Constraints on Agricultural Productivity

The development of sustainable production systems suitable for areas with low-quality soil and water resources rests on an appreciation of the constraints on agricultural productivity in the humid tropics (National Research Council, 1982; Savage, 1987). Agriculture is fundamentally a process of converting solar energy, through photosynthesis, into useful biomass. Biological productivity requires solar energy, water, and nutrients. These are abundantly available in the humid tropics, but this productive potential is not reflected in the performance of agricultural systems, which is typically poor. Intensive farming in temperate zones converts 2 percent of photosynthetically active incident solar energy to dry matter; in the humid tropics, the conversion rate is no more than 0.2 percent (Holliday, 1976). This relative inefficiency is a reflection of both socioeconomic and environmental constraints. This discussion focuses on the latter.

CLIMATE

Water can be a limiting factor in the humid tropics, despite periods of abundant rainfall (Juo, 1989; MacArthur, 1980). Many high-rainfall areas have dry periods of sufficient length to adversely affect plant growth. Water shortages often occur where the soils have low water-holding capacities, but they can also affect areas with more favorable soil environments. A few days without rain can seriously impinge on biological productivity. For example, Omerod (1978) compared rainfall distribution and water retention in London, England, with those in Lagos, Nigeria. Although the total rainfall (1,820 mm) in Lagos was 220 percent higher than that in London, the probability of drought was much higher in Lagos because of the erratic distribution of rainfall in Lagos in contrast to the relatively uniform distribution in London. Also important were the relative rates of evaporation, leaching, and runoff (higher in Lagos) and the water-holding capacity of the soils (much lower in Lagos).

The combination of high temperatures and humidity in the humid tropics restricts the types of crops and animals that can be raised and favors the spread of pests and diseases. The heat and humidity can also affect farmers and others involved in the production process, in that the hottest and wettest weather often coincides with the difficult tasks of land preparation and planting (Juo, 1989). Finally, climatic conditions in the humid tropics also result in high postharvest losses to pests and spoilage, and pose special problems for storage, transportation, and processing.

SOILS

The soils of the humid tropics vary from region to region (Table 1-6) and have special requirements, limitations, and possibilities for agricultural use. They are subject to several constraints, including low nutrient reserves, aluminum toxicity, high phosphorus fixation, high acidity, and susceptibility to erosion. These constraints, and the methods that have evolved to overcome them, vary among soil types and from region to region. Ideally, the soil, along with considerations of topography and water availability, should determine the

TABLE 1-6 General Distribution of Major Types of Soils in the Humid Tropics, in Percent

General Soil Grouping	Humid Tropic America	Humid Tropic Africa	Humid Tropic Asia	World's Humid Tropics
Acid, infertile soils (Oxisols and Ultisols)	82	56	38	63
Moderately fertile, well-drained soils (Alfisols, Vertisols, Mollisols, Andepts, Tropepts, Fluvents)	7	12	33	15
Poorly drained soils (Aquepts)	6	12	6	8
Very infertile sandy soils (Psamments, Spodosols)	2	16	6	7
Shallow soils (lithic Entisols)	3	3	10	5
Organic soils (Histosols)	—	1	6	—
Total	100	100	100[a]	100

[a]Numbers do not total to 100 due to rounding.

SOURCE: National Research Council. 1982. Ecological Aspects of Development in the Humid Tropics. Washington, D.C.: National Academy of Sciences.

optimal or ideal use of the land and its level of sustainability (Serrão and Homma, Part Two, this volume; Vincent and Hadi, Part Two, this volume). A significant challenge to researchers is how to maintain soil fertility in a sustainable manner (Ehui, Part Two, this volume).

Oxisols, found mostly in tropical Africa and South America, are used for shifting cultivation, subsistence farming, low-intensity grazing, and intensive agriculture (such as sugarcane, soybeans, and maize). In Asia, they are highly suited to producing tree fruit and spice crops. Due to extreme weathering, very low nutrient reserve, and a limited ability to hold soil nutrients, a number of nutrients in the ecosystems containing Oxisols are within living or dead plant tissue. However, these soils do have excellent physical properties and can be suitable for a wide range of uses if nutrient limitations are addressed.

Misconceptions About Humid Tropic Soils

Despite evidence to the contrary, the belief persists that the soils of the humid tropics are incapable of supporting sustainable agriculture and forestry. This belief is based on three main misconceptions about tropical soils: laterite formation, low soil organic matter content, and the role of nutrient recycling in agricultural systems.

LATERITE FORMATION

It has often been claimed that most soils of the humid tropics, when cleared of forest cover, will degrade irreversibly, ultimately forming brick-like layers known as laterite. Advances in the classification and mapping of soils show that areas in which laterite formation is a real threat are very limited and predictable (Sanchez and Buol, 1975). Only 6 percent of the Amazon region, for example, has soft plinthite in the subsoil, the substance capable of hardening into laterite if exposed by erosion. These soils occur in flat, poorly drained lands, where the danger of erosion is minimal. However, arid and semiarid regions of West Africa contain large areas of lateritic soils, especially in the West African Sahel.

Hardened laterite of geologic origin occurs in scattered areas in the humid tropics, where it serves as excellent road-building material. Low-cost roads in the Peruvian Amazon, which is essentially devoid of laterite formations, are inferior to those of the Brazilian state of Pará, where laterite outcrops occur. The laterite formation hazard, still frequently mentioned in the literature, is therefore of minimal importance

Ultisols are found mostly in regions with long growing seasons and ample moisture for good crop production. They are the most abundant soils of humid tropic Asia and are also present in Central America, the Amazon Basin, and humid coastal Brazil. Unlike Oxisols, they exhibit a marked increase of clay content with depth. They also usually contain high levels of aluminum, which is toxic to plants and severely restricts rooting in most crops. However, many Ultisols respond well to fertilizers and good management practices, and are commonly used in both shifting cultivation and intensive cultivation systems.

The agricultural production potential of Oxisols and Ultisols is improved if they are properly managed. For example, judicious applications of fertilizer can supplement their limited natural nutrient

---

as a constraint in the humid tropics. Where natural laterite outcrops occur, they are an asset to development.

SOIL ORGANIC MATTER

Organic matter content in soils of the humid tropics compares favorably with soils of temperate forests. Studies indicate that organic carbon and total nitrogen levels in tropical forest soils are somewhat higher than those found in temperate forest soils. No differences in organic matter content have been found between soils of the tropics and soils of the temperate region in uncultivated, forested ecosystems, or between Oxisols (abundant tropical soils found mostly in Africa and South America) and Mollisols (prairie soils of the U.S. Great Plains). With land clearing and continuous cropping, however, the organic matter content of soils of the humid tropics declines rapidly, because of continuously high temperatures throughout the year (Jenkinson and Ayanaba, 1977).

In most forested tropical ecosystems, soil organic matter is concentrated in the topsoil. Even though root growth within tropical forests is concentrated in the topsoil, many roots exploit the usually deep reddish subsoils for water and nutrients. In savannah Oxisols, however, soil organic matter is found in substantial quantities to a depth of 1 m or more.

NUTRIENT CYCLING

Another commonly held view is that tropical moist forests essentially feed themselves, since their soils are poor in nutrients. Some nutri-

continued

stores. In Ultisols, calcium (used to build cell walls) and magnesium (the essential ingredient in chlorophyll) are in short supply and are found primarily in the topsoil, where they have presumably been cycled by vegetation. In some Oxisols, phosphorus, which affects plant growth in many ways, is commonly so low that crops cease growth when they deplete the phosphorus contents of their seeds (Lathwell and Grove, 1986). These soils usually produce crops for only a few years before soil nutrients are exhausted or leached from the soil profile. At this point, farmers must either move to another location, restore nutrients to the soils through rotations or the application of manure or mineral fertilizers, or allow the land to revegetate before replanting.

Deforestation often leaves soils in a depleted state. Most tropical moist forests grow on an unpromising soil base, generally Ultisols

---

ent cycling studies that include the entire soil profile indicate a considerable portion of the ecosystem's nitrogen and phosphorus stocks may be located in the soil (Jordan, 1985; Sanchez, 1979). However, additional research is required to determine more accurately the content and availability of these nutrients in the biomass versus in the soils.

The high efficiency of tropical forest nutrient cycles has long been recognized (Nye and Greenland, 1960; Sanchez, 1976). Agricultural systems generally operate in the same way, with one major exception: biomass is not removed from natural ecosystems, but crop harvests in agroecosystems can remove large quantities of biomass and constitute the main pathway of nutrient loss. In grain crops, about 40 percent of the carbon, 60 percent of the nitrogen, and two-thirds of the phosphorus in crops are removed with the harvest, while most of the potassium, calcium, and magnesium remain in the crop residues (Sanchez et al., 1989). In an agricultural or forestry system, nutrients lost through harvesting must be balanced with nutrient inputs in the form of fertilizers, manures, or biological nitrogen fixation.

In agricultural systems dominated by annual crops, the flow of nutrients from soil to crop occurs seasonally and must be extremely rapid if high yields are to be attained. As crop residues are returned to the soil, they are broken down by soil fauna and flora into simple components, which are then available for uptake by the next crop. Losses from the system can occur if crop residues are removed from the field, if soil is lost through erosion, or if soluble nutrients remain in the soil with no crop growth during periods of heavy rain. The use of crop or animal residues as fuel can be a major source of nutrient (and carbon) loss from the system.

that are washed by heavy rains. Calcium and potassium are leached from the soil by rain. Iron and aluminum form insoluble compounds with phosphorus and, if present in high concentrations, will decrease the availability of phosphorus to plants. When forests are removed, rapid degradation in soil fertility can occur because of the dependence of these soils on nutrient cycling by deep-rooted plants (Buol et al., 1980).

Inceptisols, young soils of sufficient age to have developed distinct horizons, comprise the third most widespread soil type in the humid tropics. Three major kinds occur: Aquepts (poorly drained), Andepts (well drained, of volcanic origin), and Tropepts (well drained, of nonvolcanic origin). Among the Inceptisols, Aquepts are dominant in humid tropic America and Africa, and Tropepts are dominant in humid tropic Asia. Most of the Aquepts, or wet Inceptisols, are of high to moderate fertility and support dense human populations. In tropical America, they occur in the older alluvial plains along the major rivers and inland swamps of the Amazon Basin. About half have high potential for intensive agriculture. In Africa, large areas of wet Inceptisols (known locally as hydromorphic soils) long remained undeveloped because of human health hazards, although many of these hazards have been overcome and settlement has advanced. In Asia, many of the Tropept soils are used for lowland rice production. More than 90 percent of the world's rice is grown and consumed in Asia (where about 55 percent of the earth's people live). Inceptisols of volcanic origin (Andepts) are important in the volcanic regions of Asia, in parts of Central and South America, and in parts of Africa. They are generally fertile and have excellent physical properties.

Entisols are soils of recent development that do not show significant horizon layers. Within this soil type, well-drained, young alluvial soils (Fluvents) not subject to periodic flooding are considered among the best soils for agriculture in the world. Fluvents account for only 2.7 percent of the soils of the humid tropics and most are already cultivated; about two-thirds (25 million ha) are found in Asia where they are under intensive lowland rice production. Where forests remain on these soils, their preservation will be difficult due to their high agricultural potential.

BIOLOGICAL FACTORS

Biological constraints on agriculture in the humid tropics include insect and other pests, pathogens, and weeds; a lack of improved germplasm for the common crops of the region; and the loss of domestic and wild biodiversity. The hot and humid climate provides ideal conditions for pests and diseases. The growing season is essentially continuous and facilitates the development of persistent pests. Losses of crops to pests in the humid tropics are great. Preharvest losses are estimated to be 36 percent of yield, and postharvest losses are estimated to be 14 percent (U.S. Agency for International Development, 1990). The impacts of fungal, viral, and bacterial pathogens in developing countries have been studied less than those of insects, but the most comprehensive studies suggest that losses caused by pathogens are about equal to those caused by insects (Edwards et al., 1990). Weed growth is often so prolific and hard to control that it is thought to be the most important cause of yield depression (MacArthur, 1980; Sanchez and Benites, 1987).

Improved varieties of the major food crops grown by the inhabitants of forests in the humid tropics are generally lacking (especially in Africa). Rice, cassava, sweet potatoes, and cocoyams are the principal foods of indigenous populations (Juo, 1989). Root crops, in particular, have received far less attention from plant breeders than have the more conventional cereal crops. At the same time, local varieties and landraces of staple crops, many of which are highly adapted to local climatic and topographic conditions, are disappearing.

The loss of germplasm and species diversity is usually regarded as a consequence of development in the forests of the humid tropics. This loss can be seen as a serious constraint on long-term rural and agricultural well-being. The organisms within humid tropic agroecosystems provide vital services as pollinators, plant symbionts, seed dispersers, decomposers, pest predators, and disease control agents. These benefits can be diminished or lost as the diversity within agroecosystems decreases. Many local human populations also depend on nearby biological resources for food, fodder, pharmaceuticals, and other needs. Globally, tropical moist forests are the source of germplasm for many food and industrial crops. The local and global potential for using yet untapped plants and animals will remain unknown if their tropical habitats perish (Iltis, 1988). Opportunities for realizing local economic benefits through sustainable uses of biological resources could also be lost.

The Path to Sustainable Agriculture

Over the centuries, agricultural systems and techniques evolved to meet the special environmental conditions of the humid tropics. These include paddy rice systems; terrace, mound, raised-bed, and drained field systems; and a variety of agroforestry, shifting cultivation, home garden, and modified forest systems. Although these traditional systems are diverse in their particular adaptations, they share many traits: high retention of nutrients; maintenance of vegetative cover; a high level of diversity of crops and crop varieties; complex cropping patterns and time frames; and the integration of domestic and wild animals within the agroecosystem.

Shifting cultivation (also known as swidden, slash-and-burn, or slash-and-mulch agriculture) remains in wide use throughout the humid tropics. It is practiced on about 30 percent of the world's arable soils and provides sustenance to more than 300 million people and additional millions of migrants from other regions (Andriesse and Schelhaas, 1987). As traditionally practiced, shifting cultivation protects the resource base through efficient recycling of nutrients, conservation of soil and water, diversification of crops, and the incorporation of long fallow periods in the cultivation cycle. Fallows accumulate nutrients in their biomass and control weeds.

Traditional shifting cultivation systems are being disrupted, modified, and replaced as population pressures rise and as migrants unfamiliar with the humid tropics or indigenous land use practices attempt to farm newly cleared land. Typically, this results in shortened fallow periods, fertility decline, weed infestation, disruption of forest regeneration, and excessive soil erosion.

Monocultural systems have been successfully introduced over large areas of the humid tropics. Some of the more fertile soils already support monocultural production of coffee, tea, bananas, citrus fruits, palm oil, rubber, sugarcane, and other commodities produced primarily for export. However, the social and economic characteristics of monocultural crop and plantation systems are of concern in many countries where they are important land uses. While they provide productive employment, they often outcompete and, thus, discourage investment in domestic food crop production. At the same time, they occupy most of the high-quality agricultural land, although this is less true in the Asian humid tropics. They often entail concentrated ownership of large areas of land (either in the private sector or by the government), creating social and political instability, especially in densely populated countries. Where these land ownership patterns are pervasive, small-scale farmers who wish to continue farming have no other option but to move toward primary forests and marginal lands (rice farmers are an important exception in that rice production is carried out largely on long-established small farms). Fluctuations in world market prices of the commodities these systems produce, as well as the fertilizers and pesticides on which they depend, make monocultural production more vulnerable to political and macroeconomic trends than small-scale farming. This is evident, for example, in Cuba, where a high proportion of agriculture is devoted to sugar production.

The environmental characteristics of monocultural systems also raise important questions about their sustainability. The production and processing methods they employ are significant sources of pollution in many areas (Vincent and Hadi, Part Two, this volume). A high degree of biodiversity loss is incurred in establishing and maintaining monocultures. The fertile alluvial soils of the humid tropics are, in fact, so valuable for raising crops that the distinct and highly diverse lowland forests they once supported have virtually disappeared (Ewel, 1991). Because monocultures in the tropics concentrate species that under natural conditions were widely dispersed, they are more susceptible to pathogens and other pests than the same species in traditional mixed-crop systems or in natural forests. However, oil palm, rubber, sugarcane, and tea can be stable when grown in monocultures.

Despite these problems, monocultural systems are an important part of the mosaic of land uses in the humid tropics. With modifications, including reduced use of pesticides, enhanced recycling of nutrients, and more equitable distribution of productive land, these systems may continue to serve as important sources of food and agricultural production. Some monocultural crops, such as coffee, cacao, and rubber, have been produced in diversified small-landholder systems, making them more desirable both socially and environmentally. In the future, the challenge will be to better manage both the highly productive lands that are already in intensive use and the less productive lands that are used by many small-scale farmers. In advancing toward sustainability, a nation's agricultural system will need to be diverse to take advantage of available markets, to use more effectively its available natural and cultural resources, and to balance social, economic, and environmental needs.

The wide array of specific practices associated with sustainable agriculture includes the following:

- Low-impact land clearing techniques;
- Mulches, cover crops, and understory crops;
- Fertilizers and other soil amendments;
- No- and low-tillage planting techniques;
- Increased use of legumes as food crops, as cover crops, and in fallows;
- Improved fallow management techniques;
- Greater use of specially bred and alternative crops, grasses, shrubs, and trees (especially those tolerant of acidic, salinized, and high-aluminum soil conditions);

- Contour cropping and terracing;
- Biocontrol and other integrated pest management strategies;
- A variety of agroforestry systems that mix crops, trees, livestock, and other components; and
- Intercropping, double cropping, and other mixed cropping methods that allow for more efficient uses of on-farm resources.

Sustainable practices to improve productivity and conserve soil, water, and biotic resources can provide farmers with alternatives to continued clearing of forests. Based on recent research in Peru, for example, it is estimated that for every 1 ha of land put into sustainable soil management technologies by farmers, 5 to 10 ha per year of forest could be spared (Sanchez, 1991; Sanchez et al., 1990). The potential of sustainable agricultural practices to reduce deforestation will depend on the location. For example, the sustainable use of secondary forest fallows provides a viable alternative to primary forest clearing. Many of the degraded or unproductive pastures or croplands resulting from poor management practices can also be reclaimed.

The particular methods that are most appropriate in any given locality will vary both within and among the world's humid tropic regions. Local needs and opportunities, ecological circumstances, economic opportunities, and social and cultural mores, as well as the status of land and water resources, will determine which methods are most suitable. Sustainable agricultural systems cannot, in this sense, be imported. Although specific technologies can be more freely introduced, they must be adopted to the inherent opportunities and limitations of local agroecosystems.

The transition to more sustainable agricultural and land use systems is not without difficulty, particularly in the early stages. In many cases, substantial initial investments of time, labor, and money are required (for example, to construct terraces or to reforest steep slopes). In some cases, the transition requires significant changes in current farming practices and land uses (for example, restrictions on the burning of biomass). Against these short-term effects must be weighed the long-term benefits of these investments and changes. They include the following:

- Reduced pressure on primary forests and the mitigation of deforestation's effects;
- Preservation of species and germplasm diversity within the agroecosystem;
- Reduction in the amounts of carbon dioxide and other greenhouse gases released into the atmosphere;
- Conservation of soil, nutrients, and water resources;

- Increased productivity and a more stable food supply;
- Greater economic and social stability at local and national levels;
- Infrastructural developments that benefit small farms and local communities;
- Greater equity between farmers in resource-rich and resource-poor areas; and
- Increased training and employment opportunities for small-scale farmers, landless workers, and other people in rural areas.

THE NEED FOR AN INTEGRATED APPROACH

Improved land use in the humid tropics will require an approach that recognizes the characteristic cultural and biological diversity of these lands, respects their complex ecological processes, involves local people at all stages of the development process, and promotes cooperation among biologists, agricultural scientists, and social scientists. The easing of rigid disciplinary boundaries is of special importance in the humid tropics. During the past century, ecologists and other biologists have endeavored to understand the properties and dynamics of tropical forest ecosystems. Only recently, however, have they begun to transfer these insights to the study and management of tropical agricultural systems (Altieri, 1987; Gliessman, 1991a).

Most public sector agricultural research and development programs in the humid tropics have focused for the past 3 decades on developing and transferring technologies to maximize the production of cereal grains and a limited number of root and pulse crops. These technologies have led to high productivity in areas with good soil and water resources, and they have contributed substantially to national food self-reliance in Asia. Many efforts in Latin America and Africa have been directed toward increasing export earnings. Livestock production technologies have been improved, but not as part of small-scale integrated farming systems. Only recently has the agricultural development community begun to expand its programs to incorporate additional social and environmental considerations, and to devote more attention to the needs of small-scale farmers in resource-poor areas (Consultative Group on International Agricultural Research, 1990; National Research Council, 1991a).

Critics of the commodity-oriented approach hold that it has been limited by an inability to embrace all the factors and processes that influence the stability, productivity, and maintenance of tropical agroecosystems. In focusing scientific attention and development programs on particular crops and agroecosystem components, it has tended to neglect the range of physical and biotic interactions that influences crop production, the ecosystem-wide impacts of intensive production practices, the role of the crop in achieving better balanced and more equitable systems of land use, and the long-term social and economic aspects of cropping systems that require purchased inputs (National Research Council, 1991a,b).

This commodity-oriented approach has also been criticized for paying too little attention to small farms in resource-poor areas, the diverse crops and animals on which they depend, and the performance of traditional agricultural systems (Dahlberg, 1991). Many traditional resource management techniques and systems, often dismissed as primitive, are highly sophisticated and well suited to the opportunities and limitations facing farmers in the tropics. Traditional land use systems have begun to receive greater attention as the primary goal of agricultural research and development in the humid tropics shifts from maximizing short-term production and economic returns to maintaining the long-term health and productivity of agroecosystems. As noted above, their durability, adaptability, diversity, and resilience often provide critical insights into the sustainable management of all tropical agroecosystems. While most of these systems have been greatly modified or abandoned due to economic and demographic pressures, some could, with modification, contribute significantly to the stability and productivity of agriculture in many humid tropic countries. By combining the expanding scientific knowledge of tropical forest ecosystems and the empirical experience of farmers and agricultural scientists, the conceptual foundations of sustainable land use can be strengthened. By applying this knowledge back to the land, many farmers can better provide for their own needs as well as those of society and the ecosystems in which they live (Gliessman, 1990).

Agroecology—the application of ecological concepts and principles to the study, design, and management of sustainable agricultural systems—is one possible starting point in developing a more integrated approach. Agroecology tries to understand how physical conditions, soils, water, nutrients, pests, biodiversity, crops, livestock, and people act as interrelated components of agroecosystems, emphasizing the structure and function of the system as a whole. Agriculture is treated not as an independent sector or industry but as a critical element in achieving broader social and economic goals (Gliessman, 1991b). This emphasis allows particular production processes and resource management practices to be understood in their ecological as well as sociocultural contexts. It attempts to enable researchers, resource managers, development officials, and others to understand how multiple ecological, social, economic, and policy factors collectively determine the performances of agricultural systems (Conway, 1985; Gliessman, 1985, 1991a; Gliessman and Grantham, 1990).

The agroecological approach, if it is to become effective, will require interdisciplinary cooperation not only among tropical ecologists, biologists, foresters, and agricultural scientists but also among anthropologists, economists, political scientists, and other social scientists. Integrated investigations of this type can help ensure that the biophysical and agronomic components of the agroecosystem are to be considered alongside the historical, sociological, economic, political, and other cultural components (Edwards, 1987; Francis, 1986; Grove et al., 1990). However, the institutional structures and scientific environment for accomplishing this goal have yet to evolve.

MOVING TOWARD SUSTAINABILITY

Many obstacles impede progress toward sustainable land use in the humid tropics. To break the cycle of resource decline, people must be able to meet their needs in ways that are socially, economically, and environmentally viable on a long-term basis. Most of the fertile lands in the humid tropics are already being intensively used. Continued conversion of primary forests offers increasingly marginal gains. The only other alternatives are to enhance, through improved management, the stability and productivity of those lands currently devoted to agriculture, and to rehabilitate previously deforested lands that are now degraded or abandoned. Both strategies are needed. Together with continuing forest protection efforts, they can make land use as a whole more sustainable throughout the humid tropics.

There are no easy methods for reversing resource degradation, and no one land use method alone will suffice. Rather, agricultural sustainability will involve a variety of land uses, each of which requires a different strategy and a different degree of management intensity. These diverse efforts, however, rest on several basic realizations:

• Over the next several decades all land resources in the humid tropics must be more effectively managed to reverse current trends.
• Success depends not only on making each land use more sustainable but also on coordinating an appropriate mixture of land uses and management strategies for each region.
• Land use systems must maintain flexibility and allow time for natural processes of ecosystem recovery and change.

Building on these premises, a combination of improved land management techniques and innovative policy reforms can contribute to a better quality of life for the people of the humid tropics, and to more effective conservation of the natural resources on which they depend. Although there are lands in the humid tropics that are, and will continue to be, devoted exclusively to production agriculture, sustainability necessarily involves a spectrum of land uses, including low-intensity shifting agriculture, mixed cropping and agroforestry systems, perennial tree plantations, and managed pastures and forests, as well as restoration areas, extractive reserves, and strict forest reserves. Agricultural and nonagricultural land uses can in this way be coordinated to enhance sustainability at the field, landscape, watershed, regional, and even global scales. Operationally, this will entail the adoption of sustainable agricultural technologies on intensively managed lands; the restoration of cleared, degraded, and abandoned lands to biological and economic productivity; improved fallow and secondary forest management; and the protection and careful use of the remaining primary forests.

2

Sustainable Land
Use Options

In response to a combination of socioeconomic, agronomic, and environmental concerns, many scientists and policymakers are encouraging the implementation of sustainable agricultural systems (Altieri, 1987; Christanty et al., 1986; Consultative Group on International Agricultural Research, 1989; International Rice Research Institute, 1988; Ruttan, 1991; Vosti et al., 1991). Definitions of sustainable agriculture vary widely. For the purposes of this report, sustainable agriculture includes a broad spectrum of food and fiber production systems suited to the varied environmental conditions in the humid tropics. These systems attempt to keep the productive capacity of natural resources in step with population growth and economic demands while protecting and, where necessary, restoring environmental quality.

This chapter provides a basis for identifying the technical and policy changes needed to make land use in the humid tropics more sustainable (see Chapters 3 and 4). It discusses a variety of land use options that can be used to formulate plans for restoring abandoned and degraded lands and for preserving natural resources, including the primary forest. These land use options are defined and presented here under 12 descriptive categories ranging from highly managed intensive cultivation to forest reserves. These categories represent sets of activities commonly practiced in the humid tropics, but not necessarily found or applicable in all regions or to both upland and lowland areas. Although these categories do not include all land use

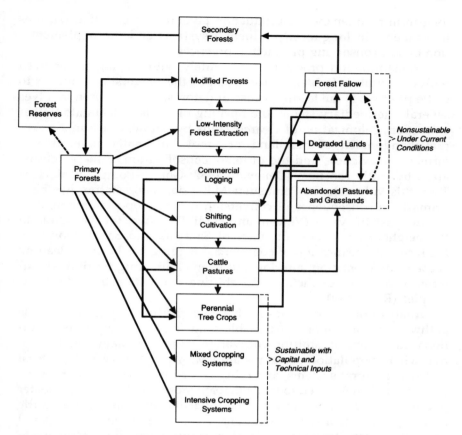

FIGURE 2-1 Examples of land transformation in the humid tropics.

activities in the humid tropics, they represent land uses with great potential for stabilizing forest buffer zone areas, reclaiming cleared lands, restoring degraded and abandoned lands, improving the productivity of small farms, and providing rural employment.

Examples of sustainable and nonsustainable uses are shown in Figure 2-1. Uses that reduce or eliminate forest cover have a broad range of requirements for capital and technical inputs, such as fertilizers and pesticides. Where social and economic conditions encourage resource depletion and short-term economic gain, however, land uses shift toward shorter and shorter production and harvest cycles, often leading to complete loss of economic production potential and abandonment. This pattern can be avoided if conditions encourage long-term maintenance of production potential—a goal that requires investments in long-term production systems and the implementation of soil-conserving production practices.

Transformation processes vary widely within a region, or even within a country. In Mexico, for example, the conversion of forests to cattle pastures is the leading cause of deforestation. It often involves several intermediary steps: the opening of roads to facilitate timber extraction, colonization of cleared lands by landless peasants, eventual abandonment of these lands or removal of small communities of farmers by eviction, and the ultimate consolidation of these "clean" areas by cattle ranchers (Denevan, 1982; Gómez-Pompa et al., Part Two, this volume). In Peninsular Malaysia, deforestation has been primarily a consequence of conversion to tree crop plantations during the past 100 years (Vincent and Hadi, Part Two, this volume). In the neighboring Malaysian states of Sarawak and Sabah, however, the recent intensification of commercial logging has been the leading cause of deforestation, altering and even eliminating traditional patterns of resource extraction and shifting cultivation by indigenous peoples (Rush, 1991).

Analysis of the processes of change is the first step in finding the pathways toward more sustainable land uses. For example, traditional low-intensity shifting cultivation systems remain a viable option where population pressures are low. Agroforestry, agropastoral and silvopastoral systems, and other labor-intensive mixed cropping systems are better suited to lands that are more fragile or under greater population pressure. More capital-intensive systems such as cattle ranching, perennial crop operations, forest plantations, and upland agricultural crop systems, while often environmentally destructive in the past, can present important opportunities for land restoration and improved land management. To be viable, they require secure land tenure, long-term investment, market access, and appropriate technologies.

No one system will simultaneously meet all the requirements for sustainability, fit the diverse socioeconomic and ecological conditions within the humid tropics, and alleviate the pressures that have brought about rising deforestation rates. The biological, social, and economic attributes of the land uses described in this chapter are summarized in Chapter 3 and technical and research needs are discussed. The order in which these land uses are presented corresponds broadly to the degree to which they change the composition and structure of primary forests. Figure 2-2 is a generalized depiction of changes to primary forests as they relate to agricultural land uses.

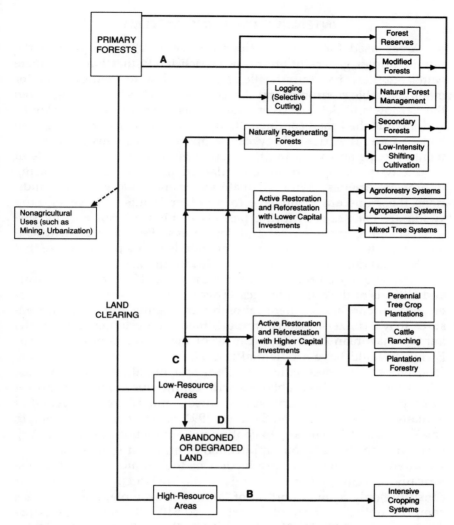

FIGURE 2-2 Pathways to sustainable agriculture and forestry land use. Management of land resources for sustainability depends on social and political forces as well as technological and economic development at local and national levels. National policy plays a significant role, particularly when maintaining various forest types (pathway A). Market forces determine the use of resource-rich areas following clearing (pathway B). The more critical pathways follow the clearing of resource-poor areas with less fertile soils. In some cases, with appropriate market incentives, sustainable use may evolve with modest public support (pathway C). Where the land resource has become severely degraded, more aggressive public sector involvement, such as incentives and subsidies, may be required (pathway D).

INTENSIVE CROPPING SYSTEMS

Areas used for intensive (high-productivity) agriculture in the humid tropics generally are resource-rich lands that have adequate water supplies, naturally fertile soils, very low to modest slope, or other favorable environmental characteristics. These areas range from the flat lowland delta or river valley areas to gently rolling uplands, and include the broad continental, high rainfall plains of the Amazon and of Central Africa. They can support input-intensive management systems and yield multiple harvests of crops at high levels of productivity. Crops are usually planted in rapid sequence, using improved varieties. With adequate water and good growing conditions the crops are responsive to fertilizer inputs. However, crop yields are constrained during periods of high rainfall and by seasonal flooding in some river and delta areas. Pest management usually prevents economic loss but often entails heavy pesticide use that can have adverse environmental and health impacts.

Intensive agriculture is agronomically feasible for most Oxisols and Ultisols of the humid tropics. This alternative may interest farmers near urban areas where favorable marketing infrastructure ensures that fertilizer-based continuous food crop production is viable. Large Amazonian cities import most of their food from other areas. Farmers would have a potential comparative advantage in growing food crops near these cities. In Peru and Brazil, respectively, sustained yields have been obtained with continuous cropping trials for 41 crops (17 years) in Yurimaguas Ultisols and 17 crops (8 years) in Manaus Oxisols (Alegre and Sanchez, 1991; Sanchez et al., 1983; Smyth and Cravo, 1991). The key to continuous production is effective crop rotations and the judicious application of lime and fertilizers.

Intensive agricultural production in the humid tropics has historically concentrated on the highly fertile lowlands. These lowlands constitute only a small portion of land. For example, lowland areas comprise only 20 percent of the estimated 510 million ha of the Amazon located within the national territory of Brazil (Serrão and Homma, Part Two, this volume). They account for between 10 and 40 percent of the total land areas of Southeast Asian countries (Garrity, 1991). In some river bottom and delta areas, annual flooding and receding water cycles deposit enriching organic and inorganic sediments. However, these flooded areas represent an even smaller portion of the total land base.

Soil characteristics coupled with water availability make these areas especially suitable for the intensive production of high-value food crops. Paddy rice production in Southeast Asia is one well-known example. Other intensive systems include terrace, mound, and drained-field systems of Africa, Asia, Central and South America, and the Pacific (Wilken, 1987a,b). These systems combine water control for drainage and irrigation through intricate systems of ditches, dikes, and shaping of the land. They provide harvests of high quality and quantity, and they are fairly predictable in their ability to provide consistent harvests from year to year.

The Development of Intensive Agriculture

Because of their high agricultural potential, resource-rich areas were the first to be developed, with early investment in roads, electricity, irrigation, and other infrastructural features. From the standpoint of national investment, these areas produced the greatest return per dollar. With few exceptions, most had been deforested and converted to high-productivity agriculture by the 1960s. Exceptions include malaria-infested portions of Nepal and Thailand, much of Mindanao in the Philippines, and large areas of inaccessible forestland in Brazil and Central Africa. These remaining areas may still be converted because of their value to agricultural production. Given social and economic pressures, the maintenance of forested areas can probably be justified only on the basis of preserving biodiversity. In most Asian countries, the few forested areas remaining on highly productive soils represent a small portion of total land area.

Internationally supported research and development in the 1960s and 1970s focused on realizing the high-production potential of these resource-rich lands. International agencies perceived an increasingly critical need for food and recognized the potential for existing scientific understanding and research methods to contribute to meeting this need. The international agricultural research centers (IARCs), such as the International Rice Research Institute (IRRI) in the Philippines, Centro Internacional de Agricultura Tropical (CIAT, International Center for Tropical Agriculture) in Colombia, and the International Institute of Tropical Agriculture (IITA) in Nigeria, were purposely situated in high-productivity tropical environments. The crop varieties that were developed had the genetic potential to respond to physical and managerial inputs under favorable soil and water environments. The widespread application of these new agricultural technologies gave rise to the green revolution. The agencies' focus also influenced the selection of areas with high-development potential and the placement of research centers within them (Dahlberg, 1979).

As a result of this concentrated investment in research and development, information and technology are readily available for high-productivity areas, both for individual crops and for high-intensity cropping systems (Chandler, 1979; DeDatta, 1981; International Irrigation Management Institute, 1987; Sanchez, 1976). Much of the information pertains to the major cereal, pulse, and other vegetable crops grown on a more intensive scale.

By the mid-1970s, most of the available highly productive land in the humid tropics was devoted to cultivating input-responsive crop varieties, and increases in individual crop yields began to level out (especially for Asian rice production). Attention turned to increasing annual area yields through more effective farming systems. From this early work came a broad range of research literature on farming systems methodologies for intensive cropping systems (Bureau of Agricultural Research of the Philippines, 1990; Harwood, 1979; International Rice Research Institute, 1975; Sanchez et al., 1982; Sukmaana et al., 1989). In the 1980s several of these research efforts shifted to particular types of cropping systems, such as wheat and rice rotations in the northern portion of the humid tropic zone (Harrington, 1991). It has been only recently, as researchers turned their attention to the rolling uplands and steeply sloping areas in Asia and to the

---

Intensification in Sustainable Agricultural Systems

Intensification is essential to developing sustainable agricultural systems in the humid tropics and elsewhere, but it can have various meanings in different contexts. Intensification in sustainable agricultural systems generally refers to the fuller use of land, water, and biotic resources to enhance the agronomic performance of agroecosystems. While intensification may involve increased levels of capital, labor, and external inputs, the emphasis here is on the application of skills and knowledge in managing the biological cycles and interactions that determine crop productivity and other aspects of agroecosystem characteristics.

This approach differs from that which has guided agricultural systems in the industrial countries in recent years. Over the past 5 decades, these systems have sought to maximize yields per hectare or per unit of labor through the development and dissemination of relatively few high-yielding crop varieties and through increased use of external inputs such as fuel, fertilizers, and pesticides. This model of agricultural development stresses intensification through progressively specialized operations and the substitution of capital and purchased inputs for labor. In general, it has entailed loss of diversity (in crop germplasm, cropping patterns, and agroecosystem biota) and high cash production costs.

reclamation of degraded pastures in Latin America, that on-farm, integrated animal systems have been studied (Amir and Knipscheer, 1987; Serrão and Toledo, 1990).

As farming system research became an important aspect of agricultural intensification efforts, researchers introduced socioeconomic considerations more systematically into their studies (Bonifacio, 1988; Hansen, 1981; Lovelace et al., 1988). Intensive farming systems were then increasingly studied with respect to their use of geophysical resources within different social and economic environments. Methodologies were developed to address more complex systems and their interactions in fragile and resource-limited environments, where changing land use patterns often have major social implications.

Intensive cropping systems face critical challenges. Questions are being raised about the ability of these systems to respond to the food needs of expanding populations. For several decades, lowland crop production has benefitted from the availability of improved varieties and hybrids, better agricultural chemicals, and mechanized farm equipment. For example, two to three crops of lowland rice with growing seasons of three to four months can now be produced

---

In meeting the concurrent goals of increased productivity and reduced environmental risk, intensification can occur in both temporal and spatial dimensions. Farmers can intensify the use of the resources available to them at different times by using more diverse rotations and optimal harvesting schedules. They can intensify the use of resources spatially by adopting techniques and growing crops that take fuller advantage of available sunlight, moisture, nutrient reserves, and biotic interactions, both aboveground (for example, through mixed cropping) and belowground (for example, through the use of legumes and deep-rooted tree crops). Optimum resource use in hilly areas of heterogeneous slope, soil type, and water resources requires a diversity of systems and system components.

In both the spatial and temporal dimensions, intensification through diversification involves the selection of crops, livestock, inputs, and management practices that foster positive ecological relationships and biological processes within the agroecosystem as a whole. These choices vary according to local environmental conditions and socioeconomic needs and opportunities. Improved agroecosystem performance is often sought through mixed cropping systems, while all internal resources (and necessary external inputs) are carefully managed to improve productive efficiency.

Rice terraces in the upper watershed area of the Solo River, Indonesia, are carefully tended to cultivate every available portion of land through the use of many different agronomic land use types, which are shown here in a single landscape. Population pressure on arable land is high in this area of Central Java. Credit: Food and Agriculture Organization of the United Nations.

each year. However, the growth in yield rates for cereal crops in Asia is increasing more slowly than demand (Harrington, 1991). Fallow periods that formerly allowed for the accumulation of nutrients and the suppression of pests have essentially been removed from the crop rotation sequence, their role being assumed by applications of purchased chemical inputs. Furthermore, pressures from pests and diseases are increasing as the area devoted to the cultivation of new varieties increases in size (Fearnside, 1987a).

In many countries, lowland areas that are relied on for producing staple and cash crops are in danger of becoming unfit for crop production as a result of improper management. The inappropriate use of high-productivity technologies is being implicated in various forms of natural resource degradation, including nutrient loading from fertilizers, water contamination from insecticides and herbicides, and waterlogging and salinization of land (Harrington, 1991). Loss of lowland cropland could seriously impair the capacity of countries in the humid tropics to meet future food demands.

The pressure to meet the subsistence needs of populations is causing governments to convert additional lowland as well as upland areas. In Indonesia for example, as transmigration programs continue, previously unmodified wetland ecosystems are being considered for cultivation of irrigated, monoculture rice or for mixtures of coconut plantations with secondary crops, which are grown to meet local needs rather than for cash or market (Kartasubrata, Part Two, this volume). In some areas, the high risk of malaria, schistosomiasis, and other diseases remains a significant barrier to the use of lowland areas. At present, these health concerns are greatest in the humid tropics of Africa and Asia.

Programs and Research Activities

To the extent that productivity in lowland areas declines and forested upland areas are environmentally degraded for future food production, sustainability in the humid tropics is placed at risk. These concerns are becoming the focal points of the preservation programs and research efforts of regional and international agricultural research centers. Efforts are being made to preserve lowland areas that have unique qualities. The Chitwan National Forest in Nepal is one of the few lowland rain forests successfully protected from development pressure. It constitutes a rich source of biological diversity in undisturbed Asian lowland, high-productivity ecozones. Further development of the Chitwan area for agriculture has so far been rejected.

Throughout the humid tropics, efforts are also being made to curtail soil erosion on intensively cultivated sloping lands. In the 1980s the Philippine Department of Agriculture initiated the Sloping Agricultural Land Technology Program, which proposed an intercropping system to produce permanent cereal crops with minimal or no fertilizer use. Between hedgerows of *Leucaena leucocephala,* a commonly grown fodder source for cattle, rows of woody perennial crops, such as coffee, were planted in contour strips alternating with several rows of food crops. Versions of this cropping system, using various plant species, provide farmers with a diverse income source and fertility-enhancing soil mulch. They can also reduce by as much as 90 percent the amount of soil lost under conventional cropping practices on open fields (Garrity, 1991).

More generally, agriculture production programs and research agencies that have traditionally focused on intensive cropping systems are reevaluating and redirecting their efforts. The IARCs of the Consultative Group on International Agricultural Research (CGIAR) now focus not only on increasing yields of intensive agriculture in favorable environments, such as irrigated lowlands, but also on developing programs to increase productivity and sustainability of cropping and livestock systems in less fertile, marginal environments, like sloping and hilly uplands (Consultative Group on International Agricultural Research, 1990).

The CGIAR has not defined the limits of the IARCs' research activities on issues of sustainability. Rather, those decisions are made by each center. For example, the CGIAR has not advocated the rehabilitation of degraded lands as a central priority of its system. However, most centers acknowledge that an increased percentage of arable land in their mandate areas has been degraded or removed from production and some have begun initiatives to address this issue (Consultative Group on International Agricultural Research, 1990).

Some centers, such as the IRRI and Centro Internacional de Mejoramiento de Maíz y Trigo (International Maize and Wheat Improvement Center), have emphasized sustainable agriculture through reallocation of internal resources, while others, such as the CIAT, IITA, and International Livestock Center for Africa, have developed explicit goal and mission statements. The International Center for Research in Agroforestry focuses its resource management agenda on mitigating tropical deforestation, land depletion, and rural poverty through improved agroforestry systems. In addition, several centers have increased the role of social science research to address the human and socioeconomic constraints on improved natural resource management practices (Consultative Group on International Agricultural Research, 1990). Perhaps the most important aspect of this increased attention will be the ability to share with resource-poor areas the institutional capacity, field research methodologies, and scientifically trained human resources of the IARCs, which had been developed primarily for agriculture on resource-rich lands.

Implications for Forest Boundary Stabilization

The ability of areas with high-quality soil and water resources in Asia to absorb more people engaged in agriculture is limited. These lands have been cleared and settled for many years, even centuries, often predating colonialism. Labor use levels are stable after the increases caused by the green revolution technologies of the 1960s and 1970s. Food production is increasing, but often at a rate not sufficient to keep up with national demand. The few remaining forest areas on these high-potential soils are unique in their genetic diversity and require extreme measures for protection. For the most part, the presence of these few remaining forests is testimony to the effectiveness of protection policies.

In the Americas and in Africa, significant forest areas remain. As roads are built, however, these areas are increasingly threatened with the possibility of land conversion. The short-term economic benefits of logging and the subsequent availability of these highly productive soils make the prospect of further agricultural expansion almost inevitable.

SHIFTING CULTIVATION

Shifting cultivation is one of the most widespread farming systems in the humid tropics, and it is often labeled as the most serious land use problem in the tropical world (Grandstaff, 1981). Shifting cultivation is usually defined as an agricultural system in which temporary clearings are planted for a few years with annual or short-term perennial crops, and then allowed to remain fallow for a period longer than they were cropped (Christanty, 1986). Conditions that limit crop yields, such as soil fertility losses, weeds, or pest outbreaks, are overcome during the fallow time, and after a certain number of years the area is ready to be cleared again for cropping (Sanchez, 1976).

While most shifting cultivation consists of various slash-and-burn methods, areas with high amounts of rainfall can use a slash-and-mulch system, which has less adverse effects on the environment. In warm wet conditions, relatively rapid decomposition of the mulch provides nutrient recycling benefits unavailable through burning, while

An example of slash-and-burn clearing of tropical rain forest. Credit: James P. Blair © 1983 National Geographic Society.

protecting the soil surface and increasing the amount of organic matter in the soil (Thurston, 1991).

As long as the human population density is not too high and fallow periods are long enough to restore productivity, shifting cultivation can be ecologically sound and can efficiently respond to a variety of human needs (Christanty, 1986). These systems are especially well suited for producing basic foodstuffs and meeting subsistence and local market needs.

However, in many of the areas where shifting cultivation had formerly been practiced successfully for centuries, population and poverty pressures have forced the shortening of the fallow period and field rotation cycle and the loss of productivity. Unless there are substantial social and economic changes, short-term cycles will continue and more lands will be cleared.

Although shifting cultivation generates limited income, few alternative cropping systems are ecologically feasible for many marginal lands. In most developing countries of the tropics, the expansion of cropping systems that depend on purchased inputs, especially those that are imported, are not economically feasible on these lands. Therefore, ways must be found to reduce the intensity of shifting cultivation if stabilization is to occur, yields are to be sustained, and the pressure on primary forests is to diminish.

Stabilization Guidelines

The length of the fallow period is the most critical factor for the long-term sustainability of shifting cultivation systems (Christanty, 1986). Shifting cultivation becomes more intensified with the combined pressures of rapidly increasing human populations, demands for income above subsistence levels, and the growing demand for cash crops. As the cropping period lengthens, the conditions that maintain a productive soil deteriorate. On much of the hilly, steep land where deforestation for cropping is occurring, erosion becomes a serious problem, soil nutrients are lost, and weedy vegetation quickly invades. Stabilization can only be achieved by allowing for an effective rest or fallow, accompanied by a series of improvements during the cropping period that lessen erosion and help maintain a fertile soil.

Guidelines for stabilizing shifting cultivation include the following:

• Respect local knowledge on cropping practices, use of local varieties, use of fire, soil management, and manipulation of the fallow period.

• Develop systems that strictly adhere to crop and fallow practices that maintain soil fertility. The length of time required before eventually recropping an area depends on local conditions, such as rainfall, soil conditions, and crop type, and can range from a few years to 30 or 40 years (Ruthenberg, 1971). Stable population levels and land tenure conditions are needed to maintain this system.

• Develop and refine organic matter management practices that improve soil and water conservation during the cropping period in order to reduce fertility loss, improve crop yields, and hasten the recovery of the system during the following fallow. The key to success is to maintain a continuous ground cover at all times during the cropping cycle. This can be achieved through minimum tillage, mulching, cover cropping, and multiple cropping (Amador and Gliessman, 1991).

• Diversify cropping systems to intensify the production of useful species, thus lessening the need for additional plantings. Diversification can be achieved through a variety of multiple cropping arrangements (Francis, 1986), such as introducing perennials or tree species into annual cropping systems. This approach usually requires market access for nonstaple food products, as the system is moved toward a perennial crop base.

• Develop managed fallow systems by intentionally introducing fallow plants that accumulate nutrients in their biomass at a faster rate than the natural fallow (Sanchez, 1976) and permit the harvest of useful or edible materials from the second growth vegetation (Sanchez and Benites, 1987).

By stabilizing shifting cultivation systems at a level of production that sustains yields, meets the needs of the local people, and respects the importance of an adequate fallow, both ecological and social benefits are obtained. Soil erosion, fertility loss, and invasion by weeds are minimized, and people are more likely to remain in one location. Research institutions as well as policymakers should realize that stabilized shifting cultivation systems are most appropriate in more remote and economically limited areas. With proper incentives, and research to develop alternatives, stabilized and diverse shifting cultivation systems could become effective buffers against further encroachment into tropical forests (Sanchez et al., 1990).

Managed Fallows and Forests in Mexico: An Example

The use of managed fallows and forests is one method by which productivity is maintained in stable shifting cultivation systems. Tropical farmers in Mexico typically plant or protect trees found along the edges of or scattered through their agricultural fields. Many of the trees are nitrogen-fixing species and their abundance may reflect centuries of human selection and protection (Flores Guido, 1987). Nitrogen-fixing trees provide most of the nitrogen required to maintain soil fertility under intensive high-yield cultivation. The use of legume trees as shade trees for cacao is a pre-Hispanic practice still used today and it has been extended to coffee production (Cardos, 1959; Jiménez and Gómez-Pompa, 1981). Shaded coffee plants produce less annually, but the shade adds many years to the useful life of the plants.

Other agroforestry techniques for managing agricultural plots (predominately used for corn production) include selecting and protecting useful trees on the cultivation site. After a year or two of intensive cultivation these plots are left to fallow. The protected trees can serve as a seed source and as habitat for birds and other seed dispersers and pollinators. During this time, postcultivation crops, which consist of perennial cultivated or volunteer crops, continue to be produced and harvested. Some species of shrubs and trees are planted, thereby providing a continuous source of products as well as influencing the composition of regenerating stands (Wilken, 1987b). Species selected for protection are determined by the interest, knowledge, and needs of the farmer, a factor which explains the high biological diversity found in fallows and in old secondary forests.

The way in which trees are cut when the plot is cleared also affects their survival. Coppicing involves cutting trees or shrubs close to ground level so they will regrow from shoots or root suckers rather than seed. Coppicing with a high trunk remaining improves survival and is a key factor in the successional process. Although only 10 percent of the trees may be coppice starts, they may account for more than 50 percent of biomass during the recovery phase depending on the type of forest (Illsley, 1984; Rico-Gray et al., 1988).

The distinction between an agricultural plot and the adjacent mature forest in the humid tropics may not be as clearly evident as in temperate regions. Rather than being separate categories of vegetation, milpas (small cleared fields) and mature forest patches are different stages of the cyclical process of shifting agriculture. Even mature vegetation is part of a more extensive management system that includes sparing trees in the milpa and protecting and cultivating useful plant species during the regrowth of the forest patch. These forest patches, along with other uncut areas where the mature vegetation is protected or where useful tree species have been encouraged or transplanted, are considered here to be forest gardens, managed forests, or modified forests.

The conservation of a strip of forest along the trails and surrounding the milpas is also important. This strip plays an important role in regeneration on fallowed lands (Remmers and de Koeyer, 1988), provides shade for travel by foot to distant fields, and maintains a habitat for wildlife. Links between patches of forest also may have a key role in maintaining deer, birds, and other game valued as food by local people.

Low-Input Cropping: A Transition Technology

Low-input cropping is a management option that has evolved as a transition technology between shifting cultivation and several sustainable options (Sanchez, 1991). It enables farmers to substantially increase short-term crop production while preparing themselves and their land for sustained land use alternatives. This option is applicable to farmers on acid, infertile soils in rural areas with limited capital and marketing infrastructure. Its principal features are the following: clearing of secondary forest fallows by slash and burn; use of acid-tolerant upland rice and cowpea cultivars in rotation, with only grain removal to minimize nutrient export; no use of fertilizers, lime, or external organic inputs; establishment of legume fallows when weed competition and nutrient deficiencies make cropping unfeasible; and elimination of fallows by slash and burn after 1 year, shifting to other management options such as grass-legume pastures, agroforestry, or mechanized continuous cropping (Sanchez and Benites, 1987).

Current results indicate the initial cropping cycle lasts 2 or 3 years and there is progressive reduction in cycle length after each legume fallow. The system is considered transitional because of two major constraints: nutrient depletion and weed encroachment. Ongoing investigations seek to prolong the duration of low-input cropping by broadening the base of acid-tolerant cultivars and species; increasing knowledge about components of the nutrient depletion process; and improving weed management through crop rotations, plant density, and frequency and time of legume cover crop fallows.

AGROPASTORAL SYSTEMS

Farming systems that combine animal and crop production vary across regions and agroecological zones. In Asia the animal components of small farming operations vary with cropping systems (McDowell and Hildebrand, 1980; Ruthenberg, 1971). In lowland rice farming areas, buffalo provide (1) traction for cultivating fields and (2) milk and meat that are consumed domestically or sold in markets. Cattle, fowl (mainly chickens and ducks), and swine are also commonly raised on these farms. Feeds include crop residues, weeds, peelings, tops of root crops, bagasse, hulls, and other agricultural by-products. In highland areas, swine, poultry, buffalo, and cattle are raised in combination with rice, maize, cassava, beans, and small grains. Livestock is less important on farms dominated by multistory gardens, which may occasionally include cattle, sheep, and goats. Feed is typically cut and carried from croplands. Livestock animals are also of some importance on tree crop farms where they either graze freely in pastures, are tethered to clean specific areas, or are fed with tree cuttings.

The cropping systems of tropical humid Africa are dominated by rice, yams, and plantains (McDowell and Hildebrand, 1980; Ruthenberg, 1971). Goats and poultry are the dominant animals. Sheep and swine are less abundant, but still common. Feeds include fallow land forage, crop residues, cull tubers, and vines. The small farms of Latin

America typically include crop mixtures of beans, maize, and rice (McDowell and Hildebrand, 1980; Ruthenberg, 1971). Cattle are common and maintained for milk, meat, and draft. Swine and poultry are raised for food or for sale. Pastures, crop residues, and cut feeds support animal production.

The literature dealing with agropastoral systems is scarce due to the lack of directed research and development efforts. Much of it was contributed by farming systems research (for example, Harwood, 1979; McDowell and Hildebrand, 1980; Shaner et al., 1982). The variety of agropastoral systems and the complexity of mixtures and interactions have discouraged systematic research and development. As farm diversification, soil and pasture management, and crop nutrient management become increasingly important to sustainable land use, these closely integrated systems should receive greater attention. Presently, most knowledge of agropastoral systems in the humid tropics resides with the native populations that manage them.

Features and Benefits of Agropastoral Farms

The close interaction between crops and livestock is the most striking feature of agropastoral farms. The structure of agropastoral farming systems is defined by the mix of crop and animal components, the extent of each, use of on-farm resources, interactions among the components, flows of energy and nutrients, and the individual contribution of each component to farm productivity (Harwood, 1987).

For example, in humid areas of Asia, land characteristics are a major determinant of crop and livestock components (Garrity et al., 1978). Heavy rains and fine textured soils make the lowlands most suitable for rice and a few other crops. Swine are raised by shifting cultivators, but the interaction between the animals and crops is largely unstructured. On more permanent farms, swine are typically raised in close association with vegetables that are produced for market (Harwood, 1987). In the humid areas of Africa, pests and diseases severely restrict the distribution of ruminants and people (Jahnke, 1982).

Agropastoral farming systems are usually highly diverse (Harwood, 1987). In most, several crops are produced on the same land within a single growing season or period, as in relay cropping or rotation systems, or within the same space simultaneously, as in intercropping systems. Rotations and polycultures are effective in controlling pests, diseases, and weeds (Altieri, 1987; Kass, 1978). They can also make nutrient cycles more efficient, protect soils from erosion, and influence the composition of the biota in and on the soil (Grove et al.,

1990). Mixed systems appear to enhance productivity and stability, which may account for their widespread appeal.

Other benefits accrue from agropastoral systems. In effect, the incorporation of livestock into farming systems adds another trophic level to the system. Animals can be fed plant residues, weeds, and fallows with little impact on crop productivity. This serves to turn otherwise unusable biomass into animal protein, especially in the case of ruminants. Animals recycle the nutrient content of plants, transforming them into manure and allowing a broader range of fertilization alternatives in managing farm nutrients. The need for animal feed also broadens the crop base to include species useful in conserving soil and water. Legumes are often planted to provide quality forage and serve to improve nitrogen content in soils.

Beyond their agroecological interactions with crops, animals serve other important roles in the farm economy. They produce income from meat, milk, and fiber. Livestock increase in value over time, and can be sold for cash in times of need or purchased when cash is available (McDowell and Hildebrand, 1980).

Incorporation of animals into cropping systems requires increases in management and labor inputs in contrast to crop farming. Farmers also need to gather and process large amounts of information. For example, decisions and actions must occur according to complex time schedules and the flow of labor and materials must be coordinated.

Requirements for Greater Sustainability

The high degree of sustainability of agropastoral systems is a consequence of the efficient use of on-farm resources. But these farms are not isolated from external influences. Markets must be available if the economic benefits of livestock are to be realized. Labor must be available to fulfill the additional demands of the mixed system. Knowledge must be preserved and communicated to assure that managerial skills are maintained. These farmers must be protected from policy distortions that cause them to alter their mixed systems in ways that decrease their sustainability (for example, incentives to exceed the animal carrying capacity of their resources).

If the agropastoral farming systems employed by small-scale farmers are to be improved and promoted within the humid tropics, institutional and policy changes are required. Research institutions must address the complexity of these systems and undertake studies to improve them. Project sponsors must recognize that such research is new and may require continuous and perhaps long-term support.

Educational outreach programs will be needed to promote improvements. Because traditional extension programs rarely focus on integrated management or small farms, changes are also required in these institutions. Governments need to avoid policies that cause small-scale farmers to abandon their mixed systems, and they must formulate policies that encourage and reward the protection of natural resources and environmental quality. A greater understanding of the interactions between national policies and local incentives would help assure that appropriate policies are developed.

CATTLE RANCHING

The conversion of tropical rain forests to open pastureland for cattle ranching is governed by socioeconomic and political pressures existing in each country. This section discusses the potentials and limitations of pasture-based cattle raising, with emphasis on regions where cattle ranching has greater importance.

Cattle are herded in Brazil on land cleared from tropical rain forest. Credit: James P. Blair © 1983 National Geographic Society.

Cattle Pastureland in Asia

Cattle raising on pasturelands takes place in Southeast Asian countries, mainly in Indonesia (Kartasubrata, Part Two, this volume), the Philippines (Garrity et al., Part Two, this volume), and Thailand (Toledo, 1986), but it is not a significant factor in increasing deforestation since crop (mainly rice) production systems are dominant. Cattle and buffalo constitute the main work force for many farm operations. They are also used for meat and dairy production. Generally their forage consists of stubble in the dry season and herbaceous vegetation that grows during the rainy season on dikes and rice fields, along the roadside, and in marginal areas of community pastures.

In some countries vast expanses of originally forested land are increasingly being converted to low-forage-value savannah grasslands of *Imperata cylindrica* due to intensive shifting agriculture on acid and infertile soils (Garrity et al., Part Two, this volume). In the Philippines, the human population of more than 5 million that subsists on shifting agriculture exert persistent pressure on formerly forested land that, due to frequent burning, is steadily being converted to *I. cylindrica* (Sajise, 1980). This same situation has been documented in Indonesia by Kartasubrata (Part Two, this volume). In parts of India, Bangladesh, and Nepal, overgrazing on communal lands is a major factor in productivity decline and soil erosion in the absence of incentives or institutions to control land access.

Cattle Pastureland in Africa

Livestock production in the humid zone of Africa is not important as an economic activity. Although some land is being cleared for cattle pasture, much of this land is not suitable for pasture beyond a few years because of soil erosion and low fertility (Brown and Thomas, 1990). Many cattle in equatorial Africa are also vulnerable to the effects of trypanosomiasis, which can cause poor growth, weight loss, low milk yield, reduced capacity for work, infertility, abortion, and often death. Annual losses in meat production alone are estimated to be $5 billion. This economic cost is compounded by losses in milk yields, tractive power, waste products that provide natural fuel and fertilizer, and secondary products, such as hides (International Laboratory for Research on Animal Diseases, 1991). Projects to eradicate the tsetse fly, which transmits the disease, are expensive and the use of large amounts of chemicals damages the environment (Goodland et al., 1984; Linear, 1985).

Some of the African breeds of cattle are genetically resistant to the effects of trypanosome infection, but they generally do not possess favorable production traits (International Laboratory for Research on Animal Diseases, 1991). Milk and meat yields are much lower than those of the European breeds, which are not tolerant to the disease and do not thrive in infested areas. However, the total efficiency of an animal is most important for African farmers, who need livestock that can produce milk, blood, and meat under poor range conditions and that can be used as draft animals (Brown and Thomas, 1990).

Dwarf sheep and goats tolerant to trypanosomiasis are more prevalent in the humid zone of equatorial Africa. Compared with cattle, these smaller ruminants have greater resistance to drought conditions, faster breeding cycles, and lower feed requirements. They are kept around the farmers' homes, are usually sedentary or restricted in movement to short distances, and often compete with food crops for space, soil, water, and nutrients (Sumberg, 1984). Research is being conducted into tsetse vector control, epidemiology, trypanosome biology, host resistance, and drug applications (International Laboratory for Research on Animal Diseases, 1991; International Livestock Center for Africa, 1991). Work is also under way on the use of bushy legumes, such as *Leucaena leucocephala* and *Gliricidia sepium,* as a high-quality forage for goats and sheep and as mulch material because of their high-nitrogen content for crop production (International Livestock Center for Africa, 1991).

Cattle Pastureland in Latin America

The socioeconomic and ecological importance of cattle raising in Latin America is based on several factors, some of which are the following:

- Biological and soil-related constraints on agriculture;
- Low human population density;
- Lack of infrastructure for transporting agricultural inputs and consumable products;
- Tax incentives and lines of credit for cattle ranching in some countries;
- Priority ranking and protection by Latin American governments;
- Cultural traditions that give cattle ranchers respect and status regardless of production and profit; and
- High levels of regional and international demands for meat.

Another important factor is the ability of cattle to transport themselves to markets by walking long distances, regardless of road and weather conditions. As a result labor requirements are lower—an especially significant consideration along the Amazonian frontier, where transportation of agricultural products is often difficult (Gómez-Pompa et al., Part Two, this volume; Serrão and Homma, Part Two, this volume; Serrão and Toledo, In press; Toledo, 1986).

In the Brazilian Amazon, Central America, and Mexico, cattle raising is a leading cause of forest conversion. In Central America, between 1950 and 1975, the pasture areas developed from deforested primary forest doubled; so did the cattle population. In the Brazilian Amazon, generous tax incentives and credits led to more than 112 big projects of farming and cattle ranching between 1978 and 1988. They were linked to development policies supported by international loans—an investment of more than $5 billion (De Miranda and Mattos, 1992).

In Andean countries, such as Colombia, Ecuador, and Peru, active colonization is also moving toward Amazonian forested areas. In the Peruvian Amazon, production systems that involve deforestation are found mostly in small areas (less than 100 ha) and consist of shifting agriculture, plantations, and cattle raising for meat and milk production (Toledo, 1986).

In general, cattle raising on previously forested land, whether large or small ventures, has often been uneconomical due to the decreasing productivity and stocking rates of pastures. This deterioration combined with the relative growth in herd size requires ranchers to convert more forestland to cattle production. The result has been a form of large-scale "shifting pasture cultivation" where the ecological damage, in terms of losses in biomass, biodiversity, soil, and water and possible changes in the climate, can be high (Salati, 1990; Serrão and Homma, 1990; Serrão and Toledo, 1990).

In the Peruvian Amazon, soil-plant-animal research has focused on developing pastures for dual-purpose (beef and milk) production in small landholdings where farmers will also grow crops and trees. Technology from CIAT's Tropical Pastures Program, developed primarily in savannah ecosystems, was adapted to humid tropic conditions. Legume and grass ecotypes were screened for their performance under acid soil conditions and subsequently evaluated for their persistence and compatibility when subjected to various grazing intensities. A grazing trial in Yurimaguas is the longest running replicated trial to test an acid-tolerant, grass-legume mixture in the humid tropics (Ayarza et al., 1987). If legume-dominated pastures prove to be sustainable, a new concept for cattle production may emerge in the humid tropics. New studies are also under way to gain further insight on nutrient cycling and to refine management practices.

Pasture Degradation: A Common Feature

Pasture degradation is the primary problem that cattle raising faces in the humid tropics. Although it is a common problem throughout the humid tropics, pasture degradation has been most evident in Latin America. Toledo and Ara (1977) and Serrão et al. (1979) identified the phenomenon and described the degradation process. The main cause of declining pasture productivity is low soil fertility and, more specifically, low soil phosphorus and nitrogen availability. Low fertility is a particularly important constraint on grass species that require more nutrients, such as *Digitaria decumbens, Hyparrhenia rufa,* and *Panicum maximum.*

During the past 25 years, particularly in Latin America, commonly used grasses that demand more nutrients have been gradually replaced by less demanding species. For example, *Brachiaria decumbens* can grow satisfactorily despite low soil fertility and has been rapidly adopted. However, because of high susceptibility to spittlebugs (*Aneolamia* spp., *Deois* spp., *Mahanarva* spp., and *Zulia* spp.), pastures of *B. decumbens* rapidly degrade (Calderón, 1981; Silva and Magalhães, 1980). Within the past 15 years, *B. humidicola,* which is more tolerant of low-fertility conditions, has been increasingly adopted in the Brazilian Amazon due to its supposed tolerance to the spittlebug (Silva, 1982). However, at the commercial production level, it has proved to be susceptible to this insect pest at high levels of infestation and has shown limited productivity potential due to its low nutritional value and poor palatability compared with other more nutritious forages (Salinas and Gualdrón, 1988; Tergas et al., 1988).

Cattle ranchers also face the serious problem of weed invasion, considered by many to be a cause of degradation and by others to be a secondary effect of the loss in competitive capacity and productivity of sown forage species. When the forest is cleared to establish pastures, available forage species are planted. Normally, the first year of establishment is successful and grazing begins. Depending on soil fertility, tolerance to biotic factors (insects, diseases, and weeds), and the quality of management, pastures can increase in productivity and stabilize at a level that is both economically favorable and ecologically justifiable. In practice, however, pastures commonly degrade rapidly, weed species invade, and a secondary forest begins to develop. If grazing pressure continues and effective weed control and burning are not carried out, biomass continues to decline and the pasture becomes a derived "native" ecosystem of generally low productivity and quality (Serrão and Toledo, In press).

Reclamation of Degraded Pasture
on Deforested Lands

Low agronomic sustainability characterizes pasturelands in their first cycle, that is, when they are first formed using available grasses after the clearing of the primary forest and mature secondary forest (Serrão et al., 1979; Serrão and Toledo, In press). As a result, large tracts of degraded pasturelands have become unproductive and eventually have been abandoned. This situation is more typical of Latin America than elsewhere, especially in the Brazilian Amazon, where in the past two decades between 5 million and 10 million ha of pasturelands have reached advanced stages of degradation (Serrão and Homma, Part Two, this volume).

If appropriate technology were applied to about 50 percent of the areas deforested for cattle raising production in the Brazilian humid tropics, it would be possible to produce animal protein and other agricultural products for the region's growing population (now close to 18 million people) at least until the year 2000 (Serrão and Homma, Part Two, this volume). In other words, from a technological viewpoint, Brazil could meet its crop and cattle production needs during the 1990s without further deforestation.

Cattle-raising development efforts should concentrate on degraded and abandoned first-cycle pasturelands (that is, those that are formed after the clearing and burning of a primary forest or a mature secondary forest). Scientific understanding of pasture reclamation through mechanization, improved forages, fertilization, and weed control is becoming increasingly available. New reclamation technologies, building on years of research, are being used in the Brazilian Amazon, with varied success (Serrão and Homma, Part Two, this volume; Serrão and Toledo, 1990).

However, several factors impede adoption of these relatively high-input technologies. Subsidies, which few developing countries can afford, are often required to make adoption of these technologies economically feasible, especially in the early stages of reclamation. Moreover, reclaimed pastures are based on a few forage species and cultivars with limited adaptability to the naturally poor and acid soil conditions or to the prevailing biotic pressures. Consequently, reclaimed pastures, although generally more stable than first-cycle pastures, are still prone to degradation. Their stability depends on relatively high investments for maintenance fertilization, grazing management, and weed control.

The Appropriate Pasture Technology
for Sustainability

The development of sustainable pasture-based production systems in acid, low-fertility soils in deforested lands of the humid tropics should be based on the following:

- Adaptation of forage grasses and legumes to the environment.
- Efficient nitrogen fixing and nutrient cycling.
- Well-established and well-managed pastures of grasses and legumes that can efficiently recycle the relatively small quantities of nutrients in the modified ecosystem.
- Intensification of pasture production using appropriate technology to increase pasture sustainability, thus reducing the pressure for more deforestation.
- Research on stable pasture-crop and pasture-tree systems that are biologically, socioeconomically, and ecologically more efficient than pure herbaceous open pastures.

To be sustained, pasture-based cattle production operations must be technically and socioeconomically manageable. That is, the farmer should have the financial resources and knowledge necessary for successfully operating on a sustainable basis.

Intensified pasture-based cattle production systems, together with crops and trees, can play an important ecological and socioeconomic role in reclaiming already deforested and degraded lands. The integration of annual crops with pastures that are established using residual crop fertilization can sometimes pay for upgrading the soil environment and further improve the soil's physical and chemical conditions through effective nitrogen fixation and nutrient recycling. Multipurpose trees can pump nutrients to the upper-soil layers, fix nitrogen, and provide supplemental animal feed, shade, and income. These integrated systems can be very efficient in using and conserving natural resources in the humid tropics, but they must be adapted to the environment, internally compatible, and relevant to farmers' needs.

Diversified and integrated pasture-based animal-crop-tree systems in deforested lands are found throughout the humid tropics, and are generally associated with small- and medium-sized farm operations. In many cases, however, they lack high levels of sustainability (Veiga and Serrão, 1990). Research is needed to understand, and develop management principles to optimize, the productivity and sustainability of agrosilvopastoral systems. Research is also needed on selecting multipurpose trees for poor soils and on developing markets for well-adapted native timbers and fruit trees.

AGROFORESTRY SYSTEMS

Agroforestry, the combined cultivation of tree species and agricultural crops, is an ancient and still widespread practice throughout the world. It encompasses a variety of land use practices and systems, some of which are presented individually in this chapter. This section presents a general overview of the principles of agroforestry and their implications for maintaining or developing sustainable agriculture and forestry practices.

In agroforestry systems, woody and herbaceous perennials are grown on land that also supports agricultural crops or animals. The mixture of these components, in the form of spatial arrangement or temporal sequence, enhances ecological stability and production sustainability. This integration allows the components to complement one another in their use of resources and in the timing of that use. Perennials have deeper roots and higher canopies than those of annuals, allowing better management of above- and belowground resources. Under ideal conditions:

- Nutrients recycled from the subsoil to the surface by deep-rooted perennials can be used by annuals.
- Leguminous perennials fix atmospheric nitrogen that can be used by annuals.
- There is minimal competition for water because of differences in depth from which the roots of annuals and perennials extract water from the soil.
- Some perennials produce allelopathic compounds that can suppress weeds.
- Differences in the structure of perennials and annuals, leading to a multistory canopy, reduce competition for light among plants.

Agroforestry systems have the potential to improve production and to enhance the agronomic and ecological sustainability of resource-poor farmers in the humid tropics. In practice, however, the potential benefits of agroforestry systems can be harnessed only through skillful and labor-intensive management of compatible systems. There are no simple blueprints of a universally applicable system that can harness all potential benefits possible under ideal conditions. Thus, a wide range of agroforestry systems has been designed to alleviate agronomic, ecologic, or managerial constraints.

Types of Traditional Agroforestry Systems in the Humid Tropics

Agroforestry is not a new concept in the humid tropics. Several types of traditional agroforestry systems exist, but no standard classification system is available to categorize them. Nair (1989) proposed a classification system based on structural, functional, agroecological, and socioeconomic factors (Figure 2-3). These broad categories are interrelated, and not necessarily mutually exclusive. In agroforestry land use systems, three basic components are managed by people: the tree (woody perennial), the herb (agricultural crops, including pasture species), and the animal. Based on their structure and function, agroforestry systems can be classified into the following three categories:

- *Agrisilviculture* is the use of crops and trees, including shrubs or vines. It includes shifting cultivation, forest gardens, multipurpose trees and shrubs on farmland, alley cropping, and windbreaks as well as integrated multistory mixtures of plantation crops.
- *Silvopastoral systems* are combinations of pastures (with or with-

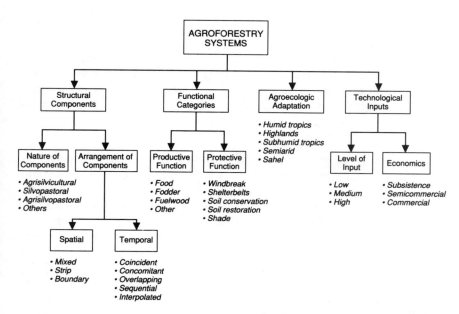

FIGURE 2-3 Characteristics of traditional agroforestry systems used in the humid tropics.

out animals) and trees. They include cut-and-carry fodder production, living fences of fodder trees and hedges, and trees and shrubs grown on pastureland.

• *Agrisilvopastoral systems* are those that combine food crops, pastures (with or without animals), and trees and include home gardens and woody hedges used to provide browse, mulch, green manure, erosion control, and riverbank stabilization.

Other types of agroforestry systems include apiculture (beekeeping) using honey-producing trees, aquaculture whereby trees lining fishponds provide leaves as forage for fish, and multipurpose woodlots that serve various purposes such as wood, fodder, or food production and soil protection or reclamation.

Principal types of agrisilvicultural systems traditionally used in the humid tropics are:

• *Rotational agroforestry.* In traditional shifting cultivation, trees and wood species are naturally regenerated over a period of 5 to 40 years and rotated with annual crops that are cultivated from 1 to 3 years. Improved tree species can be grown in place of native vegetation to achieve better soil conditions. This technique is used in multipurpose woodlots (where diverse mixtures of trees are used), home gardens (where trees and crops are grown close to the house), and compound farms (where trees, animals, crops, and the farmer's dwelling are in a fenced area).

• *An intercropping system.* Annual and perennial groups of plants are grown within the same land management unit. This system enables continuous production of food and tree products with a minimum need for restorative or idle fallow. Typical examples of intercropping systems include alley cropping and boundary planting of trees and wood hedges.

Two examples of the successful use of agroforestry systems by resource-poor farmers in the tropics are found in the Philippines and Rwanda (Lal, 1991a). In the Philippines, many small-scale farmers took up cash-crop-tree farming to produce pulpwood, poles, timber, charcoal, or fuelwood in the 1960s (Spears, 1987). The program gained significant momentum in 1972 when the Paper Industries Corporation of the Philippines (PICOP) entered into an agreement with the Development Bank of the Philippines to develop a loan scheme for small-scale tree farmers with titled or untitled land. Provision was made for part of the farm area to be maintained under food crops. PICOP guaranteed a minimum purchase price, but allowed farmers to sell wood to other outlets if they could get better prices. Within 10 years, the program covered 22,000 ha and supported 3,800 farmers, about 30 percent of whom had taken advantage of the credit program. A key to the program's success has been the high financial returns from tree growing. Adequate market incentives and security of land tenure were the basic factors responsible for acceptance by farmers.

The second example involves restoring eroded land in Rwanda using an agrisilvopastoral system. At Nyabisindu, a complex system of trees, animals, and crops was developed using the community's existing knowledge. Trees and hedges, yielding fruit, wood, and fodder, were used as protective ground cover against soil erosion. Extensive use was also made of perennial crops to further stabilize the soil (Dover and Talbot, 1987).

In Amazonian Ecuador, a sustainable system has been developed to raise sheep in association with cassava and contour strips of *Inga edulis*, which is a deep-rooted leguminous fuelwood tree. After the cassava is harvested, a perennial leguminous ground cover, *Desmodium* sp., is planted between the trees to enrich the soil. Sheep graze on the ground cover (Bishop, 1983).

Keys to the success of these projects included building on traditional knowledge, involving farmers in the choice of species, and providing economic incentives greater than those of traditional systems. Resource conservation and land restoration were additional benefits to the local community.

The viability and sustainability of these systems can be attributed to some combination of the following factors:

• A reduced fallow period and a greater ability to cultivate on a long-term basis, thereby eliminating the need to move to new land;

• Reduced use of chemical fertilizers and other fossil-fuel-based inputs due to enhancement of soil organic matter and improvement in soil fertility;

• Improved soil structure and physical properties (for example, better sizes of pores and channels in the soil that allow better water penetration and drainage);

• Decreased risks of soil degradation from accelerated erosion and other degenerative processes;

• Increased production and a rise in economic status from subsistence to partially commercialized farm; and

• Decreased need for clearing new land.

Improved Agroforestry Systems

Scientists and policymakers generally are eager to improve traditional agroforestry systems by enhancing productivity and ecological compatibility. Ways of improving these systems include using better trees and woody shrubs and creating an orderly arrangement of trees, crops, and livestock.

IMPROVED TREES AND WOODY SHRUBS

Several trees and woody shrubs are used in traditional and natural fallow systems. Some commonly used species include *Acioa baterii*, *Afzelia bella*, *Alchornea cordifolia*, *Anthonotha macrophylla*, and *Gliricidia sepium* (Okigbo and Lal, 1977). Some improved species have several advantages in an agroforestry system, including their ability to fix nitrogen, grow fast, tolerate soil acidity, and withstand regular coppicing. Commonly recommended tree species are listed in Table 2-1. However, validation for and adaptation to specific local systems are essential. More must be known about the agronomic and ecological bases of the mixtures to increase their attractiveness and usefulness to farmers.

Multipurpose trees can also be grown on cropland or pastureland.

TABLE 2-1 Commonly Recommended Species for Agroforestry Systems in the Humid Tropics

Species	Growth Characteristic(s)	Uses
Acioa baterii	Fast-growing shrub	Alley cropping, nitrogen fixation
Albizia falcata	Tree grows to 30 m	Erosion control, nitrogen fixation
Albizia lebbeck	Tree grows to 25 m	Erosion control, nitrogen fixation
Anthonotha macrophylla	Fast-growing shrub	Alley cropping, nitrogen fixation
Calliandra calothyrsus	Fast-growing shrub to 8 m, on acid soils	Alley cropping, nitrogen fixation
Cassia siamea	Shrub grows to 8 m, vigorous coppicing	Fuelwood, nitrogen fixation, lumber
Erythrina spp.	Tree grows to 20 m, often thorny, coppices well	Live fences, nitrogen fixation, fuelwood, fodder
Flemingia macrophylla	Shrub grows to 3 m	Alley cropping, nitrogen fixation
Gliricidia sepium	Fast-growing tree to 20 m, vigorous coppicing	Alley cropping, nitrogen fixation, forage, fodder, staking material,
Inga spp.	Nitrogen-fixing shrub, acid-tolerant	Alley cropping, nitrogen fixation
Leucaena leucocephala	Tree grows to 20 m, fast growing on nonacid soils, vigorous coppicing	Fodder, fuelwood, erosion control, nitrogen fixation, alley cropping, staking material
Pangomia pinneta	Small tree grows to 8 m	Erosion control, live hedges
Sesbania spp.	Fast-growing low tree	Erosion control, nitrogen fixation

TABLE 2-2 Net Primary Production of
Biomass for Commonly Recommended
Multipurpose Tree Species in the Humid
Tropics

Species	Net Primary Production of Biomass (kg/ha/yr)
Acacia auriculiformis	3,000–4,000
Acacia mangium	2,500–3,500
Albizia falcata	4,000–5,000
Alchornea cordifolia	2,000–3,000
Calliandra calothyrsus	2,500–3,500
Cordia alliodora	2,500–3,500
Dalbergia latifolia	4,000–5,000
Erythrina poeppigiana	4,000–6,000
Gmelina arborea	1,500–5,000
Leucaena leucocephala	3,000–5,000

They may be planted randomly or according to systematic patterns
on embankments, terraces, or field boundaries. They provide a variety of products including fruit, forage, fuelwood, fodder, shade, and
fence and timber material. Some commonly recommended multipurpose trees are listed in Table 2-2. Once again, local adaptation to and
validation for site-specific systems are essential.

ARRANGEMENT OF TREES, CROPS, AND LIVESTOCK

Rather than using a random and difficult-to-mechanize system of
growing trees with annuals or animals, mixtures can be grown in an
improved spatial or temporal arrangement. In an agrisilvicultural
system, for example, trees can be grown in alternate rows or strips,
as contour hedges to control erosion, or on field boundaries. These
orderly arrangements can facilitate the use of animal power and of
mechanization of farm operations, save labor, and enhance economic
and ecological benefits.

Alley cropping is a common example of a spatial arrangement.
Food crops are grown in alleys formed by contour hedgerows of trees
or shrubs (Kang et al., 1981). Trees and shrubs can be pruned to
prevent shading of the food crops and to provide nitrogen-rich mulch
for crops and fodder for livestock. Shrubs and trees also act as windbreaks, facilitate nutrient recycling, suppress weed growth, decrease
runoff, and reduce soil erosion (Ehui et al., 1990).

The most common trees for alley cropping are fast-growing, multipurpose, nitrogen-fixing trees. Tree species with the potential for use with nonacid tropical soils include *Acioa baterii, Alchornea cordifolia, Gliricidia sepium,* and *Leucaena leucocephala.* Species for acid soils include *Acioa baterii, Alchornea cordifolia, Anthonotha macrophylla, Calliandra calothyrsus, Cnestis ferruginea, Dialium guineense, Erythrina* spp., *Flemingia congesta, Harungana madagascariensis, Inga edulis, Nuclea latifolia,* and *Samanea saman.* Hedgerows of *Cassia* spp., *G. sepium,* and *L. leucocephala* can be established from seed. Other species are established from seedlings or stem cuttings. However, the use of stem cuttings often results in a patchy stand with a high rate of mortality. Trees established from stem cuttings are also easily uprooted because of poor root system development.

When successfully established, alley cropping systems can produce two or more products, such as food grains, fodder, mulch, fuelwood, and staking and building materials, and can increase or maintain soil structure. However, the beneficial effects of these systems depend on many factors, such as the tree species, area of land allocated to trees, hedgerow management, crop management, soil type, and prevalent climate. In areas with nonacid soils, satisfactory yields of cereals can be attained with the added benefit of erosion control (Kang et al., 1984; Lal, 1989). These systems are also labor intensive (Lal, 1986), therefore they are adapted primarily to areas of high population density and modest to low labor cost.

Advantages and Disadvantages of Agroforestry

Given a compatible association of trees and annual crops, agroforestry systems are likely to sustain economic productivity without causing severe degradation of the environment. Because of the low fertility of most upland tropical soils, some degradation is inevitable with any cultivation system. The rate and risks of such degradation are lower with agroforestry than with annual crop rotations. Soil organic matter, pH, soil structure, infiltration rate, cation exchange capacity, and the base saturation percentage are maintained at more favorable levels in agroforestry systems due to reduced losses to runoff and soil erosion, efficient nutrient recycling, biological nitrogen fixation by leguminous trees, favorable soil temperature regime, prevention of permanent changes in soil characteristics caused by drying, and improved drainage because of roots and other biochannels (Lal, 1989).

It is important to note, however, that trees have both positive and negative effects on soils. Negative effects include growth suppression caused by competition for limited resources (nutrients, water, and light) and by allelopathic effects. Mismanagement of trees (through, for example, improper fertilizer application or inadequate water control) can also cause soil erosion, nutrient depletion, water logging, drought stress, and soil compaction.

Economic evaluation is an important tool to assess a technology. Labor-intensive alley cropping can be economical under severe cash constraints and where hired labor is available at relatively low cost. The available data on alley cropping indicate that the system cannot sustain production without supplemental inputs of chemical fertilizers if high yields are desired. In fact, soil degradation and attendant yield reductions can occur even with the fertilizer application (Lal, 1989, 1991a).

Erosion control is a definite advantage of closely spaced contour hedgerows of *L. leucocephala* or other shrubs, but it can also be achieved through cover crops, grass strips, or no tillage. Nonetheless, the erosion preventive effects of *L. leucocephala* hedgerows must also be considered in evaluating the economic impact of an alley cropping system.

Data on soil properties indicate that intensive cultivation resulted in decreases in soil organic matter content, total nitrogen, pH, and exchangeable calcium, magnesium, and potassium in all systems including alley cropping and control (Lal, 1989). This drastic decline in soil fertility was observed in relatively fertile soils (Alfisols). The relative rates of decline, however, were somewhat less in alley cropping than with plow-based control. These results are also supported by data on acidic tropical soils in Yurimaguas, Peru (Szott, 1987). Szott observed significantly more calcium, magnesium, phosphorus, and potassium in the upper 15 cm of soil with control without trees treatments than with alley-cropping treatments. Fertilized control without trees significantly exceeded all other treatments in topsoil calcium and magnesium. The pH values were also significantly greater in the fertilized control.

Research Priorities

The agronomic aspects and biophysical processes of agroforestry using traditional cropping systems need to be more fully evaluated. For example, farmers using traditional systems commonly space their plants more widely apart than farmers using improved systems, and hence grow fewer plants per unit area. More scientific data are needed on interactions among plant species, specifically in relation to competition for water, nutrients, and light, and on the suppression of growth of one species by another species' release of toxic substances.

Major distinctions also should be made for research on acidic versus nonacidic soils. Too often soils and their constraints are ignored when designing or evaluating agroforestry systems. The ability of agroforestry systems to enhance nutrient availability on infertile soils is very limited compared with systems on fertile soils. On both, however, agroforestry systems can play an important role in reducing nutrient losses. Although litter production and quantities of nutrients recycled in litter are greater on fertile than on infertile soils, management techniques for accelerating nutrient fluxes through pruning hold promise for increasing plant productivity on infertile soils. More information is needed on the magnitude of and controls on belowground litter production and how it can be managed. Litter decomposition and soil organic matter dynamics in agroforestry systems might most easily be manipulated by managing woody vegetation to produce organic residues of a certain quality and to regulate soil temperature and moisture. More attention needs to be paid to specific soil organic matter pools, their importance in nutrient supply and soil structure, how they are affected by soil properties, and how they can be managed (Szott et al., 1991).

In addition to understanding the agronomic and biophysical aspects of agroforestry systems, the social, ecological, and economic elements require more attention. The economic feasibility of agroforestry systems needs to be assessed at the farm level. Human ecology and sociology play an important role in the acceptance and spread of technologies, as do the specific sociopolitical and institutional constraints.

Agroforestry can be a sustainable alternative to shifting cultivation. However, systems suited to many major soils and ecological regions of the tropics have yet to be developed. For example, alley cropping has shown some advantages in Alfisols but not in other soils and harsh environments. Further research is needed to develop systems performance indicators and to document ecological viability of agroforestry systems across a range of biophysical conditions.

MIXED TREE SYSTEMS

Mixed tree systems, also known as forest or home gardens and mixed tree orchards, constitute a common but understudied form of agriculture. These systems involve the planting, transplanting, sparing, or protecting of a variety of useful species (from tall canopy trees to ground cover and climbing vines) for the harvest of various forest products, including firewood, food for the household and marketplace, medicines, and construction materials. Commercially, for example, cacao plantations in Latin America are commonly intercropped with maize and bananas or plantains. The components of home gardens and many other traditional systems are selected for high productivity and minimum effort. Weeding and pest control efforts are reduced by using a combination of shade, domesticated animals, and plant species. These household plots also serve as sites for conducting small-scale crop experimentation and for cultivating seedlings before transplanting them to agricultural plots.

Typical cultivation and management practices include integrating the placement and planting times of tree species so that different products can be collected and harvested throughout the year. The heterogeneity of mixed tree systems provides a protective upper canopy that protects lower canopy and ground species from seasonal torrential rains and direct tropical sunlight. In harsh tropical environments, this practice allows the production of delicate economic species, such as cacao. In addition, the upper canopy helps maintain relatively constant moisture and temperature levels and contributes to soil regeneration (Niñez, 1985; Soemarwoto et al., 1985).

Types of mixed tree systems range from intensive systems such as home gardens, where the trees are planted along with other useful species directly adjacent to a dwelling, to more extensive systems of natural forest management, such as the artificial forests described by Alcorn (1990). Orchards sometimes integrate pastureland with trees (including timber species) for livestock production combined with annual and perennial crops (Altieri and Merrick, 1987; Fernandes et al., 1983; Russell, 1968). Mixed tree systems can also be found in the fallow fields of shifting cultivators, where useful tree species are spared or planted in the cleared agricultural plot and the subsequent forest regeneration is managed to encourage forest patches that provide desired products (Caballero, 1988; Soemarwoto and Soemarwoto, 1984). Many farmers also conserve a strip of mature vegetation between or surrounding their agricultural plots (Pinton, 1985). Research and historical accounts throughout the tropics indicate that mature forests are often composed of patches dominated by species that have been encouraged, spared, or planted by past and present human inhabitants (Gómez-Pompa and Kaus, 1990).

Indigenous groups of small-scale farmers are predominately responsible for maintaining and cultivating mixed tree areas in tropical regions, without subsidies or international expertise. In contrast, single species tree plantations, such as for coffee, cacao, rubber, or oil palm production, have been encouraged and managed for large-scale production through foreign or agribusiness investments (see below). Smaller scale production in single species plantations has typically been supported by bank credits, government-funded agricultural extension programs, and international development agencies (Niñez, 1985). These monoculture tree plantations can be fairly lucrative if they come into production when international market demands are strong. Production processes can me mechanized, thus reducing labor needs and maintenance costs. Capital investment requirements, however, are high.

Little research has been undertaken to understand the dynamics of mixed tree systems or their comparative productivity to plantation systems over the long-term. Social, economic, and ecological evaluations of mixed tree systems versus single species tree plantations are necessary before appropriate land use or investment recommendations can be made for any region.

Past and Present Forest Management

Limited studies have begun to reveal the complexity of crop and tree interactions. For the most part, these studies involve time-tested selections and local experimentation with tree species.

Mitigating Climate Change Through Sustainable Land Use

To what degree can the adoption of sustainable land uses in the humid tropics help to offset increasing concentrations of greenhouse gases in the atmosphere? Research on climate change and land use in the tropics has focused mostly on the impact of deforestation and other forms of forest conversion on greenhouse gas emissions and accumulation. Few studies have attempted to quantify the potential of sustainable land uses to mitigate these impacts. In terms of greenhouse gases, the most important feature of sustainable land use systems in the humid tropics is their potential to reduce atmospheric carbon dioxide concentrations by accumulating carbon on land. The land use systems described in this chapter can affect atmospheric carbon concentrations by (1) reducing the incidence of forest conversion, and hence the release of carbon; and (2) serving as carbon sinks, withdrawing carbon from the atmosphere and storing it in biomass and, to a lesser degree, in the soil.

This suggests a crude formula for estimating the total potential impact of sustainable land uses on greenhouse gas levels: the total impact equals the amount of carbon sequestered by adopting sustainable land uses *plus* the amount of carbon allowed to remain in undisturbed forests as a result of reduced conversion *plus* the impact of sustainable land uses on emissions of other greenhouse gases. In this equation, the amount of carbon sequestered by adopting sustainable land use op-

Managed forest patches or groves may have been one of the first forms of agriculture. Fruit and nut trees were important sources of food for early humans. Knowledge of areas with abundant tree species having edible fruits was essential information for survival (Harlan, 1975). These same areas may have also provided important sites for "garden hunting" of frugivorous animals (Linares, 1976).

The "management" of forests by early humans is considered to be an important evolutionary step. Recent ethnoecological, archaeobotanical, and paleobotanical studies have indicated that ancient management practices have influenced the present-day abundance and presence of certain species, such as *Annona* spp., *Byrsonima* spp., *Carica* spp., *Ficus* spp., *Manilkara* spp., *Quercus* spp., and *Spondias* spp. (Gómez-Pompa, 1987a,b; Harlan, 1975; Hynes and Chase, 1982; Kunstadter, 1978; Posey, 1990; Roosevelt, 1990; Turner and Miksicek, 1984). Various types of mixed tree gardens coupled with other agricultural systems, such as shifting cultivation, were able to maintain high-density populations (Lentz, 1991).

---

tions would be determined by multiplying the area of land suited to each land use option by the potential carbon sequestration capacity (in both vegetation and soils) of each option (Houghton et al., In press). Thus, sustainable land uses can retain more carbon on land in two ways: by reducing the total area of converted forestland and by reducing the total amount of biomass removed in the process of conversion.

Few of the factors in this "formula" have been investigated systematically, and none of the factors have been determined with a high degree of accuracy. Houghton (1990b) compared current land use and potential forest area in the tropics and concluded that, over the next century, reforestation efforts could reverse the net flux of carbon and withdraw almost as much carbon (about 150 Gt) from the atmosphere as would be released if current land use trends continue unchecked. Houghton et al. (In press) examined the potential of plantations, secondary forests, and agroforestry systems to accumulate carbon and concluded that, in the tropics as a whole, these systems have the potential to recover between 80 and 180 Pg of carbon (and up to 250 Pg if the recovery of soil carbon is factored in). The potential for carbon accumulation was shown to be highest in tropical Africa (40 percent of the potential total), followed by Latin America (39 percent) and Asia (21 percent). Agroforestry systems were shown to have the highest potential to accumulate carbon, followed by plantations and fallow and secondary forests. Precise figures of the carbon storage capacities of different land use systems are lacking. A rough comparison of capacities is presented in Table 3-1.

In many humid tropic areas these managed forest systems still play a key role in human subsistence. For example, the Bora people from Brillo Nuevo, eastern Peru, subsist largely on various varieties of manioc interspersed with an assortment of trees, usually peach palm (*Bactris gasipaes*), uvillia (*Pourouma cecropiifolia*), star apple (*Pouteria caimito*), macambo (*Theobroma bicolor*), guava (*Psidium* spp.), barbasco (*Lonchocarpus* spp.), and coca (*Erythroxylum coca*) (Denevan et al., 1984). The Guaymí Indians from Soloy, Panama, and the Cabecar Indians of the Telire Reserve, Costa Rica, live from the products derived from the palm *Bactris gasipaes*, which provides food and drink from its fruit and beverage from its roots (Hazlett, 1986).

More than 200 fruit tree species are found in the humid tropics today. Many of the tree fruits of Southeast Asia evolved from wild rain forest species and were gathered for thousands of years prior to the advent of agriculture (Frankel and Soulé, 1981). For example, in village gardens in the Trengganu mountains of Peninsular Malaysia, Whitmore (1975) found 26 fruit tree species being cultivated. Of these, 12 were identical to the same species growing in the wild, 6 were improved selections from the wild, 5 were indigenous but were not found in the forest, and 3 were from the New World. Historically, important tree species in Asia include the breadfruit tree (*Artocarpus* spp.) and the coconut (*Cocos nucifera*). The avocado (*Persea* spp.), cacao (*Theobroma* spp.), and the breadnut tree (*Brosimum* spp.) have played a central agricultural role in many regions of the Americas, as have the oil palms in Africa. Most of these species have been cultivated in mixed tree orchards, and efforts are being made to change them into single species plantations.

The survival and presence of mixed tree areas in the tropics today, despite external pressure for monoculture production, are largely due to the many advantages they provide their caretakers. Their structure, composition, and management can be adjusted to local environmental and social conditions. Introduced economic species can be mixed with native species. Both household and market production can be included in system management, which can respond rapidly to changing demands in local, regional, or international markets. In the Mexican state of Yucatán, small-scale fruit production, usually from home gardens, supplies much of the diverse selection of fruits found in the local markets. Mixed trees in Mexico are also producers of important international commodities such as coffee, cacao, and vanilla. In West Sumatra, mixed tree areas, known as *parak*, constitute 50 to 88 percent of the cultivated land of different villages and are important suppliers of popular fruits for the region such as durian (*Durio zibethinus*) as well as international products such as cinnamon, nutmeg, and coffee (Michon et al., 1986). Throughout Indonesia and Malaysia, the cultivated durian trees (*Durio* spp.) are grown from seeds or seedlings gathered from the adjacent forests or selected from the best cultivated fruits (Budowski and Whitmore, 1978; Michon et al., 1986; Whitmore, 1975).

The wide range of products and functions of mixed trees, combined with an increased resource base, help minimize economic risk for the farmer. Farmers derive steady income from fruit trees and cash crops without a high cost of production (Soemarwoto and Soemarwoto, 1984). Since these orchards are polycultures, they can be harvested throughout the year and provide both food and income for villagers. These orchards require low-cost inputs and part-time labor, of which the labor source is mostly family members (women, the elderly, and children) in the case of home gardens. By spreading out cultural and management requirements over the year, these systems can also reduce peak workloads and ensure a more stable subsistence and cash economy.

The ecological advantages of mixed tree systems have allowed their regeneration over centuries of use, and are thereby instrumental in the design of sustainable agriculture systems and biodiversity conservation in the humid tropics. The potential benefits and advantages of mixed tree systems were recognized by Smith (1952) over 40 years ago. These advantages include the potential for more efficient use of resources both above- and belowground, with roots from 50 to 60 m deep on some trees and canopies reaching 50 to 70 m high. The multistory canopies characteristic aboveground is also reflected belowground. The roots of the upper canopy trees are able to penetrate to the deepest strata of the subsoil; roots of the smaller tree and bush species occupy the intermediate layers; and shallow rooting annual and perennial plants form just below the surface (Douglas and Hart, 1984). Minerals and nutrients extracted from the different strata are interchanged between the various root systems by burrowing activities of various soil organisms. From the veins of the highest trees in the subsoil, water may be drawn up and made available to the shallower rooted plants. Aboveground, the plant density reduces solar rays and provides a filtering system for rainwater, while the fallen leaves help contribute to soil regeneration (Douglas and Hart, 1984; Niñez, 1985). These characteristics enable these systems to foster environmental rehabilitation and improve living conditions on marginal or degraded lands (Boonkird et al., 1984).

Mixed tree systems can also provide improved habitats for wildlife, control erosion, mitigate landslides, and reduce the risks of soil deterioration and runoff. The complexity of these managed ecosystems may be higher than the natural system since they combine the natural functions of a forest system in a small space, sometimes with domestic animals, with a high diversity of useful species to fulfill the socioeconomic needs of the household. These systems also foster in situ conservation by local residents, which enables wild, rare, and endangered species to continue evolving within the ecology of the entire habitat and permits an artificial selection of great diversity of size, shape, color, and taste variants (Wilkes, 1991).

Mixed Tree Systems Throughout the World

Agroforestry systems using mixed trees are common forms of small-scale production for farmers throughout the world (see Alcorn [1990] and Brownrigg [1985] for detailed descriptions and references). In Indonesia, the best known forest gardens are the home gardens, or *pekarangan*, a typical feature of the rural landscape. They are cultivated and managed areas surrounding a house on which mixtures of plant species are generally sown (Soemarwoto and Soemarwoto, 1982). The *pekarangan*, like most traditional home gardens in the tropics, conserves many important plant and animal landraces. These Indonesian home gardens also produce cash fruit crops, such as the durian (*Durio zibethinus*) and rambutan (*Nephelium lappaceum*), in addition to providing areas for other customary sources of income such as livestock production. Coconut and bamboo cultivation are also common.

Home gardens in Mexico are plots of land that include a house surrounded by or adjacent to an area for raising a variety of plant species and sometimes livestock. They are also known as kitchen gardens, dooryard gardens, *huertos familiares*, or *solares*. The home garden is representative of a household's needs and interests, providing food, fodder, firewood, market products, construction material, medicines, and ornamental plants for the household and local community. Many of the more common trees are those same species found in the surrounding natural forests, but new species have also been incorporated, including papaya (*Carica papaya*), guava (*Psidium* spp.), banana (*Musa* spp.), lemon (*Citrus limon*), and orange (*Citrus aurantium*). In light gaps or under the shade of trees, a series of both indigenous and exotic species of herbs, shrubs, vines, and epiphytes are grown. Seedlings from useful wild species brought into the garden by the wind or animals are often not weeded out and are subsequently integrated into the home garden system.

One of the most striking features of present-day Maya towns in the Yucatán Peninsula is the floristic richness of the home gardens. In a survey of the home gardens in the town of Xuilub, 404 species were found (Herrera Castro, 1991) where only 1,120 species are known for the whole state (Sosa et al., 1985). Home gardens also provide diverse environments where many wild species of animal and plants can live (Herrera, 1991), although the diversity of species depends on the size of the gardens and the degree of management. Estimated average family plots range from 600 m^2 to 6,000 m^2 (Caballero, 1988; Herrera, 1991). Taking into consideration that most households in rural communities of the Yucatán Peninsula have some type of home garden, local traditional practices of orchard management have already contributed to the forest cover in the peninsula and have the potential for contributing more.

On Java, home gardens occupy from 15 to 75 percent of the cultivated land (Stoler, 1978). More than 600 species are known to be grown in Indonesian home gardens (Brownrigg, 1985). In a hamlet of 40 families near Bandung, Soemarwoto and Soemarwoto (1982) reported more than 200 of species of plants. A comparative study conducted by Soemarwoto and Soemarwoto (1984) of the production and nutritional value of three predominant agricultural systems—home gardens, *talun-kebun* (another agroforestry system), and rice fields—demonstrated that their production levels did not vary greatly. However, for nutritional value, the home gardens and *talun-kebun* were better sources for calcium, Vitamin A, and Vitamin C than rice fields.

Other important agroforestry systems within Indonesia are similar to the *pekarangan*. Mixed tree plantations occur on uninhabited private lands, usually associated with shifting cultivation. They are dominated by perennial crops under which annual crops are cultivated (*kebun campuran*) or where spontaneously grown trees and perennial crops occur (*talun-kebun*) (Wiersum, 1982).

The forest gardens of Sri Lanka are another example of important mixed tree systems. Unlike the forest gardens of Indonesia and Mexico, these gardens are built on the degraded grassland hillsides of the Sri Lankan highlands (Everett, 1987). Located immediately around the houses, they may account for nearly 50 percent of private land use (Everett, 1987). The types and allocations of plants reflect local knowledge of the ecological needs of each species.

The Bari garden system, found in the tropical forest region of Catatumbo, Colombia, depicts a gradual change in the size of the vegetation between the house location and the surrounding forest. Crops similar to those depicted by the first missionaries in 1772 are cultivated in these home gardens. They include plantains (*Musa* spp.), sugarcane (*Saccharum officinarum*), cassava (*Manihot esculenta*), sweet potato (*Ipomoea batatas*), yam (*Dioscorea trifida*), pineapple (*Ananas* spp.), cotton (*Gossypium* spp.), and chiles (*Capsicum* spp.) (Pinton, 1985). This garden system offers a self-supportive and practical adaptation to economic and environmental changes (Pinton, 1985), and may represent a technique for adoption by other poor farmers in the region.

The management of fallow succession in cultivated fields is also a common technique used by farmers all over the world. The planting, sparing, protecting, transplanting, or coppicing of trees interspersed with annual crops in the cultivated plots results in the establishment of a productive mixed tree system years after the annual crops are gone. The Bora Indians of Peru plant seeds and seedlings of fruit trees along with manioc (Denevan et al., 1984). Seedlings of useful species are also spared, others are protected, or the trunks coppiced. As the trees mature and the cultivation of manioc and other annuals diminishes, the cleared plot develops into an "orchard fallow" and eventually merges with the surrounding mature vegetation. The process may take 35 years or more.

Small-scale farmers in Peru have created systems with valuable economic species through a process of managed fallowing (Padoch et al., 1985). After clearing the standing vegetation on a plot, much of the slash is burned for charcoal. Tree crops, often with high commercial value, are planted with annual and semiperennial crops and gradually predominate production in the plot.

Protected forest patches are also found in inhabited areas throughout the tropics. Old and uncut forest sections are protected by the Lua' of Thailand. Gathering is allowed in these areas, but the cutting of trees is prohibited by village rules (Kunstadter, 1978). The forest fields of the Kayapó in Brazil represent a well-known managed forest system (Posey, 1984), where useful plants are concentrated and encouraged in patches of forest near where the Kayapó travel or hunt.

A recently established system in Peru indicates the potential of local management for forest protection and use. An organization of nonindigenous farming villages in northeast Peru has established several communal forest reserves where extraction is allowed but regulated (Pinedo-Vásquez et al., 1990). The trees are used for their fruit, construction material, artisan material, and medicinal purposes.

The Role of Mixed Tree Systems in Tropical Forest Conservation

Mixed tree systems represent one of the most promising land use options available for integrating tropical forest conservation with production. The cultivation techniques already exist, local residents are already knowledgeable in cultivation practices, and the local to international markets already demand their products. The individual variation found in the different orchards contributes to forest species diversity, and orchard expansion results in more local reforestation. Mixed tree systems may be one of the few agroforestry systems that can meet household, economic, and conservation goals in the humid tropics.

Research on traditional farming systems in many areas of the world suggests that complex polycultures with trees have many advantages for the local economy over modern systems of extensive annual monocultures (see Alcorn [1990]). Unfortunately, international promotion of various local tree-based systems, from home gardens to managed forests, has not been accompanied by strong, interdisciplinary research programs to guide and assess their efficacy. This also holds for mixed tree systems. The complex forest management practices required by these systems do not fit under either conventional forestry or agriculture. Most of the research on traditional resource management in the humid tropics has been undertaken by individual researchers in separate, unintegrated disciplines. Little research has been undertaken by foresters, and agroforestry in general remains an unconventional discipline in the international scientific community. Funding to date has been minimal, often because of the obvious and reasonable caution exhibited by funding agencies to invest in unresearched, unquantified ventures. To present a viable and comprehensive plan for forestry programs that is integrated with conservation and development concerns, several research objectives need to be met:

• Baseline information on the species composition, spatial and temporal structure, age, and maintenance of present mixed tree systems in the humid tropics;

• Long-term monitoring of ecological relationships and comparisons to adjacent natural forest vegetation and to single-species plantation systems;

• Documentation and integration of traditional, technical, local, and international experience with mixed tree systems;

• Comparative production and marketing assessments of both mono- and polycultural systems to determine long-term sustainability and stability for small scale-producers; and

• Establishment of demonstration plots to design more efficient agroforestry systems that are based on ecological and economic productivity.

This type of extensive, comparative research may help to uncover the principal reasons behind poor resource management by both small- and large-scale producers in the humid tropics, and may identify the pitfalls for conventional forestry development programs. It may also illuminate the reluctance of small-scale farmers to alter their agricultural production systems. Poverty and the actions of local farmers are often blamed for tropical deforestation. Mixed tree systems, however, show that local farmers can and do manage agroecosystems on a sustainable basis. As such, they represent an existing, locally accepted alternative for biodiversity conservation and sustainable agriculture in the humid tropics. Further research, however, is needed to recognize and document their contributions to forest conservation and restoration.

PERENNIAL TREE CROP PLANTATIONS

Perennial tree crop plantations can be a useful means of converting deforested or degraded land into a system that is both ecologically and economically sustainable. They are part of a broader category of plantation agriculture that includes short rotation crops, such as pineapple and sugarcane, as well as tree crops, such as bananas

A cacao plantation was carved out of the tropical rain forest in Malaysia. Credit: James P. Blair © 1983 National Geographic Society.

and rubber. This section discusses their role in economic development and sustainable agriculture. Plantation forestry, which involves lumber, pulpwood, and fuelwood production or environmental protection, is discussed later.

Plantation Crops and Economic Development

The role of plantations in the agricultural and economic development of countries in the humid tropics has been controversial (Tiffen and Mortimore, 1990). In the 1950s plantations were considered a part of the modern sector and capable of absorbing capital investment, generating new employment opportunities, and serving as a source of foreign exchange earnings (Lewis, 1954). This positive view of the economic efficiency of plantation agriculture was often accompanied by an erroneous perception that small-scale tropical farmers were unresponsive to economic incentives and unwilling to adopt new production practices. Yet, this attitude was often attributable to the high risk or impracticality of new technologies. The hesitancy of farmers may also have been a reflection of ineligibility for credit programs, lack of access to the necessary infrastructure and markets, distrust due to previously failed rural development programs, or incompatibility with local socioeconomic structures.

As plantation systems came under greater scrutiny, they were often associated with colonial exploitation, or viewed as primary sources of persistent regional poverty (Beckford, 1972; North, 1959). These criticisms were often based on the fact that after the plantations were established and in production, and transport and processing facilities in place, little further development, diversification, or intensification could occur. The rigid production system offered few opportunities to absorb additional labor, and was held responsible for the persistence of low wages.

By the 1980s, many developing countries and assistance agencies were taking a more balanced view of both the efficiency and equity of plantation and small-landholding systems. It was recognized that the plantation system of organization often had substantial advantages in establishing highways, markets, processing facilities, and other infrastructure needs and in mobilizing required financial, managerial, and research resources. It was also recognized that in areas characterized by effective physical and institutional infrastructure, small-landholdings often achieved levels of productivity comparable with or higher than plantations. Under conditions of rising wage rates, small-landholding production often remained profitable, while the profitability of plantation production declined. For at least some crops the plantation may be an intermediate stage in the transition toward more extensive mixed cropping systems. The traditionally sharp distinction between small-landholding and plantation crops, defined by technical requirements for sustainable production, gave way to a realization that every plantation crop is produced successfully by small-landholdings in some countries or regions.

Plantation crops are sometimes equated with tropical export crops such as rubber or palm oil, or even with cash crops, as distinguished from subsistence or food crops, such as rice, maize, and cassava. In practice, however, a crop such as coffee or sugarcane may be grown for local consumption as well as for the export market. Tiffen and Mortimore (1990) suggest the following characteristics of plantation crops:

- They are tropical products (bananas, rubber) or subtropical products (tea, oranges, sugar) for which an export market exists.
- Most require prompt initial processing.
- Whether exported or sold domestically, the crop is funneled through a few local marketing or processing centers before reaching the consumer.
- They typically require large amounts of fixed capital investment (for example, for establishing the plantation and for constructing processing facilities).
- They generate some activity for most of the year, so that economic efficiency is not incompatible with a large permanent labor force.
- Monocropping is characteristic, since it is simpler than polycultures and makes the development of standardized management practices and marketing channels possible.

These characteristics imply a limited capacity to make short-term responses to changes in either the price of the product or purchased inputs such as chemicals, transportation, or labor. In the past, when local financial markets in the tropics were relatively underdeveloped, larger production units with access to developed country financial markets had substantial advantages. However, when tropical countries became independent, and their ties to central capital markets atrophied, the plantation sector in several former colonial economies declined. Other contributing factors have included the transfer of plantation management to the public sector, which occurred with tea plantations in Sri Lanka; the exploitation of producers by marketing boards through export taxes and resulting low producer prices; and other disincentives, such as the maintenance of overvalued exchange rates to protect import-substituting industrialization (Bates, 1981).

These considerations probably represent more severe constraints on perennial tree crop estates than on plantation crops in general. One implication is that adverse economic conditions, whether market or policy generated, affect tropical tree crop production more slowly because of the long-term nature of the investment. However, these conditions, if they extend over long periods, can result in the deterioration of production capacity and the depreciation of infrastructure, and these impacts may be long-lasting. An adverse economic environment, largely the result of government policy, resulted in the deterioration of oil palm production in several East African countries in the 1960s and 1970s (Bates, 1981). In contrast, more favorable economic policy and support for productivity enhancing research, land development, and infrastructure enabled Peninsular Malaysia to achieve world leadership in oil palm production while production was declining in West Africa.

Environmental Effects

The establishment of plantations can have substantial negative environmental consequences in the absence of effective public policies and private management. These effects include the following:

- The conversion of natural forest into plantations will always lead to loss of species diversity on the affected land. The seriousness of the loss depends on the amount of land that is converted to plantation relative to the total forestland in the same agroecological zone.
- The conversion of natural forest into plantations may be accompanied by substantial soil erosion. The extent of erosion will differ according to the method used for land clearing and the production systems used for each plantation crop. Typically, the establishment of rubber or oil palm plantations causes more erosion than establishment of coconut or cacao plantations. The land clearing methods used by small-landholdings often generate less erosion than the methods used by the larger plantations or by government settlement schemes. The latter are more likely to entail extensive clearing of established smaller plantations using heavy machinery.
- Because nutrients are removed from the soil when crops are harvested, production levels can only be sustained with systematic fertilizer application (Tiffen and Mortimore, 1990). These nutrients must be replaced if yields are not to decline. On a per hectare basis, there are wide differences among crops in the level of nutrients removed from the soil. Rubber, for example, imposes a relatively small nutrient drain, while oil palm imposes a high drain (Tiffen and Mortimore, 1990).

These negative effects can be mitigated by conservation practices, such as the use of leguminous ground cover, mulches, intercropping, and terracing. For example, rubber and oil palm plantations can produce stable or increasing yields on a long-term basis in Peninsular Malaysia (Vincent and Hadi, Part Two, this volume). Rubber has been grown on some sites for nearly 100 years, and oil palms for more than 70. Yields of both crops continue to increase, mostly due to the extensive use of agrichemicals and other purchased inputs and the development of higher yielding varieties by the Rubber Research Institute of Malaysia and the Palm Oil Research Institute of Malaysia (Pee, 1977). However, these practices require relatively high levels of both research and extension efforts to achieve sustainable production. Improved management, planting, and harvesting techniques, fertilization, pest control, and (for rubber) use of chemicals that stimulate higher flows of latex have also been important (Vincent and Hadi, Part Two, this volume).

The adoption of sustainable plantation management methods (especially if they prove highly profitable) may not forestall the expansion of these (and other) systems into undisturbed forests. In Peninsular Malaysia, the productivity of rubber and palm plantations led to their rapid expansion. In recent years, however, industrialization has led to more off-farm employment and greater rural labor shortages, thereby decreasing agricultural expansion. The phase of land development marked by conversion of forests to plantations appears to be closing rapidly in Peninsular Malaysia (Vincent and Hadi, Part Two, this volume).

Investments for Sustainability

The slow growth in demand for most perennial tree crop products can be partially offset by technical change leading to lower production costs. In the 1950s and 1960s, a profound "export pessimism" constrained research and development investment in the tree crop sector in several developing countries. Malaysia was one of the few postcolonial economies that continued to make the research investment needed to enhance the competitiveness of its tree crop economy against industrial synthetic substitutes, as in the case of rubber, and against competing producers of tropical tree crop products, as in the case of oil palm and cacao (Ruttan, 1982). In contrast, the regional research system for tropical tree crops in the former British colonies in West Africa fell into disrepair in the 1960s and 1970s. In the former French colonies of West Africa, the regional research institutions remained viable, with substantial support from France into the early 1980s.

The first requirement for maintaining and enhancing the sustainability of tropical tree crop production systems is to strengthen national agricultural research systems in the tropics. The second major challenge is to broaden the research agenda on tropical tree crop production to place greater emphasis on the management of tree crop systems for sustainability and on the policy environment needed to enhance sustainable development of land and labor productivity (National Research Council, 1991a).

PLANTATION FORESTRY

Tropical tree plantations cover about 11 million ha of land and are composed of many tree species (Brown et al., 1986). Although plantations do not constitute a natural biome and are in fact a heterogeneous mix of managed ecosystems, they have many common characteristics. For example, most tropical tree plantations were established after the 1960s and are thus fairly young (Food and Agriculture Organization and United Nations Environment Program, 1981; Lanly, 1982). Moreover, most plantations occur in subtropical and premontane environments; few examples of successful plantations are found in the lowland wet tropics (Lugo et al., 1988). Plantations are usually established on damaged or deforested lands for sawn wood, veneer, and pulpwood production (industrial plantations), environmental protection (nonindustrial plantations), or for supplying fuelwood (energy plantations). Common genera in plantations worldwide include *Acacia, Eucalyptus, Pinus, Swietenia,* and *Tectona.*

The literature on plantation forestry in the tropics is copious. Most studies deal with species adaptability and trials, spacing studies, and other aspects of plantation culture. A number of books summarize the state of knowledge on tropical tree plantations (for example, Bowen and Nambiar [1984], Evans [1982], Lamprecht [1989], and Zobel [1979]). More recent studies have examined plantation biomass accumulation (Lugo et al., 1988), the role of plantations in the global carbon cycle (Brown et al., 1986), the use of plantations for rehabilitating damaged lands (Lugo, 1988), and ecological comparisons of plantations and tropical secondary forests (Cuevas et al., 1991; Lugo, 1992). These studies show that plantation productivity is a function of climate and soil factors. The highest yields are usually the result of intensive management, high technological inputs (such as genetic improvement of varieties), and intensive care of plantings (Cuevas et al., 1991; Lugo, 1992). Without constant maintenance, plantations will not remain as monocultures and can gain plant species at rapid rates. This tendency toward diversification can be used to rehabilitate damaged lands, to foster ecosystems for native species (Lugo, 1988), or to serve as habitat for wildlife (Cruz, 1987, 1988).

Plantation function reflects the behavior of the planted species, as demonstrated in their cycling of nutrients and in organic matter dynamics. In a comparative study of native forests paired to plantations of similar age, for example, Lugo (1992) found that Caribbean pine (*Pinus caribaea*) plantations consistently accumulated more litter (dead and decaying bark, leaves, branches, and other plant material) than the native forest. Aboveground nutrient use efficiency was higher in the plantation because it had greater aboveground biomass production with less uptake of nutrients from the soil. However, native forests consistently outproduced the plantation in belowground root production and biomass. The net effect of these differences was that total primary productivity in the paired forests was equal (Cuevas et al., 1991). In contrast, the functions of mahogany (*Swietenia macrophylla*) plantations are more similar to those of the natural forests.

Findings from about 70 comparisons between plantations and paired native forests (Lugo, 1992) revealed that generalizations about plantation structure and function cannot be made without adequate study of the many climatic, soil, biotic, or temporal characteristics of the ecosystem. The age of the plantation, for example, is an important variable that explains many of the characteristics of these human-dominated ecosystems. With age, tree stands accumulate more species, biomass, and nutrients. The forest's impact on soil fertility, organic matter, and other characteristics is also age dependent, the cumulative effects becoming more apparent as plantations mature and successional processes proceed.

From a managerial point of view, plantations are flexible ecosystems because they can be designed and used for a multiplicity of purposes, ranging from food production and land rehabilitation to wildlife habitat and mixed uses (Figure 2-4). They contribute an important tool for land managers who are striving to diversify the productive capacity of the land (Wadsworth, 1984). The major drawbacks of plantations relate to cost, knowledge requirements, and the length of time required before products are ready for market. Like any intensive land use, plantation establishment and care require high investments, although costs are generally lower than those required for food crop production. Knowledge of species adaptability and site factors is critical to avoid costly failures, particularly in moist tropical conditions. Failures can result from insect or disease outbreaks, poor species response to local conditions, or catastrophic events, for example.

Most tropical countries have identified tree species that grow well in sites available for tree plantation establishment, and adaptability

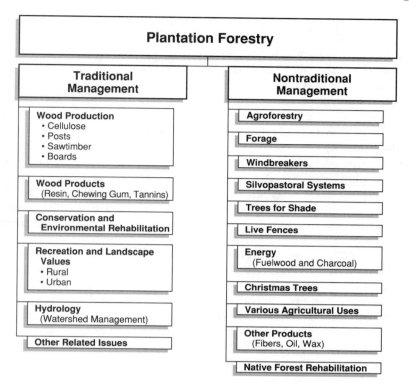

FIGURE 2-4   Uses of tropical forestry plantations.

trials are advanced in those countries with established forest management agencies. In agrarian societies, plantation forestry is a required management option for addressing many human needs, including fuelwood and charcoal production and land rehabilitation. It can be applied at the village level, where human labor and degraded land are usually available but where wood products require much time to gather and transport.

Success in plantation forestry programs depends on strong outreach efforts, well-operated nurseries, and timely human interventions in all phases of plantation establishment (that is, site preparation, planting, tree care, and adequate protection of young trees, particularly in their early stages when they are vulnerable to grazing, fires, or other accidents that can destroy them). The benefits of a well-established program are many and long-lasting because plantations can be very productive, improve soil conditions, and provide many tangible and intangible benefits associated with forest cover. Yet, those benefits will not materialize if information is not transmitted effectively to practitioners in the field and if economic incentives are inadequate. The yields and benefits from these systems of production thus depend, in part, on the efficacy of extension services as well as the financial returns to plantation owners.

Plantation research is widely practiced in the tropics. Tropical foresters have been very successful in establishing tree plantations in most tropical conditions, documenting growth rates, identifying hazards, and improving the use of superior seed. More recently, reports on the biomass and nutrient aspects of plantation management have been published (Cuevas et al., 1991; Lugo, 1992; Wang et al., 1991). Because plantation forestry requires site-specific knowledge to assure long-term success, research on all aspects must continue to be supported. In addition, much of the information, particularly concerning the function of plantation forests, has not been synthesized. Such a synthesis should seek common principles of management and forest response that can be extrapolated widely. Moreover, as the uses of plantations diversify into nonwood products, it is important to widen the number of species planted and learn about lesser-known species that have been ignored in traditionally wood-oriented research. Other new research areas include the establishment of plantations in diverse landscapes and for a variety of other purposes, such as to graze animals, to plant crops, to recycle wastes, and to serve as wildlife habitat.

REGENERATING AND SECONDARY FORESTS

The development of sustainable agriculture and land use systems in the humid tropics requires an understanding of the forest regeneration process and the factors that influence it. Regenerating forests can be viewed as a transitional land use option, preparing tracts of deforested land for more intensive management, or as a permanent land use itself, maintaining forest cover and maturing into secondary (and eventually primary) forest. Secondary forests, which have often been dismissed as inferior to primary forests and less important from a conservation standpoint, also possess many ecological and economic benefits (Table 2-3). Furthermore, primary forests cannot be restored without the development, first, of secondary forests. In this sense, secondary forests should also be considered a viable land use option.

Factors Affecting Forest Regeneration

The rate of forest regeneration is inversely related to the scale of forest clearing and the intensity and duration of use prior to abandonment (Brown and Lugo, 1990; Uhl et al., 1990a). Forest cover

TABLE 2-3   Products and Benefits Derived from Secondary Forests

Products and Benefits	Reference(s)
Fruits, medicinal plants, construction materials, and animal browse	Sabhasri (1978)
Valuable timber species (e.g., *Aucoumea klaineana, Cordia alliodora, Swietenia macrophylla*)	Richards (1955), Budowski (1965), Rosero (1979)
Uniform raw materials with respect to wood density and species richness	Ewel (1979)
Woods low in resins and waxes, which facilitates their use	Ewel (1979)
Biomass production at a fast rate	Ewel (1979)
Ease of natural regeneration	Ewel (1979)
Ability to support higher animal production and serve as productive hunting grounds	Ewel (1979), Posey (1982), Lovejoy (1985)
Habitat for greater numbers of vertebrates, which may enhance tourism	Lovejoy (1985)
Tree species with properties often sought by foresters for establishing plantations	Ewel (1979)
Generally more accessible to markets than remaining primary forests	Wadsworth (1984)
Availability as foster ecosystems for valuable late secondary species	Ewel (1979)
A useful template for designing agroecosystems	Ewel (1986)
Restoration of site productivity and reduction of pest populations	Ewel (1986)

SOURCE: Brown, S., and A. E. Lugo. 1990. Tropical secondary forests. J. Trop. Ecol. 6:1–32.

returns relatively rapidly following clearing, burning, and immediate abandonment (Uhl et al., 1988). Previously forested lands that are repeatedly burned, grazed over long periods, or tilled and scraped with heavy machinery may remain treeless for many years following abandonment, especially where soils have been extensively damaged and nutrient reserves have been depleted (Nepstad et al., 1991; Uhl et al., 1990a).

SHORT-TERM FACTORS

Short-term differences in forest regeneration rates can be traced to the failure of tree seedlings and sprouts to establish themselves on abandoned lands (Nepstad et al., 1991; Toky and Ramakrishnan, 1983). Factors that impede establishment include:

- Lack of seed or residual tree roots in the soil that can give rise to new tree stems;
- Lack of fruiting shrubs and small trees to attract seed-carrying birds and bats into abandoned fields;
- Abundant seed- and seedling-eating ants or rodents in the abandoned fields; and
- An aggressive weed community that suppresses the growth of other plants through high root length density, competition for water and available nutrients, or allelopathic influences.

Young tree seedlings in abandoned fields are also subject to higher temperatures, higher vapor pressure deficits, and lower soil moisture availability than seedlings established in natural treefall gaps, where many forest tree species regenerate (Nepstad et al., 1991). Little is known about the ability of forest tree seedlings to tolerate these extreme physical conditions.

Grasses impede forest regeneration in many areas. They do not provide perches or fleshy fruits to attract the birds and bats that carry tree seeds into abandoned fields. (However, they do provide excellent habitat for seed-eating rodents and leafcutter ants, and their dense root systems effectively compete for soil nutrients and water.) In the Amazon Basin, abandoned fields with long histories of repeated burning or grazing are sometimes occupied by *Paspalum* spp., *Hyparrhenia rufa,* and other grass species that resist tree establishment and forest regeneration for many years (Nepstad et al., 1991; Serrão and Toledo, 1990). In Southeast Asia, roughly 200,000 km^2 of tropical forest have been replaced by the aggressive grass, *Imperata cylindrica* (Barnard, 1954; Jensen and Pfeifer, 1989). Land use practices that eliminate on-site sources of new trees (buried seeds and residual tree roots) and allow a dense cover of grasses to develop may lead to long-term deforestation.

LONG-TERM FACTORS

Once trees are established in abandoned fields, and aggressive weed communities are weakened by the shade of overtopping saplings, forest regeneration can proceed. The total leaf area of the original forest is often recovered within the first few years of regeneration. Fine root distribution, although poorly studied, appears to be reestablished within the first 5 to 10 years of regeneration (Nepstad et al., 1991). Recovery of the biomass and nutrient stocks of the original forest, however, may take much longer. In a study of forest regeneration following abandonment of shifting cultivation plots in the Venezuelan Amazon, Saldarriaga et al. (1988) found that the accumulation of biomass and nutrients in regrowing forests reached a plateau at about 60 years following abandonment. This plateau probably arises from two factors. First, the rapidly growing pioneer species that comprise the young, regrowing forests are often short-lived. As they begin to die, biomass and nutrient accumulation is slowed until the density and size of slower growing, longer-lived trees increase. Second, biomass and nutrient accumulation may slow as reserves of essential soil nutrients, which probably did not limit tree growth soon after field abandonment, become scarce. Long-term recovery of the biomass and nutrient stocks of the original forest may depend on the rate at which nutrients arrive in the ecosystem through rainfall (Buschbacher et al., 1988; Harcombe, 1977) and the rate at which nutrients are released through the weathering of primary soil minerals, if they are present.

Fire

The most important factor affecting forest regeneration is fire. Abandoned agricultural lands are most fire-prone when they are overtaken by weeds that quickly dry out after rainfall and provide abundant fuel close to the ground. In eastern Amazonia, and presumably in Southeast Asia, grass-dominated abandoned fields can be ignited within a few days of rain events (Uhl and Kauffman, 1990; Uhl et al., 1990b). The high flammability of grasses is one of the greatest threats to successful forest regeneration on abandoned agricultural lands, and probably explains the persistence of vast tracts of *I. cylindrica* on previously forested land in Southeast Asia. As tree establishment and growth proceed, fire susceptibility declines but continues to threaten forest regeneration. Young secondary forests in the eastern Amazon can be ignited within 10 days of dry-season rain events and are far more flammable, because organic fuels on the ground dry out faster than in the primary forest (Uhl and Kauffman, 1990; Uhl et al., 1990b).

Susceptibility to fire is also a function of the geographical distribution of agricultural and forestlands. Young forests that lie along roads or are adjacent to agricultural lands are at much higher risk than those surrounded by a matrix of primary or late-secondary forests.

Acceleration of Forest Regeneration

The best techniques for accelerating forest regeneration are based on knowledge of the specific barriers to tree establishment and tree growth. In grass-dominated fields, forest regeneration may be fostered by protecting the site from fire and, where necessary, freeing

A view of secondary forest in the foreground with primary forest in the
background. Secondary forest is the regrowth after major disturbance, such
as logging or fire. Credit: James P. Blair © 1983 National Geographic Society.

tree seedlings from the competitive cycle. In Southeast Asia, tree seedlings are liberated by matting down neighboring stems of the *Imperata* grass. In eastern Amazonia, fire suppression alone permits the rapid growth of tree clusters that attract seed dispersal agents and ameliorate harsh local climate conditions (Nepstad et al., 1991). The acceleration of biomass and nutrient accumulation is more difficult to achieve and, omitting the use of fertilizers, may be best accomplished by planting within young secondary forests those trees that are effective at acquiring nutrients from acid infertile soils. Active reforestation programs using appropriate mixes of native species can be useful at initial as well as advanced stages of regeneration.

On some sites, the growth rates of available native species may be inadequate. In these cases, forest rehabilitation can be accelerated with fast-growing exotic tree species. These can quickly restore forest environments, modify site conditions, and allow native forest species to regenerate in their shade. In this way, the plantings serve as a "foster ecosystem" for native forests (Lugo, 1988).

THE ROLE OF SECONDARY FORESTS

Most regenerating forests, if not cleared again or managed as part of an agricultural system, will eventually mature into secondary forests. The total area of secondary forests in the tropics has been increasing rapidly. In 1980, secondary forests accounted for 40 percent of the total forest area in the tropics and increased at an annual rate of 9 million ha (Food and Agriculture Organization and United Nations Environment Program, 1981). The diverse ecological characteristics within this large area have created different types of secondary forests (Table 2-4). However, these young forests share several characteristics: their biomass and nutrients quickly accumulate; they are dominated by pioneer species; they experience rapid turnover of their component species; and their appearance changes rapidly.

Indigenous people learned to use the characteristics of secondary forests to their advantage (Clay, 1988; Rico-Gray et al., 1991). Rather than just occupying space and repairing soil fertility, forest fallows in shifting cultivation cycles became elements of complex land use patterns. Most species within secondary forests had some use or value to indigenous people (Barrera et al., 1977; Gómez-Pompa, 1987a,b; Rico-Gray et al., 1985). Over time, forest composition was modified to meet specific needs (Gómez-Pompa et al., 1987).

Secondary forest vegetation must be evaluated as part of the complex mosaic of tropical landscapes and the human activities within them. A typical landscape in the humid tropics is a mixture of land uses,

TABLE 2-4   Ecological Characteristics of Secondary Forests

Ecological Characteristics	Reference(s)
Fast growth rates and short life spans	Budowski (1965)
Higher numbers of reproductively mature individuals per species than in mature forests	Zapata and Arroyo (1978)
Conditions suitable for recolonization of mycorrhiza after agriculture	Ewel (1986)
Short life cycles that are adapted to timed cycles of human use of land	Gómez-Pompa and Vásquez-Yanes (1974)
Many tree seeds that are widely dispersed	Budowski (1965), Gómez-Pompa and Vásquez-Yanes (1974), Opler et al. (1980)
Seeds can remain viable in soil for several years	Gómez-Pompa and Vásquez-Yanes (1974), Lebron (1980)
Ability to germinate and grow well on impoverished soils, which suggests low-nutrient requirements	Gómez-Pompa and Vásquez-Yanes (1974)

each representing a different intensity of human intervention, with scattered secondary forests in different stages of recovery from previous uses. The task is to maintain the overall primary productivity of the land, keep human activities at stable and acceptable levels, and protect biodiversity. Properly managed secondary forests are critical for attaining these goals because they can supply forest products, repair site fertility, and maintain a high level of native biodiversity. They are also important for research into agroecosystem functions. Agroecosystems that mimic secondary forests hold promise for achieving improved agricultural production without permanent damage to sites (Ewel, 1986; Hart, 1980).

Within the land use mosaic described in this report, different types of social organizations and institutions are required. For example, in landscapes composed of primary forests, humans are generally organized as shifting cultivators or hunters and gatherers. In highly degraded landscapes, humans must either migrate to new lands or depend on external sources of fertilizers and other inputs to rehabilitate damaged ecosystems. Secondary forests, because of their diverse ecological and social attributes, offer many opportunities for improving production. However, to take advantage of these opportunities, policymakers, conservationists, agriculturalists, and development officials must focus on their potential, and not just on what has been lost with the primary forests. Intensified management of secondary forests can increase yields of some products, but output cannot be sustained without increased attention, improved technology, and fuller knowledge of forest ecosystem processes (Wadsworth, 1983, 1984, 1987a).

NATURAL FOREST MANAGEMENT

Natural forest management offers a promising alternative to the depletion of commercial timber resources within primary and secondary tropical moist forests. It involves controlled and regulated harvesting, combined with silvicultural and protective measures, to sustain or increase the commercial value of subsequent stands, and it relies on natural regeneration of native species. On the spectrum of sustainable land use options, natural forest management occupies a position between strict forest protection and higher intensity production systems that require permanent clearing or conversion of forests. Although varied in their approaches and methods, all natural forest management systems seek to protect forest cover, ensure the reproduction of commercially important species, and derive continuing economic benefits from the forests.

Only a small percentage of the world's timber-producing tropical forest is managed. A 1982 survey of 76 countries possessing tropical forests found that of 210 million ha being logged only 20 percent was being managed (Lanly, 1982; Moad, 1989). In the Asia-Pacific region, where most of the world's managed tropical forests are found, less than 20 percent of production forests receive systematic silvicultural treatments (Food and Agriculture Organization and United Nations Environment Program, 1981). Only 0.2 percent of the world's moist tropical forests is being managed for sustained timber production, according to recent estimates (Poore et al., 1990).

Forest Management in the Humid Tropics

The ecological complexity of tropical moist forests places special constraints on applying forest management practices, especially those developed in temperate zone forests (Buttoud, 1991). Silvicultural practices in the humid tropics must consider the high degree of tree species diversity, the vulnerability of tropical forest soils, and the regeneration biology of leading commercial tree species.

The high degree of diversity in tropical moist forests complicates the harvesting, extraction, marketing, and regeneration of forest trees. In any given area of tropical forest, only a minority of tree species is commercially marketable. In Suriname, for example, about 50 tree species comprising between 10 and 20 percent of the total forest tree species diversity are commercially harvested (de Graaf, 1986). Even in Southeast Asian forests, where logging focuses on the dipterocarp and other closely related trees, only about 100 species are exploited (about 2,500 tree species are native to the Malay Peninsula alone).

Tropical forest soils are easily damaged by the mechanized processes of timber harvesting and extraction and by the larger scale of forest clearing that mechanization allows. These impacts include soil compaction and erosion, higher soil temperatures, desiccation, loss of soil biodiversity, removal of aboveground nutrient reserves (especially phosphorus), and lower nutrient retention capacity.

The capacity to manage tropical forests effectively is limited by a lack of understanding of forest regeneration processes (Lugo, 1987). The reproductive requirements of many leading commercial tree species are neglected under current management systems. Some species require specialized pollinators and dispersers that are not considered in management plans. Many timber trees depend on persistent seedling populations for regeneration, making them highly vulnerable to understory disturbance (Moad, 1989).

Timber extraction affects all of these characteristics, altering the structure, function, and species diversity of the forest. Because tropical forests are so diverse, most commercial logging that occurs in the humid tropics involves selective extraction. Selective harvesting may provide the basis for more sustainable management systems, but most extraction methods, as currently practiced, extensively damage other forest trees, the regenerative capacity of the forest, and forest soils. Genetic depletion, and even extinction, can occur if harvesting is excessive. Uncontrolled selection also opens forests to illegal harvesting of timber and wildlife and increases the susceptibility of forests to fire. Finally, the decline in economic value of forested land that follows extraction fosters further conversion, especially through agricultural expansion and settlement.

Management Systems

Natural forest management systems offer mixed benefits and costs. They are suited to areas with less productive soils and afford greater protection of soil and water resources than land uses that require permanent large-scale clearing. Although they simplify the structure and composition of primary forests, and hence result in lost biological diversity, these systems allow the forests to retain a greater degree of diversity than that provided by more intensive agricultural, agroforestry, or plantation systems (Buschbacher, 1990). On-site carbon storage rates are high, and because much of the extracted wood is intended for construction and other permanent uses, the carbon can remain sequestered. Long-term nutrient loss through removal of biomass may serve as the ultimate limitation on the sustainability of managed forests, but these losses can be minimized through careful logging operations. A degree of risk is inevitably incurred in the opening of access roads. Even where selective timber harvesting is feasible and well regulated, postharvest management may not be, which sets the stage for more intense forms of forest conversion.

The socioeconomic attributes of natural forest management are also variable. Compared with plantation and agroforestry systems, natural forest management systems are less labor intensive, require fewer capital inputs, and yield forest products at relatively low levels. At the same time, they create more employment opportunities per investment unit than do cattle ranches (Goodland et al., 1990). If planned and undertaken with care, they can provide employment and income for forest dwellers and protect cultural integrity. For this reason, local participation is especially critical. Several reviews of sustainable forestry methods and natural forest management systems have been published in recent years (Moad, 1989; Office of Technology Assessment, 1984; Schmidt, 1987; Wadsworth, 1987a,b; Wyatt-Smith, 1987). Natural forest management systems are usually grouped into three broad categories: uniform shelterwood systems, strip shelterwood systems, and selection systems.

UNIFORM SHELTERWOOD SYSTEMS

Uniform shelterwood systems are designed to produce even-aged stands rich in timber species (Office of Technology Assessment, 1984). Under these systems, all marketable trees within a given area are harvested during the initial phase of management. Subsequent silvicultural operations further open the forest canopy, allowing seedlings and saplings of commercially valuable species to thrive. Logging is monocyclic, taking place once at the end of each rotation.

The foremost example of uniform shelterwood systems is the Malayan Uniform System (MUS), first developed in the lowland dipterocarp forests of the Malay Peninsula after World War II (Buschbacher, 1990) and commonly practiced from the early 1950s to the 1970s. After the initial harvest, forests were managed according to a 60-year rotation cycle of regeneration, periodic low-intensity silvicultural interventions (for example, removal of vines and elimination of noncommercial species, defective stems, and competing stems), and reharvesting. The aim of this system was to produce a relatively uniform growth of young *Shorea* spp. It offered acceptable rates of regeneration and appeared to be biologically sustainable. However, the widespread conversion of the lowland forests to oil palm and rubber plantations and other more intensive agricultural systems almost completely removed these forests, obviating the need to manage them. Hence, the MUS was not in practice long enough for second rotation cuts to be made. Today the MUS is practiced in a modified form, with an emphasis on selective management systems.

The Malaysian experience illustrates difficulties in the transferability of the MUS to other regions. The uniform system, as developed in Malaysia, was most applicable in fertile, lowland forests with high seedling densities. Attempts to transfer the MUS to nearby hill forests were generally unsuccessful due to less predictable seedling production, greater topographic effects on tree species composition and abundance, and greater damage to regenerating seedlings during logging operations (Gradwohl and Greenberg, 1988; Lee, 1982). As a result, uniform systems appear silviculturally appropriate only when an adequate stock of seedlings of desirable species exists prior to harvesting and a large enough proportion of commercially valuable species exists in the original forest canopy to justify complete canopy removal (Buschbacher, 1990).

The Tropical Shelterwood System (TSS), analogous to the Malayan system, was tested and introduced in several African countries in the 1940s, but results were less promising. Seedlings in the African forests were less abundant and distributed less uniformly, requiring more extensive and more frequent interventions to open the forest canopy. This led to greater infestation by weed trees and vines, higher labor costs, and ultimately poor regeneration of the desired species (Asabere, 1987). Plantation and other more intensive land uses, as well as intensified logging, precluded further systematic development of uniform shelterwood systems suitable to Africa.

STRIP SHELTERWOOD SYSTEMS

Strip shelterwood (or strip clearcut) systems are still largely in the experimental phase, but they show high potential for small-scale, sustainable management of tropical forests. In these systems, narrow strips of forest are cleared on a rotating basis, and regeneration occurs by seed dispersal from adjacent undisturbed forest and by stump sprouting. Careful harvesting plans and operations are designed to simulate the natural processes of tropical forest gap formation and regeneration (Hartshorn, 1989). The rotation schedule allows equal areas of forest to be harvested annually, the size of the cuts determined by the total area of managed forest and the period required for regeneration (Moad, 1989).

Extraction operations are carefully planned to minimize environmental damage. Local topographic and ecological conditions determine the size, location, and orientation of strips. Access roads are designed to minimize erosion and compaction and to protect areas of adjacent undisturbed forest, which is critical for regeneration. The use of heavy machinery is minimized and draft animals are often used to remove sawn logs. Logs are cleaned on site, and the slash (the bark, leaves, and branches of the harvested trees) is left to decompose rather than be burned or removed, allowing more retention of nutrients.

The most extensive test of a strip shelterwood system has taken place in the Palcazú valley of eastern Peru. Demonstration strips were first harvested in 1985. Initial postharvest inventories indicate abundant regeneration, with twice the tree species diversity of the preharvest strip (Hartshorn, 1990). This project has also placed high priority on social and economic considerations in its design. Project planners and indigenous communities work closely to coordinate harvesting, processing, and marketing operations; to distribute project benefits; and to ensure sustainable management of the communal forestlands (Buschbacher, 1990; Hartshorn, 1990).

The success of strip shelterwood systems depends on the ability of early successional stage trees to establish themselves rapidly in forest gaps, grow quickly, and produce marketable wood (Moad, 1989). Consequently, strip systems may be less applicable in Asian forests, where most timber trees, including the dipterocarps, are unlikely to regenerate rapidly on cleared sites. The potential for use is higher in the humid tropics of West Africa and Latin America, where suitable tree species and genera are more abundant. Further research may establish how variables, including the regenerative biology of tree species, postharvest silvicultural treatments, and the size, location, and frequency of cuts, can be altered to suit local conditions. For example, studies conducted at the Bajo Calima Concession in Colombia suggest the need to adjust the size and rotation schedule of cuts as well as the extent and placement of forest reserves to allow nonpioneer tree species, many of which have large seeds and depend on dispersal by birds and mammals, to regenerate (Faber-Langendoen, 1990).

SELECTION SYSTEMS

Most forests managed for timber in the humid tropics employ selection (or polycyclic felling) systems. In selection systems, trees are removed on a limited basis from mixed-age forests in a series of fellings, rather than in one large harvest (Wyatt-Smith, 1987). Less timber is extracted from the forest during each harvest, but harvesting occurs more frequently than in monocyclic systems. Two or more cuts, generally on a cycle of 25 to 35 years, take place in the course of a single rotation.

Selection systems were developed in response to site limitations, low regeneration rates, high labor costs, and other difficulties associated with even-aged forest management (Buschbacher, 1990). Variations include the Modified Selection System, employed in Ghana in the 1950s; Malaysia's Selective Management System (which began to replace the MUS in the early 1970s); and the Selective Logging System in Indonesia and the Philippines. Other polycyclic systems have been implemented or tested in Australia, Cameroon, India, Mexico, Myanmar, Nigeria, the Philippines, Trinidad, Uganda, and other humid and subhumid tropical countries (World Bank, 1991). Relatively little attention has been given to research and development of polycyclic systems appropriate for the Amazon Basin (Boxman et al., 1985; Rankin, 1985). The Celos Management System, recently developed on an experimental basis in Suriname, has yielded favorable early results in terms of minimizing ecological impacts and providing relatively high economic returns (Anderson, 1990; de Graaf and Poels, 1990).

Selection systems rely on the advanced regeneration of young, pole-sized trees to produce the subsequent timber crop (in contrast to shelterwood systems, which rely on seedling establishment). In some selection systems, advanced regeneration is promoted through improvement (or liberation) thinning (Moad, 1989). Improvement thinning usually involves the poisoning or girdling of less economically valuable trees and vines that compete with the most promising understory trees. Thinning removes 15 to 30 percent of the total number of stems and can reduce the time required to second harvest from 45 to 30 years, or as much as 33 percent (Buschbacher, 1990; Moad, 1989). Thinning has been employed most extensively in Southeast Asian forests, but it has also been tried in Côte d'Ivoire, Gabon, Ghana, Nigeria, Suriname, and Zaire. In most of these cases, however, the practice has been curtailed due to inadequate funding and a shortage of trained personnel (Moad, 1989).

In practice, successful selection systems still face significant obstacles. Tree regeneration and growth rates are often inadequate to meet projected rotation goals, and economic pressures force forest managers to shorten cutting cycles. High-grading (the unregulated extraction of only the most valuable trees) is prevalent throughout the tropics, but less so in the Southeast Asian dipterocarp forests. Poor planning of felling and transport operations results in excessive reduction in forest cover and damage to soil and water resources. Especially critical is damage to seedlings and pole-sized trees, on which successful forest regeneration depends. Improvement thinning and other silvicultural treatments are hindered by a lack of economic incentives and trained personnel and by ineffective government control and enforcement of forestry operations (Wyatt-Smith, 1987).

Constraints on Sustainable Forestry

It is not yet possible to find a natural tropical forest that has been successfully managed for the sustainable production of timber, because no management system has yet been maintained through multiple rotations (Poore et al., 1990). Some critics dismiss sustainable forestry in the humid tropics as a "myth" on the grounds that it remains unproved, provides low yields and slow economic returns, and is liable to be superseded by more disruptive or lucrative land use practices (see Spears [1984]). Others respond that natural forest management has been proved to be feasible on technical grounds, but it has generally failed for social and economic reasons (Anderson, 1990; Buschbacher, 1990). Forestry in the humid tropics may be sustainable, but it will require changes in logging practices, in the economics of the forestry sector, and in the land use policy environment (Goodland et al., 1990; Poore et al., 1990).

Past experience suggests a combination of silvicultural and socioeconomic factors behind the lack of successful implementation. On most sites, the key silvicultural constraint on sustained timber production is inadequate regeneration of seedlings, saplings, and pole-sized trees (Wyatt-Smith, 1987), usually resulting from excessive damage during logging operations. In other cases, biological constraints, such as weed and vine infestation, lack of seed dispersers, and lack of trees with appropriate regeneration capabilities, are more important. Socioeconomic factors include insufficient tenure provisions; lack of local involvement in management decisions and project benefits; ineffective regulation, supervision, and monitoring of forestry activities and methods; and the inability of forest managers to control land use over the long term (Buschbacher, 1990; Moad, 1989). The economic viability of sustainable forestry systems is hindered by a lack of adequate information on the resource base and potential markets, by international market forces that focus on a few tree species that are difficult or expensive to regenerate, by incentive policies that favor short-term timber exploitation, and by the undervaluation of timber products, nontimber products, and other forest services (World

Bank, 1991). In many cases, these are the same forces that hinder implementation of other sustainable land use systems described in this chapter.

MODIFIED FORESTS

As a land use option, modified forests can only be considered viable where the human population remains low and the extractive activities of forest dwellers is limited. By studying these ecosystems and societies, researchers gain insights into the processes of landscape change in the humid tropics and human influences on those processes.

Indigenous people often modify the structure and composition of primary forests. Technically, a primary forest is one without human influence (Ford-Robertson, 1971). Even in the least disturbed forests, however, human influence is evidenced by the presence of stumps, charcoal in the soil profile, artifacts, or exotic species.

Indigenous people also modify forests by altering the frequency of native species or the size of wildlife populations in ways that are difficult to detect. Only through detailed study and long-term analysis can the effects of people be detected. For example, Maya cultures apparently managed forests for food, fiber, medicines, wood, resins, and fuel, thereby modifying the species composition of large areas of Central American landscapes long believed to be primary forest (Barrera et al., 1977; Gómez-Pompa et al., 1987; Rico-Gray et al., 1985). The human-modified forest is almost impossible to segregate from pristine primary forest.

It is clear that even limited human presence can change the structure of forest ecosystems. It is doubtful, however, that forest processes, such as rates of primary productivity or the velocity and efficiency of nutrient cycles, are significantly altered. The key point is that wherever humans interact with natural forest ecosystems, forest modification is unavoidable. It is equally clear that there are thresholds beyond which modification is incompatible with the conservation of forest resources.

In practice modified forests are likely to be most appropriate where indigenous peoples and local communities retain secure tenure over large areas of forestland and where strong national policies support and protect these cultural groups and their ways of life. In recognizing modified forests for what they are—ecosystems that have been managed in subtle but sophisticated ways to provide their human inhabitants with sustainable livelihoods—their value as primary forests is not diminished. Rather, they acquire even greater sociocultural value as models and examples of successful human interaction with tropical moist forests.

FOREST RESERVES

Although a complete examination of the role and value of forest reserves is beyond the scope of this report, they need to be considered in devising comprehensive land use strategies in the humid tropics. The lack of secure protection for primary forests and wildlands diminishes the potential for sustainable agriculture, land use, and development throughout the tropics. These lands provide the biotic foundation on which human activity can be sustained and enhanced, and they protect the biological legacy of the humid tropics along with its many values.

Protected forests now constitute a small fraction of the tropical landscape—about 3 percent in Africa, 2 percent in Asia, and 1 percent in South and Central America (Nations, 1990). The protection mechanisms are as diverse as the number of countries and organizations that strive to protect forest ecosystems. They include biosphere reserves, wildlife preserves, national parks, national forests, refuges, sanctuaries, extractive reserves, privately owned lands, and land trusts. These efforts, however, require stronger political and financial support, especially for law enforcement, local community involvement, land acquisition, and effective reserve management. Without this support, the contribution these lands can make toward sustainable land use more generally is undermined (MacKinnon et al., 1986).

At this point, biologists cannot accurately determine the amount of land to preserve for optimal protection of biological diversity. No single standard exists for determining the amount or location of lands that should be set aside. However, long-term ecological studies are under way to understand the dynamics of species loss in tropical forests so that reserves of adequate size and configuration may be established (McNeely et al., 1990; Myers, 1988; Reid and Miller, 1989).

Many social and ecological factors endanger forest reserves. Conservation biologists are concerned with the sizes and shapes of reserves, global climate change, and the fragmentation of forest habitats by roads and other developments as some of the most urgent ecological factors that determine the integrity of reserves (Diamond, 1975; Harris, 1984; Peters and Lovejoy, 1992). Research on the effects of these and other factors on reserve function and effectiveness is a high priority (Ecological Society of America, 1991; Soulé and Kohm, 1989). Social forces that affect forest reserves revolve around the growing human pressures on reserve boundaries and resources, and

This border of a 10-ha (25-acre) reserve near Manaus, Brazil, illustrates the edge effect. Trees and other vegetation that form a barrier between natural and disturbed vegetation often experience a reduced vigor and are challenged or replaced by species that are well adapted to colonizing newly disturbed or cleared areas. In this case, the reserve is separated by only a few meters from agricultural fields of cassava (*Manihot esculenta*). The reserve is part of a project to determine the minimum critical size of ecosystems. Credit: Douglas Daly.

the difficulties associated with granting protection status without providing proper institutional, educational, and on-site support.

Much interest has focused on extractive reserves as a solution to deforestation in tropical areas. A discussion of its potential as well as environmental, social, economic, and research issues follows.

Defining a Role for Extractive Reserves

Extractive reserves can be among sustainable land uses in the humid tropics. They are forest areas where use rights are granted by governments to residents whose livelihoods customarily depend on extracting rubber latex, nuts, fruits, medicinal plants, oil seeds, and other forest products (Browder, 1990). These rights enable people to use and profit from land resources not legally belonging to them. Extractive reserves protect traditional agricultural practices and the forestlands on which they depend.

The development and long-term viability of extractive reserves face significant social, economic, and ecological obstacles. Under some circumstances, extractive reserves can contribute to sustainability in the humid tropics as components within more comprehensive land use strategies. Expectations, however, need to be tempered by a better understanding of their real potential and inherent limits.

The concept of extractive reserves originated in the mid-1980s as rubber tappers gained support in the state of Acre in western Brazil (Allegretti, 1990). Since then, the national government has designated 14 reserves, covering 3 million ha, within the Brazilian Amazon. The National Council of Rubber Tappers is trying to obtain reserve status for 100 million ha, or about one-fourth, of the Brazilian Amazon (Ryan, 1992).

Other efforts to establish extractive reserves are occurring both within and beyond the Amazon Basin. In Guatemala, for example, half of the 1.5 million ha in the Maya Biosphere Reserve has been allocated for traditional extraction of chicle, a gum derived from the sapodilla tree (*Achras zapota*), and the leaves of the xate (*Chamaedorea* spp.), which are used as ornamentals (Ryan, 1992). Interest has been further stimulated by studies indicating the economic value and potential of nontimber forest products (Balick and Mendelsohn, 1992; Peters et al., 1989a,b).

In weighing extractive reserves as a land use option, it is important to recognize that the primary goal in establishing reserves in the Brazilian Amazon has not been to protect biological diversity or tropical forests, but to secure reforms in land tenure and land use (Browder, 1990; Sieberling, 1991). Because opportunities for extraction are most advantageous where marketable species—especially tree species—are found in relatively high concentrations, extractive reserves are less likely to be located in the most species rich areas of the humid tropics (Browder, 1992; Peters et al., 1989a). In effect, reserves often will serve to maintain and protect biological diversity, forest cover, and the environmental services that intact tropical moist forests provide, but these functions are incidental to their social and economic benefits, and thus subject to changing socioeconomic conditions.

Commercial extraction is less intrusive than other forms of forest conversion, but it does alter forest ecosystems. In general, little research has focused on the long-term impacts of commercial extraction on the function and composition of tropical moist forests or on the ability of forests to sustain harvests of fruits, nuts, or other products (Ehrenfeld, 1992). Impacts can vary depending on the type of product extracted, the scale and methods of extraction, and the nature of the forest in which extraction occurs. Commercial extraction can result in degradation if large quantities of biomass (or small quantities of key ecosystem components) are removed, or if harvesting techniques cause excessive damage. In addition, researchers have noted the tendency to exploit extracted forest products to the point of depletion, for example, in the case of wild fruits and palm hearts in Peru and rattan in parts of Southeast Asia (Bodmer at el., 1990; DeBeer and McDermott, 1989; Vasquez and Gentry, 1989). At the species level, changes in population levels may affect the reproductive biology of extracted species and the status of associated plant and animal populations. Enrichment planting—the enhancement of populations of economically advantageous species by artificial means—may reduce species diversity within the forest as a whole. At the genetic level, market forces may result in the selection of specific individuals or traits, altering genetic variability within the species. Extractive reserves, depending on the scope and effectiveness of their management strategies, may amplify or minimize all of these effects.

The economic viability of extractive reserves is compromised, in both the long and short term, by a variety of factors. The economic base of most extractive reserves will be narrow. Existing reserves in the Brazilian Amazon depend primarily on production of rubber and Brazil nuts, and thus depend on volatile market conditions and subsidy policies (Browder, 1990; Ryan, 1991). Other factors complicate the sustainability of trade in extracted products. In most cases, viable commercial markets must be developed. The perishability of many tropical products may limit the ability to create or supply distant markets. Many products will not be conducive to standardized production because of highly varied harvest, transport, packaging, and storage needs.

Where markets for products do exist, extraction is vulnerable to increased competition from domesticated and synthetic sources. Extraction from wild sources is labor intensive, thus inviting artificial cropping and plantation systems (Browder, 1990). For example, Brazil nuts are being produced on plantations in Brazil. Finally, the capacity of extractive activities to improve standards of living may be limited as profits are absorbed by intermediaries before they reach harvesters (Browder, 1992; Ryan, 1992).

These biological and economic constraints should not obscure the social benefits that extractive reserves can provide (Sieberling, 1991). Most extractors in the humid tropics are poor and must contend with limited economic opportunities, threatened or inequitable land and resource rights, and unresponsive political structures. Most of them also engage in subsistence agriculture and depend on extractive activities for primary or supplementary income as well as food, fiber, and medicines. As the Brazilian experience has shown, the process of organizing, advocating, and managing extractive reserves can stimulate local participation and affect other areas of need, including health and extension services, housing, education, tenure reform, and marketing and infrastructure development. As the extractive reserve concept develops, it will provide valuable lessons for rural development efforts.

Extractive reserves should not be viewed as the solution to either deforestation or sustainable development in the humid tropics. They can, in the immediate future, stimulate needed land reforms, supply income and employment for limited local populations, protect some forestlands from more intensive forms of conversion, and provide important models of sustainable forest use. They cannot, however, meet the long-term needs of the growing numbers of shifting cultivators arriving at the forest frontier, provide full income or economic independence for the rural poor, preserve areas of the humid tropics that are especially diverse, or restore lands that are already in advanced stages of degradation. They may provide an important complement to other land uses, but they are not a substitute for forest reserves or for better managed agroecosystems, restoration areas, or more comprehensive and equitable land use strategies.

The record in creating and managing extractive reserves suggests several key guidelines for their further development. First, the limits and opportunities of extractive reserves should be clearly recognized. Designation should be initiated and supported by local people and communities, and the intended beneficiaries should be involved at all development stages. Government commitment—financial, political, and technical—is needed during the initial stages of reserve establishment and over time. As demographic, economic, and ecological conditions change, reserve management goals and methods need to remain flexible. Economic strategies should initially stress opportunities to develop known products, but they should also emphasize the need to diversify with time, to secure local benefits through value-adding processes, to work with all local resource users, and to reinvest in reserve operations (Clay, 1992). Local forest management skills need to be strengthened, with particular emphasis on improved extension services and increased interaction between biologists and extractors.

Research should seek to clarify the social, economic, and ecological factors that influence the long-term viability of extractive reserves and activities. Specific biological research is needed on commercially important species, their reproductive biology and ecological functions, and the impacts of extraction on forest composition, structure, and function (Ehrenfeld, 1992).

3

Technological Imperatives
for Change

It is apparent from the wealth of materials surveyed that the causes of forest conversion and deforestation vary with the characteristics of the natural resource base, the level of national or local development, demographics, institutional philosophy and policy, and the resulting social and economic pressures on land resources. The appropriateness of solutions for sustainable resource use depend on these same determinant factors, only some of which are subject to change and to management. Solutions are thus highly time- and place-dependent. The focus of the discussion and recommendations in this chapter is on the assessment of land use options and on the factors limiting their broad implementation.

The committee has found that publicly supported development efforts are confined to a range of land use choices that is too narrow. Use of some systems is being supported in places where they are clearly nonsustainable, while other potentially highly productive systems for some environments are being neglected. The study has identified sustainable land use options suitable for a broad range of conditions in the humid tropics. That so many instances of diverse production systems was found is not surprising; that they appear to have such broad applicability across the humid tropics is of great development interest.

KNOWLEDGE ABOUT LAND USE OPTIONS

Land uses have different goals and involve varying degrees of forest conversion, management skill, and investment. They confer different biophysical, economic, and social benefits. Geographic and demographic factors define their opportunities and constraints. Consequently, trade-offs are involved in choosing among them.

A Comparison of Land Use System Attributes

To be readily usable by development planners, land use systems should be defined according to their environmental, social, and economic attributes, and described in detail. The place and role for each system, which will depend on the level of national or local development, should be identified along with conditions required for their implementation and evolution.

Throughout the humid tropics, intensive cropping systems now occupy most of the resource-rich lands—those with fertile soils, little slope, and adequate rainfall or irrigation for crop growth during much of the year. The potential for continued increases in productivity on these lands through genetic improvement is uncertain, although it is probable for some crops in some regions. In addition, opportunities exist to reduce losses from pests and diseases and to cut back on the use of pesticides through better application of integrated pest management. Modest improvements in health and nutritional benefits may come through additional crop diversification and reduction in pesticide use. Changes in other social and economic attributes are likely to be very gradual.

More efforts are being made to identify and measure the attributes of agroecosystems that can serve as indicators of sustainability (Dumanski, 1987; Ehui and Spencer, 1990). Physicochemical, biological, social, cultural, and economic factors are being used to analyze system performance and potential. Many aspects of agricultural sustainability are difficult to categorize and quantify. In applying information that is quantifiable, issues of scale are critical (Consultative Group on International Agricultural Research, 1989, 1990).

Table 3-1 provides a framework for comparing the attributes and potential contributions to sustainability of land use systems. It is a tool that researchers, resource managers, policymakers, and development planners and practitioners can use in devising land use strategies.

The biophysical attributes in Table 3-1 include the nutrient cycling capacity of the system, the capacity of the system to conserve soil and water, the resistance of the system to pests and diseases, the level of biological diversity within the system, and the carbon flux

TABLE 3-1 Comparison of the Biophysical, Social, and Economic Attributes of Land Use Systems in the Humid Tropics[a]

Land Use Systems	Nutrient Cycling Capacity[b] L	M	H	Soil and Water Conservation Capacity L	M	H	Stability Toward Pests and Diseases[g] L	M	H	Biodiversity Level[h] L	M	H	Carbon Storage L	M	H
Intensive cropping															
High-resource areas[c]		X[e]	X[f]		X		X	O		X			X		
Low-resource areas	X	O		X	O		X	O		X	O		X		
Low-intensity shifting cultivation	X			X	O		X				X		X	O	
Agropastoral systems		X			X	O	X				X			X	
Cattle ranching	X[d]				X	O	X			X	O[i]		X	O	
Agroforestry		X		X			X			X	O			X	
Mixed tree systems		X		X	O		X				X			X	O
Perennial tree crop plantations		X		X	O		X			X				X	
Plantation forestry		X		X	O		X	O		X				X	O
Regenerating and secondary forests		X		X	O		X				X			X	O
Natural forest management		X		X			X			X				X	O
Modified forests	X			X			X			X					X
Forest reserves	X			X				X			X				X

NOTE: The letters L (low), M (moderate), and H (high) refer to the level at which a given land use would reflect a given attribute.

[a]In this assessment, "X" denotes results using the best widely available technologies for each land use system. The "O" connotes the results of applying best technologies now under limited-location research or documentation. The systems could have the characteristics denoted by "O" given continued short-term (5- to 10-year period) research and extension.

[b]The capacity to cycle nutrients from the soil to economically useful plants or animals and replenish them without significant loss to the environment.

[c]Those areas having fertile soils with little slope and few, if any, restrictions to agricultural land use. They have adequate rainfall or irrigation during much of the year for crop growth.

[d]High efficiency of recycling but low levels of nutrient removal through harvesting.

[e]Present technologies may develop high flow with high crop production, but they often entail high nutrient loss. Future technologies hold promise for greater containment and efficiency.

[f]Lowland, flooded rice production has both high nutrient flow and very high efficiency of recycling and of nutrient containment.

Social Attributes									Economic Attributes								
Health and Nutritional Benefits[j]			Cultural and Communal Viability[k]			Political Acceptability[l]			Required External Inputs[m]			Employment Per Land Unit			Income		
L	M	H	L	M	H	L	M	H	L	M	H	L	M	H	L	M	H
X	O				X			X		X			X				X
X	O		X	O		X	O		X			X	O		X	O	
	X		X			X			X				X			X	
X	O		X	O		X	O		X			X	O				X
X	O		X	O				X	X			X					X
X			X	O		X	O		X			X	O		X	O	
X	O		X	O		X	O		X	O		X	O		X	O	
X			X					X		X^n		X					X
X			X					X	X	X^n		X					X
X			X			X			X			X			X		
X			X	O		X	O		X			X			X		
X	O		X			X			X	O		X			X		
X				X		X			X			X			X		

[g]Indicates the natural ability to maintain pests and diseases below economic threshold levels in tropical ecosystems.

[h]Refers to the diversity of plant and crop species which, in turn, fosters diversity of flora and fauna both above and below the ground.

[i]Assumes diversity of plant species under well-managed grazing systems, which may include tree species in silvipastoral systems.

[j]To farms and their local communities.

[k]The ability to survive as a land use system and to provide income, employment, and needed goods in communities under continued and increasing population pressure. The systems must make optimum use of local resources and encourage acceptable levels of local equity.

[l]Politically desirable at levels above the local community (that is, county, region, province, state, or national level). At higher government levels it is assumed that generating cash flow through national or international channels usually takes precedence, but with the well-being of local communities having increasing consideration.

[m]Levels of external inputs appropriate to maintain optimal production with best available technologies. These levels, particularly of pesticides, may not be environmentally sustainable in the long term.

[n]Includes capital investment for establishment.

and storage capacity of the system. They serve to characterize the relative complexity, efficiency, and environmental impacts of the various land uses. Perennial tree crop plantations, for example, are generally monocultural systems, and less biologically diverse than primary forests. The biological simplicity of these plantations renders them more susceptible to insect pests and microbial and fungal diseases (Ewel, 1991). Perennial tree plantations, however, have a higher capacity for nutrient cycling than annual crop systems, and are better able to conserve soil and water due to the presence of a permanent, often stratified, vegetative cover. Plantations, due to the large biomass of the trees, also store about 10 times more carbon than do annual crops. The carbon storage capacity of plantations, however, is less than primary or mature forests (Dale et al., Appendix, this volume; Houghton et al., 1987). Once a forest matures, the storage and release of carbon achieves equilibrium; carbon dioxide sequestered through new growth equals that discharged from the oxidation of decaying old growth.

Important social attributes of these land use systems include health and nutritional benefits, cultural and communal viability, and political acceptability. Health and nutritional benefits reflect the capacity of a system to offset problems associated with intensive agrochemical use, heavy metal contamination, degraded water resources, high disease vector populations, and other public health concerns, as well as the capacity of the system to provide local people with a variety of food products at adequate levels. Cultural and communal viability refers to the ability of production systems to be adapted to local cultural traditions and to enhance community structures. Similarly, the ability of a system to ensure and enhance social welfare could be taken as a measure of its political acceptability.

Among the economic attributes that should be taken into account in comparing land use systems are the level of external inputs (such as fertilizer and equipment) required, the amount of employment generated, and the amount of income generated. Precise assessments of these economic attributes are especially difficult to derive. All can vary widely, even within a given type of land use, depending on the management practices employed, the impact of market fluctuations (or, in some cases, the lack of accessible markets), the type of crops grown, and other variables. The approximations in Table 3-1 are intended only to offer a sense of the relative economic costs and benefits across the spectrum of land use systems. Agroforestry systems, for example, require little fertilizer (although initial amendments may be required on degraded lands). With modification, they can be designed to generate moderate levels of employment or income on a per unit area basis. Perennial tree plantations require considerable chemical inputs and labor to maintain productivity, but generate more employment and income than agroforestry systems on a per unit area basis.

None of these features alone will determine the viability or sustainability of a given system. Rather, each system entails positive and negative attributes that must be viewed in the context of local biophysical, economic, and social opportunities and constraints. Furthermore, these attributes, and others not included, interact in complex ways to determine the rate and direction of change within an agroecosystem, and, more widely, within landscapes and regions. Hence, different systems will be appropriate and sustainable for different locations depending on the level of development and the relative availabilities of land, labor, and capital.

Table 3-1 also assumes the use of the best available technologies. For example, areas with poor soil and water resources have received far less attention from rural and agricultural development programs, but increasing population and development pressures and the need for greater cash income are forcing conversion of these areas to more productive and intensive land uses. For many of these areas, the newly researched and demonstrated technologies for mixed cropping systems show considerable promise. Low-input transitional technologies have potential for stabilizing erosion and lengthening the rotation cycle in low-intensity shifting cultivation areas, which are under severe stress to produce more by shortening the fallow period (Sanchez and Benites, 1987).

In all attribute categories, intensive cropping, agroforestry systems, agropastoral systems, mixed tree plantations, and, to some extent, modified forests offer significant benefits. This is particularly true in countries where industrial expansion or tourism is creating markets for high-value fruit, spice, and fiber products produced as woody perennial species or for animal products that can be integrated into small farm systems. Mixed perennial and annual crop systems (agroforestry) have a relatively high capacity to conserve soil and water, good nutrient cycling characteristics, and moderately high levels of diversity, which in turn provides enhanced protection against pests and diseases. They are suited to small-scale, labor-intensive settings and require modest capital to initiate. These land uses rate high in social and political acceptability in that they promote social well-being and generate income.

Cattle ranching, perennial tree crop plantations, and plantation forestry offer some desirable biophysical attributes but somewhat fewer social benefits. Although they require higher capital investments, they can be politically desirable from the viewpoint of national investment strategy because they usually generate products for export. They conserve production resources if well managed, but because they involve government or private ownership of large tracks of land, local people may not view such extensive land use as socially desirable where employment and incomes are low. Overgrazing and poor management practices may reduce or even destroy long-term soil productivity. With the use of the best available technologies, however, the biophysical attributes of each of these systems can be improved to acceptable levels of sustainability for a wide range of conditions.

Forest reserves and secondary forests have excellent biophysical attributes, but their social and economic acceptability at the local level is often low, especially where population pressure is great. While secondary and managed natural forests can be moderately productive, members of the local community need to share in, and gain from, the management of forest resources. Forest reserves have national value and may generate considerable local benefits if tourism and other low-impact uses are properly managed.

Indigenous Knowledge and Production Systems

The vast body of indigenous knowledge on land use systems must be recorded and made available for use in national development planning.

The need for widely adaptable sustainable land use systems in the humid tropics has brought increased attention to traditional systems of agricultural production and land management, and indigenous knowledge of tropical resources. Until recently the long history of agricultural adaptations among indigenous people was neglected as researchers focused on transferring modern crop production models and techniques perfected in the temperate zones. Many traditional forms of land management, including stable shifting agriculture, agroforestry, home gardens, and modified forests, are being lost along with the forests and the cultures in which they evolved. It is important that these systems be investigated and understood. Research can offer insights into many aspects of traditional systems: their structure, genetic diversity, species composition, and functioning as agroecosystems; their social and economic characteristics; the decision-making processes of the farmers and forest dwellers who manage them; their impact on local communities and ecosystems; and their potential for wider application.

Likewise, indigenous knowledge of local plants and animals is being lost as traditions of intergenerational training are eroded. This loss, of special interest to ethnobotanists and conservation biologists, needs to become a matter of general concern for all who are interested in the foundations of sustainable land use. The study of traditional uses of plants and animals may suggest new ways to diversify farming operations, to take advantage of natural forest resources, and to gain financial returns for the protection of biological diversity. The research process can have additional benefits by fostering collaborative relationships between researchers and indigenous peoples, and providing the groundwork for successful local development projects.

Traditional systems and indigenous knowledge will not yield panaceas for land use problems in the humid tropics. Researchers need to evaluate both the benefits and drawbacks of traditional systems, with the aim of understanding the ability of these agroecosystems to meet regional environmental needs and to help alleviate poverty (Gómez-Pompa et al., Part Two, this volume). Traditional ways of making a living in humid tropical environments, refined over many generations by intelligent land-users, provide necessary insight into managing tropical forests, soils, waters, crops, animals, and pests. Many of the practices, products, and processes inherent in these traditional approaches can provide lasting benefits within more modern agricultural systems.

LAND USE DESIGN AND MANAGEMENT CONSIDERATIONS

Agricultural development involves a wide range of land use design and management considerations. If land use activities or interventions are planned and undertaken at the wrong level or scale, these efforts can hinder rather than enhance sustainability. To development appropriate land use designs, geophysical diversity, population pressures, and socioeconomic needs must be fully examined. Development activities need to be highly detailed and finely tuned to local conditions. This, in turn, requires community and farmer input and control. Centralized operations at the regional or national level cannot provide the attention to detail that is needed. It may be necessary, however, to establish guidelines and long-term plans for erosion or pollution control through more centralized institutions.

At the national and regional levels, general land use characteristics need to be appraised and monitored in forming national policy, allocating development resources, and fashioning broad resource use guidelines. General land use planning requires data on soil type, topography, forest cover, and other geographic factors, as well as data bases on demographic and other socioeconomic factors. Data must be available in adequate quantity and quality for central planning.

Successful planning at the community level includes cultural, social, political, and economic factors at the local level (Chambers et al., 1989; National Research Council, 1991a). It is also essential that communities have access to a wide range of land use options. In most communities, knowledge of a variety of land use systems is limited. Descriptive literature is either inappropriate or unavailable, and no procedure is in place for local people to gain direct access to sources of information. Development specialists tend to promote the particular system in which they were trained or that donors have mandated. The fact that they are specialists usually precludes broad-based training in or knowledge of integrated resource management.

Activities undertaken at the farmer level should focus on the constraints that farmers face in adopting appropriate systems, including insecure and inadequate land tenure, lack of credit and economic incentives, and lack of access to technology and required inputs (often planting materials).

Sustainability and the Integration of Land Uses

A scientific basis for designing and selecting land uses, and their combinations, must be developed.

In moving toward more sustainable means of agricultural production and resource conservation in the humid tropics, land uses need to be integrated so their interactions are mutually reinforcing. In other words, the land use options used by a community must not only make optimum use of the resource base, but complement each other in nutrient flow, biodiversity, and in meeting the range of community needs. Progress toward this goal could be hastened if:

• The attributes and long-term environmental and socioeconomic effects of various land uses were better understood;

• The biological and agricultural characteristics of humid tropic landscapes, watersheds, or other areas amenable to areawide management plans were more fully ascertained and useful land use classification systems were developed; and

• Appropriate land use planning and development efforts, involving people and institutions at the farm, community, regional, and national levels, were further advanced.

The spatial and temporal integration of land uses is fundamental to sustainable agriculture and the conservation of natural resources. Spatial arrangements are defined by the area being considered. For example, on the farm they can refer to cropping patterns and terrain management, such as terracing. In a larger area, they can pertain to

At an elevation of about 1,800 m (6,000 ft) in the Cameron Highlands of Malaysia, villagers grow vegetables on terrace farms that are situated on land cleared of tropical forests. Credit: James P. Blair © 1983 National Geographic Society.

different types of land uses in close proximity. The spatial arrangement of various land uses affects biophysical factors (for example, the presence of pollinators and pest predators or the rate of soil erosion within a watershed) as well as socioeconomic factors (for example, the availability of markets and reliable infrastructure) within the agroecosystem. Much of the theoretical groundwork and applied research on land use spatial patterns and relationships has been developed in terms of biogeography, forestry, landscape ecology, and conservation biology (Harris, 1984; Hudson, 1991; MacArthur and Wilson, 1967). Increased interaction between agricultural researchers, planners, and scientists from these related disciplines would allow greater insight into the best arrangement of land uses.

As difficult as it is to determine the appropriate mix of land uses within a region, country, or specific site, sustainability also requires the temporal arrangement of land uses and their integration over time. Time frames have always been taken into account in traditional shifting cultivation systems and are incorporated, for example, in the design of rotation schedules. Expanded time frames should also be considered in planning long-term shifts in land use. In practice, regenerating forests and various low-input cropping and agroforestry systems may serve primarily as preparatory or transitional land uses (Sanchez, 1991).

While there is no exact way to determine which mix of systems will be most appropriate at any one place or point in time, the need to consider issues of scale in making decisions is critical. Land use scenarios should be considered at various geographical scales, from the farm to the landscape to the watershed to the region. An agricultural technology may offer sustainable productivity at the farm level, but have adverse social, economic, and environmental effects on the surrounding landscape (Okigbo, 1991). An individual farmer, for example, may benefit by capturing a significant proportion of a water source for irrigation. If, however, that source provided water for domestic use by downstream users, supported other downstream economic activities, or was critical to the stream's ecological functions, the individual benefit would have potentially serious community-wide effects.

Conversely, the success of a particular technology will be influenced by its ability to adapt to the components, processes, and relationships within the larger agroecosystem. Terracing, for example, is most often undertaken on steeply sloping lands to reduce the effective slope on which farming occurs. Successful implementation, however, depends on the physical, social, and economic characteristics of the larger ecosystem. The type of terraces, their height, closeness to each other, and the extent of terracing must be suited to the specific conditions of the ecosystem. These considerations of scale are especially important in weighing the information presented in Table 3-1 and the policy issues discussed in Chapter 4.

Improved resource use also requires an appreciation of changing demographics. For example, traditional shifting cultivation has been the most sustainable form of agriculture in many areas of the humid tropics. It may remain a suitable land use system where population levels are low and stable. However, to prescribe its continued (or expanded) use in areas lacking a sufficient land base would diminish the sustainability of the area as a whole. As population density increases or decreases, the appropriate role of shifting cultivation will change. The conditions that define this role are not easily predicted.

Many resource management problems in the humid tropics reflect the inability of institutions to address land use problems and potential solutions in an integrated manner (Lundgren, 1991). Most institutions involved in research, education, training, resource management, and international aid and development are structured according to various components of land use systems—for example, soils, water, crops, forests, range, livestock, and fisheries. Many have focused on managing one or two components for maximum productivity, without considering other consequences.

To overcome professional and institutional divisions, local and national agencies in the humid tropics need to foster cross-sector communication and action. Integrated management requires closer cooperation among hydrologists, soil scientists, agronomists, foresters, livestock and fishery managers, conservation biologists, cartographers and geographic information specialists, economists, sociologists, and other professionals. It also requires close cooperation between resource professionals, farmers, and other rural residents.

At the same time, local resource management activities need to be viewed within a broader context. The land management problems that undermine agroecosystem sustainability—soil erosion and sedimentation, nutrient depletion, declining water quality and availability, the loss of biological diversity, pest outbreaks, and destructive floods and fires—should be addressed through coordinated responses at scales larger than the field or local village level. Solutions require critical understanding of how the mosaic of land types and land uses within a given landscape or watershed supports or destabilizes local physical, biological, and ecological functions. This broader scale is also needed to address social and economic aspects of land use in a manner that extends beyond the local community (Okigbo, 1991).

Achieving an optimal mix of land uses will not be easy anywhere in the humid tropics. In any given area, this mix will vary according to the status of forest resources, climatic factors, topography, soil characteristics, levels of biological diversity, population pressures, indigenous populations, current land uses, and other considerations. To encourage optimal use of the land, zoning may be necessary. Decisions about major categories of land use can best be made at the national level; more specific decisions about land use must be made at lower levels. For example, in countries that retain large areas of primary forest, such as Brazil and Zaire, extractive reserves and natural forest management will be more important than in countries where deforestation is well advanced (see Part Two, this volume). Countries with high-population density, poor soils, and large areas of degraded lands will seek to allocate more space for labor-intensive restorative agroforestry systems than countries with fertile soils suitable for more intensive forms of crop agriculture. Countries that also contain large areas outside the humid tropics will need to coordinate land allocations across ecological boundaries.

It is highly important to involve local farmers, forest dwellers, and communities in zoning decisions at every step in the process. Just as the basic aim of land tenure reforms is to give people a stake in land and land uses, so should the zoning process seek to give citizens a greater voice in determining the land's future.

All countries have areas of special biological interest, whether these contain rare or endemic species, unusually high levels of biological diversity, or remnants of primary forest. The amount, types, and location of land that can and should be protected need to be determined. The design of forest reserves needs to be coordinated with agroecological zoning to avoid, to the extent possible, the effective destruction of habitat through isolation and fragmentation, to establish effective buffer zones and corridors, and to provide opportunities for integrated management. This is especially important in areas where forest reserves provide critical environmental services, such as the protection of upland watersheds.

The criteria used to evaluate land resources will themselves vary from country to country. In many cases, ongoing research will be required to delineate more precisely basic land attributes such as levels of biological diversity, susceptibility to erosion, potential for different agroforestry systems, and the state of forest regeneration in deforested areas. Remote sensing and geographical information systems can make the agroecological zoning process more efficient. Clearly this is one area where international support should be given to national resource agencies to strengthen their capacities.

Land Use Patterns and Land Classification

Land use classification systems that include geophysical, biological, and socioeconomic determinants must be developed for each country. Their evolution must involve the local communities that will ultimately be responsible for resource use. National priorities and ability to provide resources and infrastructure must also be considered.

Biological, geophysical, and climatic characteristics (including natural vegetation type, soil type and condition, slope, slope aspect, water availability, rainfall, humidity, light, wind, storm type, and storm intensity and frequency) determine land suitability for different types and combinations of agricultural and forestry systems. Ultimately, social, economic, and institutional conditions will determine the actual patterns of land use and the productivity levels within a landscape. Where human population density is low, more land tends to be used for agriculture, and the variety of land uses tends to be limited. As population pressure on the land rises, the variety of land

This typical agricultural landscape in the lowlands of Tabasco, Mexico, shows a mixed-crop land use system adapted to various field conditions by local farmers. Intercropped maize and beans occupy the better soil in the foreground; rice grows in a wet area in the middle; cassava has recently been harvested from the poorer, more well-drained soils around the house; maize grows in the background to the right on soils enriched by annual flooding but high enough for cropping during the rest of the year; a multistoried mixed tree plantation occupies the background where a diversity of timber and fruit trees provide shade for cacao trees below. Credit: Stephen Gliessman.

uses increases as people take fuller advantage of natural resources and of each production niche.

Land use patterns may become extremely diverse and complex if productivity and sustainability are demanded of all available land. For example, in a village at a lower mountain elevation, farmers may work the valley-bottom floodplains, the gentle to steeply sloping mountain soils (which may be too steep to terrace or may have cooler northern exposures), dry hilltops, and eroded gullies or stone outcroppings. They must take into account climate—heavy seasonal rains and the possibility of summer thunderstorms with hail, which restricts the growing of tree fruit. If land pressures in the village are high and markets are available, appropriate land use systems could include lowland rice with winter crop rotation, terraced rice, terraced mixed upland crops, growth of animal fodder on terrace faces, agroforestry, mixed forest plantings, highland grazing, animal feed gathered from nearby vegetation, and extractive reserves. This entire range of systems types can be found, for example, in many villages in Southeast Asia. Although population pressures in these villages may be high, the social, political, and institutional conditions permit the blending of national, regional, community, and farmer interests in adjusting land uses to the geophysical environment.

Such adjustments could be encouraged if methods of land use classification that better incorporated biological, geophysical, climatic, and socioeconomic characteristics were developed. Few if any countries in the humid tropics have programs for detailed and systematic evaluations of natural resources of the type and at the scale necessary for assessing management options (Lal, 1991a). Similarly, there is no general classification system of ecological zones, of agricultural production potential, or of agricultural land use patterns that can provide an adequate framework for global-scale analysis of forestry and agriculture in the humid tropics (Lal, 1991a; Okigbo, 1991; Oram, 1988).

Existing land classification schemes do provide important baseline information. The soil and geophysical classifications of the Food and Agriculture Organization of the United Nations, for example, can be used to determine land use potential and environmental fragility, and to map and quantify the area within various categories of land use (Food and Agriculture Organization, 1976). Holdridge's classification of life zones based on climatic data is an important tool for understanding plant species adaptability and comparing forest system properties, and may be of value in indicating the potential of management options most appropriate for different lands (Holdridge, 1967; Lugo and Brown, 1991).

In general, these and other land classification systems have not been designed to incorporate socioeconomic factors, such as human population density and access to roads, or important biological factors, such as the degree of biodiversity. Inventories that might yield basic data for improved land use classification systems have usually been conducted on a partial basis, have focused only on resources of known commercial value, and have been hindered by a lack of strong institutional support (Latin American and Caribbean Commission on Development and Environment, 1990). As a result, science cannot calculate with precision the areas suitable for various land uses.

Maintenance of Biomass

The ability of a land use system to maintain high residual biomass in the form of wood, herbaceous material, or soil organic matter should be a primary requirement for restoring degraded or abandoned lands.

Biomass in the form of wood, herbaceous material, or soil organic matter significantly effects local and global ecological systems. It is essential for sustaining soil structure and fertility, recycling rainfall, and preventing soil erosion and floods. For example, a healthy stand of rain forest produces high levels of cloud recharge. About three-fourths of the rainfall is evaporated either directly from the soil and from the surface of leaves or from transpiration by plants, and roughly one-fourth runs off into streams, returning to the ocean (Salati and Vose, 1984).

Plant biomass, above and below ground, also plays a role in air quality and potential climatic changes. Through photosynthesis, trees use carbon dioxide in the atmosphere to produce the oxygen necessary to support life. Terrestrial soils are the largest reservoir of carbon, containing two times as much carbon as green plants (Lal, 1990). The clearing of forests releases carbon into the atmosphere that had previously been stored in trees and soils.

Through proper agricultural management techniques, some land uses have the potential for increasing the storage of soil carbon and the production of biomass. When fallow periods are long enough, carbon and other nutrient levels are maintained under shifting cultivation. A by-product of plantation cropping of fast-growing forests is the carbon fixation both in the standing forests and in their root systems. However, research is needed to determine which types of systems and combinations of plant and animal species are most effective in different regions.

Monitoring Systems and Methodologies

Resources should be available for linking national monitoring agencies with global satellite-based data sources so these agencies can refine, update, and verify their data bases for tracking land use changes and effects.

Monitoring systems and methodologies must be improved to trace land use changes and their effects. For example, only within the past 2 decades in the United States has it become possible to estimate the magnitude of soil loss and its effect on productivity. In most countries of the humid tropics, only rudimentary data on soil loss are available (World Resources Institute, 1992). The same holds for data on groundwater pollution, salinization, sedimentation rates, levels of biological diversity, greenhouse gas emissions, and other environmental phenomena (Ruttan, 1991). In addition to collecting these data, this effort should include assessments of the social effects of environmental change on human populations, especially the health of individuals and communities. It is also important that monitoring add to the knowledge of, and ability to quantify, the impact of agricultural practices on the levels of greenhouse gases.

Broad-scale environmental phenomena are inherently difficult to quantify. This problem is exacerbated in the humid tropics by the escalating rate of deforestation. Accurate data on the spatial extent of, and biogeochemical processes associated with, deforestation and land use in the humid tropics are critical. Such data are especially important for research on global climate change, which relies heavily on computer models. International data bases employing satellite-generated information have improved monitoring capacities, but they should be more effectively linked with national monitoring systems. In many cases, these international data bases cannot be accessed at the national level. As a result, major discrepancies occur between the international and national data on basic questions such as the extent of forest cover and the rate of deforestation. Where data are available, their utility can be impaired by a lack of standard definitions and land use classifications.

The Global Environment Monitoring System (GEMS) of the United Nations Environment Program is an example of international efforts toward making data more readily available to resource planners and other analysts who might use them to advise development decision makers. The GEMS has activities related to air and water quality in 142 countries. However, due to inadequate financial resources, the coverage and quality of data have been weakened (World Bank, 1992).

ECOLOGICAL GUIDELINES FOR SYSTEMS MANAGEMENT

Systems options are selected, as discussed above, through stakeholder negotiation based on geophysical resources, social needs, markets, and the range of social and economic conditions. The target systems then evolve from existing conditions to higher productivity through progressive changes. The degree to which these systems increase in ecological sustainability, particularly in a fragile soil environment, depend largely on the following six biologically based elements:

• *The degree to which nutrients are recycled.* Productivity within a system is directly related to the magnitude of nutrient mobilization and flow. Sustainability is directly related to the efficiency of nutrient use and to the reduction of nutrient loss, either to ground or surface water or to the atmosphere.

• *The extent to which the soil surface is physically protected.* Soil loss through water transport or wind erosion must be minimized. It should be protected from oxidation or other chemical deterioration through protective plant cover. Physical deterioration, compaction, and loss of structure through rainfall can be equally damaging, reducing productive potential. Continuous crop or crop residue cover from appropriately managed systems is crucial to maintenance of productive potential.

• *The efficiency and degree of utilization of sunlight and soil and water resources.* With increasing limitations on the extent of natural resources in many populous countries, the selected agricultural systems must be managed for optimal use, including continuous crop cover, good crop and animal genetic potential, minimal pest damage, and optimal nutrient supply.

• *A small offtake (harvested removal) of nutrients in relation to total biomass.* This factor is especially important on the more fragile soils. Where soils are erosive, have poor nutrient status, or are otherwise chemically or physically fragile, the maintenance of high biomass systems is critical.

• *Maintenance of a high residual biomass in the form of wood, herbaceous material, or soil organic material.* A carbon source for both energy and nutrient retention is critical to the support of biomass in the soil and to crop and animal productivity.

• *The structure and preservation of biodiversity.* The efficiency of nutrient cycling and the stability of pests and diseases in the system depend on the amount and type of biodiversity as well as its temporal and spatial arrangement (structural diversity). Traditional systems, particularly those in marginal production environments, often have significant stability and resiliency as a result of structural diversity. Research is only now beginning to quantify these effects.

TECHNICAL NEEDS COMMON TO ALL LAND USE OPTIONS

Three scientific areas, interwoven throughout the report, are an essential part of every land use option and its application to any given environment. The degree to which a land use is sustainable often depends on the success in dealing with pest management, nutrient cycling, and water management.

Pest Management

Plant and animal protection is crucial to the productivity of any land use system. Although many land uses have an inherent stability or resiliency with regard to pests and diseases, additional steps may be needed to protect plants and animals from damage due to insects, weeds, pathogens, or nematodes. Pest-induced losses to crops before

A farmer in Zaire tends his coffee trees. Coffee is one of the country's most profitable crops and is well suited to mixed crop small farms. Credit: James P. Blair © 1983 National Geographic Society.

harvest can be as high as 36 percent in developing countries (U.S. Agency for International Development, 1990). Current efforts to manage losses emphasize the use of chemical pesticides. Heavy, widespread use, however, can lead to detrimental effects on nontarget organisms, water contamination, pesticide resistance, and chemical residues on food. Chemical control for some important pests and pathogens may also not be economically viable.

The development of economically and environmentally sound solutions to these problems is central to the issues of resource sustainability and achieving agricultural production goals. Research suggests that knowledge about natural biological processes in the crop and animal production environment may lead to management approaches or new products that, alone or combined with the careful use of chemicals, are effective against pests. For example, integrated pest management (IPM) is an ecologically based strategy to control pest populations and minimize crop loss through biological, cultural, and chemical means. It relies on natural mortality factors, such as pest predators, weather, and crop management, and reduces the need for pesticide use. Its adoption, however, is hindered by technical, institutional, socioeconomic, educational, and policy constraints in developing countries.

Technical constraints, such as knowledge of the controlling factors of the pest, the ability to manage predator populations, and difficulty in making the necessary crop management changes, are beginning to be overcome. In Indonesia, IPM was successfully used to control the rice plant hopper (Kenmore, 1991). Biological control methods tailored to the crop and pest were effectively used in Africa to control damage from the cassava mealybug and cassava green mite (Herren, 1989).

Nutrient Cycling

High productivity requires the enhanced movement of nutrients from soil to crops and trees, or from crops to animals and returning to crops. The lack of nutrients is often the most limiting factor on low-fertility soils. As productivity increases, however, nutrient flow and containment become increasingly critical, posing significant risk to water quality. Surface runoff containing phosphorus and nitrogen enriches water and accelerates the aging of lakes, whereby aquatic plants are abundant and oxygen is deficient. Nitrate buildup in water at levels above 10 parts per million poses serious health risks to humans.

High residual biomass systems are efficient in the extraction, use, and recycling of nutrients. Yet, even with perennial tree plantations, the fertilization needed for optimum yields can lead to loss to the environment unless appropriate cover crop and other measures are taken for their containment (Vincent and Hadi, Part Two, this volume).

Integrated nutrient management to reduce nutrient losses is thus critical to all systems. The magnitude of loss will vary with location, topography, cropping system, and other site-specific factors. Increases in soil fertility can be gained through the integration of livestock with tree and food crops, tillage practices and mulching, alley cropping, crop rotation, cover crops, and mixed cropping (Cashman, 1988; Francis, 1986; Gliessman, 1982; Lal, 1987; Part Two, this volume).

Water Management

Water is an increasingly precious and limiting resource in all systems. Its quantity and quality play a vital role in the functioning of natural ecosystems and in economic development. Water resources are shared by all life forms in the environment. Its multiple use in hydroelectric generation, irrigation, fish and shellfish production, and waste disposal requires an integrated approach to its management.

Quality of the water resource is determined both by its purity and by the variability in stream flow or aquifer level. Land use in catchment areas is a critical determinant of both aspects of downstream water quality (Bjorndalen, 1991; Lundgren, 1985). Degradation of upstream areas leads to cycles of declining productivity and poverty both in the directly affected area as well as for downstream irrigation, fisheries, tourism, and other uses. Management of water resources is a cross-cutting issue that can serve as a focal point for a development program's organization, institutional structure, and impact assessment. It can only be addressed in an integrated fashion, beginning with selection of land use options appropriate not only to the geophysical setting but to the social and economic environment (Lal and Rassel, 1981). Watershed-level management capacity is required for all successful land use development planning.

COMMODITY-SPECIFIC RESEARCH NEEDS

Major public sector support is needed for research on basic food and feed grain commodities, both in genetic improvement and in management technologies. An appropriate economic environment must be maintained to continue and expand private sector technology development in the capital-intensive, vertically integrated industries, such as poultry, hogs, fish, and silk production, and in the development of appropriate inputs. Above all, farmer-collaborative networks for integrative technology adaptation and dissemination are needed. These are discussed in Chapter 4.

4

Policy-Related Imperatives
for Change

If agricultural technologies and land use options exist that can make agricultural development in the humid tropics more sustainable, then why have they not been more widely adopted? Why has deforestation or forest conversion not been more effectively managed? There are no simple answers to these complex questions, as illustrated by the country profiles in Part Two of this volume. For countries in the humid tropics to make real progress toward sustainability, the broad range of social, economic, and political factors that affect land use patterns must be recognized and considered throughout the development process. Progress will depend not only on the availability of improved land use techniques, but on the creation of a more favorable environment for their further development, implementation, and dissemination. These changes must be achieved through the national and international institutions that determine the character of public policy.

The goal of the committee's policy-related recommendations is to meet human needs, at individual, national, and international levels, without further undermining the long-term integrity of tropical soils, waters, flora, and fauna—the foundations of sustainable development. The countries of the humid tropics will need to take the lead if these efforts are to succeed. The countries beyond the humid tropics will need to extend their support and be willing to make their own sacrifices. All countries will need to share the conviction that success is possible, and offer their commitment to its realization.

The strategy for change outlined here to promote sustainability in the humid tropics involves efforts to (1) manage forests and land resources more effectively and (2) encourage sustainable agriculture. Most reform efforts emphasize the removal of policies that have led to accelerated deforestation rates in recent years. Until now, however, these reforms have not focused on stabilizing and rehabilitating already deforested lands, nor have they served to guide small-scale farmers toward more sustainable agricultural production systems through forest conversion strategies.

Sustainable agriculture will not automatically slow forest conversion or deforestation in the humid tropics. However, the combination of forest management and the use of sustainable land use options will provide a framework within which each country can achieve an equilibrium appropriate to its development stage and natural resource use requirements. These systems can help to offset the impacts of heightened economic and demographic pressures on intact primary and secondary forests by improving the management of agricultural systems, diversifying crop production systems, stabilizing shifting agriculture on steep lands and in forest margins, and restoring degraded and abandoned lands.

At the same time, however, the ability to enhance the performance and profitability of croplands, pastures, mixed systems, or plantations may encourage further migration into and conversion of undisturbed forests. The combination of improved land productivity and further population growth could also result in higher land prices, causing small-scale farmers to migrate to cheaper lands at the forest frontier.

Pressures to extend sustainable agricultural systems to undisturbed forest will remain, especially where timber profits are high or population growth is rapid. In some areas, such as parts of Africa, Brazil, and Venezuela, additional conversion of forests to agricultural, or nonagricultural, uses may be necessary and appropriate based on national environmental and food needs. In all situations, however, technical innovations must be accompanied by policies that guide their applications and protect undisturbed forests.

Both the causes and consequences of nonsustainable land use in the humid tropics are global in nature. Action by, and coordination among, all countries will be required to effect change. Accordingly, the actions recommended here are wide ranging. Some apply primarily to the policies and activities of industrial nations, while others focus on developing countries within the humid tropics. All countries, however, stand to gain from multinational cooperation.

The changes discussed in this report focus primarily on low productivity lands worked by small-scale farmers and on forested or recently deforested lands. An improved policy environment should, however, consider the role that highly productive agricultural lands and input-intensive agroecosystems play in protecting forests and stabilizing degraded lands. If the productivity of these areas can be increased in a sustainable manner, part of the pressure for expansion into marginal areas may be reduced.

MANAGING FOREST AND LAND RESOURCES

Sustainable agricultural practices cannot be expected to take hold in the humid tropics as long as development policies and economic forces continue to encourage more expedient uses of land resources. Governments, international development agencies, and other organizations have begun to address this fundamental problem, but the acceleration of deforestation rates through the 1980s indicates the need for a stronger commitment to reform. Recent analyses have described in detail national and international policies and their impact on tropical resources (for example, Barbier et al., 1991; Binswanger, 1989; Hurst, 1990; Leonard, 1987; Repetto and Gillis, 1988). This growing body of analysis points to the need for policymaking bodies at the local, national, and international levels to reexamine their roles and responsibilities in determining the future welfare of tropical land resources and the people who depend on them.

Reviews of Existing Policies

Policy reviews under way at local, national, and international levels must be broadened to consider the negative effects that policies have had on sustainable land use.

In response to escalating rates of deforestation and increased awareness of local, regional, and global effects, many international and bilateral development agencies have reassessed their forest policies. These include the Dutch Development Corporation, the Inter-American Development Bank, the Asian Development Bank, the Finnish International Development Agency, the World Bank, the Food and Agriculture Organization (FAO) of the United Nations, and the International Tropical Timber Organization (Spears, 1991). Most of the recent forest policy statements of these agencies focus on the forest resource itself, and analyze the changing market conditions, institutional and social forces, and policies that have encouraged forest conversion and deforestation. Few focus on the need for agricultural sustainability in responding to deforestation in the humid tropics.

Land has been set aside for the Surui in Aripuana Park, Rondônia, Brazil. The Surui people resent the influx of settlers who, they say, destroy the forest and take the land. This small cluster of houses is nestled in a mixed garden setting surrounded by a partially converted forest. Credit: James P. Blair © 1983 National Geographic Society.

Government agencies within humid tropic countries have also adjusted policies in response to global environmental concerns and their own socioeconomic and environmental priorities. In recent years, for example, Brazil has removed the financial incentives that promoted conversion of forestland to large-scale cattle ranches and has put into place programs to encourage sustainable agricultural development (Serrão and Homma, Part Two, this volume). Brazil and Colombia have recently recognized the claims of indigenous people to large forest areas and have given them greater responsibility for managing these areas. As new national policies are instituted, however, they can sometimes have unintended effects. In the wake of severe floods in 1988, for example, the government of Thailand instituted a ban on commercial logging. The subsequent rise in timber prices led to increased illegal cutting and failed to check the forces behind forest encroachment by shifting cultivators (Myers, 1989).

Efforts to review policies that contribute to deforestation are a prerequisite to sustainable land use in the humid tropics, and they merit expansion. At national and regional levels, policy reviews should respond to the specific biophysical, social, and economic circumstances that affect land use patterns within countries and regions. These reviews should also focus on the in-country effects of international trade, lending, and debt-reduction policies. At the international level, the review process will vary from institution to institution, depending on its size and objectives and the range of its activities.

Although the policy review process will necessarily vary, the following considerations are generally applicable.

• Given the complexity of the socioeconomic and ecological aspects of land use in the humid tropics, reviews should be undertaken by multidisciplinary teams.

• Economic policies that encourage large-scale logging and agricultural clearing should be identified and evaluated in terms of their externalized costs, social and ecological costs, and availability of transport infrastructure, such as roads and bridges. For example, the fees charged loggers for the right to cut standing timber seldom come close to the costs of replacing the volume removed with wood grown in plantations (World Bank, 1992). In general, these policies discourage long-term interest and investments in forest management (both in the public and private sectors), undervalue the full economic and environmental benefits of conserving primary forests, and hinder the adoption of sustainable land use alternatives.

• New methods of assessing and assigning value to the forests should be sought. Reviews should assist in recognizing the full range of the forests' economic benefits, the key environmental services they provide, the potential for sustainable use of their resources, the opportunity costs involved in forest conversion, and the rights of future generations to forest services and products. When possible, values should be expressed in standard economic terms, such as financial costs and returns, with cost and benefit streams discounted to a common base. Those that cannot, such as aesthetic values and environmental services secured through conserving biological diversity, should nonetheless be explicitly noted in all economic analyses (Barbier et al., 1991; Norgaard, 1992; Randall, 1991).

• Reviews should seek opportunities to integrate more fully the activities of the forest and agricultural sectors, and to incorporate an interdisciplinary perspective in development, research, and training programs.

• Ways should be found to integrate infrastructure, land use, and development policies. Well-conceived infrastructure development policies can fail if sustainable land use technologies and systems are not in place to support them.

The Negative Impacts of Land Use Policies

Despite increasing recognition of the importance of tropical forest resources, the exploitation of tropical forests to meet short- to medium-term development objectives still takes precedence over most long-term uses in many countries. Economic analyses and policies have failed to recognize the full market and nonmarket values of forest conservation and sustainable land uses. Thus many of the potential benefits of forest conservation are not realized. Similar economic factors have contributed to, and continue to support, deforestation in temperate zone countries, including the United States (Repetto and Gillis, 1988).

Often national economic and land use policies contribute to this dilemma by directly or indirectly promoting the inefficient and nonsustainable conversion of forests to other uses. In many cases, the policies of international development agencies have encouraged these moves, especially as developing countries try to reduce their burden of outstanding debts. Areas with the highest rates of deforestation in recent decades include the Brazilian Amazon, the Philippines, Malaysia, and Côte d'Ivoire. A variety of economic incentives has encouraged exploitation of forest resources, including tax incentives and credits for land clearing, subsidized credit, timber pricing procedures, price interventions, land subsidies and rents, concessions, tenure, and property rights.

• *Tax incentives and credits.* The adverse effects of tax policies on forests and forest management procedures throughout the humid tropics are well documented (Binswanger, 1989; Browder, 1985; Repetto and Gillis, 1988). Policy mechanisms that have been identified include direct tax rate incentives and credits, tax holidays (tax-free periods), and tax subsidies (differential rates based on land use). The role of tax incentives and credits in stimulating conversion of forests to cattle ranches in the Brazilian Amazon in the 1970s and 1980s has been well documented (Browder, 1988; Hecht, 1982; Hecht et al., 1988; Serrão and Homma, Part Two, this volume). Tax policies have also been identified as important factors in encouraging destructive logging operations in Indonesia, Côte d'Ivoire, Malaysia, and other countries where timber extraction is a leading cause of deforestation (see Part Two, this volume).

- Finally, as institutions reexamine their priorities, they should recognize the need to better coordinate their efforts and institutional commitments. Lack of coordination—within national governments, among international organizations, and between international agencies and governments—often gives rise to conflicting resource policies. International development agencies, in particular, should seek opportunities to coordinate policies in support of conservation and sustainable development objectives in the humid tropics. The Tropi-

---

- *Subsidized credit.* Subsidized credit has been used to stimulate large-scale commercial investment in forests. These credit policies have induced excessive timber harvest rates and conversion to ranching, large-scale farming, and other competing land uses (Repetto and Gillis, 1988). By contrast, small-scale farmers in many humid tropic countries have limited access to credit. Consequently, they are unable to invest in the improvements needed to make their operations more economically and environmentally sound over the long term.
- *Timber pricing procedures.* Pricing policies have led to the undervaluing of tropical timbers. The price of timber in national and international markets reflects the costs of logging, milling, and transport, but not the foregone environmental benefits, goods and services, and other indirect or nonmarket-related forest values, such as aesthetics or siltation of reservoirs due to erosion. Timber is thus made available in the market at prices that do not reflect the full social and environmental costs.
- *Price interventions.* In some countries, domestic price interventions, especially price supports for products grown on converted lands, have contributed to the loss of primary forests. In Indonesia, for example, price interventions have stimulated the conversion of forests to palm oil and other tree crop plantations.
- *Land subsidies and rents.* Low rent and fee collections by governments have encouraged excessive logging. Rent collections in the form of royalties are not responsible for excess rates of deforestation since rent does not affect allocation decisions (Hyde and Sedjo, In press). Rather, the problem of excessive deforestation rates is rooted in subsidies for land clearing and in insecure tenure. Subsidies, in the form of direct payments or tax concessions, provide incentives to deforest areas that would otherwise remain uncleared. Subsidies are also used to support activities that require forest clearing (for example, livestock pastures and certain crops).
- *Concessions.* A logging concession is an agreement between the government and the logger that establishes the terms and conditions for the harvest of trees on public lands. Concession agreements may stipulate, for example, the size of trees to be cut, the area to be cut, the

continued

cal Forestry Action Plan, developed by FAO, the United Nations Development Program, the World Bank, and the World Resources Institute, is an example of such a coordinated effort (Food and Agriculture Organization, 1987).

Anticipated corrections as a result of these reviews include: reforms in tax, credit, and subsidy policies that remove incentives to

---

period in which harvesting occurs, and any precautions that are to be taken in harvesting operations. Currently, the logging concession system contributes to tenure insecurity in many countries. Although the absence of tenure long enough to allow trees to grow to harvestable size inhibits conservative land use, this issue is often absent in tropical forest and land management policymaking. In many parts of the tropics, timber concessions are granted for relatively short periods of time. In Southeast Asian forests, for example, the harvest interval typically recommended in forest management systems is 35 years, while most concessions are granted for 20 years at most. Consequently, the incentive to adopt a long-term perspective is weak. The concession holder is expected, and is often required by law, to undertake reforestation. The anticipated termination of tenure, however, inhibits reforestation efforts. Thus, institutions in charge of forest management are generally biased against sustainable forestry techniques, even where such techniques are proved effective and where growth and cost conditions could support sustainable management.

• *Tenure.* Tenure is a key determinant of the status of small farmers and indigenous groups in the humid tropics. Where tenure is insecure, exploitation for short-term gains is more common. Long-term investment is discouraged because the potential investor has no assurance of retaining tenure to obtain benefits over a longer time period. Where tenure considerations have been disregarded, as for example in the Philippines (Vincent and Hadi, Part Two, this volume), disastrous timber harvest practices have ensued and reforestation efforts have suffered.

• *Property rights.* In many developing countries, the lack of secure tenure is compounded by confusion over the question of rights to various land resources. In many cases, property rights apply differentially to the forestland, the trees, and other forest products. The rights to land, timber, and minor forest products are frequently attenuated or in apparent conflict with one another (Fortmann and Bruce, 1988). This is a particular problem where tree and forest tenure are divorced from local land tenure (Gómez-Pompa et al., Part Two, this volume). Local people may hold traditional rights to harvest nuts, fruits, firewood, and other minor forest products from communal forests while the state retains title to the trees. The state may then sell or otherwise grant maximize timber production and that encourage more sustainable forest management techniques; international trade and financing reforms that can bring more realistic prices to tropical timber while reducing wasteful harvesting methods; clarification of property rights and support for local and indigenous land tenure; and changes in concession agreements to prompt greater investment in long-term forest management and reforestation efforts.

---

concessions for harvest or may empower its forest management agency to do so (Lynch, 1990; Rush, 1991).

In many tropical countries, political corruption contributes to deforestation and other forms of natural resource degradation. Close ties between commercial timber interests and politicians have encouraged the exploitation of forest resources for political purposes (Rush, 1991). For example, the awarding of noncompetitive timber concessions through military and government contacts has significantly contributed to the rapid rates of deforestation in Indonesia, Malaysia, and the Philippines (Garrity et al., Part Two, this volume; Repetto, 1988b). Under these circumstances, the tenure rights of indigenous people are often disregarded. Forestry regulations and guidelines affecting extraction techniques, rotation schedules, the environmental impacts of logging and processing operations, and reforestation requirements are ineffective due to lack of enforcement (Repetto, 1988b). Forest encroachment, poaching, timber smuggling, and other illegal logging practices become important problems (McNeely et al., 1990). In addition, corruption has allowed private timber interests to have undue influence on government subsidies, tax policies, the location of infrastructure development projects, and the distribution of land, aid, and credit.

The impact of corruption can be seen in the case of the Philippines. Rush (1991) notes that access to timber concessions and other state-owned natural resources has played an important role in the political patronage system. Garrity et al. (Part Two, this volume) identify "large-scale corruption" as a distinguishing characteristic of the Philippine government during the late 1970s, when deforestation rates were particularly high. In many cases, timber operators in the Philippines have themselves held political office, making it impossible to enforce policies that would result in lower profit margins (Baodo, 1988). At the same time, deficiencies in community organization, training, and cooperative management at the local level have allowed forest regulations to be abused (Garrity et al., Part Two, this volume). Although the impact of political corruption on resource management is especially evident in South and Southeast Asia, where timber extraction has been especially lucrative, the same forces operate throughout the humid tropics as well as in temperate regions.

A housing development of about 15,000 units was carved out of the Brazilian rain forest. The tight clustering of houses reduces the use of land but dramatically decreases opportunities for low-income families to have mixed gardens and keep animals so critical to their well-being. Credit: James P. Blair © 1983 National Geographic Society.

Planning of Major Infrastructure Projects

Impact assessments of infrastructural development projects should be broadened to anticipate changes in land use systems and subsequent social effects.

Infrastructural development projects, usually undertaken with the backing of international development agencies, have caused widespread forest degradation in the humid tropics. The construction of mines, dams, railroads, highways, and logging roads directly and indirectly affects large areas of primary forest, leading to changes in land use. Larger areas are affected by soil, air, and water pollution, soil erosion and sedimentation, disruption of hydrological systems, forest fragmentation, and other associated consequences.

Until recently, these social and environmental costs were rarely considered. The international development organizations that provide much of the support for these projects—such as the World Bank, other major development banks, and some bilateral donors—now require impact assessments. In many cases, however, these assessments have failed to prevent or mitigate adverse impacts. For example, high-sedimentation rates threaten the viability of dam projects throughout the humid tropics: Eljon in Honduras, Chixoy in Guatemala, Ambuklao in the Philippines, and Arenal in Costa Rica. Development proceeded without effective provisions for sustainable agriculture, watershed management, protection of adjacent forestland, forest restoration and rehabilitation, pollution control, and other mitigation measures. In the future, environmental provisions should seek to prevent land degradation by requiring that sustainable land use practices accompany infrastructure development projects from the outset.

The social impacts of these projects have also been inadequately addressed by governments and international agencies. Local communities and indigenous people are often displaced or disrupted despite their tenure or property rights. Moreover, infrastructural development often precedes or takes place simultaneously with resettlement and colonization projects, yet settlers are rarely provided with adequate tenure, tools, financing, or knowledge needed to use these lands sustainably. The result frequently is the perpetuation of the pattern of resource decline. Poor farmers gain access to primary forests, yet they continue to farm in a manner that depletes resources and keeps them impoverished.

These adverse social impacts need to be anticipated and, where necessary, mitigated. Development projects that entail relocation or resettlement should recognize the need for sustainable land use systems (and effective land use restrictions) in the surrounding cleared lands, forests, and watersheds. The tenure rights of indigenous people and colonists should be secured prior to major infrastructural development projects. Land titling is not always an issue in these cases, but where questions of ownership and usufruct rights exist they should be resolved before projects proceed. This approach was taken, for example, in the Pichis-Palcazu Project in Peru. Land titling and property boundary surveys were undertaken prior to road construction, allowing the native Amuesha-Campa communities as well as settlers to gain secure tenure before the influx of new migrants occurred.

National Resource Management Agencies

The mission of national resource management agencies as custodians of national forest and land resources should be redefined to focus more atten-

tion on achieving a balance among resource users. The strengthening of resource management agencies is a key area for cooperation among the governments of tropical countries and the international assistance agencies.

Throughout the humid tropics, national resource management institutions, particularly forest agencies, are often nonexistent or weak. Where they do exist, they receive limited political and financial support. While agricultural agencies generally receive the greater portion of financial support, they in turn allot little funding to forest-related activities or to research and development in sustainable agriculture (Okigbo, 1991; Repetto, 1988a; Villachica et al., 1990). Few national or state resource bureaucracies are capable of effective protection and stewardship of the resources under their jurisdiction, or of supporting basic or applied research in forest ecology, agroecology, farming systems, indigenous knowledge, or other areas relevant to sustainable land use. In some countries, effective agencies may need to be built from the bottom up through long-term investments. In others, where strong agency structures are already present, they may need to be better integrated.

The structure of resource management agencies is usually determined by discrete resource categories, such as agriculture, forestry, and environmental protection. As a result, the division of responsibilities—in legal jurisdiction and in scientific research, training, extension, and development programs—has made integrated management difficult. In these cases, it may be most effective to invest in training and continuing educational opportunities in the environmental sciences for agency personnel.

Biodiversity

Biodiversity should be conserved through both the establishment of forest reserves and the inclusion of broad genetic diversity as a basis for sustainable land use systems.

The development of sustainable agriculture and the protection of biodiversity are not two different undertakings, but allied aspects of conservation as a whole in the humid tropics. The diversity of soil organisms, plant and animal genetic material, pest and disease control agents, plant pollinators, symbionts, and seed dispersers underlies the functioning and productivity of tropical agroecosystems as well as managed forests (Edwards et al., 1991; Grove et al., 1990; Lal, 1991b; Pimentel, 1989). Improved management on more intensively used lands may ease the pressure to develop forested areas rich in biodiversity.

The establishment and effective management of forest reserves should be seen as part of the development process. All lands can and should contribute to sustainable development. This alone justifies allocating resources for preserving lands and improving their stewardship. Moreover, some uses are permissible in reserves under certain circumstances and may warrant encouragement as part of a strategy of sustainable development. Examples would include scientific research and educational activities, low levels of extractive activities, recreation and ecotourism, and modest efforts to interpret the scenic and natural values embodied in these reserves. When properly planned and managed, these uses do not endanger the primary forest values that the reserves were created to protect.

Policies that simultaneously emphasize the goals of conserving biodiversity and implementing sustainable agricultural systems—especially policies aimed at improving the quality of life for small-scale farmers and local communities through conservation measures—need further development and additional support. From an agricultural and rural development perspective, the benefits of this integrated approach are substantial. Direct economic benefits can be realized through the identification of new products for local use and export. Investments in biodiversity research by industrialized nations can serve to transfer financial resources to countries in the humid tropics and strengthen local research institutions. The establishment, management, and maintenance of germplasm banks can protect local genetic resources, bring farmers and researchers closer together, and provide local employment. Biodiversity research (involving, for example, soil organisms and insect populations) can offer new insights and techniques for agroecosystems. Rural communities can provide services for visitors to national parks and biological reserves. The establishment of reserves and buffer zones can also protect the tenure rights, resources, traditional management methods, and knowledge of indigenous cultural groups.

Often the benefits of biodiversity conservation accrue outside the local community. For example, germplasm from the humid tropics has improved crop productivity in the temperate zone. These contributions must be recognized and efforts made to obtain benefits for the local population. Incentives to identify important natural areas, and to protect and manage them, should be made available to farmers and communities. Creative partnerships between local people and research organizations, management agencies, funding agencies, nongovernmental organizations, and private enterprises can help to ensure that the benefits and costs of conservation are fairly distributed (Altieri, 1989; Brush, 1989).

Global Equity Considerations

The adoption of sustainable agriculture and land use practices in the humid tropics should be encouraged through the equitable distribution of costs on a global scale.

Industrialized countries have a responsibility to assume a proportionate share of these costs, to compensate the countries of the humid tropics for foregoing the short-term economic benefits of resource depletion, and to provide incentives for conservation measures that provide global benefits (Sachs, 1992; Swanson, 1992). Strategies for cost-sharing have already been devised to promote reductions in global atmospheric carbon emissions (through, for, example, carbon taxes and permit trading). At the same time, economic analyses are beginning to explore the means by which environmental costs and benefits may be reflected more accurately in markets and incorporated into international development, trade, and lending policies (see, for example, Costanza and Perrings, 1990; Norgaard, 1992). These innovative cost-sharing and valuation methodologies are becoming increasingly important in achieving a broad range of environmental and development goals, and should be supplemented with other foreign assistance mechanisms that promote equity at the global level.

The World Bank (1991) emphasizes three broad areas of assistance through which the international community can facilitate the transfer of resources and the conservation of tropical resources: technical assistance, research, and institution building; financing; and international trade reforms. Within these categories, a number of specific measures can be adopted. Direct transfers of funds allow the countries of the tropics to decide how to allocate these funds. Other forms of transferral may better meet other, more specific needs. For example, debt-for-nature swaps, which have been arranged with Brazil and several other countries, may be most important in countries with high foreign debt burdens. Investments in institutions or carefully planned infrastructure projects may be more beneficial in countries where these institutions and projects are weak. Improved access to markets and better terms of trade can serve to promote new products and to achieve more equitable trading patterns. Innovative partnerships and exchanges—scholarships and stipends for students in resource management, collaborative research enterprises, private investments in new products from the tropics, and funding for programs in public health and community development—link conservation and development activities. The objective in all of these examples is the same: to use the financial and institutional resources of developed countries in encouraging the conservation of natural resources and the development of human resources in developing countries.

SUPPORTING SUSTAINABLE AGRICULTURE

Changing policies that contribute to deforestation and natural resource degradation in the humid tropics will not by itself encourage the adoption of sustainable agricultural systems. The fact that land use alternatives exist does not ensure they will be widely adopted by farmers. International and national institutions need to support these alternatives at all phases of development, dissemination, and implementation. Without support, sustainable agricultural practices are likely to be adopted only slowly and erratically.

The overarching need throughout the world's humid tropics is to implement land use systems that simultaneously address social and economic pressures and environmental concerns. In areas where short-fallow shifting cultivation is the leading proximate cause of defores-

Cacao is an important cash crop in many developing countries. Pictured is a plant growing in Africa. Credit: Food and Agriculture Organization of the United Nations.

tation and land degradation, the primary goal should be to encourage shifting cultivators to adopt alternatives to low-yielding, slash-and-burn agriculture. In areas where other causal factors are important, actions should reflect the potential of sustainable agricultural systems to reduce these pressures and mitigate their effects. In all areas, much greater emphasis needs to be given to the rehabilitation, restoration, and reforestation of degraded and abandoned lands.

Efforts to support sustainable agriculture can be grouped into three categories:

- Providing an enabling environment;
- Providing incentives and opportunities; and
- Strengthening research, development, and dissemination.

Within these categories, a wide range of reforms and initiatives need to take place at the local, national, and international levels.

Providing an Enabling Environment

National governments in the humid tropics should promote policies that provide an enabling environment for developing land use systems that simultaneously address social and economic pressures and environmental concerns.

Many small-scale farmers and forest dwellers in the humid tropics are unable to adopt sustainable practices due to local socioeconomic and infrastructural constraints. The policy initiatives described here are intended to provide guidance for removing basic obstacles and providing opportunities for sustainable practices to take hold. Essential components of an enabling environment include assurance of resource access through land titling or other tenure-related instruments, access to credit, investment in infrastructure, local community empowerment in the decision-making process, and social stability and security.

LAND TITLING AND OTHER LAND TENURE REFORMS

More than any other factor, the status of land tenure will determine the destiny of land and forest resources in the humid tropics. This conclusion holds true for all classes of local land users—native peoples and forest dwellers as well as more recent settlers and small-scale farmers.

Indigenous forest dwellers retain their traditional territories in many parts of the humid tropics, but their territorial rights are seldom secure. In many cases, the government agencies that hold jurisdiction over resources have not even acknowledged the presence (much less the claims) of native peoples. Because many indigenous territories overlap areas of commercial concessions, these groups often face arbitrary displacement or destruction of their homelands (Lynch, 1991). Their tenure systems, many of which are based on common property systems of management, are often incompatible with national laws and difficult to delineate and protect. As a first step in the process of bringing sustainable and equitable land use to the humid tropics, the legitimacy of these territorial rights and tribal domains, and their value in forest conservation and development programs, should be recognized.

For hundreds of millions of small-scale farmers and other resource users in the humid tropics whose livelihoods depend on access to land and forest resources, tenure issues are fundamental to their choice of land use practices and to their future welfare. Lacking secure tenure, farmers and other small-scale resource users have little incentive to conserve, manage, improve, or invest in land resources. Deprived of the benefits of local resources, they must often overexploit those to which they do have access. Lack of tenure also contributes to mutual animosity among small-scale users, large landowners, government officials, and resource bureaucracies, and hence to a diminished public capacity to respond to the need for resource conservation.

The mechanisms by which insecure tenure results in resource degradation vary widely throughout the tropics. In some areas, inappropriate tenure arrangements, such as inequitable share-cropping requirements or lack of secure ownership, force farmers into short-term behavior—encroachment onto marginal lands, cultivation of steep slopes, and intensified cycles of shifting cultivation. Often the process is more passive; lacking secure tenure, farmers are discouraged from investing in terracing, agroforestry systems, timber plantations, tree crops, and other long-term land improvements. Moreover, they are often unable to make investments because they require credit to do so, and credit, if available, is extended only to those who have tenure and can pledge their land as security. Breaking this cycle is particularly important in countering the tendency of shifting cultivators to enter new areas and in removing the obstacles to the reclamation of abandoned lands. Ownership of land is often transitory in areas where shifting cultivation is widely practiced. Few farmers in these areas are able or willing to invest in alternatives to slash and burn, which typically involve planting trees in agroforestry and other mixed systems, if they do not have secure tenure (Sanchez, 1991).

In all these cases, tenure arrangements that provide long-term access to land resources are prerequisites to efficient land-use decision making and to the implementation of sustainable land-use systems. This need is beginning to be recognized in national mandates and allocation legislation in many tropical countries—Brazil, Colombia, Indonesia, Peru, and the Philippines, among others—but these moves are often difficult to enforce. Economic and political elites who benefit from existing tenure arrangements are resistant to change. In addition, many national forestry and other resource agencies are actively opposed to these policy changes, fearing that recognition of tenure will eliminate the role of foresters and other government agency officials. This fear, for example, has impeded progress in the Philippine government's efforts to delineate indigenous territorial boundaries (Garrity et al., Part Two, this volume; Lynch, 1991).

The importance of tenure provisions is also beginning to be recognized and incorporated in the programs of bilateral and international development agencies, human rights and conservation organizations, and other nongovernmental organizations (NGOs) (Plant, 1991; World Resources Institute, 1990b). Perhaps most significant, local and indigenous people themselves are more aware of their stake in tenure disputes and of their protection under international law (Lynch, 1991). In addition to immediate support for efforts to improve the status of tenure for small-scale farmers and indigenous people, development agencies should support much-needed research in the social sciences on a wide variety of tenure issues: accurate, country-specific demographic surveys of the number and distribution of people in forests and forest margins; forms of tenure and their connection to land use, agricultural productivity, and conservation practices; traditional means of resolving tenure and resource disputes within and between local communities; the role of women in various tenure systems; the changes in tenure that have accompanied modern settlement and forest conversion; and conflicts between traditional and modern tenure systems.

Even as research continues to illuminate the important connections between tenure reform and sustainable land use, national governments in the humid tropics should endeavor to resolve tenure disputes and to anticipate and prevent future conflicts. Territorial boundaries should be delineated and land title granted prior to infrastructural development projects and resettlement programs. This is especially important in areas where migrants are encroaching on areas traditionally used for extraction (as, for example, in the rubber tapping regions of Acre in Brazil) or on tribal lands (as in the Yanomani lands of Brazil and Venezuela, where in the last decade gold mining has resulted in a rush of new settlers). Such conflicts are never easily resolved once they develop, and the best strategy is to build preventive measures into all development planning.

ACCESS TO CREDIT FOR SMALL-SCALE FARMERS

Lack of access to credit is a major constraint that shifting cultivators and other small-scale farmers face in improving their resource use. While some sustainable land use practices can improve productivity even in the absence of credit, most will require long-term investments, since the costs of implementation will not be recovered in the short term. In areas where the chemical and nutrient limitations of soils were traditionally overcome through slash-and-burn cycles, credit for initial soil preparations can be critical in the period of transition to sustainable systems. Credit is essential in areas where soil amendments, seeds, tree stock, tools, and other purchased inputs are needed to initiate land rehabilitation and the conversion of destructive shifting agriculture or cattle ranching to more stable systems. The provision of both credit and secure tenure is especially important in rehabilitating badly degraded lands, where the rebuilding of "biological capital" requires substantial investments of time and money.

Credit mechanisms and structures should vary to suit local social and land use conditions, and innovative arrangements should be encouraged. The Grameen Bank in Bangladesh, a community-based cooperative development bank that makes small grants and loans, is one example (World Resources Institute, 1990b). Innovation in credit programs, however, must entail careful planning to ensure they promote flexibility in land use and do not lock farmers and other landowners into nonsustainable practices. The objective is to give small-scale farmers the means to adjust their operations and adopt new practices that encourage the local rehabilitation, sustainable use, and conservation of resources.

INVESTMENT IN INFRASTRUCTURE

National and international infrastructure investment policies have often encouraged access to and through primary forests. In the future, infrastructural development's primary aim should not be to advance deforestation, but rather to support more appropriate land uses on already cleared lands.

Strengthening the connections between the small farm and the market can be an efficient and cost-effective means of stimulating the diversified activities on which sustainable land use largely depends (Brannon and Baklanoff, 1987; Gómez-Pompa et al., Part Two, this volume). This implies the provision of reliable roads, bridges, and railroads, suitable processing equipment, adequate storage facilities, and improved marketing mechanisms (particularly for new products). Improved transportation networks are needed not only to allow farmers to market their products, but to enhance their access to necessary inputs, including information through extension services and other means. Storage facilities are needed to protect tropical products, many of which are highly perishable, from spoilage and postharvest pest problems. Processing equipment is needed to convert products into more readily marketable forms (often with value-added benefits to local economies), and to develop new products for local use as well as national and even international export. Improved marketing mechanisms and facilities can create additional opportunities for traditional and newly developed products.

LOCAL DECISION MAKING

If sustainable land use practices are to be successfully introduced, they must be responsive to the concerns and needs of small farms and rural communities and adaptable to local social, economic, and political conditions (Chambers et al., 1989; Edwards, 1989). The annals of development agencies contain many cases of well-intended projects that have failed due to inadequate farmer and community participation in project development, planning, and management. Farmers who do not have a stake or perceived self-interest in developing a locally suitable agroforestry project or mixed cropping system will not be committed to its success. Where local people participate in the planning process, and receive immediate benefits, the results can be striking (Gómez-Pompa et al., Part Two, this volume).

Local responsiveness calls for modifications in conventional approaches to development planning. Especially under the highly variable conditions of communities in the humid tropics, top-down strategies that emphasize only the transfer of technologies from centralized research stations to farmers are prone to assume or overlook key biophysical, social, political, or cultural factors that determine the local acceptance of land use practices. National and international development agencies, policymakers, and institutions need to involve local communities from the inception of planning on all projects, beginning with a realistic appraisal of the problems, needs, desires, and opportunities that farmers and communities face (Chambers et al., 1989; Gómez-Pompa and Bainbridge, 1991). These assessments need to take into account the status of local natural resources and community needs, using this information to plan and implement better coor-

An agroforestry program, sponsored by the Food and Agriculture Organization in a deteriorated region of Madagascar, incorporates the cultivation of trees with other agricultural production for integrated rural development. Pictured is an improved breed of chicken. Credit: Food and Agriculture Organization of the United Nations.

dinated development programs. In many parts of the humid tropics, for example, education and public health services can be better integrated with sustainable land use goals. The development agencies can play a critical role by providing technical assistance in community planning.

The social forestry programs that have been implemented in several Southeast Asian countries provide important working models for the increased participation of local farmers and communities in the humid tropics. Social forestry programs work with local communities to provide training and incentives for reforestation, forest protection, the local use of forest products, and the implementation of plantation and agroforestry projects on private and communal lands. By 1987, some 10,000 households, representing 10,000 ha of forestland, had become involved in Indonesia's Social Forestry Program, with the ultimate goal being the rehabilitation of 270,000 ha of degraded forestland (Kartasubrata, Part Two, this volume). In the Philippines, the Community Forestry Program has met with early success in its efforts to give upland farmers and forest dwellers greater access to, and responsibility for, local forest resources (Garrity et al., Part Two, this volume).

Programs such as these hold great promise, but they are also confronted with many difficulties: the reluctance of governments to undertake needed reforms in land tenure; insufficient funds for needed subsidies and appropriate infrastructure; poor coordination among resource agencies; corruption and abuse in program administration; a lack of personnel with the necessary mix of skills in forestry as well as training in management and community development; a lack of tried and tested, locally adaptable agroforestry technologies; and a shortage of technicians willing to work with farmers (Garrity et al., Part Two, this volume). These deficiencies should not diminish the importance of social forestry and other experimental efforts to communicate the needs of small-scale farmers and foster the participation of local communities. Rather, international agencies and national governments should carefully review the record of these initial successes and failures, and work together to build programs that anticipate problems through the closer involvement of the users—the small-scale farmers and forest dwellers.

Providing Incentives and Opportunities

National governments in the humid tropics and international aid agencies should develop and provide incentives to encourage long-term investment in increasing the production potential of degraded lands, for settling and restoring abandoned lands, and for creating market opportunities for the variety of products available through sustainable land use.

In many cases, the steps already outlined will provide the conditions under which more sustainable agriculture can take hold and evolve. In these instances, the economic and environmental benefits of alternative practices and products are obvious and accrue quickly enough to induce individual farmers and local communities to make the necessary investments of time, labor, and money. In other cases, however, additional steps may be needed to stimulate investment and action.

INCENTIVES TO ENCOURAGE INVESTMENTS IN LAND IMPROVEMENT

The most promising methods of sustainable land management are often financially marginal in the short term. Some require terracing and other land improvement investments. Others may include the use of perennial crops that entail long establishment times and high start-up costs. These costs may be especially prohibitive where the lands themselves are difficult to work (for example, uplands, steep slopes, poorer soils, soils that have been badly degraded in the process of clearing, and areas overtaken by tenacious weed or grass species). While various sustainable systems and agricultural practices hold great promise in stabilizing and improving these lands, the immediate financial returns may be inadequate to attract farmers and investments. Reform will be particularly difficult when decreases, rather than increases, in the productivity of the land are required. In such cases, alternative employment opportunities are a most probable solution, but it may be necessary to provide direct subsidies to compensate landholders while they allow their properties to stabilize.

Policy devices that have encouraged deforestation in the past—tax abatement, credits, pricing policies, concessions, and subsidies—can be revised to induce small-scale farmers and other landholders to adopt sustainable agricultural practices. Optimally, national development agencies and international aid agencies would work together toward this goal. With a consortia of researchers, NGOs, and other institutional interests, they would identify the lands of greatest need, gauge local community conditions, coordinate appropriate land use and conservation measures, and help provide the financial backing for investment programs.

To attain the most efficient use of limited funds, it will be necessary to determine where natural regeneration is proceeding most acceptably and investments can be delayed or used most sparingly, and where human needs are more pressing and regenerative processes require "boosting." As regeneration and economic development proceeds, the mix of land use inputs is likely to change and so too will the mix of appropriate incentives. Thus, for example, labor-intensive agroforestry systems that might be highly suitable in low-wage countries may be less financially viable in high-wage countries. Some degree of anticipation of the consequences of changing economic and agroecological conditions is prudent. The necessary initial steps, however, remain clear: provide local farmers and communities in the humid tropics with incentives to improve their current land use practices and restore degraded lands.

INCENTIVES TO ENCOURAGE REHABILITATION OF ABANDONED LANDS

The incentives and investments just described will mainly affect lands that are already inhabited but in a degraded state. Special measures must also be taken to rehabilitate completely abandoned lands. Throughout the humid tropics, land abandonment has often followed deforestation. This pertains in particular to those lands that have been heavily exploited for timber and cattle ranching in recent decades. Over vast areas, these lands have simply been logged and then abandoned. In others they have been purposely cleared for (or converted after logging to) agricultural uses that have proved, for one reason or another, nonsustainable. The growth of secondary forests will take decades.

In either case, there are definite strategic and logistical advantages to focusing on abandoned lands. If small-scale farmers can be helped to return abandoned lands to productivity, these lands can absorb populations, provide local employment opportunities, ease the pressure to extend deforestation, and stabilize soils and watersheds. Moreover, most abandoned lands retain at least rudimentary transportation and market infrastructures that can be improved with proper investments. Securing tenure on abandoned lands is a critical step in their rehabilitation, but special concessions may be required to attract farmers, especially landless shifting cultivators, to these areas.

Abandoned lands are heterogenous. The methods and goals of restoration vary, and so must incentive strategies. Lands that have been overtaken, for example, by *Imperata cylindrica* and other invader species may require incentives to induce tree planting and fire protection efforts as small landowners convert to agroforestry and perennial crop systems. Lands where the nutrients have been depleted and ash inputs are low require fertilizers. Tillage operations are needed on seriously compacted lands. Abandoned or degraded pastures in the Brazilian Amazon and elsewhere will require incentives for intensified management through improved forages, fertilization, crop introduction, and weed control. In areas where fuelwood needs are acute, reforestation with fast-growing trees may be the highest priority. Where commercial logging has opened steep slopes, the immediate need is for vegetative cover; where some cover has been restored, additional terracing or contouring may be needed.

Needs will not only vary from region to region, but also within regions. Depending on local tenure arrangements, it may be necessary to target subsidies, tax concessions, and other incentives toward villages and communities instead of (or in addition to) individual landowners. This is especially important in situations where the stabilization of entire watersheds is critical, and points to the importance of landscape-level planning in treating abandoned lands.

Additional incentives, not specifically aimed at site rehabilitation, are nonetheless necessary for restoring abandoned lands. Local and regional market incentives may be needed to stimulate demand for products raised on these lands. Subsidies are usually required to build tree nurseries and processing facilities. Government agencies often retain exclusive responsibility for tree nursery management, thus discouraging private investment. However, privatization can be a desirable means of stimulating investment. Incentives for investment in collaborative research, demonstration areas, and education and extension projects may also be needed to build the local knowledge base.

MARKETS FOR AGRICULTURAL AND FOREST PRODUCTS

In developing market opportunities, it may be difficult for new products to compete with established humid tropical crops such as rubber, cacao, and oil palm. Opportunities may exist, however, to produce a wide variety of lesser-known crops and other products if market outlets for them can be developed. These can be incorporated into many land use systems as alternative crops in more intensive cropping systems, as trees in agroforestry systems, as restoration agents (particularly through the use of acid-tolerant cultivars), and as harvested products from extractive reserves.

Examples of potentially important products include the peach palm (*Bactris gasipaes*); achiote (*Bixa orellana*), a colorant; guaraná (*Paullinia cupana*), a flavoring for soft drinks; Brazil nut (*Bertholletia excelsa*); and fruits used in juice concentrates and other food products. The growing industrial and service economies of Asia, for example, are providing enormous market potential for forest products. This is only a partial list of food products from the Amazon Basin. Many other potential crops exist elsewhere in the humid tropics, including a wide variety of fruits and spices in humid tropic Asia. Medicines, resins, oils, latex, gums, fibers, and other materials have the potential to reach wider markets. Efforts to establish a specific international market niche for new products can take advantage of the developed world's changing values as reflected by its rising interest in environmental issues. Reliance will likely need to be placed on public institutions for market intelligence, establishment of grades and standards, and possibly the creation of a means of addressing the risks, such as insurance, protection from pests and pathogens, and genetic improvement. Market development is best undertaken by the private sector.

Development programs should be prepared to foster awareness and cooperation among private and public sectors concerned with sustainable land use (Kartasubrata, Part Two, this volume). For-profit firms can serve an important function by stimulating new investment and enterprises at the local level, and their responsible participation should be encouraged through an appropriate mix of rewards, incentives, and disincentives. As interest in the conservation of tropical forests has grown, so have examples of creative, collaborative investment. Recently, for example, Merck and Company, a U.S.-based pharmaceutical firm, entered into an agreement with the government of Costa Rica to "prospect" native flora and fauna for natural chemical compounds with commercial potential. By providing $1 million over a 2-year period, Merck has acquired exclusive rights to screen materials collected by Costa Rica's Instituto Nacional de Biodiversidad (National Biodiversity Institute, INBio). These funds and others from U.S. and European universities, foundations, and governments will establish INBio's chemical prospecting activities. This effort is designed to make the forests pay for themselves and to acquire the technology needed to screen natural compounds. Other arrangements to conserve the country's biodiversity include the exchange of patent rights for royalties.

It is also important that the research underlying market development be undertaken as an interdisciplinary endeavor, and that it directly involve farmers and forest dwellers. Economists, social scientists, and natural scientists should collaborate with each other and with farmers to determine the best means of introducing new products and to assess their long-term impacts on farm performance, farmer income, community development, genetic diversity, and ecosystem composition and function.

STRENGTHENING RESEARCH, DEVELOPMENT, AND DISSEMINATION

New partnerships must be formed among farmers, the private sector, nongovernmental organizations, and public institutions to address the broad needs for research and development and the needs for knowledge transfer of the more complex, integrated land use systems.

The successful adoption of sustainable agricultural systems and practices requires a strong network for research, development, and dissemination of information.

New Methodologies for Research and Development A comprehensive, interdisciplinary approach to research, education, and training is fundamental to developing and managing the complex, sustainable agroecosystems of the humid tropics. The land at greatest risk of degradation is of modest production potential due to slope, limited availability of water, and soils that are low in fertility and highly erosive. These lands require technologies quite different from high-productivity lands, which have received the bulk of agricultural development attention.

The international community has given substantial support for research to increase the productivity of major crops such as rice and maize, and for research on tropical soils, livestock, chemical methods of pest control, human nutrition, and other aspects of agriculture in developing countries (McCants, 1991). Much less research has been directed toward smallholder agroforestry systems, tree crops, improved fallow and pasture management, low input cropping, corridor systems, biocontrol and other methods of integrated pest management, and other agricultural systems and technologies appropriate to higher risk land types. Research in these areas has begun to receive greater attention. The activities, for example, of the International Center for Research in Agroforestry in Kenya and of the Centro Agronómico Tropical de Investigación y Enseñanza (Tropical Agriculture Research and Training Center) in Costa Rica have recently been expanded. Additional support for similar initiatives is needed.

It is important that the research knowledge base be expanded geographically and adapted to particular climatic, biotic, soil, and socioeconomic situations. Specific research needs for different land use options vary. All, however, require validation research and effective means of gathering and disseminating information. More on-farm testing and research should involve the rehabilitation, sustainable use, and management of recently cleared, degraded, and abandoned lands. This work should focus on the potential of these lands to support intensive agriculture as well as less intensive agroforestry and forest management systems. Sustainable agricultural technologies exist for these lands, but they require much more refinement and usually yield low rates of return to capital, management, and labor. No-tillage agriculture, for example, could be used on steep slopes throughout the tropics, but economical and environmentally sound methods of weed control are needed.

As new methodologies for research are developed, they can build on the efforts of existing methods. Studies of productivity constraints will continue to be necessary, but effective solutions to the agricultural problems of farmers on marginal lands are unlikely to be found solely through experiment station and laboratory research. As basic agronomic research continues, there is increasing need for studies that emphasize the experience and experimentation of farmers. On-farm studies themselves often suggest questions for further laboratory-based research (Chambers et al., 1989).

Participation of Nongovernmental Organizations   The diversity and complexity of agroecological, political, social, and economic conditions throughout the humid tropics require a degree of sensitivity and microadaptation that large, centralized development agencies, especially those operating at the international level, do not and cannot efficiently provide.   Only locally based organizations can handle the complexity that arises out of local conditions, while serving as conduits for the flow of information to, from, and among local farmers and communities.

The burgeoning of nonprofit private voluntary organizations (PVOs) and NGOs in the developing world is a response to this need.   While many of these organizations focus specifically on conservation and agricultural development, many others with an interest and a stake in land use issues lack the experience, resources, and personnel to follow up on their concerns.   National and international development agencies need to foster the productive involvement of local NGOs as intermediaries between themselves, national government agencies, universities, and local communities in support of the methods and goals of sustainable land use.

In particular, NGOs can assume a prominent role in training and education at the community level, in partnership with (or in the absence of) official extension services.   NGOs can also serve as vital links in improved communication networks, connecting local farmers with researchers, agency administrators, aid officials, and other development workers.   Perhaps most important, local NGOs are likely to be more effective than external organizations in shaping environmentally and socially acceptable land use policies based on local needs and priorities.

The organizations that comprise the NGO and PVO community are highly diverse (National Resource Council, 1991a).   Some are international, others indigenous; some are community based, others are national associations; some consist of poor farmers, while others are well-funded urban institutions.   Relatively few, however, have extensive research and extension capabilities in sustainable agriculture or resource management.   For this reason, those groups that are in place and prepared to assume greater responsibilities involving land use issues should receive increased support for technical training.

Support for training may take the form of direct funding or innovative collaborative linkages with other organizations having needed expertise.   NGO linkages with established agricultural development institutions, such as the international agricultural research centers and national agricultural research systems, have been limited by mutual distrust or by a lack of collaborative mechanisms.   As institutional barriers are overcome, and new mechanisms developed, development projects increasingly bring together a wide array of public and private organizations.

In 1984, for example, the Cooperative for American Relief Everywhere, the New York Zoological Society, the U.S. Agency for International Development (USAID), and the Ugandan Forestry Department initiated the Village Forest Project in southwestern Uganda. The goal of the Village Forest Project is to improve living conditions for local farmers through the introduction of agroforestry techniques while simultaneously reducing pressures on the Kibale Forest Reserve, a protected area of moist lowland forest (Cooperative for American Relief Everywhere, 1986; Struhsacker, 1987). The International Center for Research in Agroforestry provides on-site technical assistance. The Sustainable Agriculture and Natural Resource Management program of USAID is attempting to bring the same collaborative spirit to a full range of sustainable resource management issues in developing countries (National Research Council, 1991a).

Dissemination of Information Through Extension Services The implementation of sustainable agriculture systems and practices in the humid tropics will require the active involvement of extension services. Decentralization, local adaptation, and innovation are key to the successful adoption and refinement of these systems, and extension services can be adapted to meet these needs. Working together with NGOs and others in the private sector, extension personnel can link farmers, researchers, resource agencies, community officials, and development officials. Through them, agencies should promote relevant research findings, develop demonstration projects and networks, and disseminate the information, management practices, plant materials, and tools necessary for the wider application of sustainable agricultural systems. Information, however, must flow both ways: extension workers should assist researchers in identifying the socioeconomic, environmental, and agronomic constraints that small farms and rural communities face.

Sustainability begins with an approach that is attuned to these environmental, social, and cultural realities, to local belief systems, and to traditional methods and knowledge. Accordingly, future extension services need to adopt an interdisciplinary approach. Extension personnel may require exposure to and training in aspects of land use and the environmental sciences that they have not previously received, including forestry and agroforestry, land use planning and zoning, and the conservation of biological diversity. In addition, the social aspects of rural development must become a more prominent part of all extension services. Rural women, in particular, will need to be involved more actively in extension activities.

Education and training programs at all levels can benefit from adopting similar interdisciplinary approaches. Educational materials incorporating research findings need to be developed for use in schools and communities at all levels. Where training for work with natural resources is unavailable in-country, support should be provided for scientists and resource managers to receive graduate and postgraduate training in countries where appropriate programs are available, with the requirement that scientists return to their countries of origin to work.

OTHER POLICY AREAS AFFECTING LAND USE

This report is principally concerned with the implementation of improved agricultural techniques and the rehabilitation of degraded lands. However, other areas of public policy significantly affect sustainability in the humid tropics. These include political and social stability, population growth, greenhouse warming, and alternative energy sources.

Political and Social Stability

In the humid tropics, as elsewhere, long-term patterns of land use and the status of land resources are determined, in part, by the degree of stability within the society and its political institutions. The problems of resource management, and of deforestation in particular, cannot be separated from the issues of urban poverty, social justice, economic inequity, ineffective administration, deteriorating urban infrastructure, political corruption, agrarian reform, human rights abuses, and other pressing social concerns. Environmental degradation often reflects the desperate competition for access to resources under unstable social conditions, and unless these conditions are addressed, it will be impossible to make progress toward sustainable development (Latin American and Caribbean Commission on Development and Environment, 1990; Rush, 1991).

Under unstable conditions, both urban and rural populations are less likely to be concerned with long-term environmental health and more likely to engage in activities that yield short-term benefits. Declining environmental conditions, in turn, increase the degree of social and political instability. In the extreme case of warfare, traditional patterns of resource use can be grossly disrupted, and entire agricultural, wetland, and forest areas degraded through clearing, defoliation, burning, draining, and bombing. Large expanses of land throughout Southeast Asia and Central America have experienced this fate over the past three decades (Office of Technology Assessment, 1984). The self-reinforcing cycle of social instability and environmental degradation fundamentally undermines the conditions necessary to sustainable use of resources: the mixture of technological innovation, education and access to information, long-term investment, policy reform, political empowerment at the local level, and economic and demographic stability.

Population Growth

There is little hope of accomplishing sustainable land use unless population growth is brought under control. The world's population is expected to increase by a billion people each decade well into the twenty-first century, with the developing nations of the tropics accounting for most of this growth. Because underdevelopment and poverty are directly related to higher fertility rates, any strategy for resource conservation in the humid tropics must entail strong policies to reduce poverty, an effort that could take many years.

Short-term problems of population distribution are commonly solved by resettlement. However, this approach to reducing local population pressures typically results in a host of new social and environmental problems. In many countries of the humid tropics, national resettlement policies and programs have resulted in large numbers of settlers moving into primary forests. This has occurred, for example, in the Philippines in the 1950s and 1960s, and more recently under the large-scale resettlement programs in Indonesia and Brazil (see Part Two, this volume). In other cases, such as Mexico, areas slated for colonization programs have first been prepared for settlement by the commercial extraction of valuable timber (Gómez-Pompa et al., Part Two, this volume). Whenever possible, resettlement policies should provide opportunities for transmigrants to develop abandoned lands and other less sensitive ecosystems.

Greenhouse Gas Emissions

Over the next several decades, sustainable agriculture and land use systems in the humid tropics can play an important role in efforts to stabilize and possibly reduce greenhouse gas concentrations. Evolving policies need to recognize, encourage, and reward actions that allow this potential to be realized. International climate change negotiations and agreements should proceed with greater emphasis on the benefits of sustainable land uses in the humid tropics. The overriding goal should be to provide incentives and opportunities for improved land use at the local level.

Current international policy discussions on carbon dioxide emissions must consider more systematically the ability of sustainable land use systems in the humid tropics to reduce atmospheric carbon dioxide concentrations by slowing deforestation, withdrawing carbon and storing it in plant biomass and soil, and providing alternative sources of energy. Changes in land use offer a practical means of removing large quantities of a greenhouse gas from the atmosphere through human intervention (Intergovernmental Panel on Climate Change, 1990b). Yet, even the best economic models and analyses involving the abatement of carbon dioxide concentrations focus primarily on the costs of reducing industrial emissions. Most do not factor in the positive contributions that sustainable land uses in the humid tropics offer (Darmstadter, 1991).

However, this potential should not be overstated. Improved land use in the humid tropics alone cannot offset the impact of industrial emissions of carbon dioxide. The capacity to sequester carbon through land use changes should not imply an abdication of the responsibility of developed countries to bring emissions under control. Support for land use changes that have local benefits can also provide global benefits, but not in the absence of policy changes that affect industrial emissions.

At the international level, the question of equity will continue to be a critical factor in the success of efforts to mitigate global warming. Although many in the international community share a deep sense of purpose and responsibility within the arena of global climate change, the attitudes, positions, and interests involved vary greatly, and international agreements will not be easy to forge or to enforce (Morrisette and Plantinga, 1991). However, the movement toward sustainable agriculture and land use in the humid tropics can serve as a focal point for shared actions based on common concerns. There is much room for collaboration and cooperation among the industrial nations of the north and the developing countries of the south in providing the people, the knowledge, the tools, and the political and financial support that are needed to transform the potential climate-related benefits of sustainable agriculture into reality.

Alternative Energy Sources

Many people in developing countries use wood and charcoal as their principal energy sources. Within the humid tropics, rising demand has increased wood gathering. In these areas, alternate energy sources and national energy strategies that reduce the use of wood to sustainable levels are needed to help relieve the pressures on forested lands. More research should be devoted to fuelwood plantations; alternative sources of wood (for example, sawmill wastes) for charcoal production and more efficient production processes; improved kilns, stoves, and furnaces as well as solar technologies; and sustainable extraction practices.

In general, moist forests are less affected by fuelwood demand than drier forest types, but there are important exceptions. In Zaire, for example, fuelwood accounts for 75 to 90 percent of the total national energy budget, and fuelwood gathering is the leading cause of deforestation (Barbier et al., 1991). According to projections to the year 2000, 5.5 million ha of forestland in Zaire would need to be depleted each year to meet increasing fuelwood requirements (World Bank and United Nations Development Program, 1983; Ngandu and Kolison, Part Two, this volume). Forests near large urban areas and surrounding industrial development projects that require charcoal are especially susceptible to heavy exploitation. The Grande Carajas project in the eastern Amazon, for example, is projected to produce and consume 1.1 million metric tons of charcoal annually in its iron and cement operations. Eucalyptus plantations will meet some of this demand, but nearby forests are likely to be affected as well (Fearnside, 1987b; Gradwohl and Greenberg, 1988).

References

Alcorn, J. B. 1990. Indigenous agroforestry systems in the Latin American tropics. Pp. 203–213 in Agroecology and Small Farm Development, M. A. Altieri and S. B. Hecht, eds. Boca Raton, Fla.: CRC Press.

Alegre, J.C., and P.A. Sanchez. 1991. Central continuous cropping experiment in Yurimaguas: Y-101. Pp. 249–251 in TropSoils Technical Report, 1988–1989. Raleigh, N.C.: TropSoils Management Entity, North Carolina State University.

Allegretti, M. H. 1990. Extractive reserves: An alternative for reconciling development and environmental conservation in Amazonia. Pp. 252–264 in Alternatives to Deforestation: Steps Toward Sustainable Use of the Amazon Rain Forest, A. B. Anderson, ed. New York: Columbia University Press.

Altieri, M. A. 1987. Agroecology: The Scientific Basis of Alternative Agriculture. Boulder, Colo.: Westview.

Altieri, M. A. 1989. Rethinking crop genetic resource conservation: A view from the south. Conserv. Biol. 3:77–79.

Altieri, M. A., and L. C. Merrick. 1987. In situ conservation of crop genetic resources through maintenance of traditional farming systems. Econ. Bot. 41(1):86–96.

Amador, M. F., and S. R. Gliessman. 1991. An ecological approach to reducing external inputs through the use of intercropping. Pp. 146–159 in Agroecology: Researching the Ecological Basis for Sustainable Agriculture, S. R. Gliessman, ed. New York: Springer-Verlag.

Amir, P., and H. C. Knipscheer. 1987. On-Farm Animal Research/Extension and Its Economic Analysis. Los Baños, Philippines: Southeast Asian Regional Center for Graduate Study and Research in Agriculture.

Anderson, A. B. 1990. Deforestation in Amazonia: Dynamics, causes, and alternatives. Pp. 3–23 in Alternatives to Deforestation: Steps Toward Sustainable Use of the Amazon Rain Forest, A. B. Anderson, ed. New York: Columbia University Press.

Andriesse, J. P., and R. M. Schelhaas. 1987. A monitoring study of nutrient cycles in soils used for shifting cultivation under various climatic conditions in tropical Asia. Agric. Ecosyst. Environ. 19:285–332.

Asabere, P. K. 1987. Attempts at sustained yield management in the tropical high forests of Ghana. Pp. 47–70 in Natural Management of Tropical Moist Forest, F. Mergen and J. Vincent, eds. New Haven, Conn.: Yale University School of Forestry and Environmental Studies.

Ayarza, M. A., R. Dextre, M. Ara, R. Schaus, K. Reátegui, and P. A. Sanchez. 1987. Producción animal y cambios en la fertilidad de suelos en asociaciones bajo pastoreo en un Ultisol de Yurimaguas. Peru. Suelos Ecuatoriales 18:204–208.

Balick, M. J., and R. Mendelsohn. 1992. Assessing the economic value of traditional medicines from tropical rain forests. Conserv. Biol. 6:128–130.

Baodo, E. L. 1988. Incentive policies and forest use in the Philippines. Pp. 165–203 in Public Policies and the Misuse of Forest Resources, R. Repetto and M. Gillis, eds. Cambridge, U.K.: Cambridge University Press.

Barbier, E. B., J. C. Burgess, and A. Markandya. 1991. The economics of tropical deforestation. Ambio 20(2):55–58.

Barnard, R. C. 1954. The control of Laland (*Imperata arundinacea* Var. Major) by fire protection and planting. Malayan Forest. 17:152–154.

Barrera, A., A. Gómez-Pompa, and C. Vazquez Yanes. 1977. El manejo de las selvas por los Mayas: Sus implicaciónes silvicolas y agrícolas. Biotica 2:47–61.

Bates, R. H. 1981. Markets and States in Tropical Africa: The Political Basis of Agricultural Policies. Berkeley: University of California Press.

Beckford, G. L. 1972. Persistent Poverty: Underdevelopment in Plantation Economies in the Third World. London: Oxford University Press.

Binswanger, H. 1989. Brazilian Policies that Encourage Deforestation in the Amazon. World Bank Environment Department Working Paper No. 6. Washington, D.C.: World Bank.

Bishop, J. P. 1983. Tropical forest sheep on legume forage/fuel wood fallows. Agroforest. Syst. 1:79–84.

Bjorndalen, J. E. 1991. An ecological approach to the inventory and monitoring of rain forest catchments in Tanzania. Pp. 97–102 in The Conservation of Mount Kilamanjaro, W. O. Newmark, ed. Gland, Switzerland: World Conservation Union.

Bodmer, R. E., T. G. Fang, and I. Moya. 1990. Fruits of the forest. Nature 343:109.

Bonifacio, M. F., ed. 1988. Working Papers on Community Based Agriculture. Quezon City, Philippines: Agriculture Training Institute.

Boonkird, S.-A., E. C. M. Fernandes, and P. K. R. Nair. 1984. Forest villages: An agroforestry approach to rehabilitating forest land degraded by shifting cultivation. Agroforest. Syst. 2:87–102.

Bowen, G. D., and E. K. S. Nambiar. 1984. Nutrition of Plantation Forests. New York: Academic Press.

Boxman, O., N. R. de Graaf, J. Hendrison, W. B. J. Jonkers, R. L. H. Poels, P. Schmidt, and R. Tjon Lim Sang. 1985. Towards sustained timber production from tropical rain forests in Suriname. Netherlands J. Agric. Sci. 33:125–132.

Brannon, J., and E. N. Baklanoff. 1987. Agrarian Reform and Public Enterprise in Mexico. Tuscaloosa, Ala.: University of Alabama Press.

Browder, J. O. 1985. Subsidies, deforestation, and the forest sector in the Brazilian Amazon. Report prepared for the World Resources Institute, Washington, D.C.

Browder, J. O. 1988. The social costs of rainforest destruction: A critique and economic analysis of the "hamburger debate." Interciencia 13(3):115–120.

Browder, J. O. 1990. Extractive reserves will not save tropics. BioScience 40:626.

Browder, J. O. 1992. The limits of extractivism: Tropical forest strategies beyond extractive reserves. BioScience 42:174–182.

Brown, H. C. P., and V. G. Thomas. 1990. Ecological considerations for the future of food security in Africa. Pp. 353–377 in Sustainable Agricultural Systems, C. A. Edwards, R. Lal, P. Madden, R. H. Miller, and G. House, eds. Ankeny, Iowa: Soil and Water Conservation Society.

Brown, S., and A. E. Lugo. 1982. The storage and production of organic matter in tropical forests and their role in the global carbon cycle. Biotropica 5:6–87.

Brown, S., and A. E. Lugo. 1990. Tropical secondary forests. J. Trop. Ecol. 6:1–31.

Brown, S., A. E. Lugo, and J. Chapman. 1986. Biomass of tropical tree plantations and its implications for the global carbon budget. Can. J. Forest Res. 16:390–394.

Brown, S., A. E. Lugo, and L. R. Iverson. 1992. Processes and lands for sequestering carbon in the tropical forest landscape. Water, Air, Soil Pollution 64:139–155.

Brownrigg, L. A. 1985. Home Gardening in International Development: What the Literature Shows. Washington, D.C.: League for International Food Education, U.S. Agency for International Development.

Brush, S. B. 1989. Rethinking crop genetic resource conservation. Conserv. Biol. 3:19–29.

Budowski, G. 1965. Distribution of tropical American rain forest species in the light of successional processes. Turrialba 15:40–42.

Budowski, G., and T. C. Whitmore. 1978. Food from the forest. Pp. 911–917 in Proceedings from the Eighth World Forest Congress, Sixth Technical Session, Djakarta, Indonesia.

Bunyard, P. 1985. World climate and tropical forest destruction. Ecologist 15:125–136.

Buol, S. W., F. D. Hole, and R. J. McCracken. 1980. Soil Genesis and Classification. Ames, Iowa: Iowa State University Press.

Bureau of Agricultural Research of the Philippines. 1990. Planning On-Farm Research. Manila: Philippine Department of Agriculture.

Buschbacher, R. 1990. Natural forest management in the humid tropics: Ecological, social, and economic considerations. Ambio 19(5):253–258.

Buschbacher, R., C. Uhl, and E. A. S. Serrão. 1988. Abandoned pastures in eastern Amazonia. II. Nutrient stocks in the soil and vegetation. J. Ecol. 76:682–699.

Buttoud, G. 1991. Expert in what? The lessons European forestry has to teach are based on hard experience. Ceres 129:31–34.

Caballero, J. 1988. The Maya homegardens of the Yucatan Peninsula: A regional comparative study. Paper presented at the First International Congress of Ethnobiology, Belém, Brazil, July 18–22, 1988.

Calderón, M. 1981. Insect pests of tropical forage plants in South America. Pp. 778–788 in Proceedings of the XVI International Grasslands Congress, J. A. Smith and V. W. Hayes, eds. Boulder, Colo.: Westview.

Cardos, A. 1959. El comercio entre los maya antiguos. Acta Antropol. 2:50.

Cashman, K. 1988. The benefits of alley farming for the African farmer and her household. Pp. 231–237 in Global Perspectives on Agroecology and Sustainable Agricultural Systems, P. Allen and D. Van Dusen, eds. Santa Cruz: University of California.

Chambers, R., A. Pacey, and L. A. Thrupp. 1989. Farmer First: Farmer Innovation and Agricultural Research. London: Intermediate Technology Publications.

Chandler, R. F. 1979. Rice in the Tropics: A Guide to the Development of National Programs. Los Baños, Philippines: International Rice Research Institute.

Christanty, L. 1986. Shifting cultivation and tropical soils: Patterns, problems, and possible improvements. Pp. 226–240 in Traditional Agriculture in Southeast Asia: A Human Ecology Perspective, G. G. Marten, ed. Boulder, Colo.: Westview.

Christanty, L., O. L. Abdoellah, G. G. Marten, and J. Iskandar. 1986. Traditional agroforestry in West Java: The pekarangan (homegarden) and kebun-talun (annual-perennial rotation) cropping systems. Pp. 132–158 in Traditional Agriculture in South East Asia, G. G. Marten, ed. Boulder, Colo.: Westview.

Clay, J. W. 1988. Indigenous Peoples and Tropical Forests. Cambridge, Mass.: Cultural Survival.

Clay, J. W. 1992. Report on funding and investment opportunities for income generating activities that could complement strategies to halt environmental degradation in the greater Amazon Basin. Paper prepared for the World Wildlife Fund, Nature Conservancy, and World Resources Institute, Washington, D.C.

Consultative Group on International Agricultural Research (CGIAR). 1989. Sustainable Agricultural Production: Final Implications for International Research. Rome, Italy: Food and Agriculture Organization of the United Nations.

CGIAR. 1990. Sustainable agricultural production. Final report of the CGIAR committee. MT/90/18. Presented at the CGIAR meeting, The Hague, Netherlands, May 21–25, 1990.

Conway, G. 1985. Agroecosystem analysis. Agric. Admin. 20:31–55.

Cooperative for American Relief Everywhere (CARE). 1986. Uganda village forestry project evaluation. Paper prepared for CARE, New York City, New York.

Costanza, R., and C. A. Perrings. 1990. A flexible assurance bonding system for improved environmental management. Ecol. Econ. 2:57–76.

Crutzen, P. J., and M. O. Andreae. 1990. Biomass burning in the tropics: Impact on atmospheric chemistry and biogeochemical cycles. Science 250:1669–1678.

Cruz, A. 1987. Avian community organization in a mahogany plantation on a neotropical island. Caribbean J. Sci. 23:286–296.

Cruz, A. 1988. Avian resource use in a Caribbean pine plantation. J. Wildlife Manage. 52:274–279.

Cuevas, E., S. Brown, and A. E. Lugo. 1991. Above- and belowground organic matter storage and production in a tropical pine plantation and a paired broadleaf secondary forest. Plant Soil 135:257–268.

Dahlberg, K. A. 1979. Beyond the Green Revolution: The Ecology and Politics of Global Agricultural Development. New York: Plenum.

Dahlberg, K. A. 1991. Sustainable agriculture—Fad or harbinger? BioScience 41(5):337–340.

Darmstadter, J. 1991. Estimating the cost of carbon dioxide abatement. Resources 103:6–9.

DeBeer, J. H., and M. J. McDermott. 1989. The Economic Value of Non-timber Forest Products in Southeast Asia. Amsterdam: Netherlands Committee for the International Union for the Conservation of Nature.

DeDatta, S. K. 1981. Principles and Practices of Rice Production. New York: John Wiley & Sons.

de Graaf, N. R. 1986. Natural regeneration of tropical rain forest in Suriname as a land-use option. Netherlands J. Agric. Sci. 35:71–74.

de Graaf, N. R., and R. L. H. Poels. 1990. The Celos Management System: A polycyclic method for sustained timber production in South American rain forest. Pp. 116–127 in Alternatives to Deforestation: Steps Toward Sustainable Use of the Amazon Rain Forest, A. B. Anderson, ed. New York: Columbia University Press.

De Miranda, E. E., and C. Mattos. 1992. Brazilian rain forest colonization and biodiversity. Agric. Ecosystems Environ. 40:275–296.

Denevan, W. M. 1982. Causes of deforestation and forest and woodland degradation in tropical America. Paper commissioned by the Office of Technology Assessment, U.S. Congress, Washington, D.C.

Denevan, W. M., J. M. Treacy, J. B. Alcorn, C. Padoch, J. Denslow, and S. Flores Paitan. 1984. Indigenous agroforestry in the Peruvian Amazon: Bora Indian management of swidden fallows. Interciencia 9(6):346–357.

Diamond, J. 1975. The island dilemma: Lessons of modern biogeographic studies for the design of natural reserves. Biol. Conserv. 7:129–146.

Douglas, J. S., and R. A. J. Hart. 1984. Forest Farming. Boulder, Colo.: Westview.

Dover, M., and L. M. Talbot. 1987. To Feed the Earth: Agro-ecology for Sustainable Development. New York: World Resources Institute.

Dudal, R. 1980. Soil-related constraints to agricultural development in the tropics. Pp. 23–37 in Priorities for Alleviating Soil-Related Constraints to Food Production in the Tropics. Los Baños, Philippines: International Rice Research Institute.

Dumanski, J. 1987. Evaluating the sustainability of agricultural systems. Pp. 195–205 in Africaland: Land Development and Management of Acid Soils in Africa. Proceedings No. 7. Bangkok, Thailand: International Board for Soil Research and Management.

Ecological Society of America. 1991. The sustainable biosphere initiative: An ecological research agenda. Ecology 71:371–412.

Edwards, C. A. 1987. The concept of integrated systems in lower input/sustainable agriculture. Amer. J. Alt. Agric. 2(4):148–152.

Edwards, M. 1989. The irrelevance of development studies. Third World Quart. 11(1):116–135.

Edwards, C. A., R. Lal, P. Madden, R. H. Miller, and G. House, eds. 1990. Sustainable Agricultural Systems. Ankeny, Iowa: Soil and Water Conservation Society.

Edwards, C. A., H. D. Thurston, and R. Janke. 1991. Integrated pest management for sustainability in developing countries. Pp. 109–133 in Toward Sustainability: Soil and Water Research Priorities for Developing Countries. Washington, D.C.: National Academy Press.

Ehrenfeld, D. 1992. The business of conservation. Conserv. Biol. 6:1–3.

Ehrlich, P. R., and E. O. Wilson. 1991. Biodiversity studies: Science and policy. Science 253:758–762.

Ehui, S. K., and D. S. C. Spencer. 1990. Indices for measuring the sustainability and economic viability of farming systems. RCMP Research Monograph No. 3. Ibadan, Nigeria: International Institute of Tropical Agriculture.

Ehui, S. K., B. T. Kang, and D. S. C. Spencer. 1990. Economic analysis of soil erosion effects in alley cropping, no-till, and bush fallow systems in southwestern Nigeria. Agric. Syst. 34:349–368.

Evans, J. 1982. Plantation Forestry in the Tropics. Oxford: Clarendon.

Everett, Y. 1987. Seeking Principles of Sustainability: A Forest Model Applied to Forest Gardens in Sri Lanka. Master's thesis. University of California, Berkeley.

Ewel, J. J. 1979. Secondary forests: The tropical wood resource of the future. Pp. 53–60 in Simposio Internacional sobre las Ciencias Forestales y su Contribucion al Desarrollo de la America Tropical, M. Chavarria, ed. San Jose, Costa Rica: Concit, Interciencia, and SCITEC.

Ewel, J. J. 1986. Designing agricultural ecosystems for the humid tropics. Annu. Rev. Ecol. Syst. 17:245–271.

Ewel, J. J. 1991. Yes, we got some bananas. Conserv. Biol. 5(3):423–425.

Faber-Langendoen, D. 1990. Natural Rain Forest Management at the Bajo Calima Concession, Colombia. Forestry Research Report No. 6. Bogotá: Smurfit Cartón de Colombia.

Fearnside, P. M. 1987a. Rethinking continuous cultivation in Amazonia. BioScience 37(3):209–214.

Fearnside, P. M. 1987b. Deforestation and international economic development projects in Brazilian Amazonia. Conserv. Biol. 1:214–221.

Fernandes, E. C. M., A. Oktingati, and J. Maghembe. 1983. The Chaga home gardens: A multi-storied agro-forestry cropping system on Mt. Kilimanjaro, Northern Tanzania. Agroforest. Syst. 1(3):269–273.

Flores Guido, J. S. 1987. Yucatan, tierra de las leguminosas. En Revista de la Universidad Autónoma de Yucatán, Númbero 163 (Oct./Nov.).

Food and Agriculture Organization (FAO). 1975. Soil Map of the World. Vol. IV, South America. Paris: United Nations Educational, Scientific, and Cultural Organization.

FAO. 1976. Framework for Land Evaluation. Soils Bulletin No. 32. Rome, Italy: Food and Agriculture Organization of the United Nations.

FAO. 1977. Soil Map of the World. Vol. V, Africa. Paris: United Nations Educational, Scientific, and Cultural Organization.

FAO. 1978. Soil Map of the World. Vol. VI, South Asia, and Vol. IX, Southeast Asia. Paris: United Nations Educational, Scientific, and Cultural Organization.

FAO. 1987. The Tropical Forestry Action Plan. Rome, Italy: Food and Agriculture Organization of the United Nations.

Food and Agriculture Organization and United Nations Environment Program. 1981. Forest Resources of Tropical Africa, Asia, and the Americas. Rome, Italy: Food and Agriculture Organization of the United Nations.

Ford-Robertson, F. C., ed. 1971. Terminology of Forest Science, Technology Practice and Products. Washington, D.C.: Society of American Foresters.

Forest Resources Assessment 1990 Project. 1990. Interim report on Forest Resources Assessment 1990 Project. Item 7 of the Provisional Agenda presented at the Tenth Session of the Committee on Forestry of the Food and Agriculture Organization of the United Nations, Rome, Italy, September 24–28, 1990.

Forest Resources Assessment 1990 Project. 1991. Second interim report on the state of tropical forests. Prepared for the Tenth World Forestry Congress, Paris, France, September 17–26, 1991.

Forest Resources Assessment 1990 Project. 1992. Third interim report on the state of tropical forests. Prepared for the Food and Agriculture Organization of the United Nations, Rome, Italy.

Fortmann, L., and J. W. Bruce, eds. 1988. Whose Trees: Proprietary Dimensions of Forestry. Boulder, Colo.: Westview.

Francis, C. A., ed. 1986. Multiple Cropping Systems. New York: Macmillan.

Frankel, O. H., and M. E. Soulé. 1981. Conservation and Evolution. New York: Cambridge University Press.

Garrity, D. P. 1991. Sustainable land use systems for the sloping uplands of Southeast Asia. Paper presented at the Asia Environment and Agriculture Officers Conference, Colombo, Sri Lanka, September 9–13, 1991.

Garrity, D. P., H. G. Zandstra, and R. Harwood. 1978. A classification of Philippine upland rice growing environment for use in cropping systems research. Philippine J. Crop Sci. 3(1):25–37.

Gliessman, S. R. 1982. Nitrogen distribution in several traditional agroecosystems in the humid tropical lowlands of southeastern Mexico. Plant Soil 67:105–117.

Gliessman, S. R. 1985. An agroecological approach for researching sustainable agroecosystems. Pp. 184–199 in The Importance of Biological Agriculture in a World of Diminishing Resources, H. Vogtmann, E. Boehnke, and I. Fricke, eds. Witzenhausen, Germany: Verlagsgruppe.

Gliessman, S. R. 1990. Understanding the basis of sustainability for agriculture in the tropics: Experiences in Latin America. Pp. 378–390 in Sustainable Agricultural Systems, C. A. Edwards, R. Lal, P. Madden, R. H. Miller, and G. House, eds. Ankeny, Iowa: Soil and Water Conservation Society.

Gliessman, S. R. 1991a. Agroecology: Researching the ecological basis for sustainable agriculture. Pp. 3–10 in Agroecology: Researching the Ecological Basis for Sustainable Agriculture, S. R. Gliessman, ed. New York: Springer-Verlag.

Gliessman, S. R. 1991b. Quantifying the agroecological component of sustainable agriculture: A goal. Pp. 366–370 in Agroecology: Researching the Ecological Basis for Sustainable Agriculture, S. R. Gliessman, ed. New York: Springer-Verlag.

Gliessman, S. R., and R. Grantham. 1990. Agroecology: Reshaping agricultural development. Pp. 196–207 in Lessons of the Rainforest, S. Head and R. Heinzman, eds. San Francisco: Sierra Club Books.

Gómez-Pompa, A. 1987a. On Maya silviculture. Mex. Stud. 3(1):1–17.

Gómez-Pompa, A. 1987b. Tropical deforestation and Maya silviculture: An ecological paradox. Tulane Stud. Zool. Bot. 26:19–37.

Gómez-Pompa, A., and D. A. Bainbridge. 1991. Tropical forestry as if people mattered. In A Half Century of Tropical Forest Research, A. E. Lugo and C. Lowe, eds. New York: Springer-Verlag.

Gómez-Pompa, A., and A. Kaus. 1990. Traditional management of tropical forests in Mexico. Pp. 45–64 in Alternatives to Deforestation: Steps Toward Sustainable Use of the Amazon Rain Forest, A. B. Anderson, ed. New York: Columbia University Press.

Gómez-Pompa, A., and A. Kaus. 1992. Taming the wilderness myth. BioScience 42:271–279.

Gómez-Pompa, A., and C. Vázquez-Yanes. 1974. Studies on secondary succession of tropical lowlands: The life cycle of secondary species. Pp. 336–342 in Proceedings of the First International Congress of Ecology. The Hague: International Association of Ecology.

Gómez-Pompa, A., J. S. Flores, and V. Sosa. 1987. The "Pet Kot": A man-made tropical forest of the Maya. Interciencia 12(1):10–15.

Goodland, R. J. A., C. Watson, and G. Ledec. 1984. Environmental management in tropical agriculture. Boulder, Colo.: Westview.

Goodland, R. J., E. O. A. Asibey, J. Post, and M. Dyson. 1990. Tropical moist forest management: The urgency of transition to sustainability. Environ. Conserv. 17:303–318.

Gradwohl, J., and R. Greenberg. 1988. Saving the Tropical Forests. London: Earthscan.

Grainger, A. 1988. Estimating areas of degraded tropical lands requiring replenishment of forest cover. Int. Tree Crops J. 5:3–6.

Grandstaff, T. B. 1981. Shifting cultivation. Ceres 4:28–30.

Green, G. M., and R. W. Sussman. 1990. Deforestation history of the eastern rain forests of Madagascar from satellite images. Science 248:22–25.

Grove, T. L., C. A. Edwards, R. R. Harwood, and C. J. Pierce Colfer. 1990. The role of agroecology and integrated farming systems in agricultural sustainability. Paper prepared for the Forum on Sustainable Agriculture and Natural Resource Management, November 13–16, 1990, National Research Council, Washington, D.C.

Hansen, G. E., ed. 1981. Agriculture and Rural Development in Indonesia. Boulder, Colo.: Westview.

Harcombe, P. A. 1977. Nutrient accumulation by vegetation during the first year of a tropical forest ecosystem. Pp. 347–378 in Recovery and Restoration of Damaged Ecosystems, J. Cairns, K. L. Kickson, and E. E. Herricks, eds. Charlottesville: University Press of Virginia.

Harlan, J. 1975. Crops and Man. Madison, Wis.: American Society of Agronomy and Crop Science Society of America.

Harrington, L. W. 1991. Is the green revolution still green? Staple crop systems under stress. Paper presented at the Conference on Asia's Environment and Agriculture Initiatives: Integrating Resources for Sustainability and Profit, U.S. Agency for International Development, Colombo, Sri Lanka, September 9–13, 1991.

Harris, L. 1984. The Fragmented Forest: Island Biogeography Theory and the Preservation of Biological Diversity. Chicago: University of Chicago Press.

Harrison, S. 1991. Population growth and deforestation in Costa Rica, 1950–1984. Interciencia 6:83–93.

Hart, R. D. 1980. A natural ecosystem analog approach to the design of a successful crop system for tropical forest environments. Biotropica 2(Suppl.):73–82.

Hartshorn, G. S. 1989. Application of gap theory to tropical forest management: Natural regeneration on strip clearcuts in the Peruvian Amazon. Ecology 70:567–569.

Hartshorn, G. S. 1990. Natural forest management by the Yanesha Forestry Cooperative in Peruvian Amazonia. Pp. 128–138 in Alternatives to Deforestation: Steps Toward Sustainable Use of the Amazon Rain Forest, A. B. Anderson, ed. New York: Columbia University Press.

Harwood, R. R. 1979. Small Farm Development: Understanding and Improving Farming Systems in the Humid Tropics. Boulder, Colo.: Westview.

Harwood, R. R. 1987. Agroforestry and mixed farming systems. Pp. 273–301 in Ecological Development in the Humid Tropics: Guidelines for Planners, A. E. Lugo, J. R. Clark, and R. D. Child, eds. Morrilton, Ark.: Winrock International.

Hazlett, D. L. 1986. Ethnobotanical observations from Cabecar and Guaymí settlements in Central America. Econ. Bot. 40:339–352.

Hecht, S. B. 1982. Cattle Ranching in the Brazilian Amazon: Evaluation of a Development Strategy. Ph.D. dissertation. University of California, Berkeley, Calif.

Hecht, S., and A. Cockburn. 1989. The Fate of the Forest. London: Verso.

Hecht, S. B., R. B. Norgaard, and G. Possio. 1988. The economics of cattle ranching in eastern Amazonia. Interciencia 13(5):233–239.

Herren, H. R. 1989. The biological control program of IITA: From concept to reality. Pp. 18–30 in Biological Control: A Sustainable Solution to Crop Pest Problems in Africa, J. S. Yaninek and H. R. Herren, eds. Ibadan, Nigeria: International Institute of Tropical Agriculture.

Herrera Castro, N. 1991. Los Huertos Familiares Mayas en el Oriente de Yucatán. Master's thesis. Universidad Nacional Autónoma de México, México, D.F.

Holdridge, L. R. 1967. Life Zone Ecology, rev. ed. San Jose, Costa Rica: Tropical Science Center.

Holliday, R. H. 1976. The efficiency of solar energy conversion by the whole crop. Pp. 127–146 in Food Production and Consumption, A. N. Duckman, J. G. W. Jones, and E. H. Roberts, eds. Amsterdam: North Holland Publishing.

Houghton, J. T., G. J. Jenkins, and J. J. Ephraums, eds. 1990. Climate Change: The IPCC Scientific Assessment. Cambridge, U.K.: Press Syndicate of the University of Cambridge.

Houghton, R. A. 1990a. The global effects of tropical deforestation. Environ. Sci. Technol. 24:414–422.

Houghton, R. A. 1990b. The future role of tropical forests in affecting the carbon dioxide concentration of the atmosphere. Ambio 19(4):204–209.

Houghton, R. A., R. D. Boone, J. R. Fruci, J. E. Hobbie, J. M. Melillo, C. A. Palm, B. J. Peterson, G. R. Shaver, G. M. Woodwell, B. Moore, D. L. Skole, and N. Myers. 1987. The flux of carbon from terrestrial ecosystems to the atmosphere in 1980 due to changes in land use: Geographic distribution of the global flux. Tellus 39B(1–2):122–139.

Houghton, R. A., J. D. Unruh, and P. A. LeFebvre. In press. Current land use in the tropics and its potential for sequestering carbon. Global Biogeochem. Cycles.

Hudson, W. 1991. Landscape Linkages and Biodiversity. Washington, D.C.: Defenders of Wildlife.

Hurst, P. 1990. Rainforest Politics: Ecological Destruction in South-East Asia. London: Zed Books.

Hyde, W. F., and R. Sedjo. In press. Managing tropical forests: Reflections on the rent distribution discussion. Land Econ.

Hynes, R. A., and A. K. Chase. 1982. Plant, sites and domiculture: Aboriginal influence upon plant communities in Cape York Peninsula. Archaeol. Oceania 17:38–50.

Illsley, C. 1984. Vegetación y producción de la milpa bajo roza, tumba y quema en el ejido de Yaxcaba, Yucatán, México. Tesis Profesional. Universidad de Michoacana de San Nicolas de Hidalgo, Morelia, Michoacán, México.

Iltis, H. H. 1988. Serendipity in the exploration of biodiversity: What good are weedy tomatoes? Pp. 98–105 in Biodiversity, E. O. Wilson, ed. Washington, D.C.: National Academy Press.

Intergovernmental Panel on Climate Change. 1990a. Proceedings of the Conference on Tropical Forestry Response Options to Global Climate Change. Washington, D.C.: U.S. Environmental Protection Agency.

Intergovernmental Panel on Climate Change. 1990b. Greenhouse Gas Emissions from Agricultural Systems, Vols. 1 and 2. Washington, D.C.: U.S. Environmental Protection Agency.

International Irrigation Management Institute. 1987. Irrigation Management for Diversified Cropping. Colombo, Sri Lanka: Kandy.

International Laboratory for Research on Animal Diseases. 1991. ILRAD 1990: Annual Report of the International Laboratory for Research on Animal Diseases. Nairobi, Kenya: International Laboratory for Research on Animal Diseases.

International Livestock Center for Africa (ILCA). 1991. ILCA 1990: Annual Report and Program Highlights. Addis Ababa, Ethiopia: International Livestock Center for Africa.

International Rice Research Institute (IRRI). 1975. Proceedings of the Cropping Systems Workshop. Los Baños, Philippines: International Rice Research Institute.

IRRI. 1988. Sustainable Agriculture; Green Manure in Rice Farming. Los Baños, Philippines: International Rice Research Institute.

Jahnke, H. E. 1982. Livestock Production Systems and Livestock Development in Tropical Africa. Kiel, Germany: Kieler Wissenschaftsverlag Vauk.

Jenkinson, D. S., and A. Ayanaba. 1977. Decomposition of carbon-14 labeled plant material under tropical conditions. J. Soil Sci. Soc. Am. 41(5):912–915.

Jensen, C. L., and S. Pfeifer. 1989. Assisted natural regeneration: A new reforestation approach for DENR? An appraisal prepared for the Department of Environment and Natural Resources (DENR). Manila: Government of the Philippines and U.S. Agency for International Development.

Jiménez, E., and A. Gómez-Pompa. 1981. Estudios Ecológicos en el Agroecosistema Cafetalero. Xalapa, México: Instituto Nacional de Investigaciones sobre Recursos Bióticos.

Jordan, C. F. 1985. Nutrient Cycling in Tropical Forest Ecosystems. New York: Wiley Interscience.

Juo, A. S. R. 1989. New farming systems development in the wetter tropics. Exp. Agric. 25:145–163.

Kang, B. T., G. F. Wilson, and L. Sipkens. 1981. Alley cropping maize and *Leucaena* in southern Nigeria. Plant Soil 63:165–179.

Kang, B. T., G. F. Wilson, and T. L. Lawson. 1984. Alley Cropping: A Stable Alternative to Shifting Cultivation. Ibadan, Nigeria: International Institute of Tropical Agriculture.

Kangas, P. 1990. Deforestation and diversity of life zones in the Brazilian Amazon: A map analysis. Ecol. Modelling 49:267–275.

Kass, D. C. L. 1978. Polyculture Cropping Systems: Review and Analysis. Cornell International Agriculture Bulletin No. 32. Ithaca, N.Y.: Cornell University.

Kenmore, P. E. 1991. Indonesia's Integrated Pest Management—A Model for Asia. Manila, Philippines: Food and Agriculture Organization of the United Nations, Rice Integrated Pest Control Program.

Kunstadter, P. 1978. Ecological modification and adaptation: An ethnobotanical view of Lua' swiddeners in northwestern Thailand. Anthropol. Papers 67:169–200.

Lal, R. 1986. Soil surface management in the tropics for intensive land use and high and sustained production. Adv. Soil Sci. 5:1–108.

Lal, R. 1987. Managing soils of sub-Saharan Africa. Science 236:1069–1076.

Lal, R. 1989. Agroforestry systems and soil surface management of a tropical Alfisols, I-V. Agroforest. Syst. 8:1–6,7–29,97–111,113–132,197–215,217–238.

Lal, R. 1990. Introduction and overview of greenhouse gases from tropical agriculture. P. VI-9 in Greenhouse Gas Emissions from Agricultural Systems, Vol. 1: Summary Report. Washington, D.C.: U.S. Environmental Protection Agency.

Lal, R. 1991a. Myths and scientific realities of agroforestry as a strategy for sustainable management of soils in the tropics. Adv. Soil Sci. 15:91–137.

Lal, R. 1991b. Soil research for agricultural sustainability in the tropics. Pp. 66–90 in Toward Sustainability: A Plan for Collaborative Research on Agriculture and Natural Resource Management. Washington, D.C.: National Academy Press.

Lal, R., and E. W. Rassel, eds. 1981. Tropical Agricultural Hydrology: Watershed Management and Land Use. Chichester, U.K.: John Wiley & Sons.

Lamprecht, H. 1989. Silviculture in the tropics. Robdorf, Germany: TZ-Verlagsgesellschaft.

Lanly, J. P. 1982. Tropical Forest Resources. Forestry Paper No. 30. Rome, Italy: Food and Agriculture Organization of the United Nations.

Lathwell, D. J., and T. L. Grove. 1986. Soil-plant relationships in the tropics. Annu. Rev. Ecol. Syst. 17:1–16.

Latin American and Caribbean Commission on Development and Environment. 1990. Our Own Agenda. New York: Inter-American Development Bank and United Nations Development Program.

Lean, G., D. Hinrichsen, and A. Markham. 1990. Atlas of the Environment. New York: Prentice-Hall.

Lebron, M. L. 1980. Physiological plant ecology: Some contributions to the understanding of secondary succession in tropical lowland rainforest. Biotropica 12(Suppl.):31–33.

Lee, H. S. 1982. The development of silvicultural systems in the hill forests of Malaysia. Malayan Forest. 45:1–8.

Lentz, D. L. 1991. Maya diets of the rich and poor: Paleoethnobotanical evidence from Copán. Latin Amer. Antiquity 2(3):269–287.

Leonard, H. J. 1987. Natural Resources and Economic Development in Central America. New Brunswick, N.J.: Transaction Books.

Lewis, W. A. 1954. Economic development with unlimited supplies of labor. Manchester School Econ. Soc. Stud. 22(May):139–191.

Linares, O. F. 1976. "Garden hunting" in the American tropics. Human Ecol. 4(4):331–349.

Linear, M. 1985. The tsetse war. Ecologist 15:27–35.

Lovejoy, T. E. 1985. Rehabilitation of degraded tropical forest lands. Environmentalist 5:1–8.

Lovelace, G. W., S. Subhadhira, and S. Simaraks. 1988. Rapid Rural Appraisal in Northeast Thailand: Case Studies. Thailand: Khonkaen University.

Lugo, A. E. 1987. Tropical forest management with emphasis on wood production. Pp. 169–189 in Ecological Development in the Humid Tropics, A. E. Lugo, J. R. Clark, and R. D. Child, eds. Morrilton, Ark.: Winrock International.

Lugo, A. E. 1988. The future of the forest: Ecosystem rehabilitation in the tropics. Environment 30(7):16–20,41–45.

Lugo, A. E. 1991. Cities in the sustainable development of tropical landscapes. Nature Resources 27(2):27–35.

Lugo, A. E. 1992. Comparison of tropical tree plantations with secondary forests of similar age. Ecol. Monogr. 62(1):1–41.

Lugo, A. E., and S. Brown. 1991. Comparing tropical and temperate forests. Pp. 319–330 in Comparative Analyses of Ecosystems, J. Cole, G. Lovett, and S. Findlay, eds. New York: Springer-Verlag.

Lugo, A. E., and S. Brown. In press. Tropical forests as sinks of atmospheric carbon. Forest Ecol. Manage.

Lugo, A. E., S. Brown, and J. Chapman. 1988. An analytical review of production rates and stemwood biomass of tropical forest plantations. Forest Ecol. Manage. 23:179–200.

Lundgren, B. O. 1991. Commentary. Agroforest. Today 3(3):6–7.

Lundgren, L. 1985. Catchment Forestry in Tanzania. A revised report for the joint Tanzania/Nordic Forestry Sector Review 1985. Nairobi, Kenya: Regional Soil Conservation Unit, Swedish International Development Authority.

Lynch, O. J. 1990. Whither the People? Demographic, Tenurial, and Agricultural Aspects of the Tropical Forestry Action Plan. Washington, D.C.: World Resources Institute.

Lynch, O. J. 1991. Community-based tenurial strategies for promoting forest conservation and development in South and Southeast Asia. Paper presented at the Asia Environment and Agriculture Officers Conference, U.S. Agency for International Development, Colombo, Sri Lanka, September 10–13, 1991.

MacArthur, J. D. 1980. Some characteristics of farming in a tropical environment. Pp. 19–29 in Farming Systems in the Tropics, H. Ruthenberg, ed. New York: Oxford University Press.

MacArthur, R. H., and E. O. Wilson. 1967. The Theory of Island Biogeography. Princeton, N.J.: Princeton University Press.

MacKinnon, J. R., K. MacKinnon, G. Child, and J. Thorsell. 1986. Managing Protected Areas in the Tropics. Gland, Switzerland: International Union for the Conservation of Nature and Natural Resources.

McCants, C. B. 1991. Contributions of international agricultural research centers, Agency for International Development, Food and Agriculture Organization, and U.S. Department of Agriculture to sustainable agriculture and gaps in the information base. Paper presented at the Forum on Sustainable Agriculture and Natural Resource Management, Washington, D.C., November 13–16, 1990.

McDowell, R. E., and P. E. Hildebrand. 1980. Integrating crop and animal production: Making the most of resources available to small farms in developing countries. New York: Rockefeller Foundation.

McNeely, J. A., K. R. Miller, W. V. Reid, R. A. Mittermeier, and T. B. Werner. 1990. Conserving the World's Biological Diversity. Gland, Switzerland and Washington, D.C.: International Union for the Conservation of Nature and Natural Resources, World Resources Institute, Conservation International, World Wildlife Fund-US, and the World Bank.

Michon, G., F. Mary, and J. Bompard. 1986. Multistoried agroforestry garden system in West Sumatra. Agroforest. Syst. 4:315–338.

Moad, A. S. 1989. Sustainable forestry in the tropics: The elusive goal. Paper presented at the Workshop on the U.S. Tropical Timber Trade: Conservation Options and Impacts, New York City, April 14–15, 1989.

Morrisette, P. M., and A. J. Plantinga. 1991. The global warming issue: Viewpoints of different countries. Resources 103:2–6.

Myers, N. 1984. The Primary Source: Tropical Forests and Our Future. New York: W.W. Norton.

Myers, N. 1988. Threatened biotas: Hotspots in tropical forests. Environmentalist 8(3):1–20.

Myers, N. 1989. Deforestation Rates in Tropical Forests and Their Climatic Implications. London: Friends of the Earth, U.K.

Nair, P. K. R. 1989. Classification of agroforestry systems. Pp. 39–52 in Agroforestry Systems in the Tropics, P. K. R. Nair, ed. Boston: Kluwer.

National Academy of Sciences. 1980. Research Priorities in Tropical Biology. Washington, D.C.: National Academy of Sciences.

National Research Council (NRC). 1982. Ecological Aspects of Development in the Humid Tropics. Washington, D.C.: National Academy Press.

NRC. 1991a. Toward Sustainability: A Plan for Collaborative Research on Agriculture and Natural Resource Management. Washington, D.C.: National Academy Press.

NRC. 1991b. Toward Sustainability: Soil and Water Research Priorities for Developing Countries. Washington, D.C.: National Academy Press.

NRC. 1992. Conserving Biodiversity: A Research Agenda for Development Agencies. Washington, D.C.: National Academy Press.

Nations, J. 1990. Protected areas in tropical forests. Pp. 208–216 in Lessons of the Rainforest, S. Head and R. Heinzman, eds. San Francisco: Sierra Club Books.

Nepstad, D. C., C. Uhl, and E. A. S. Serrão. 1991. Recuperation of a degraded Amazonian landscape: Forest recovery and agricultural restoration. Ambio 20(6):248–255.

Niñez, V. 1985. Introduction to household gardens and small scale food production. Food Nutr. Bull. 7(3):1–5.

Norgaard, R. 1989. Three dilemmas in environmental accounting. Ecol. Econ. 1(4):303–314.

Norgaard, R. 1992. Sustainability and the Economics of Assuring Assets for Future Generations. Policy Research Working Paper WPS-832. Washington, D.C.: World Bank.

North, D. C. 1959. Agriculture in regional economic growth. J. Farm Econ. 51(Dec.):943–951.

Nye, P. H., and D. J. Greenland. 1960. The Soil Under Shifting Cultivation. Technical Communication 51. Harpenden, U.K.: Commonwealth Bureaux of Soils.

Office of Technology Assessment, U.S. Congress. 1984. Technologies to Sustain Tropical Forest Resources. OTA-F-24. Washington, D.C.: Office of Technology Assessment, U.S. Congress.

Okigbo, B. N. 1991. Development of Sustainable Agricultural Production Systems in Africa: Roles of International Agricultural Research Centers and National Agricultural Research Systems. Ibadan, Nigeria: International Institute of Tropical Agriculture.

Okigbo, B. N., and R. Lal. 1977. Residue mulches, intercropping and agrisilvicultural potential in tropical Africa. Pp. 54–69 in Proceedings of the IFOAM Conference, Montréal, Canada. Montreal: International Federation of Organic Agriculture Movements.

Omerod, W. E. 1978. The relationship between economic development and environmental degradation: How degradation has occurred in West Africa and how its progress might be halted. J. Arid Environ. 1:257–279.

Opler, P. A., H. G. Baker, and G. W. Frankie. 1980. Plant reproductive characteristics during secondary succession in neotropical lowland forest ecosystems. Biotropica 12(Suppl.):40–46.

Oram, P. A. 1988. Building the agroecological framework. Environment 30(9):14–17,30–36.

Padoch, C., J. Chota Inuma, W. de Jong, and J. Unruh. 1985. Amazonian agroforestry: A market-oriented system in Peru. Agroforest. Syst. 3:47–58.

Pee, T. Y. 1977. Social Returns from Rubber Research in Peninsular Malaysia. Ph.D. dissertation. Michigan State University, East Lansing.

Peters, C. M., M. J. Balick, F. Kahn, and A. B. Anderson. 1989a. Oligarchic forests of economic plants in Amazonia: Utilization and conservation of an important tropical resource. Conserv. Biol. 3:341–349.

Peters, C. M., A. H. Gentry, and R. O. Mendelsohn. 1989b. Valuation of an Amazonian rainforest. Nature 339:656–657.

Peters, R. L., and T. Lovejoy, eds. 1992. Consequences of Greenhouse Warming for Biological Diversity. New Haven, Conn.: Yale University Press.

Pimentel, D. 1989. Ecological systems, natural resources, and food supplies. Pp. 2–31 in Food and Natural Resources, D. Pimentel and C. W. Hall, eds. San Diego, Calif.: Academic Press.

Pinedo-Vásquez, M., D. Zarin, and P. Jipp. 1990. Use-values of tree species in a communal forest reserve in northeast Peru. Conserv. Biol. 4(4):405–416.

Pinton, F. 1985. The tropical garden as a sustainable food system: A comparison of Indians and settlers in northern Colombia. Food Nutr. 7(3):25–28.

Plant, R. 1991. Land Rights for Indigenous and Tribal Peoples in Developing Countries: A Survey of Law and Policy Issues, Current Activities, and Proposals for an Inter-agency Programme of Action. Geneva, Switzerland: United Nations International Labour Office.

Poore, M. E. D., P. Burgess, J. Palmer, S. Rietbergen, and T. Synnot. 1990. No timber without trees: Sustainability in the tropical forest. London: Earthscan and the International Tropical Timber Organization.

Population Reference Bureau (PRB). 1988. 1988 World Population Data Sheet. Washington, D.C.: Population Reference Bureau.

PRB. 1991. World Population Data Sheet 1991. Washington, D.C.: Population Reference Bureau.

Posey, D. A. 1982. Keepers of the forest. Garden 6:18–24.

Posey, D. A. 1984. A preliminary report on diversified management of tropical forest by the Kayapó Indians of the Brazilian Amazon. Adv. Econ. Bot. 1:112–126.

Posey, D. A. 1990. The science of the Mebengokre. Orion Summer 9(3):16–23.

Randall, A. 1988. What mainstream economists have to say about the value of biodiversity. Pp. 217–223 in Biodiversity, E. O. Wilson, ed. Washington, D.C.: National Academy Press.

Randall, A. 1991. Nonuse benefits. In Measuring the Demand for Environmental Quality, J. B. Braden and C. D. Kolstad, eds. New York: Elsevier.

Rankin, J. M. 1985. Forestry in the Brazilian Amazon. Pp. 360–392 in Key Environments: Amazonia, G. T. Prance and T. E. Lovejoy, eds. New York: Pergamon.

Raven, P. 1988. Biological resources and global stability. Pp. 3–27 in Evolution and Coadaptation in Biotic Communities, S. Kawano, J. H. Connell, and T. Hidaka, eds. Tokyo: University of Tokyo Press.

Reid, W. V., and K. R. Miller. 1989. Keeping Options Alive: The Scientific Base for Conserving Biodiversity. Washington, D.C.: World Resources Institute.

Remmers, G., and H. de Koeyer. 1988. El "Tolche" en Pixoy. Master's thesis. University of Wageningen, The Netherlands.

Repetto, R. 1988a. The Forest for the Trees? Government Policies and the Misuse of Forest Resources. Washington, D.C.: World Resources Institute.

Repetto, R. 1988b. Overview. Pp. 1–41 in Public Policies and the Misuse of Forest Resources, R. Repetto and M. Gillis, eds. Cambridge, U.K.: Cambridge University Press.

Repetto, R., and M. Gillis, eds. 1988. Public Policies and the Misuse of Forest Resources. Cambridge, U.K.: Cambridge University Press.

Richards, P. W. 1955. The secondary succession in the tropical rain forest. Sci. Prog. London 43:45–57.

Rico-Gray, V., A. Gómez-Pompa, and C. Chan. 1985. Las selvas manejadas por los mayas de Yohaltun, Campeche, México. Biotica 10:321–327.

Rico-Gray, V., J. G. García-Franco, A. Puch, and P. Sima. 1988. Composition and structure of a tropical dry forest in Yucatan, Mexico. Int. J. Environ. Sci. 14:21–29.

Rico-Gray, V., A. Chemas, and S. Mandujano. 1991. Uses of tropical deciduous forest species by the Yucatecan Maya. Agroforest. Syst. 14:149–161.

Roosevelt, A. 1990. The historical perspective on resource use in tropical Latin America. Pp. 30–64 in Economic Catalysts to Ecological Change. Gainesville, Fla.: Center for Latin American Studies, University of Florida.

Rosero, P. 1979. Some data on a secondary forest managed in Siquirres, Costa Rica. In Workshop: Agroforestry Systems in Latin America, G. de las Salas, ed. Turrialba, Costa Rica: Centro Agronómico Tropical de Investigación y Enseñanza.

Rush, J. 1991. The Last Tree: Reclaiming the Environment in Tropical Asia. New York: Asia Society.

Russell, W. M. S. 1968. The slash and burn technique. Natur. Hist. 78(3):58–65.

Ruthenberg, H. 1971. Farming Systems in the Tropics. Oxford, U.K.: Clarendon.

Ruttan, V. W. 1982. Agricultural Research Policy. Minneapolis: University of Minnesota Press.

Ruttan, V. W. 1991. Sustainable growth in agricultural production: Poetry, policy, and science. Paper presented at the Seminar on Agricultural Sustainability, Growth, and Poverty Alleviation: Issues and Policies, International Food Policy Research Institute, Feldafing, Germany, September 23–27, 1991.

Ryan, J. 1991. Goods from the woods. World Watch 4(4):19–26.

Ryan, J. 1992. Conserving biological diversity. Pp. 9–26 in State of the World 1992, L. Starke, ed. New York: W. W. Norton.

Sabhasri, S. 1978. Effects of forest fallow cultivation on forest production and soil. Pp. 160–184 in Economic Development and Marginal Agriculture in Northern Thailand, P. Kunstadter, E. C. Chapman, and S. Sabhasri, eds. Honolulu: University Press of Hawaii.

Sachs, I. 1992. Transition strategies for the 21st century. Nature Resources 2(1):84–17.

Sader, S. A., and A. T. Joyce. 1988. Deforestation trends in Costa Rica, 1940–1983. Biotropica 20:11–19.

Sajise, P. E. 1980. Alang-Alang (*Imperata cylindrica* [L. Blaur]) and Upland Agriculture. Pp. 35–46 in Biotrop Special Publication No. 5. Bogor, Indonesia: Biotrop.

Salati, E. 1990. Possible climate changes in Latin America and the Caribbean and their consequences. Paper presented at the Technical Meeting of Government Experts, Towards an Environmentally Sustainable Form of Development, United Nations Mission for Latin America and the Caribbean, Santiago, Chile, June 1990.

Salati, E., and P. B. Vose. 1984. Amazon Basin: A system in equilibrium. Science 225:129–138.

Salati, E., T. E. Lovejoy, and P. B. Vose. 1983. Precipitation and water recycling in tropical forests. Environmentalist 3:67–72.

Saldarriaga, J. G., D. C. West, M. L. Tharp, and C. Uhl. 1988. Long-term chronosequence of forest succession in the upper Rio Negro of Colombia and Venezuela. J. Ecol. 76:938–958.

Salinas, J. G., and R. Gualdrón. 1988. Adaptación y requerimientos de fertilización de *Brachiaria humidicola* (Rendle) en la altillanura plana de los Llanos Orientales de Colombia. Pp. 457–472 in Simpósio sobre o Cerrado, Vol. 6. Savana, Brasil: Alimentaçao e Energia.

Sanchez, P. A. 1976. Properties and Management of Soils in the Tropics. New York: Wiley Interscience.

Sanchez, P. A. 1979. Soil fertility and conservation considerations for agroforestry systems in the humid tropics of Latin America. Pp. 79–124 in Soils Research in Agroforestry, H. O. Mongi and P. A. Huxley, eds. Nairobi, Kenya: International Center for Research in Agroforestry.

Sanchez, P. A. 1991. Alternatives to slash and burn: A pragmatic approach to mitigate tropical deforestation. Paper presented at the Conference on Agricultural Technology for Sustainable Economic Development in the New Century: Policy Issues for the International Community, World Bank, Airlie, Virginia, October 23, 1991.

Sanchez, P. A., and J. R. Benites. 1987. Low-input cropping for acid soils of the humid tropics. Science 238:1521–1527.

Sanchez, P. A., and S. W. Buol. 1975. Soils of the tropics and the world food crisis. Science 188:598–603.

Sanchez, P. A., and T. T. Cochrane. 1980. Soil constraints in relation to major farming systems in tropical America. Pp. 107–139 in Priorities for Alleviating Soil-Related Constraints to Food Production in the Tropics. Los Baños, Philippines: International Rice Research Institute.

Sanchez, P. A., D. E. Bandy, J. H. Villachica, and J. J. Nicholaides. 1982. Amazon Basin soils: Management for continuous crop production. Science 216:821–827.

Sanchez, P. A., J. H. Villachica, and D. E. Bandy. 1983. Soil fertility dynamics after clearing a tropical rainforest in Peru. Soil Sci. Am. J. 47:1171–1178.

Sanchez, P. A., C. A. Palm, L. T. Scott, E. Cuevas, and R. Lal. 1989. Organic input management in tropical agroecosystems. Pp. 125–152 in Dynamics of Soil Organic Matter in Tropical Ecosystems, D. C. Coleman, J. M. Oades, and G. Uehara, eds. Honolulu: University of Hawaii Press.

Sanchez, P. A., C. A. Palm, and T. J. Smyth. 1990. Approaches to mitigate tropical deforestation by sustainable soil management practices. Devel. Soil Sci. 20:211–220.

Savage, J. M. 1987. Introduction. Pp. 3–11 in Ecological Development in the Humid Tropics: Guidelines for Planners, A. E. Lugo, J. R. Clark, and R. D. Child, eds. Morrilton, Ark.: Winrock International.

Schmidt, R. 1987. Tropical rain forest management: A status report. Unasylva 39(156):2–17.

Serrão, E. A. S., and A. K. O. Homma. 1990. The question of sustainability of cattle raising on altered areas in the Amazon: The influence of agronomical, ecological, and socioeconomic variables. Paper prepared for the World Bank, Washington, D.C.

Serrão, E. A., and J. M. Toledo. 1990. The search for sustainability in Amazonian pastures. Pp. 195–214 in Alternatives to Deforestation: Steps Toward Sustainable Use of Amazon Rain Forest, A. B. Anderson, ed. New York: Columbia University Press.

Serrão, E. A. S., and J. M. Toledo. In press. Sustained pasture-based production systems for the humid tropics. In Development or Destruction: The Conversion of Tropical Forest to Pasture, T. E. Downing, S. B. Hecht, H. A. Pearson, and C. Garcia-Downing, eds. Boulder, Colo.: Westview.

Serrão, E. A. S., I. C. Falesi, J. B. da Veiga, and J. F. T. Neto. 1979. Productivity of cultivated pastures on low fertility soils in the Amazon of Brazil. Pp. 195–225 in Pasture Production in Acid Soils of the Tropics, P. A. Sanchez and L. E. Tergas, eds. Cali, Colombia: Centro Internacional de Agricultura Tropical.

Shaner, W. W., P. F. Philipp, and W. R. Schnehl. 1982. Farming Systems Research and Development: Guidelines for Developing Countries. Boulder, Colo.: Westview.

Sieberling, L. 1991. Extractive reserves (letter). BioScience 41:285–286.

Silva, A. de B. 1982. Determinaçao de Danos de Cigarrinha-das-Pastagens Deois incompleta a Brachiaria humidicola e Brachiaria decumbens. Circular Técnica No. 27. Belém, Brasil: Empresa Brasileira de Pesquisa Agropecuária.

Silva, A. de B., and B. P. Magalhães. 1980. Insectos Nocivos às Pastagens no Estado do Pará. Boletim de Pesquisa No. 8. Belém, Brasil: Centro de Pesquisa Agropecuária do Trópico Úmido.

Smiet, A. C. 1990. Forest ecology on Java: Conversion and usage in a historical perspective. J. Trop. Forest Sci. 2:286–302.

Smith, J. R. S. 1952 (1929). Tree Crops. Westport, Conn.: Devin-Adair.

Smyth, T. J., and M. S. Cravo. 1991. Continuous cropping experiment in Manaus: M-901. Pp. 252–255 in TropSoils Technical Report, 1988–1989. Raleigh, N.C.: TropSoils Management Entity, North Carolina State University.

Soemarwoto, O., and I. Soemarwoto. 1982. Homegarden: Its nature, origin and future development. Pp. 130–139 in Ecological Basis for Rational Resource Utilization in the Humid Tropics of South East Asia, K. Awang, L. S. See, L. F. See, A. R. M. Derus, and S. A. Abod, eds. Darul Ehsan, Malaysia: Universiti Pertanian.

Soemarwoto, O., and I. Soemarwoto. 1984. The Javanese rural ecosystem. Pp. 261–270 in An Introduction to Human Ecology Research on Agricultural Systems in SouthEast Asia, A. T. Rambo and P. E. Sajise, eds. Los Baños: University of the Philippines.

Soemarwoto, O., I. Soemarwoto, Karyono, E. M. Soekartadiredja, and A. Ramlan. 1985. The Javanese home garden as an integrated ecosystem. Food Nutr. Bull. 7(3):44–47.

Sosa, V., J. S. Fores, V. Rico-Gray, R. Lira, and J. J. Ortiz. 1985. Lista florística y sinonimía Maya. Fascículo 1. Etnoflora Yucatanense. Xalapa, México: Instituto Nacional de Investigaciones sobre Recursos Bióticos.

Soulé, M. E., and K. A. Kohm. 1989. Research Priorities for Conservation Biology. Washington, D.C.: Island Press.

Spears, J. S. 1984. Role of forestation as a sustainable land use strategy option for tropical forest management and conservation and as a source of supply for developing country wood needs. Pp. 29–47 in Strategies and Designs for Afforestation, Reforestation and Tree Planting, K. F. Wiersum, ed. Wageningen, Netherlands: Pudoc.

Spears, J. 1987. A development bank perspective. Pp. 53–66 in Agroforestry: A Decade of Development, H. A. Steppler and P. K. R. Nair, eds. Nairobi, Kenya: International Center for Research in Agroforestry.

Spears, J. 1991. Summaries of multilateral and bilateral forest policy papers. Paper prepared for the World Bank, Washington, D.C.

Stoler, A. L. 1978. Garden use and household economy in rural Java. Bull. Indonesian Econ. Stud. 14(2):85–101.

Struhsacker, T. 1987. Forest resources and conservation in Uganda. Biol. Conserv. (1987):209–234.

Sukmaana, S., P. Amir, and D. M. Mulyadi. 1989. Developments in Procedures for Farming Systems Research. Jakarta, Indonesia: Agency for Agricultural Research and Development.

Sumberg, J. E. 1984. Alley farming in the humid zone linking crop livestock production. ILCA Bull. 18:2–6.

Swanson, T. M. 1992. Economics of a biodiversity convention. Ambio 21:250–257.

Szott, L. T. 1987. Improving the Productivity of Shifting Cultivation in the Amazon Basin of Peru Through the Use of Leguminous Vegetation. Ph.D. dissertation. North Carolina State University, Raleigh.

Szott, L. T., E. C. M. Fernandes, and P. A. Sanchez. 1991. Soil-plant interactions in agroforestry systems. Forest Ecol. Manage. 45:127–152.

Tergas, L. E., O. Paladines, and I. Kleinheisterkamp. 1988. Productividad animal y manejo de Brachiaria humidicola (Rendle) en la altillanura plana de los Llanos Orientales de Colombia. Pp. 499–506 in Simpósio sobre o Cerrado, Vol. 6. Savana, Brasil: Alimentaçao e Energia.

Thurston, H. D. 1991. Sustainable Practices for Plant Disease Management in Traditional Farming Systems. Boulder, Colo.: Westview.

Tiffen, M., and M. Mortimore. 1990. Theory and Practice in Plantation Agriculture: An Economic Review. London: Overseas Development Institute.

Toky, O. P., and P. S. Ramakrishnan. 1983. Secondary succession following slash and burn agriculture in north-eastern India. J. Ecol. 71:735–745.

Toledo, J. M. 1986. Pasturas en trópico húmedo: Perspectiva global. Pp. 19–35 in Simpósio do Trópico Úmido I, Vol. V. Belém, Brasil: Empresa Brasileira de Pesquisa Agropecuária.

Toledo, J. M., and M. Ara. 1977. Manejo de suelos para pastura en la selva Amazónica. Paper presented at the Reunión Taller sobre Ordenación y Conservación de Suelos de America Latina, Food and Agriculture Organization of the United Nations and Swedish International Development Authority, Lima, Peru, May 1977.

Tosi, J. 1980. Life zones, land use, and forest vegetation in the tropical and subtropical regions. Pp. 44–64 in The Role of Tropical Forests in the World Carbon Cycle, S. Brown, A. E. Lugo, and B. Liegel, eds. Washington, D.C.: U.S. Department of Energy.

Tosi, J., and R. F. Voertman. 1964. Some environmental factors in the economic development of the tropics. Econ. Geogr. 40:89–205.

Tucker, R. P. 1990. Five hundred years of tropical forest exploitation. Pp. 39–52 in Lessons of the Rainforest, S. Head and R. Heinzman, eds. San Francisco: Sierra Club Books.

Turner II, B. L., and C. H. Miksicek. 1984. Economic plant species associated with prehistoric agriculture in the Maya lowlands. Econ. Bot. 38(2):179–193.

Uhl, C., and J. B. Kauffman. 1990. Deforestation, fire susceptibility, and potential tree responses to fire in eastern Amazon. Ecology 71(2):437–449.

Uhl, C., D. Nepstad, R. Buschbacher, K. Clark, B. Kauffman, and S. Subler. 1990a. Studies of ecosystem response to natural and anthropogenic disturbances provide guidelines for designing sustainable land-use systems in Amazonia. Pp. 24–42 in Alternatives to Deforestation: Steps Toward Sustainable Use of the Amazon Rain Forest, A. B. Anderson, ed. New York: Columbia University Press.

Uhl, C., J. B. Kauffman, and E. D. Silva. 1990b. Os caminhos do fogo na Amazônia. Ciência Hoje 11(65):24-32.

Uhl, C., R. J. Buschbacher, and E. A. S. Serrão. 1988. Abandoned pastures in eastern Amazonia. I. Patterns of plant succession. J. Ecol. 76:663–681.

U.S. Agency for International Development. 1990. Pesticide Use and Poisoning: A Global View. Washington, D.C.: U.S. Agency for International Development.

Vasquez, R., and A. H. Gentry. 1989. Use and misuse of forest-harvested fruits in the Iquitos area. Conserv. Biol. 3:350–361.

Veiga, J. B., and E. A. S. Serrão. 1990. Sistemas silvopastoris e produção animal nos trópicos úmidos: A experiência da Amazônia brasileira. Pp. 37–68 in Pastagens. Piracicaba, Brasil: Sociedade Brasileira de Zootecnia.

Villachica, H., J. E. Silva, J. R. Peres, and C. M. C. da Rocha. 1990. Sustainable agricultural systems in the humid tropics of South America. Pp. 391–437 in Sustainable Agricultural Systems, C. A. Edwards, R. Lal, P. Madden, M. H. Miller, and G. House, eds. Ankeny, Iowa: Soil and Water Conservation Society.

Vosti, S. A., T. Reardon, and W. von Urff, eds. 1991. Agricultural Sustainability, Growth, and Poverty Alleviation: Issues and Policies. Feldafing, Germany: German Foundation for International Development.

Wadsworth, F. H. 1983. Production of usable wood from tropical forests. Pp. 279–288 in Tropical Rain Forest Ecosystems, F. B. Golley, ed. New York: Elsevier.

Wadsworth, F. H. 1984. Secondary forest management and plantation forestry technologies to improve the use of converted tropical lands. Paper commissioned by the Office of Technology Assessment, U.S. Congress, Washington, D.C.

Wadsworth, F. H. 1987a. A time for secondary forestry in tropical America. Pp. 189–198 in Management of the Forests of Tropical America: Prospects and Technologies, J. Figueroa Colon, F. H. Wadsworth, and S. Brenham, eds. Rio Piedras, P.R.: Institute of Tropical Forestry, U.S. Department of Agriculture.

Wadsworth, F. H. 1987b. Applicability of Asian and African silviculture systems to naturally regenerated forests of the Neotropics. Pp. 93–113 in Natural Management of Tropical Moist Forest, F. Mergen and J. Vincent, eds. New Haven, Conn.: Yale University School of Forestry and Environmental Studies.

Wang, D., F. H. Borman, A. E. Lugo, and R. D. Bowden. 1991. Comparison of nutrient-use efficiency and biomass production in five tropical tree taxa. Forest Ecol. Manage. 46(1–2):1–21.

Whitmore, T. C. 1975. Tropical Rain Forests of the Far East. Oxford: Clarendon.

Wiersum, K. F. 1982. Tree gardening and taungya on Java: Examples of agroforestry techniques in the humid tropics. Agroforest. Syst. 1:53–70.

Wilken, G. C. 1987a. Role of traditional agriculture in preserving biological diversity. Paper prepared for the Office of Technology Assessment, U.S. Congress, Washington, D.C.

Wilken, G. C. 1987b. Good Farmers: Traditional Agricultural Resource Management in Mexico and Central America. Berkeley, Calif.: University of California Press.

Wilkes, G. 1991. In situ conservation of agricultural systems. Pp. 86–101 in Biodiversity: Culture, Conservation and Ecodevelopment, M. L. Oldfield and J. B. Alcorn, eds. Boulder, Colo.: Westview.

Wilson, E. O. 1988. The current state of biological diversity. Pp. 3–18 in Biodiversity, E. O. Wilson, ed. Washington, D.C.: National Academy Press.

Wilson, E. O., and F. M. Peter, eds. 1988. Biodiversity. Washington, D.C.: National Academy Press.

Wisniewski, J., and A. E. Lugo, eds. 1992. Natural Sinks of CO_2. Dordrecht, Netherlands: Kluwer Academic Publishers.

World Bank. 1991. The Forest Sector: A World Bank Policy Paper. Washington, D.C.: World Bank.

World Bank. 1992. World Development Report 1992: Development and the Environment. New York: Oxford University Press.

World Bank and United Nations Development Program. 1983. Zaire Energy Assessment Report. Washington, D.C.: World Bank.

World Health Organization. 1990. Global Estimates for Health Situation Assessment and Projections—1990. Geneva: World Health Organization.

World Resources Institute (WRI). 1990a. World Resources 1990–91. New York: Basic Books.

WRI. 1990b. Toward an Environmental and Natural Resource Management Strategy for Asian and Near East Countries in the 1990s. Washington, D.C.: World Resources Institute.

WRI. 1992. World Resources 1992–93. Washington, D.C.: World Resources Institute.

Wyatt-Smith, J. 1987. Problems and prospects for natural management of tropical moist forest. Pp. 6–22 in Natural Management of Tropical Moist Forest, F. Mergen and J. Vincent, eds. New Haven, Conn.: Yale University School of Forestry and Environmental Studies.

Zapata, T. R., and M. T. K. Arroyo. 1978. Plant reproductive ecology of a secondary deciduous forest in Venezuela. Biotropica 10:221–230.

Zobel, B. 1979. The ecological impact of industrial forest management. Research Report No. 52. Cantón de Colombia, S.A., Investigación Forestal.

APPENDIX

Emissions of Greenhouse Gases from Tropical Deforestation and Subsequent Uses of the Land

*Virginia H. Dale, Richard A. Houghton, Alan Grainger,
Ariel E. Lugo, and Sandra Brown*

Wide-scale land use change is resulting in numerous environmental consequences: degradation of soils, loss of extractive resources, loss of biodiversity, and regional and global climate change, among others. Common land use changes are forest degradation and the conversion of forests to agricultural systems and pastures. Because many agricultural systems in the humid tropics are not sustainably managed, each year large areas of forest are cleared to provide new fertile lands. Sustainable agriculture offers one means of offsetting the global consequences of large-scale land use change.

This paper discusses the emissions of greenhouse gases associated with land use change and the potential impact that sustainable agriculture may have on these emissions. Land uses involving intensive deforestation and intensive agricultural practices increase greenhouse gas emissions; in the case of deforestation, by eliminating a

Virginia H. Dale is a research scientist in the Environmental Sciences Division at Oak Ridge National Laboratory, Oak Ridge, Tennessee; Richard A. Houghton is a senior scientist at the Woods Hole Research Center, Woods Hole, Massachusetts; Alan Grainger is a bioecographer, resource economist, modeler, and environmental policy analyst and is currently a lecturer in geography at the University of Leeds, Leeds, United Kingdom; Ariel E. Lugo is director and project leader of the Institute of Tropical Forestry, U.S. Forest Service, U.S. Department of Agriculture, Puerto Rico; Sandra Brown is associate professor of Forest Ecology in the Department of Forestry, University of Illinois, Urbana–Champaign, Illinois.

source of oxygen production, carbon dioxide (CO_2) conversion, and carbon (CO) sequestration; in the case of agriculture, by increasing sources of methane (CH_4) through rice and livestock production. The emphasis here is on CO_2, the major contributor to the greenhouse effect, and on tropical deforestation, the major land use change that accounts for the current increase in atmospheric CO_2 concentrations. The net flux of carbon from land use changes is calculated by adding the stocks of carbon per unit area for the major land uses of the world to the rates of change in land use. Therefore, this paper reviews estimated carbon content and the rates of change in the carbon content of the major land uses in the humid tropics. That discussion forms a basis for estimating the flux of greenhouse gases from land use changes. Because projections of future impacts are based on particular models, this paper presents and compares the major model structures. Lastly, it discusses how the sustainable uses of land can reduce future emissions of greenhouse gases. The last section also presents a set of priorities for future research.

EFFECTS OF LAND USE CHANGE ON GLOBAL CLIMATE

Changes in the earth's climate are predicted to cause a 0.3°C warming per decade (range, 0.2° to 0.5°C per decade), which may instigate a 6-cm rise in sea level per decade (range, 3 to 10 cm per decade) in the next century (Houghton et al., 1990). These changes are anticipated as a result of the buildup of radiatively important gases in the atmosphere. Aside from water vapor, the major biogenic gases that contribute to the greenhouse effect (greenhouse gases)—CO_2, CH_4, nitrous oxide (N_2O), chlorofluorocarbons (CFCs), and ozone—result either entirely or in part from human activities. Except for CFCs, these gases are also part of the natural cycles between ocean, land, and atmosphere. The increasing concentrations of these gases in the atmosphere, however, and the enhanced greenhouse effect that may result are due to increased emissions of these gases as a result of human activities, predominantly fossil fuel combustion and the expansion of agricultural lands (for CO_2 concentrations, see Figure A-1) (Post et al., 1990). Currently, the burning of fossil fuels is the major contributor, but historically, land use changes have had a larger impact on atmospheric greenhouse gas concentrations (Houghton et al., 1983). Agriculture and the clearance of forests for agricultural use have accounted for about 50 percent of the total emissions of carbon over the past century (Figure A-2). In the past CO_2 has accounted for more than half of all gases that contribute to the greenhouse effect and is expected to account for 55 percent over the next century (Houghton et al., 1990).

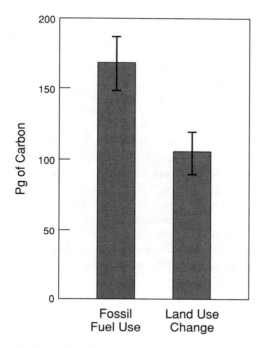

FIGURE A-1 Carbon dioxide (CO_2) released from burning of fossil fuels and expansion of agricultural lands from 1850 to 1980. The lines within the bars indicate standard deviations; Pg, petagram. Source: Dale, V. H., R. A. Houghton, and C. A. S. Hall. 1991. Estimating the effects of land-use change on global atmospheric CO_2 concentration. Can. J. Forest Res. 21:87–90. Reprinted with permission.

The annual net flux of carbon to the atmosphere from land use change is estimated to have been 0.4 to 2.6 Pg of carbon per year in 1980 (1 Pg = 10^{15} g) (Detwiler and Hall, 1988a; Houghton et al., 1987). The annual net flux of carbon as a result of fossil fuel emissions was between 5.0 and 5.5 Pg from 1980 to 1988 (Marland and Boden, 1989). Therefore, the recent contribution of CO_2 to the atmosphere from land use change in terrestrial ecosystems is between 10 and 50 percent of the flux resulting from fossil fuel emissions. If 10 percent is correct, then land use change is not a major cause of the increases in atmospheric CO_2 concentrations. Researchers must accurately identify whether the larger values are correct or whether the rate of land use change is increasing. It is also important to continue research to estimate the carbon flux resulting from the human impact on terrestrial ecosystems. The role of "undisturbed" forests also requires scientific attention because, as these forests regenerate following natural or undetected human disturbances, carbon sequestration could offset some of the emissions resulting from human activities. Note that forests classified as undisturbed have frequently been subject to some human manipulations.

This paper discusses the effects of land use change on greenhouse gas emissions and the potential impact that sustainable agriculture may have on the interaction. The emphasis is on CO_2, the major contributor to the greenhouse effect (Figure A-3), and on tropical deforestation, the major land use change involved in the current increase in atmospheric CO_2 concentrations (Dale et al., 1991). A major finding from this review is that most of the current flux of greenhouse gases to the atmosphere from the tropics is due to the conversion of forests to agricultural uses and that sustainable agricultural practices could be a significant means of controlling the expansion of deforestation. Sustainability—which exists when land can be used for a long period of time without significant declines in the

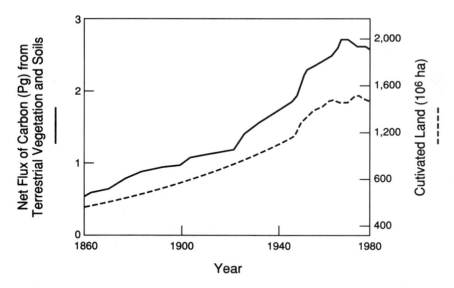

FIGURE A-2 Change in the area of cultivated land and net flux of carbon (Pg, petagram) from terrestrial sources from 1860 to 1980. Source: Houghton, R. A., J. E. Hobbie, J. M. Melillo, B. Moore, B. J. Peterson, G. R. Shaver, and G. M. Woodwell. 1983. Changes in the carbon content of terrestrial biota and soils between 1860 and 1980: A net release of CO_2 to the atmosphere. Ecol. Monogr. 53:235–262.

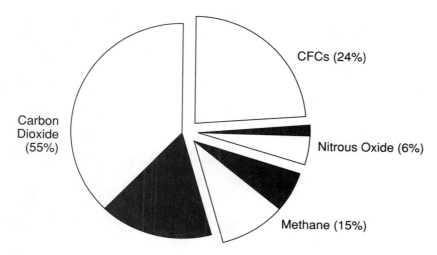

FIGURE A-3 Contributions of different gases to the greenhouse effect calculated for the 1980s. Screened segments indicate the relative contributions of deforestation and land use to the total emissions. White segments represent industrial and natural contributions. For the chlorofluorocarbons (CFCs), all the emissions are industrial. Source: Houghton, J. T., G. J. Jenkins, and J. J. Ephraums, eds. 1990. Climatic Change: The IPCC Scientific Assessment. Cambridge: Cambridge University Press.

ecologic attributes of the land—may require inputs of fertilizer, irrigation, use of machines, periods of fallow, adjacent land preserves, or other manipulations. The situation should be seen as sustainable from the viewpoints of both the landholder, who is able to make a living from the land, and the land itself, which maintains soil conditions adequate for growing agricultural or forest crops (Costanza, 1991). Therefore, it is important to evaluate the costs and benefits of particular forms of sustainable agriculture (including greenhouse gas emissions resulting from land use practices).

MAJOR LAND USE CHANGES RESPONSIBLE FOR THE FLUX OF GREENHOUSE GASES

Forests contain about 90 percent of all the carbon stored in terrestrial vegetation and are being cleared at a very rapid rate. (Table A-1 indicates the variability in estimates of deforestation, and Dale [1990] discusses the methods used to obtain the estimates.) With this clearing, the carbon previously stored in the trees and soils is being re-

TABLE A-1   Rates of Deforestation of Closed Tropical Forests by
Source of Information (in Thousands of Hectares per Year)

Region	Myers, 1980[a] (1979)	FAO and UNEP, 1981[b] (1976–1980)	Grainger, 1984[a,c] (1976–1980)	WRI, 1990[b,c] (1980s)	Myers, 1989[a,d] (1989)	FAO, 1991[e] (1981–1990)
Tropical America	3,710	4,119	3,301	10,859	7,680	7,290
Tropical Africa	1,310	1,319	1,204	1,338	1,580	4,788
Tropical Asia	2,320	1,815	1,608	2,390	4,600	4,707
Total	7,340	7,235	6,113	14,587	13,860	16,785

NOTE: FAO and UNEP, Food and Agriculture Organization of the United Nations
and United Nations Environment Program; WRI, World Resources Institute. Numbers
in parentheses are years to which deforestation data apply.

[a]Refers only to closed forests in the humid tropics.

[b]Refers to all tropical closed forests.

[c]Uses data from the Food and Agriculture Organization and United Nations Environment Program (1981) only for forests in the humid tropics.

[d]Refers to 34 countries that contain 97 percent of the world's total area of tropical humid forests.

[e]Estimates for 62 of the 76 countries in the tropics; they include almost all of the humid forests along with some dry areas (Food and Agriculture Organization, 1991). The fact that some open forests are included makes a comparison with closed forests somewhat misleading.

leased into the atmosphere. The net rate and the completeness of
carbon release depend on the fraction of the forest burned, the decomposition rate of downed wood, and the fate of forest products.
For example, when wood is burned, it quickly releases carbon into
the atmosphere, whereas wood structures retain their carbon for a
longer period of time, and some charcoal is essentially a form of
permanent carbon storage. Crops and pastures may hold 2 to 5 percent of the carbon in vegetation per unit area, compared with that
held by forest vegetation.

About half of the mass of vegetation is carbon. Estimates of
biomass come from direct measurements (Ajtay et al., 1979; Brown
and Lugo, 1982; Olson et al., 1983) or are derived from wood volumes reported in large-scale forest inventories (Brown and Lugo, 1984;
Brown et al., 1989) (Table A-2). On average, the soils of the world
contain about three times more organic carbon than is contained in
vegetation.

The net flux of carbon to the atmosphere from land use changes depends not only on stocks of carbon in forests and rates of land use change but also on uses of agricultural lands (Table A-3). Land use changes can be triggered by natural events (such as fire, hurricanes, or landslides) or by people. Because the land use changes instigated by people have the greatest effect on the net carbon flux, only those changes are discussed here. The land use changes considered below include permanent agriculture and pasture, degradation of croplands and pastures, shifting cultivation, forest plantations and tree crops, logging, and degraded forests (Table A-3).

Many surveys have addressed the causes of tropical deforestation. Myers (1980, 1984) emphasized the roles of cattle ranchers, loggers, and farmers. Grainger (1986) distinguished between the land

TABLE A-2 Carbon Stocks in Vegetation and Soils of Different Types of Ecosystems Within the Tropics (Megagrams per Hectare)

| Source and Region | Closed Forests[a] | | | | |
	Forests in Humid Tropics	Seasonal Forests	Closed Forests[b]	Open Forests or Woodlands[a]	Crops
Vegetation					
Tropical America	176, 82	158, 85	89, 73	27, 27	5
Tropical Africa	210, 124	160, 62	136, 111	90, 15	5
Tropical Asia	250, 135	150, 90	112, 60	60, 40	5
Soils[c]	100	90	NA	50	NA

NOTE: NA, not available.

[a]The first value of each pair of data is based on destructive sampling of biomass (Ajtay et al., 1979; Brown and Lugo, 1982; Olson et al., 1983); the second value is calculated from estimates of wood volumes (Brown and Lugo, 1984; Houghton et al., 1985). It is not evident which estimate is more accurate.

[b]These estimates are also based on wood volumes reported by the Food and Agriculture Organization and United Nations Environment Program (1981) and use the revised conversion factors given by Brown et al. (1989). The first value of each pair of data is for undisturbed forests; the second value is for logged forests.

[c]The values are averaged from estimates by Brown and Lugo (1982), Post et al. (1982), Schlesinger (1984), and Zinke et al. (1986).

SOURCES: Houghton, R. A. 1991a. Releases of carbon to the atmosphere from degradation of forests in tropical Asia. Can. J. Forest Res. 21:132–142; Houghton, R. A. 1991b. Tropical deforestation and atmospheric carbon dioxide. Climatic Change 19:99–118.

APPENDIX

222

TABLE A-3 Initial Carbon Stocks Lost to the Atmosphere When Tropical Forests Are Converted to Different Kinds of Land Use and the Tropical Land Use Areas, 1985

Land Use	Percentage of Carbon Lost from:		Tropical Land Use Area, 1985 (millions of ha)
	Vegetation	Soil	
Permanent agriculture	90–100	25	602[a]
Pasture	90–100	12	1,226[a,b]
Degraded croplands and pastures	60–90[c]	12–25[c]	?
Shifting cultivation	60	10	435[d]
Degraded forests	25–50[e]	?	?
Plantations	30–50[f]	?	17[d]
Logging	25[g] (range 10–50)	?	169[d]
Forest reserves	0	0	?

NOTE: For soils, the stocks are to a depth of 1 m. The loss of carbon may occur within 1 year with burning or over 100 years or more with some wood products. The question marks denote unknown information.

[a]Food and Agriculture Organization. 1987. Yearbook of Forest Products. Rome, Italy: Food and Agriculture Organization of the United Nations.

[b]Area includes pastures on natural grasslands as well as those cleared from forest.

[c]Degraded croplands and pastures may accumulate carbon, but their stocks remain lower than the initial forests.

[d]Food and Agriculture Organization and United Nations Environment Program. 1981. Tropical Forest Resources Assessment Project. Rome, Italy: Food and Agriculture Organization of the United Nations.

[e]Houghton, R. A. 1991a. Releases of carbon to the atmosphere from degradation of forests in tropical Asia. Can. J. Forest Res. 21:132–142.

[f]Plantations may hold as much or more carbon than natural forests, but a managed plantation averages one-third to one-half as much carbon as an undisturbed forest because it is generally regrowing from harvest (Cooper, C. F. 1982. Carbon storage in managed forests. Can. J. Forest Res. 13:155–166).

[g]Based on current estimates of aboveground biomass in undisturbed and logged tropical forests (Brown, S., A. J. R. Gillespie, and A. E. Lugo. 1989. Biomass estimation methods for tropical forests with applications to forest inventory data. Forest Sci. 35:881–902). When logged forests are colonized by settlers, the losses are equivalent to those associated with one of the agricultural uses of the land.

SOURCE: Unless indicated otherwise, data are from Houghton, R. A., R. D. Boone, J. R. Fruci, J. E. Hobbie, J. M. Melillo, C. A. Palm, B. J. Peterson, G. R. Shaver, G. M. Woodwell, B. Moore, D. L. Skole, and N. Myers. 1987. The flux of carbon from terrestrial ecosystems to the atmosphere in 1980 due to changes in land use: Geographic distribution of the global flux. Tellus 39B:122–139.

uses that replace forests and the underlying causes of deforestation: socioeconomic factors, environmental factors, and government policy. Fearnside (1987) divided the causes of deforestation in Brazil into proximate and ultimate causes. Repetto (1989) stressed the economic incentives set by government policies. One approach is no more correct than another, although Repetto's approach may be the most useful for determining how to change current incentives. From the perspective of sustainable agriculture, however, there is yet another approach to assigning cause to deforestation—most deforestation in the tropics has been, and still is, due to the development of new agricultural land. The expansion of agricultural land, and thus deforestation, could be reduced by adopting methods of sustainable agriculture.

Permanent Agriculture

When forests and woodlands are cleared for cultivated land, an average of 90 to 100 percent of the aboveground biomass is burned and immediately released to the atmosphere as CO_2. Up to an additional 25 percent of carbon in the 1 m of surface soils is also lost to the atmosphere (Table A-3). Most of the loss occurs rapidly within the first 5 years of clearing; the rest is released over the next 20 years.

The wood harvested for products subsequently oxidizes, but it does so much more slowly than does the wood felled for cultivated land. The material remaining above and below the ground decays, as does the organic matter of newly cultivated soil. The rates of decay vary with climate, but in the humid tropics, most material decomposes within 10 years (John, 1973; Lang and Knight, 1979; Swift et al., 1979). However, recent work has indicated that many tropical woods take up to several decades to decompose (S. Brown and A. E. Lugo, personal observations). A small fraction of burned organic matter is converted to charcoal, which resists decay (Comery, 1981; Fearnside, 1986; Seiler and Crutzen, 1980). When croplands are abandoned, the lands may return to forests at rates determined by the intensity of disturbance and climatic factors (Brown and Lugo, 1982, 1990b; Uhl et al., 1988).

Cultivation of staple food crops in fields is common in the humid tropics—as it is elsewhere in the world—and is sustainable on good soils. Rice, maize, and cassava are the principal crops. Rice is usually cultivated in flooded fields or paddies, and the productivity and sustainability of wet rice cultivation is enhanced by reducing soil acidity under anaerobic conditions. This improves nutrient availability and the fertilization capabilities of the algae, decayed stubble, and animal dung that exist in soil. Soil erosion is reduced because the soil surface is covered by water and because of the constraints on soil movement imposed by the mounds of earth surrounding the paddies.

Wet rice cultivation is one of the most sustainable land uses in the humid tropics; however, it is not universally applicable. Level, easily flooded sites are required (for example, river floodplains), although slight slopes can be accommodated by terracing. If the water above the sediment is aerobic, it can act as a sink for CH_4, another greenhouse gas. However, CH_4 is produced in copious amounts under the anaerobic conditions of the flooded fields. So, in addition to the CO_2 given off when the forest is cleared initially, there is a continuing emission of CH_4. This presents a major and not easily resolvable problem. Although control of deforestation by promoting the spread of wet rice cultivation makes sense because of its high productivity and sustainability, this might be harmful from a climate change perspective.

Pastures

The changing of forests to pastures results in a 90 to 100 percent loss of carbon from the vegetation, which is similar to that for cultivated lands (Table A-3). Because pastures generally are not cultivated, the loss of carbon from pasture soils is less than the loss from cropland soils (about 12 percent compared with 25 percent). Most studies show a loss of soil carbon (Fearnside, 1980, 1986; Hecht, 1982a), sometimes as much as 40 percent of the carbon originally contained in the forest soil (Falesi, 1976; Hecht, 1982b). However, under some conditions there appears to be no loss of soil carbon (Buschbacher, 1984; Cerri et al., 1988), and there may even be an increase (Brown and Lugo, 1990b; Lugo et al., 1986).

Theoretically, cattle ranching on planted pastures is an attractive option because it should maintain a continuous grassy cover on the soil surface and does not involve cultivation, thereby reducing soil degradation. The hydrologic and soil conservation properties of pastures observed on experimental sites are generally favorable. In practice, however, both productivity and sustainability can be low in some tropical areas, causing frequent abandonment of land. The results are a continuing need to clear more forests to provide fresh pastures; overgrazing, which causes widespread, degraded vegetative cover and changes in composition; and soil compaction from constant trampling by animals, which exposes the soil to other forms of degradation. However, in well-managed pasturelands, this pattern of events does not occur. For example, large areas of productive pasturelands that have been in use for several decades or more exist in Venezuela, Costa Rica, and Puerto Rico. The organic carbon content of the soil of well-managed pasturelands is as high or higher than that of the forests from which the pasturelands were originally derived (S. Brown, personal observation).

From a climate change perspective, there are disadvantages to pastures. First, the amount of biomass per unit area is low. Second, frequent burning of pastures to maintain productivity leads to emissions of greenhouse gases in addition to the emissions following the initial clearing. Third, cattle emit CH_4 from their guts. In this case, continuing greenhouse gas emissions are not compensated for by high sustainability, as is the case with wet rice cultivation.

Degradation of Croplands and Pastures

In many areas of the humid tropics, the abandonment of croplands is not followed by forest regeneration. Degraded croplands and pastures may accumulate carbon, but 60 to 90 percent of the carbon in the original forest and 12 to 25 percent of the soil carbon has been lost to the atmosphere (Houghton et al., 1987). Much of the land is abandoned in the first place because it has lost its fertility or has been eroded. These abandoned, degraded lands do not immediately return to forests, yet their degradation requires that new lands be cleared to keep the areas of productive croplands and pastures constant. The new lands are most frequently obtained by clearing forests.

Degraded lands are characterized by having been deforested and exposed to factors that reduced the land's productive potential (Lugo, 1988). According to Grainger (1988), the area of degraded lands in the tropics exceeds the area of unspoiled forestlands. The degraded lands have already lost a fraction of the carbon they stored initially and have the potential to serve as carbon sinks, should they be managed properly or rehabilitated by artificial or natural means.

Shifting Cultivation

The practice of traditional shifting cultivation, in which short periods of cropping alternate with long periods of fallow, during which time forests regrow, is common throughout the tropics. This form of shifting cultivation is sustainable when low population densities exist over large areas and the forests recover during the fallow phase. Shifting cultivation results in about a 60 percent loss of the original carbon in the vegetation and a 10 percent loss of the carbon in the soils when the forest is cut and burned (Houghton et al., 1987). Large amounts of soil organic carbon are lost in association with permanent agricultural systems but not in association with short-term shifting agricultural systems (Ewel et al., 1981). Deforestation for shifting cultivation releases less net carbon to the atmosphere than does deforestation for permanently cleared land because of the partial recovery of the forests (Table A-3). The length of the cycle varies considerably among regions because of both ecologic and cultural differences (Turner et al., 1977). Decay rates for the plant material left dead at the time of deforestation and accumulation rates for regrowing vegetation during the fallow periods vary by ecosystem (Brown and Lugo, 1982, 1990a; Saldarriaga et al., 1988; Uhl, 1987; Uhl et al., 1982). Less soil organic matter is oxidized during the shifting cultivation cycle than during continuous cultivation (Detwiler, 1986; Schlesinger, 1986). Under shifting cultivation, deforestation is temporary and recurrent. During the fallow stage, these areas are carbon sinks. Soils can recover their soil organic carbon at rates as high as 2 Mg/ha (1 Mg = 10^6 g) per year following abandonment of agriculture to forest succession (Brown and Lugo, 1990b) (Table A-4). However, much of the shifting cultivation today is nontraditional, and fallow periods are often shortened to the point where the land becomes so badly degraded that it is virtually useless for any agricultural activity (Grainger, 1988).

Three main types of shifting cultivation can be identified: traditional long-rotation, short-rotation, and encroaching cultivation (Grainger, 1986, In press).

TRADITIONAL LONG-ROTATION SHIFTING CULTIVATION

Traditional shifting cultivation, which is practiced on long rotations of at least 15 to 20 years and often longer, is one of the few proven sustainable land uses in the humid tropics. Cropping for 1 to 5 years is followed by a 10- to 20-year fallow period, during which time the fertility of the land (that is, the nutrient content of both soil and vegetation) regenerates and weed growth is eliminated. Although it is sustainable, this practice has low productivity and can support only a low population density. It is now restricted to fairly remote areas where competition for land is low.

SHORT-ROTATION SHIFTING CULTIVATION

Most shifting cultivation is now carried out on short rotations of less than 15 years. Rotations of 6 years are common in Asia, and

TABLE A-4 Processes that Create Carbon Sinks and Their
Potential Magnitude in the Tropical Closed-Forest Landscape

Process	Magnitude (grams of carbon/m²/year)
Biomass accumulation in forests >60–80 years old and logged forests	100–200
Biomass accumulation in secondary forest fallows 0–20 years old[a]	200–350
Biomass accumulation in plantations[b]	140–480
Accumulation of coarse woody debris[c]	
Forests >60–80 years old	20–40
Forests 0–20 years old	17–30[c]
Accumulation of soil organic carbon	
Background rates	2.3–2.5
Forest succession	50–200
Conversion of cultivation to pastureland or grassland	30–42

[a]Converted to carbon units by multiplying organic matter by 0.5.
[b]Weighted average rates across all species and age classes.
[c]Two studies described by Brown and Lugo (1990b) report an average amount
of coarse woody debris at an age of about 20 years of 8.5 percent of the
aboveground biomass; this percentage of the biomass accumulation rate was
assumed to go into coarse woody debris during the 20-year period.

SOURCE: Lugo, A., and S. Brown. In press. Tropical forests as sinks of
atmospheric carbon. Forest Ecol. Manage.

even shorter rotations are found in Africa. Rotation length is reduced in response to the need for a more settled life-style than that
led by traditional itinerant shifting cultivators (when farmers stay in
one place they use a smaller area of land and rotate crops more frequently). The amount of available land is reduced as the population
density increases or other land uses encroach onto territories where
shifting cultivation was formerly practiced. The shorter the rotation
length, the less time fertility has to regenerate and the greater the
scope for a long-term decline in soil fertility and, hence, a decline in
yields per hectare. When clearing and burning are done more frequently, there is a greater probability that the land will become infested by weeds. Weeds are just as important a cause of land abandonment as declining yields.

A number of points arise. First, because local conditions and
management practices have a crucial influence on rotation length, it
is difficult to identify a general threshold rotation at which shifting

cultivation becomes unsustainable (Young, 1989). Second, it has been argued that increases in cropping intensity in response to rising populations are usually accompanied by measures to improve productivity and sustainability (Boserup, 1965). However, some agricultural economists disagree and point out that, in practice, increasing intensity often leads to a decline in yields, increased soil degradation, and lower sustainability (Blaikie and Brookfield, 1987). Third, although the sustainability of shifting cultivation is determined by how well it sustains the yield per hectare over succeeding rotations, it can also be evaluated from a carbon budget perspective with respect to how much carbon is stored, on average, in the fallow vegetation and how much soil carbon is restored after cropping. Short rotations do not allow forests to regenerate, as is the case in traditional agricultural practices. The usual result is a low bushy vegetation technically referred to as secondary forest but commonly called forest fallow and which has a low carbon content per hectare. If agricultural sustainability declines, then the carbon stock could also fall to a low level (for example, if some robust weedy species takes hold and prevents the regeneration of woody cover).

Because short-rotation shifting cultivation is such a widespread practice, its elimination is not feasible. Instead, a major effort is required to improve its productivity and sustainability. This may involve the judicious use of fertilizers (Sanchez et al., 1983) or the development and promotion of low-input cropping practices that improve the soil (Sanchez and Benites, 1987). The latter would include the planting of trees during the fallow period as an alternative to sole reliance on natural regeneration (Juo and Lal, 1977).

ENCROACHING CULTIVATION

Encroaching cultivation, a widespread practice, is typically carried out by landless migrants. Farmers spread out in waves from roads into the forest, clearing forest and cropping land until yields are too low and weed infestation is too great to continue. They then move to an adjacent patch of forest and repeat the process. Instead of working with the nutrient cycling mechanisms of the natural ecosystem so that they can return at a later date to crop the land again, encroaching cultivators usually exhaust the fertility of the land and leave behind a scrubby wasteland. This is of little use for agriculture and renders the land incapable of supporting regenerating vegetation, which could increase the carbon stock and improve soil conditions. Thus, productivity and sustainability are both poor, and from the points of view of both deforestation and carbon budget analysis, the impact of encroaching cultivation is more akin to permanent cultivation than shifting cultivation, but with none of the former's potential advantages.

Tree Plantations

Tropical forests may also be replaced by two types of tree plantations: forest plantations and tree crop plantations, from which the output consists of food, oils, and other nontimber products. Monocultures are often, but not always, grown in both types of plantations. The forest plantation area in the tropics was only about 1 percent of the total closed-forest area in 1980 (Lanly, 1982). Forest plantations are typically established to restore cover to areas where forests are not as abundant as they once were and where both timber and fuelwood are in short supply. Plantations can contain as much carbon as the original vegetation, but they typically contain 30 to 50 percent of the carbon in the original vegetation because of short rotations (Lugo et al., 1988). The net primary productivity of plantations can be high, with values about 3 and 10 times those of secondary and mature forests, respectively (Brown et al., 1986; Lugo et al., 1988). Soil organic matter also builds up on tree plantations (Brown and Lugo, 1990a; Cuevas et al., 1991). Because of a plantation's high rate of biomass accumulation and the predominance of younger plantations, the positive impact of tree plantations on the carbon cycle in the tropics is greater than might be evident (Brown et al., 1986). Moreover, many of these plantations are established for environmental protection purposes or to rehabilitate degraded lands (about 17 percent of the total area [Evans, 1982]) and are thus likely to continue to accumulate carbon for long time periods.

Numerous tree crops are grown on plantations in the humid tropics, including oil palm, rubber, cacao, coconut, bananas, and coffee. Some plantations are very large, covering thousands of hectares; others are quite small. In all cases, however, the replacement of forest by an alternative tree cover does result in some of the factors that lead to sustainability, including maintenance of a relatively closed canopy of vegetation that covers the land and minimal disturbance of the soil. The amount of biomass per unit area is also high, but it is not equivalent to that in mature forests. Productivity is good on the best soils, and the high capital intensity of operations gives a commercial incentive to plantation operators to be careful when choosing sites. However, weed removal to increase productivity also exposes the soil to erosion, thereby diminishing sustainability. One way to overcome this problem is to intercrop the tree crops with another perennial crop or pastures—an application of the silvopastoral agroforestry system.

Logging

Logging of forests in the tropics is generally selective in that only the largest commercial trees are removed (Lanly, 1982), but there is often damage to the residual trees (Ewel and Conde, 1978). As of 1980, almost 15 percent of the closed forests had been logged. This area was increasing annually by an additional 4.4 million ha. Logging removes 10 to 50 percent of the carbon in vegetation (Houghton et al., 1987). Although logging removes living biomass, both directly for products and through transfers to dead biomass (necromass), during recovery vigorous regrowth can occur in the residual stand.

The farmers responsible for most of the deforestation in the tropics tend to prefer stands that have already been modified (usually logged) (Brown and Lugo, 1990b; Lanly, 1982). These stands are easier to cut and clear or are accessible because of road construction (Grainger, 1986). More than half of the area deforested in 1980 originated from selectively logged forests (Lanly, 1982); thus, their biomass had already been reduced.

The rate of aboveground carbon accumulation (as biomass) in tropical forests ranges widely between negative values (when stands are degrading) to more than 15 Mg/ha/year in fast-growing plantations (Lugo et al., 1988). During logging, CO_2 is released into the atmosphere from the mortality and decay of trees damaged in the harvest operations, the decay of logging debris, and the oxidation of the wood products. Logging may also cause a net withdrawal of carbon from the atmosphere if logged forests are allowed to regrow and the extracted wood is put into long-term storage, such as buildings or furniture. Long-term observations of the carbon dynamics of forest plots, either undisturbed or subjected to slight disturbances in their recent past, do not support the notion that they have steady-state levels of carbon (Brown et al., 1983; Weaver and Murphy, 1990). In all cases, tree growth plus ingrowth (trees with the minimum diameter to be included in the survey) accumulated more aboveground carbon than was lost by tree mortality. Ingrowth into tree stands tends not to be a significant carbon sink unless the stand is recovering from an acute disturbance such as intensive logging (Brown et al., In press) or a hurricane. If the land is not used following harvest, the regenerating forest probably accumulates more carbon than it releases, and in the long run the net flux of carbon may be close to zero (Harmon et al., 1990).

Rates of harvest are reported annually in the *Yearbook of Forest Products* (Food and Agriculture Organization, 1946–1987). Average extraction rates in different regions range between 8.4 and 56.9 m^3/ha of total growing stocks of 100 to 250 cm^3/ha (Lanly, 1982). About one-third of the original biomass is damaged or killed in the harvesting process (Kartawinata et al., 1981; Nicholson, 1958; Ranjitsinh, 1979). The dead material decays exponentially. The undamaged, live vegetation accumulates carbon again at rates that vary with the type of forest and the intensity of logging (Brown and Lugo, 1982; Brown et al., In press; Horne and Gwalter, 1982; Uhl and Vieira, 1989). The live vegetation then eventually dies and decomposes, returning CO$_2$ to the atmosphere. The harvested products decay at rates that depend on their end use (Food and Agriculture Organization, 1946–1987); for example, fuelwood typically decays in 1 year, paper in 10 years, and construction materials in 100 years (Houghton et al., 1987).

Degraded Forests

In addition to controlled selective logging and the extraction of other resources from forests, illicit extraction of timber products occurs in vast areas (Brown et al., 1991). This "log poaching" reduces the forests biomass and, in the process, releases 25 to 50 percent of the carbon in vegetation to the atmosphere (Houghton et al., 1987). This release of carbon has often been overlooked in estimates of carbon flux. The lowering of biomass through the illicit extraction of wood, forage, or other resources may account for some of the differences in the estimates of biomass discussed above. If the higher estimates based on direct measurement of biomass were selective of stands that showed no sign of disturbance, and if the lower estimates of biomass came from a sampling of more representative stands, the difference in estimates may be of human origin. Stands recovering from previous disturbances (young secondary forests, more than 20 years old) accumulate aboveground carbon at rates from 2.2 to 3.8 Mg/ha/year (Brown and Lugo, 1990b) (Table A-4). Depending on whether the degradation occurred long ago or recently (Brown et al., 1991; Flint and Richards, 1991), an accounting of the carbon that has been released as a result of degradation may increase estimates of carbon flux by 50 percent or more (Houghton, 1991a).

ESTIMATED FLUX OF GREENHOUSE GASES FROM LAND USE CHANGES

The estimated carbon content and rates of change of the major land uses in the tropics reviewed above can be used to estimate the flux of greenhouse gases from those land use changes. The discussion on fluxes is broken down by gas: CO_2, CH_4, N_2O, and carbon monoxide (CO).

Carbon

The specific amount of CO_2 released as a result of tropical deforestation is difficult to quantify (Table A-5). The most current estimated release of carbon from land use change in the tropics is 1.1 to 3.6 Pg for 1989. In 1980, 22 of 76 tropical countries contributed 1 percent or more to the total flux; five countries (Brazil, Indonesia, Colombia, Côte d'Ivoire, and Thailand) contributed half of the total net release (Houghton et al., 1985). In 1989, four countries (Brazil, Indonesia, Myanmar, and Mexico) accounted for more than 50 percent of the release (Houghton, 1991b).

The expansion of agricultural lands and pasturelands accounts for most of the carbon loss due to tropical deforestation (Table A-3). Losses due to forest degradation are hard to quantify because of the difficulty of identifying areas of degraded forests on a broad scale. The roles of biomass burning and carbon sinks should also be considered.

BIOMASS BURNING

Biomass burning is estimated to release 3.0 to 6.2 Pg of carbon annually (Crutzen and Andreae, 1990). This release is a gross emis-

TABLE A-5 Estimated Release of Carbon Dioxide as a Result of Tropical Deforestation

Year	Petagram (Pg) of Carbon Released as Carbon Dioxide	Reference
1980	0.9–2.5	Houghton et al., 1985
1980	0.6–1.1[a]	Molofsky et al., 1984
1980	0.4–1.6[b]	Detwiler and Hall, 1988b
1980	0.5–0.7	Grainger, 1990d
1980	0.9–2.5[c]	Hao et al., 1990
1989	1.1–3.6	Houghton, 1991b

[a]This value does not include deforestation of fallow areas, which was estimated to release 0.4 to 0.8 Pg of carbon to the atmosphere (Houghton et al., 1985).

[b]This value does not include permanent loss of fallow areas.

[c]The study did not consider long-term releases associated with decay or long-term accumulations associated with growth.

sion; however, most of the carbon released in a year is accumulated in the growth of recovering vegetation. The burning of grasslands, agricultural lands, and savannahs, however, has increased over the last century, because rarely burned ecosystems, such as forests, have been converted to frequently burned ecosystems, such as agricultural lands or grasslands or shrub lands. For example, the area of grasslands, pastures, and croplands increased by about 50 percent between 1850 and 1985 in tropical America (Houghton et al., 1991) and between 1880 and 1980 in tropical Asia (Flint and Richards, 1991; E. P. Flint and J. F. Richards, Duke University, personal communication, 1991). The area of natural grasslands actually decreased but was more than offset by increases in the combined areas of pastures and croplands. The relative increases observed in tropical America and Asia are probably greater than increases in Africa, where large areas of savannahs already existed before the last century. Burning of almost half of the world's biomass is estimated to occur in the savannahs of Africa (Hao et al., 1990). Worldwide carbon emissions from the burning of savannahs and agricultural lands have probably increased by 20 to 25 percent over the last 150 years.

The formation of charcoal as a result of burning sequesters carbon. Because carbon in charcoal is oxidized slowly, if at all, charcoal formation removes carbon from the short-term carbon cycle, resulting in long-term sequestration (Seiler and Crutzen, 1980). Each year, between 0.3 and 0.7 Pg of carbon is estimated to be converted to charcoal through fires (Crutzen and Andreae, 1990). Only about 0.1 Pg of carbon is estimated to be formed in charcoal as a result of fires associated with shifting cultivation and deforestation; however, the production, fate, and half-life of carbon in charcoal are poorly known, so the size of this carbon sink is uncertain.

TROPICAL SYSTEMS AS CARBON SINKS

The potential for vegetation to be a carbon sink depends on the balance of all natural processes of the carbon cycle and the influence of human and natural disturbances. Potential long-term carbon sinks include large trees, necromass, changes in wood density, soil organic carbon (SOC), and carbon export. A significant fraction of the net accumulation of aboveground carbon in tropical forest stands appears to occur in the continuous growth of older trees that get progressively larger with age (Brown and Lugo, 1992; Brown et al., In press). Necromass is a potential long-term carbon sink because of the relatively slow rate of wood decomposition (decades to centuries) (Table A-4) (Harmon et al., 1990). The importance of changes in wood density as a carbon sink relates to the fact that species composition changes as forests mature. Typically, mature forest species have higher density woods than do pioneer species (Smith, 1970; Whitmore and Silva, 1990), so more carbon can be stored per unit of wood volume produced by mature forest species (for example, Weaver [1987]). SOC is a long-term storage compartment for atmospheric carbon. Schlesinger (1990) recently showed that some tropical soils under forests continue to accumulate SOC over thousands of years at a rate of about 2.3 g CO/m^2/year (the flux background rate in Table A-4). However, large SOC depletions may be associated with deforestation, and the rate of recovery of SOC to initial levels is slow. Conversion of cultivated cropland to pastures also results in SOC accumulation (Lugo et al., 1986). SOC will recover under forest plantations, and some species appear to accelerate its recovery (Lugo et al., 1990a,b). Carbon export may occur when carbon is transported by rivers to oceanic systems.

Other Greenhouse Gases

Most of the carbon released to the atmosphere from land use changes is released as CO_2 (Table A-6 and Figure A-3). The emissions of CH_4, N_2O, and CO to the atmosphere are also of interest because they contribute either directly or indirectly to the heat balance of the earth and have been increasing during recent decades (Figure A-4). The accumulation of CH_4 in the atmosphere contributed about 15 percent of all gases that contributed to the greenhouse effect in the 1980s; the contribution from N_2O was about 6 percent. Although CO is not radiatively important itself, it reacts chemically with hydroxyl radicals (OH) in the atmosphere, some of which would otherwise react with, diminishing its concentration.

Land use change is a major contributor to the releases of CH_4 and N_2O. Fifty to 80 percent of the annual release of CH_4 is from land (Houghton et al., 1990). The higher estimate includes releases from natural wetlands and termites, largely natural sources. Rice paddies, ruminant animals, and biomass burning are estimated to contribute 20, 15, and 8 percent, respectively, of the total emissions of CH_4.

About 65 to 75 percent of the annual releases of N_2O are thought to come from land (Houghton et al., 1990), with soils alone contributing 50 to 65 percent. Soils may also be an important sink for atmospheric N_2O and CH_4. The magnitude of the soil sink is not known. In fact, the global budget for N_2O is not understood well enough to account for the observed increase in the concentration of N_2O in the atmosphere. An additional source is not yet accounted for.

TABLE A-6 Annual Global Emissions of Greenhouse Gases

Gas	Source	Annual Emissions[a]	Percent Emissions of Gases	Radiative Forcing Relative to CO_2 Molecule	Contribution to Greenhouse Effect in 1980s (percent) Total	Deforestation
CO_2	Industrial[b]	5.6 Pg of C	50	1	50	
	Biotic[b]	2.0–2.8 Pg of C	20–25			
	Tropical deforestation	2.0–2.8 Pg of C	20–25			13–16
CH_4	Industrial[b]	50–100 Tg of C	8–10	25	20	
	Biotic[b]	320–785 Tg of C	63–66			
	Tropical deforestation	136–310 Tg of C	26–27			8
N_2O	Industrial[b]	<1 Tg of N		250	5	
	Biotic[b]	3–9 Tg of N	75			
	Tropical deforestation	1–3 Tg of N	25			1–2
CFCs	Industrial[b]	700 Gg	100	1,000s	20	
	Biotic[b]	0 Gg	0			0
Total					95	22–26

NOTE: No natural emissions of CO_2 can be measured on a regular basis; the values of biotic and deforestation sources are identical. 1 Pg = 10^{15} g; 1 Tg = 10^{12} g; 1 Gg = 10^9 g. CO_2, carbon dioxide; CH_4, methane; N_2O, nitrous oxide; CFCs, chlorofluorocarbons; C, carbon; N, nitrogen.

[a]Data are from Ramanathan et al. (1987). The greenhouse gases considered here are only those directly released as a result of human activities. Tropospheric ozone, which is formed as a result of other emissions, contributes another 5 percent to the total. The major greenhouse gas, water vapor, is not directly under human control but will increase in response to a global warming.

[b]Biotic emissions include emissions from tropical deforestation as well as natural emissions.

[c]Relatively little CH_4 is emitted as a result of deforestation. Most of CH_4 emissions result from rice cultivation or cattle ranching, which are land uses that replace forests. Additional releases occur with repeated burning of pasturelands and grasslands.

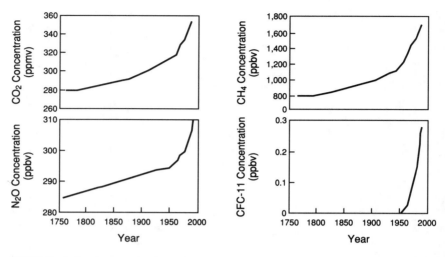

FIGURE A-4   Atmospheric concentrations of the major greenhouse gases—carbon dioxide (CO_2), methane (CH_4), nitrous oxide (N_2O), and chlorofluorocarbon 11 (CFC-11)—from 1750 to 2000; ppbv, parts per billion (by volume); ppmv, parts per million (by volume). Source: Data from Houghton, J. T., G. J. Jenkins, and J. J. Ephraums, eds. 1990. Climatic Change: The IPCC Scientific Assessment. Cambridge: Cambridge University Press.

METHANE

A small fraction of the carbon released to the atmosphere may be CH_4. The releases of CH_4 during burning are generally 2 orders of magnitude lower than those of CO_2: 0.5 to 1.5 percent of the CO_2 released (Andreae et al., 1988; Crutzen et al., 1985).   The radiative effect of a molecule of CH_4, however, is 25 times greater than that of a CO_2 molecule, so if as much as 4 percent of the carbon were emitted as CH_4, the radiative effects of CO_2 and CH_4 would be equal in the short term. Because the average residence time of CH_4 in the atmosphere is only about 10 years, whereas that of CO_2 is 100 to 250 years, the long-term greenhouse effect due to CO_2 is greater than that due to CH_4 (Table A-6) (Houghton et al., 1990; Lashof and Ahuja, 1990; Rodhe, 1990).

If the ratio of CH_4/CO_2 emitted in fires associated with deforestation is 1/100, and if 40 percent of the emissions from deforestation resulted from burning, then only about 10 Tg (1 Tg = 10^{12} g) of carbon as CH_4 would be emitted to the atmosphere directly from deforestation. This flux is based on the net flux of CO_2, however. Burning of pastures, grasslands, and fuelwood is estimated to have released 30 to 75 Tg of CH_4 annually (Cicerone and Oremland, 1988). In addition, 60 to 170 Tg of CH_4 is released from rice cultivation (Cicerone and Oremland, 1988), 27 Tg is released from natural tropical wetlands (Matthews and Fung, 1987), and 19 to 38 Tg is released from cattle ranching in the tropics (some of which occurs on natural savannahs and grasslands) (Lerner et al., 1988). Thus, a total of 136 to 310 Tg of carbon, or about 26 percent of the global emissions of CH_4, may arise from the tropics (Table A-6). The expansion of wetlands through the flooding of forests for hydroelectric dams could become a significant new source of CH_4 in the future.

NITROUS OXIDE

The gas N_2O is also emitted to the atmosphere following deforestation. Small amounts of N_2O are released during burning, but most of the release occurs in the months following a fire. Fire affects the chemical form of nitrogen in soils and, as a result, favors denitrification (Cofer et al., 1989; Levine et al., 1988). One of the by-products of denitrification is the production of nitric oxide (NO) and N_2O.

Estimates of the global emissions of N_2O are tentative. Industrial sources are thought to contribute less than 1 Tg of N_2O per year. Earlier estimates of this flux were higher, but the measurements are now thought to have been artificially high (Muzio and Kramlich, 1988). The soils of natural ecosystems are estimated to release 3 to 9 Tg of nitrogen annually as N_2O (Table A-6) (Seiler and Conrad, 1987). Fertilized soils may release 10 times more per unit area, and the soils of new pastures may release even higher amounts (Anderson et al., 1988; Levine et al., 1988). Deforestation for tropical pastures may well be a major contributor to the global increase in N_2O concentrations (Table A-6) (Luizão et al., 1989).

CARBON MONOXIDE

CO is not a greenhouse gas, but it does affect the oxidizing capacity of the atmosphere through interaction with OH and thus indirectly affects the concentrations of other greenhouse gases such as CH_4 and N_2O. Generally, CO emissions account for 5 to 15 percent of the total CO_2 emissions from burning of biomass, depending on the intensity of the burn (Andreae et al., 1988; Cofer et al., 1989; Crutzen et al., 1979, 1985). More CO is released from smoldering fires than during rapid burning or flaming. The burning associated with deforestation may thus release 40 to 170 Tg of carbon as CO. In addition, the repeated burning of pastures and savannahs in the tropics is estimated to release 200 Tg of carbon as CO (Hao et al., 1990). Together, these emissions of CO from the tropics are as large as estimates of global emissions from industrial sources (Cicerone, 1988).

Total Radiative Effect from All Gases Released as a Result of Tropical Deforestation

The global emissions (both total emissions and emissions from tropical deforestation) of the three greenhouse gases previously discussed above (CO_2, CH_4, and N_2O) are given in Table A-6. Biotic emissions include the emissions from tropical deforestation. By taking the sums of the emissions and taking into account the different radiative effects of the gases and their residence times in the atmosphere (Ramanathan et al., 1987), tropical deforestation accounts for about 25 percent of the radiatively active emissions globally (Houghton, 1990a) (Figure A-3).

ESTIMATING FUTURE IMPACTS

Estimation of the future impacts of land use changes encompasses many dimensions. The spatial scale of interest is the entire earth, but by focusing on the tropics, or key areas in the tropics, insights can be achieved. This review restricts the time dimension to the next century because critical socioeconomic and political decisions that will have major environmental repercussions will be made during that time frame. Models are an essential tool in projecting future patterns and impacts of land use change. Because no one model encompasses all of the processes of importance or all of the scales of interest, four main model types are reviewed here. One set of models emphasizes the accounting of carbon gains and losses in the landscape because of changes in land use, whereas the others integrate the socioeconomic and ecologic aspects of land use change. Together, they demonstrate approaches to determining future impacts under different scenarios of land use change, management, and population change. All of these models suffer from poor knowledge of deforestation rates, the biomass per unit area, the rate of biomass recovery, and changes in the carbon pools as a result of disturbance.

Carbon Accounting Models

The models of Moore and colleagues (1981), Houghton and colleagues (1983, 1985, 1987), Detwiler and Hall (1988a), and Bogdonoff and colleagues (1985) are based on information at the biome level, predict carbon fluxes over large regions of the earth, and account for lags in the releases or uptake of carbon. These lags result from the slowness of decay of dead plant material, soils, and wood products or the accumulation of carbon in regrowing forests following shifting cultivation and logging. The annual flux estimation includes timing of the releases and accumulations of carbon stock following land use changes. For example, 50 to 60 percent of the carbon emitted to the atmosphere in 1980 is calculated to have resulted from the deforestation during the first year after the trees are cut. The remaining net flux in 1980 resulted from the decay of vegetation and soils and from the oxidation of wood products generated by deforestation in previous years.

Deforestation and other land use changes initiate changes in vegetation and soil. In the year of deforestation, a large amount of carbon is released through burning. Afterward, the decay of soil organic matter, downed wood, and wood products continues to release carbon to the atmosphere, but at lower rates. If croplands are abandoned, regrowth of live vegetation and redevelopment of soil organic matter withdraw carbon from the atmosphere and again carbon accumulates on land. Such changes have been defined for different types of land uses and different types of ecosystems in different regions of the tropics. Annual changes in the different reservoirs of carbon (live vegetation, soils, downed wood, and wood products) determine the annual net flux of carbon between the land and atmosphere. Because ecosystems and land uses vary and calculations require accounting for cohorts of different ages, accounting (bookkeeping) models have been used for the calculations.

The accounting models developed by Houghton (1990b) allow for a projection of the effects of particular patterns of deforestation or reforestation. For example, when deforestation is based on population change, the projected rate of global deforestation more than doubles between 1980 and 2045 (Figure A-5), at which time the forests of Asia would be eliminated. Closed forests in the rest of the tropics would be eliminated by about 2065, and open forests would be eliminated 10 years later. In a reforestation scenario, land that had supported forests in the past and that is not presently used for crops or settlements is allowed to regenerate, with all logging stopping by 1991. The projected accumulation of carbon on lands abandoned by shifting cultivation would be 54 Pg; that on reforested land would be 98 Pg. The models can also be used to compare alternative assumptions on factors that affect carbon emissions. For example, the high- and low exponential curves in Figure A-5 signify the inclusion and the

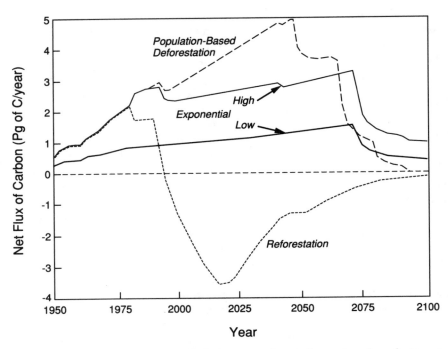

FIGURE A-5 Four projections of the annual net flux of carbon between
tropical land and the atmosphere, in petagrams (Pg) of carbon (C) per year,
based on different assumptions of deforestation and reforestation rates. Positive fluxes indicate a net release of carbon from the atmosphere. The curves
marked, Exponential High and Low, are based on two extremes of carbon
emissions associated with high and low estimates of biomass in the tropical
forest. Abrupt reductions in emissions near 2045, 2060, and 2070 result from
an elimination of forests in a major region and, hence, an abrupt reduction in
the rate of tropical deforestation. Source: Houghton, R. A. 1990b. The
future role of tropical forests in affecting the carbon dioxide concentration of
the atmosphere. Ambio 19:204–209. Reprinted with permission.

lack of inclusion of the conversion of forest fallow to permanently
cleared land and show that there is a significant difference between
the simulations in the rate of deforestation for the entire tropics.

   The accounting models have been extremely useful in projecting
the future effects of particular scenarios of land use practices. It is
largely because of the results that have been obtained from these
models that current research is focused on land use practices in the
tropics and the effects of CO_2 releases. However, these accounting
models do not incorporate feedbacks between the socioeconomic and

ecologic aspects of land use change. Therefore, the most recent models of land use integrate a variety of aspects of land use change so that the causes of deforestation and its impacts can be evaluated.

Models that Integrate Socioeconomic and Ecologic Aspects of Land Use Change

Given the socioeconomic forces that frequently initiate land use changes and, as a result, that cause major ecologic effects, modeling of the land use change process requires models that combine socioeconomic and ecologic factors.

FACTORS AFFECTING LAND USE CHANGES

Grainger (1986, 1990b, In press) has argued that land use changes can be attributed to three sets of underlying causes: socioeconomic factors, physical environmental factors, and government policies. It is assumed that national land use morphology (the relative proportions of different land uses, each with different biomass, productivity, and sustainability characteristics) changes over time in response to these underlying causes, each of which is described here briefly. Land use also has a role linked to the degree of sustainability and other factors (discussed later).

Socioeconomic factors, such as population growth and economic development, are the key driving forces causing large areas of forestland in the humid tropics to be transferred to agricultural uses. Rising populations require more land for settlement and crops to supply the increasing demand for food. Growing population densities also lead to more intensive farming practices. National population growth rates are somewhat correlated with deforestation rates in the humid tropics (Allen and Barnes, 1986; Grainger, 1986).

In terms of agriculture, economic development increases per capita food consumption and the need to grow export cash crops. Crops that are grown for export earn foreign currency, which funds continued economic growth. Both trends lead to more deforestation and, assisted by the improved access to Western technology that comes with economic development, to more chain saws, tractors, and other mechanized equipment that increase the rate at which forests can be cleared or damaged. At the same time that economic development catalyzes the spread of deforestation, however, it can also control it, by providing a greater opportunity to invest in improved agricultural techniques and technologies (for example, high-yielding crop varieties) that can grow the same amount of food on a smaller area of land than before. In addition, economic development of the industrial, service, and other non-land-based sectors can diminish the pressure for land conversion.

Physical environmental factors affect land use because deforestation is a spatial phenomenon, with land use changes occurring because of the diffusion of people, economic activity, and new techniques into forested areas from existing centers of settlement. The diffusion process is channeled by physical factors such as ease of access by rivers and roads, topography, and soil type. Some of these factors promote the expansion of agriculture; others constrain it. Many have a secondary economic component. For example, the longer the distance or more difficult the access from a given area to the nearest market, the higher the cost of transporting produce and the higher the price that needs to be charged to make cultivation of a crop economically viable.

Land use changes are also influenced by government policies. In some cases the link is direct, as in Brazil, where agricultural and regional policies have actively promoted the expansion of cattle ranching in the Amazon region. In other instances the link is indirect, through policies that either promote population growth and economic development or change the country's physical infrastructure (for example, by building the highway networks).

MODELS

Carbon flux models that emphasize supply and demand have been built at three scales. The importance of the factors instigating land use changes is scale dependent.

National-to-Global Model Estimates of future impacts of land use changes have been made by expanding the scope of a model previously built to simulate long-term trends in national deforestation rates (Grainger, 1986, 1990b). The theoretical basis of this model lies in a more complex systems model that links together land uses and the underlying socioeconomic causes of land use change outlined above (Grainger, 1986, 1990b, In press).

Because of the lack of data that can be used to initialize the many parameters in this systems model and the paucity of empirical studies of deforestation processes, a simpler national simulation model was devised, using the same principles for the quantitative simulation of possible long-term trends. In this model the area of agricultural land, and hence the rate of deforestation, is assumed to depend on (1) the population growth rate, (2) the rate of increase in food

TABLE A-7 Regional Trends in Deforestation Rates in the Humid Tropics, 1980 to 2020 (Millions of Hectares per Year)

Region	High-Deforestation Scenario					Low-Deforestation Scenario				
	1980	1990	2000	2010	2020	1980	1990	2000	2010	2020
Africa	1.6	1.5	1.2	0.9	0.9	1.0	0.9	0.7	0.6	0.4
Asia-Pacific	1.7	1.5	1.2	1.1	1.1	1.1	0.9	0.7	0.5	0.4
Latin America	3.3	3.1	2.7	2.2	1.7	2.0	1.6	1.1	0.6	0.0
Humid tropics	6.6[a]	6.1	5.1	4.3	3.7	4.1[a]	3.4	2.5	1.7	0.9

[a]Compare these values with the estimate of 5.6 million ha per year by Lanly (1982) for 1976–1980.

SOURCES: Grainger, A. 1986. The Future Role of the Tropical Rain Forests in the World Forest Economy. Oxford: Oxford Academic Publishers; Grainger, A. 1990b. Modeling deforestation in the humid tropics. Pp. 51–67 in Deforestation or Development in the Third World?, Vol. III, M. Palo and G. Mery, eds. Bulletin No. 349. Helsinki: Finnish Forest Research Institute.

consumption per capita (α), (3) the rate of increase in yield per hectare (β), and (4) the availability of forestland and agricultural land (Grainger, 1986, 1990b). Deforestation rates normally equal the additional area of farmland required each year and are assumed to become zero when the forest area per capita reaches 0.1 ha (an arbitrary limit that is estimated by empirical analysis and that corresponds to the attainment of an eventual new point of equilibrium in the national land use system). After this, increased food production can only be gained by raising the yield per hectare and obtaining extra farmland from nonforestland (Grainger, 1991).

The model simulates a decline in deforestation rates for 43 countries that contain 96 percent of the world's total area of the humid tropical forests and 92 percent of the total deforestation rate in 1980 (Table A-7). Two alternative scenarios—the high- and low-deforestation scenarios—were simulated with the simpler national deforestation model by using initial population growth rates, which were the same as those for 1970 to 1980, and growth rates in food consumption per capita (α) and yield per hectare (β), which were estimated on the basis of average regional values for 1970 to 1980 (Table A-8). In the high-deforestation scenario, the deforestation rate falls from 6.6 million ha/year in 1980 to 3.7 million ha/year in 2020. In the low-deforestation scenario, it falls from 4.1 million ha/year to just 0.9 million ha/year, ending close to zero in Latin America, where, in a number of countries, the increase in agricultural productivity made

TABLE A-8   Assumed Increases in Per Capita Food Consumption
(α) and Yield per Hectare (β) (Percentage per Year)

Region	High-Deforestation Scenario			Low-Deforestation Scenario		
	α	β	α–β	α	β	α–β
Africa	0.5	1.0	–0.5	0.0	1.0	–1.0
Asia-Pacific	1.5	1.5	0.0	1.5	2.0	–0.5
Latin America	1.5	2.0	–0.5	0.5	1.5	–1.0

SOURCES: Grainger, A. 1986. The Future Role of the Tropical Rain Forests in the
World Forest Economy. Oxford: Oxford Academic Publishers; Grainger, A. 1990b.
Modeling deforestation in the humid tropics. Pp. 51–67 in Deforestation or Develop-
ment in the Third World?, Vol. III, M. Palo and G. Mery, eds.   Bulletin No. 349.
Helsinki: Finnish Forest Research Institute.

possible a net return of land to forests (and, hence, negative deforestation rates), mostly toward the end of the simulation period. The
high-deforestation scenario predicts a reduction of about 20 percent
in total forest area, in comparison with a fall of less than 10 percent
in the low-deforestation scenario (Table A-9). These results suggest
that deforestation may not have as devastating an effect on the forests in the humid tropics as some have feared. Indeed, even if deforestation rates continue at the levels estimated for 1976 to 1980, the
overall reduction in forest area by 2020 would be only 23 percent
(Grainger, 1986, 1990b).

TABLE A-9   Regional Trends in Forest Area in the Humid Tropics,
1980 to 2020 (Millions of Hectares)

Region	High-Deforestation Scenario					Low-Deforestation Scenario				
	1980	1990	2000	2010	2020	1980	1990	2000	2010	2020
Africa	198.9	183.5	170.3	160.4	151.6	198.9	188.9	181.1	175.0	170.1
Asia-Pacific	239.4	222.8	209.5	197.5	185.8	239.4	228.9	220.8	214.8	210.2
Latin America	598.0	566.2	537.5	513.0	493.8	598.0	580.1	566.7	558.6	555.8
Humid tropics	1,036.3	972.6	917.2	870.8	831.1	1,036.3	997.9	968.7	948.5	936.1

NOTE: Totals may not be exact because of rounding.

SOURCES: Grainger, A. 1986. The Future Role of the Tropical Rain Forests in the
World Forest Economy. Oxford: Oxford Academic Publishers; Grainger, A. 1990b.
Modeling deforestation in the humid tropics. Pp. 51–67 in Deforestation or Develop-
ment in the Third World?, Vol. III, M. Palo and G. Mery, eds.   Bulletin No. 349.
Helsinki: Finnish Forest Research Institute.

Expansion of the simpler national deforestation model to simulate the net emissions of CO_2 similarly results in a decline in carbon release (Table A-10). CO_2 emissions in the model include the loss of carbon from burning of cleared vegetation and soil oxidation; the model takes into account both lags in carbon emissions and the subsequent uptake of carbon in the soil and the regenerating vegetation (as described by Grainger [1990d]). Two scenarios were simulated, corresponding to the low- and high-deforestation scenarios described above. In the low and high carbon emissions scenarios, 0.4 and 0.7 Pg of carbon, respectively, were released from the humid tropics in 1980, corresponding to estimated carbon releases of 0.5 to 0.8 Pg (Grainger [1990d]). These fell to 0.1 and 0.4 Pg of carbon, respectively, by 2020. The simulations suggest that if governments take steps to ensure that growth in agricultural productivity can outpace the rise in food consumption per capita, like the rate assumed here, then the consequent fall in deforestation rates could lead to a cut in carbon emission rates from the forests in the humid tropics of 40 to 70 percent over the next 30 years.

TABLE A-10 Two Scenarios for Future Trends in Carbon Dioxide Emissions from Deforestation in the Tropics (Petagrams of Carbon per Year), 1980 to 2020

Region	High-Deforestation Scenario					Low-Deforestation Scenario				
	1980	1990	2000	2010	2020	1980	1990	2000	2010	2020
Humid tropics										
Africa	0.192	0.187	0.158	0.131	0.125	0.132	0.122	0.902	0.756	0.609
Asia	0.211	0.193	0.164	0.157	0.151	0.138	0.120	0.949	0.712	0.557
Latin America	0.298	0.281	0.250	0.208	0.158	0.179	0.150	0.106	0.578	0.089
Total	0.702	0.661	0.572	0.495	0.434	0.449	0.391	0.291	0.205	0.126
All tropics[a]										
Africa	0.235	0.230	0.201	0.174	0.168	0.174	0.164	0.133	0.118	0.104
Asia	0.220	0.202	0.173	0.166	0.160	0.147	0.129	0.104	0.080	0.065
Latin America	0.342	0.325	0.294	0.252	0.202	0.223	0.193	0.150	0.102	0.053
Total	0.797	0.757	0.668	0.591	0.530	0.544	0.486	0.387	0.300	0.221

[a]Simulations include the following constant values of emissions (petagrams of carbon) from the dry tropics: Africa, 0.043; Asia, 0.009; Latin America, 0.044; total dry tropics, 0.096.

SOURCE: Grainger, A. 1990d. Modeling future carbon emissions from deforestation in the humid tropics. Pp. 105–119 in Tropical Forestry Response Options to Global Climate Change. Report No. 20P-2003. Washington, D.C.: Office of Policy Analysis, U.S. Environmental Protection Agency.

The omission of changes in agricultural sustainability is one of a number of structural limitations of the simpler national systems model that could, in practice, lead to higher deforestation rates in the future. As sustainability and yields decline, more deforestation is needed elsewhere unless there is some compensating increase in productivity on better soils. The model assumes that all deforestation that takes place in response to increased demand for food satisfies that demand and continues to satisfy it indefinitely. If this assumption was not justified, then deforestation rates could be higher than those simulated by the model. However, since sustainability is site and land use specific, production of a realistic simulation of the actual conditions could well require a spatial model rather than the highly aggregated model used here. This, in turn, would depend upon the results of detailed field studies of the actual effects of nonsustainability and would require that spatial data on land suitability, land use patterns, and forest cover be obtained.

Spatially Explicit Models   Spatially explicit models include such specifics as soil, vegetation, and land use practices for each model cell (unit) and can simulate feedbacks between environmental conditions, land use practices, future opportunities, and sustainability. For example, Southworth and colleagues (1991) developed a model that simulates colonization and its effects on deforestation, land use, and associated carbon losses. The model projects patterns and rates of deforestation under different immigration policies, land tenure practices, and road development scenarios and includes feedbacks between changes in soil and vegetation conditions and future opportunities for land use (Dale et al., In press).

A spatial model (Figure A-6) was used to contrast sustainable agricultural practices with the typical scheme of colonists in Rondônia, Brazil, of burning the tropical forest, planting annual and perennial crops and then pastures, and lastly, abandoning their lots. The results from simulations show how these extremes of resource management can affect carbon storage and release in the humid tropics (Southworth et al., 1991). The resulting carbon and land use profiles are markedly different at both the lot-specific and regional levels. The lot-specific net carbon loss profiles depend on the land use practices of particular families. Once a lot nears the point of full clearance, it is abandoned, starting a very gradual process of carbon recovery. In later years some of these lots have been absorbed into larger cattle ranches and are grazed. The land clearance rate is assumed to be similar to that of tenant farming (in practice, it can be expected to vary by rancher). The sustainable agriculture scenario

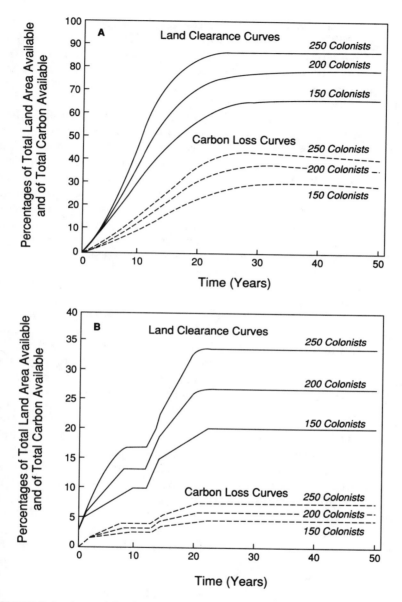

FIGURE A-6 Areawide changes in carbon release and land cleared for (A) typical colonist scenario and (B) sustainable agriculture scenario. Source: Southworth, F., V. H. Dale, and R. V. O'Neill. 1991. Contrasting patterns of land use in Rondônia, Brazil: Simulating the effects on carbon release. Int. Soc. Sci. J. 130:681–698. Reprinted with permission.

(Figure A-6) also shows significant carbon depletion during the initial 2–year phase of clearing the lot; this is followed by a stability in the carbon content of the lot during the intercropping period. Given the noticeably different results obtained from the two scenarios examined, it is worth asking what needs to be done to implement sustainable shifting cultivation in Rondônia or in other forested regions in the tropics. Activities that reduce the negative exponential decline of carbon in the simulations were represented by planting trees intermixed with annuals. Recovery of previously pastured land in this area is apparently affected more by treatment of the land than it is by soil nutrient stocks (Buschbacher et al., 1988). For example, a soil with low nutrient content but a high density of pioneer trees had twice as much biomass 2.5 years after abandonment as a site with higher soil nutrient content and few root sprouts (Buschbacher et al., 1988). In Ouro Preto, Rondônia, farmers who concentrated on perennial crops were better off in terms of material possessions and housing than the other colonists (Leite and Furley, 1985).

The introduction of economic considerations into the set of rules used by the spatially explicit model can expand understanding of the causes and consequences of particular land management scenarios. Two outcomes of this approach to dynamic microsimulation of land use changes include (1) the opportunities to experiment with "land holding capacity," or the number of people a region can be expected to sustain over a given period of time, and (2) the analysis of how ecology and socioeconomics interact to influence the spatial and temporal pattern of land use.

Abstract Spatial Consideration of Land Use Patterns Nonexplicit spatial considerations of land use practices allow researchers to analyze theoretical relationships between the socioeconomic and ecologic aspects of land use. For example, Jones and O'Neill (In press) developed a model that distributes agricultural activities (forestry or food production) within the spatial domain of a Thünen circle (an abstract area that radiates outward from a central market). The dominant process that influences spatial distribution is the costs of transporting labor and of purchasing inputs and products. The unique feature of the model of Jones and O'Neill is that the environmental impacts of the economic activity, such as soil degradation or erosion, directly reduce productivity. Thus, degradation and conservation measures to mitigate degradation become endogenous variables that influence economic decisions.

The models consider the spatial distribution of activity at equilibrium and emphasize the total spatial extent of the activity (for example, total area of deforestation) and the intensity of activity per hectare (for example, erosion potential). The models are designed to explore how changes in the economy (for example, prices, transportation costs, wages, and population) or in the ecology (for example, soil fragility) would be expected to influence the extent and intensity of the agricultural activity.

To date, a suite of seven models has been developed. These models follow the same overall conceptualization but differ in their assumptions or constraints. For example, the total population may be considered to be constant within a region, or immigration-emigration may be permitted. Three of the models explore the implications of a shifting agricultural system in which economic and ecologic parameters determine the percentage of a lot permitted to lie fallow in a given year. By using a suite of models, the individual scenario remains simple and amenable to analytic solution while a variety of scenarios covering a number of different assumptions can be explored.

SUSTAINABILITY AND THE REDUCTION OF FUTURE IMPACTS

The effect of alternative land uses and agricultural sustainability on tropical deforestation can be evaluated with spatially explicit models. However, five factors constrain the sustainability of land use in the humid tropics. (1) Site quality imposes inherent limitations on the sustainability of each type of land use practiced at a given level of intensity. Low-fertility Oxisols and Ultisols are widespread (Sanchez, 1976) and require careful husbandry to avoid degradation and fertility depletion. Sloping lands susceptible to erosion are prevalent. (2) The choice of land use for a particular site affects its sustainability in two main ways. First, the land use may be unsustainable at the level of intensity originally practiced; that is, permanent field cropping on low fertility soils on sloping lands may increase the land's susceptibility to erosion. Second, the land use may become unsustainable as its intensity of use increases. This is commonly found with shifting cultivation as rotations become shorter. (3) Socioeconomic factors constrain the sustainability of land uses. Thus, a rise in population density may result in more intensive shifting cultivation, and if rotations become too short, it may lead to soil degradation and a decline in yields and sustainability. Although economic development can enable investments that make agriculture more productive and sustainable, the lack of economic development, or poverty, means that many farmers are unable to make such investments. (4) A general erosion by the market economy of societies with people who earn their living from subsistence agricultural activities is an inevitable consequence of economic development. Permanent cultivators of cash crops focus on growing the most profitable crops and ensuring that yields are as high as possible. This may not necessarily mean that sustainability is prized, too, however. Cattle ranching in Brazilian Amazonia is a prime example of a case in which the emphasis was on short-term financial gains rather than long-term sustainability. (5) Sustainability can also decline as a result of external influences, as when one land use has an impact on another. For example, expansion of the area of permanent agriculture or logging may jeopardize shifting agriculture by affecting its area and intensity. Such expansion is possible because shifting cultivators often lack legal land rights.

The situation is complicated even more by the considerable variation in the degree of sustainability of the resulting agricultural land uses. Overintensive use of land, whether it is due to poor management, inadequate suitability of the land for a given use, inability to invest in required inputs, or socioeconomic and policy factors, can lead to a decline in soil fertility and crop yields and, hence, a decline in the amount of biomass per unit area and the carbon uptake rate. The ultimate result may be a change in land use as land is abandoned. Lack of sustainability also has another effect: more deforestation is required to increase the area of agricultural land so that overall production is maintained, and this leads to further CO_2 emissions.

A move toward sustainability is therefore a vital consideration if deforestation is to be controlled. It is also an important alternative to consider if greenhouse gas emissions are to be reduced and carbon uptake maximized. Improvements in sustainability go hand in hand with increasing productivity on selected lands. Because low-fertility soils are widespread in the humid tropics, the area of land suited to intensive agriculture under foreseeable socioeconomic conditions is limited. One solution involves increasing the productivity of intensive agriculture on only the best lands so that an increasing share of national food production can be managed. Low-intensity agriculture and forestry could then be allowed to continue elsewhere.

Agroforestry

One promising way to increase the productivity and sustainability of land use on poorer soils is to use agroforestry systems, a variety of techniques that combine the growing of herbaceous crops and trees and the raising of livestock on the same or adjacent areas of land. This simulates the multilayered structure of natural tropical forests and maintains the high level of vegetative cover needed to protect the soil from erosion (Grainger, 1980, 1991).

Agroforestry also plays a role in efforts to combat the greenhouse effect. So far most attention has been given to large-scale afforestation to increase the rate of carbon uptake by terrestrial biota (Grainger, 1990a,c; Houghton, 1990b; Sedjo and Solomon, 1989). There are practical limits, however, to how fast the establishment rate of timber plantations (for growing industrial wood or fuelwood) can be increased. Foresters identified this problem in the 1970s for reasons unconnected with the greenhouse effect: There were simply insufficient forestry personnel to plant the number of trees needed, and those trees that were planted could be cut down prematurely by local people because young plantations were poorly protected. The best solution was determined to be the establishment of social (or community) forestry programs. Personnel involved with these forestry programs support and assist with the establishment of new tree cover on communal lands or private farmlands, rather than in government forest reserves, with varying degrees of participation by local people. Tree species that satisfy local needs for food, fodder, fuelwood, and other products are chosen. Many social forestry projects involve agroforestry systems of one sort or another. Any new initiative to expand forest cover in the tropics will probably involve a combination of monoculture plantations and agroforestry systems, thereby enhancing the sustainability of land use on the poorer lands on which such activities are likely to be concentrated.

Carbon Sinks

If sustainable land use practices spread, there is a high potential for carbon sinks to increase in importance in the global carbon budget. For example, carbon accumulation in the biomass and soils of forest fallow could continue for many decades if less of this land had to be recut and burned to meet the demands of food production. At the same time, the need to continue deforesting tropical lands might be reduced, and because many of these forests are still recovering from past disturbances, they could continue to sequester carbon in biomass, necromass, and soil. Therefore, models that project future carbon releases from the tropics must consider the switch to more sustainable uses of the landscape. These uses should encompass an optimum mix of land uses (for example, mature forest, logged forests, secondary forests, and annual and perennial crops). The models should also be able to evaluate the ability of the various land uses to sequester carbon through biomass accumulation in primary and secondary forests or plantations, woody debris, and soils (Table A-4).

Priorities for Future Research

To better assess the role of land use changes on climate change, including the impact of sustainability, more research is needed to extend the scope of the data collection, analyses, and modeling approaches outlined here. Some key priorities include the following:

• There is an urgent need to undertake field studies to gain a better understanding of what constitutes sustainable systems. The critical information needed from field studies is a better understanding of what defines sustainable systems. For example, it would be useful to know the threshold rotation at which shifting cultivation becomes sustainable and what factors influence that threshold. The data should be collected within a well-defined and standardized sampling format so that comparisons can be made between different ecosystems.

• Further field and remote-sensing research into deforestation and land use change processes is needed so that many of the functional relationships can be quantified at both local and national levels.

• It is important to develop an index of agricultural sustainability that could be used in land use models and land use planning techniques. The index could be estimated on the basis of such factors as land capability, land use intensity, available investment capital, food yield per hectare, economic rate of return, transportation systems, and cultural factors.

• More development, testing, and comparisons of the models discussed above are needed. These models can explore the factors that lead to sustainability and project the regional and global repercussions of sustainable agriculture. Because each model emphasizes different aspects of the system and operates at different spatial scales, they are useful for exploring different types of questions.

• More information is needed on the biomass densities and carbon sequestration rates of alternative tropical land uses and how these vary with productivity and the degree of sustainability.

ACKNOWLEDGMENTS

W. M. Post, G. Marland, and four anonymous reviewers provided useful comments on the manuscript. J. P. Veillon made available his long-term data for the forests of Venezuela. Research was partially sponsored by the Carbon Dioxide Research Program, Atmospheric and Climatic Change Division, Office of Health and Environmental Research, U.S. Department of Energy, under contract DE-AC05–84OR21400 with Martin Marietta Energy Systems, Inc.

REFERENCES

Ajtay, G. L., P. Ketner, and P. Duvigneaud. 1979. Terrestrial primary production and phytomass. Pp. 129–182 in SCOPE 13 The Global Carbon Cycle, B. Bolin, E. T. Degens, S. Kempe, and P. Ketner, eds. New York: Wiley.

Allen, J. C., and D. F. Barnes. 1986. The causes of deforestation in developing countries. Ann. Assoc. Amer. Geog. 75:163–184.

Anderson, I. C., J. S. Levine, M. A. Poth, and P. J. Riggan. 1988. Enhanced biogenic emissions of nitric oxide and nitrous oxide following surface biomass burning. J. Geophys. Res. 93:3893–3898.

Andreae, M. O., E. V. Browell, M. Garstang, G. L. Gregory, R. C. Harriss, G. F. Hill, D. J. Jacob, M. C. Pereira, G. W. Sachse, A. W. Setzer, P. L. Silva Dias, R. W. Talbot, A. L. Torres, and S. C. Wofsy. 1988. Biomass-burning emissions and associated haze layers over Amazonia. J. Geophys. Res. 93:1509–1527.

Blaikie, P., and H. Brookfield. 1987. Approaches to the study of land degradation. Pp. 27–48 in Land Degradation and Society, P. Blaikie and H. Brookfield, eds. London: Methuen.

Bogdonoff, P., R. P. Detwiler, and C. A. S. Hall. 1985. Land use change and carbon exchange in the tropics. III. Structure, basic equations, and sensitivity analysis of the model. Environ. Management 9:345–354.

Boserup, E. 1965. The Conditions of Agricultural Growth: The Economics of Agrarian Change under Population Pressure. Chicago: Aldine.

Brown, S., and A. E. Lugo. 1982. The storage and production of organic matter in tropical forests and their role in the global carbon cycle. Biotropica 14(3):161–187.

Brown, S., and A. E. Lugo. 1984. Biomass of tropical forests: A new estimate based on forest volumes. Science 223:1290–1293.

Brown, S., and A. E. Lugo. 1990a. Effects of forest clearing and succession on the carbon and nitrogen content of soils in Puerto Rico and US Virgin Islands. Plant Soil 124:53–64.

Brown, S., and A. E. Lugo. 1990b. Tropical secondary forests. J. Trop. Ecol. 6:1–32.

Brown, S., and A. E. Lugo. 1992. Biomass estimates for tropical moist forests of the Brazilian Amazon. Interciencia 17:8–18.

Brown, S., A. E. Lugo, S. Silander, and L. Liegel. 1983. Research history and opportunities in the Luquillo Experimental Forest. General Technical Report SO-44. New Orleans: Southern Forest Experiment Station, U.S. Forest Service, U.S. Department of Agriculture.

Brown, S., A. E. Lugo, and J. Chapman. 1986. Biomass of tropical tree plantations and its implications for the global carbon budget. Can. J. Forest Res. 16:390–394.

Brown, S., A. J. R. Gillespie, and A. E. Lugo. 1989. Biomass estimation methods for tropical forests with applications to forest inventory data. Forest Sci. 35:881–902.

Brown, S., A. J. R. Gillespie, and A. E. Lugo. 1991. Biomass of tropical forests of South and Southeast Asia. Can. J. Forest Res. 21:111–117.

Brown, S., L. Iverson, and A. E. Lugo. In press. Land use and biomass changes in Peninsular Malaysia during 1972–82. In Effects of Land Use Change on Atmospheric Carbon Dioxide Concentrations: Southeast Asia as a Case Study, V. H. Dale, ed. New York: Springer-Verlag.

Buschbacher, B. 1984. Changes in Productivity and Nutrient Cycling Following Conversion of Amazon Rainforest to Pasture. Ph.D. dissertation. University of Georgia, Athens.

Buschbacher, R., C. Uhl, and E. A. S. Serrão. 1988. Abandoned pastures in eastern Amazonia. II. Nutrient stocks in the soil and vegetation. J. Ecol. 76:682–699.

Cerri, C. C., B. Volkoff, and F. Andreux. 1988. Nature and behaviour of organic matter in soils under natural forest and after deforestation, burning and cultivation in Amazonia. Paper presented at the 46th International Congress of Americanists, July 4–8, 1988, Amsterdam, Holland.

Cicerone, R. J. 1988. How has the atmospheric concentration of CO changed? Pp. 49–61 in The Changing Atmosphere, F. S. Rowland and I. S. A. Isaksen, eds. New York: Wiley-Interscience.

Cicerone, R. J., and R. S. Oremland. 1988. Biogeochemical aspects of atmospheric methane. Global Biogeochem. Cycles 2:299–327.

Cofer, W. R., J. S. Levine, D. I. Sebacher, E. L. Winstead, P. J. Riggan, B. J. Stocks, J. A. Brass, V. G. Ambrosia, and P. J. Boston. 1989. Trace gas emissions from chaparral and boreal forest fires. J. Geophys. Res. 94:2255–2259.

Comery, J. A. 1981. Elemental Carbon Deposition and Flux from Prescribed Burning on a Longleaf Pine Site in Florida. Master's thesis. University of Washington, Seattle.

Cooper, C. F. 1982. Carbon storage in managed forests. Can. J. Forest Res. 13:155–166.

Costanza, R. 1991. The ecological effects of sustainability: Investing in natural capital. In Environmentally Sustainable Economic Development Building on Bruntland, R. Goodland, H. Daly, and S. El Serafy, ed. Environment Working Paper 46. Washington, D.C.: World Bank.

Crutzen, P. J., and M. O. Andreae. 1990. Biomass burning in the tropics: Impact on atmospheric chemistry and biogeochemical cycles. Science 250:1669–1678.

Crutzen, P. J., L. E. Heidt, J. P. Krasnec, W. H. Pollack, and W. Seiler. 1979. Biomass burning as a source of atmospheric gases CO, H_2, N_2O, NO, CH_3Cl and COS. Nature 282:253–256.

Crutzen, P. J., A. C. Delany, J. Greenberg, P. Haagenson, L. Heidt, R. Lueb, W. Pollock, W. Seiler, A. Wartburg, and P. Zimmerman. 1985. Tropospheric chemical composition measurements in Brazil during the dry season. J. Atmospheric Chem. 2:233–256.

Cuevas, E., S. Brown, and A. E. Lugo. 1991. Above and below ground organic matter storage and production in a tropical pine plantation and paired broadleaf secondary forest. Plant Soil 135:257–268.

Dale, V. H. 1990. Report of a Workshop on Using Remote Sensing to Estimate Land Use Change. ORNL/TM 11502. Oak Ridge, Tenn.: Oak Ridge National Laboratory.

Dale, V. H., R. A. Houghton, and C. A. S. Hall. 1991. Estimating the effects of land-use change on global atmospheric CO_2 concentrations. Can. J. Forest Res. 21:87–90.

Dale, V. H., F. Southworth, R. V. O'Neill, and A. Rosen. In press. Simulating spatial patterns of land-use change in Rondônia, Brazil. In Some Mathematical Questions in Biology, R. H. Gardner, ed. Providence, R.I.: American Mathematical Society.

Detwiler, R. P. 1986. Land use change and the global carbon cycle: The role of tropical soils. Biogeochemistry 2:67–93.

Detwiler, R. P., and C. A. S. Hall. 1988a. Tropical forests and the global carbon cycle. Science 239:42–47.

Detwiler, R. P., and C. A. S. Hall. 1988b. The global carbon cycle. Letter. Science 241:1738–1739.

Evans, J. 1982. Plantation Forestry in the Tropics. Oxford: Clarendon Press.

Ewel, J., and L. Conde. 1978. Environmental implications of any-species utilization in the moist tropics. Pp. 107–123 in Proceedings of Conference on Improved Utilization of Tropical Forests. Madison, Wis.: Forest Products Laboratory, U.S. Forest Service, U.S. Department of Agriculture.

Ewel, J., C. Berish, B. Brown, N. Price, and J. Raich. 1981. Slash and burn impacts on a Costa Rican wet forest site. Ecology 62:816–829.

Falesi, I. C. 1976. Ecossistema de Pastagem Cultivada na Amazônia Brasiliera. Boletim Técnico No. 1. Belém, Brasil: Centro de Pesquisa Agropecuária do Trópico Úmido.

Fearnside, P. M. 1980. The effects of cattle pasture on soil fertility in the Brazilian Amazon: Consequences for beef production sustainability. Trop. Ecol. 21:125–137.

Fearnside, P. M. 1986. Brazil's Amazon forest and the global carbon problem: Reply to Lugo and Brown. Interciencia 11:58–64.

Fearnside, P. M. 1987. Causes of deforestation in the Brazilian Amazon. Pp. 37–61 in The Geophysiology of Amazonia. Vegetation and Climate Interactions, R. E. Dickinson, ed. New York: Wiley.

Flint, E. P., and J. F. Richards. 1991. Historical analysis of changes in land use and carbon stocks of vegetation in South and Southeast Asia. Can. J. Forest Res. 21:91–110.

Food and Agriculture Organization (FAO). 1946–1987. Yearbook of Forest Products. Rome, Italy: Food and Agriculture Organization of the United Nations.

FAO. 1987. 1986 Production Yearbook. Rome, Italy: Food and Agriculture Organization of the United Nations.

FAO. 1991. Interim report on Forest Resources Assessment 1990 project. Item 7 of the Provisional Agenda, Committee on Forestry, Tenth Session. COFO-90/8(a). Rome, Italy: Food and Agriculture Organization of the United Nations.

Food and Agriculture Organization and United Nations Environment Program. 1981. Tropical Forest Resources Assessment Project. Rome, Italy: Food and Agriculture Organization of the United Nations.

Grainger, A. 1980. The development of tree crops and agroforestry systems. Int. Tree Crops J. 1:3–14.

Grainger, A. 1984. Quantifying changes in forest cover in the humid tropics: Overcoming current limitations. J. World Forest Resource Management 1:3–63.

Grainger, A. 1986. The Future Role of the Tropical Rain Forests in the World Forest Economy. Oxford: Oxford Academic Publishers.

Grainger, A. 1987. The future environment for forest management in Latin America. Pp. 1–9 in Management of the Forests of Tropical America: Prospects and Technologies, J. C. F. Colon, F. H. Wadsworth, and S. Branham, ed. Washington, D.C.: U.S. Department of Agriculture.

Grainger, A. 1988. Estimating areas of degraded tropical lands requiring replenishment of forest cover. Int. Tree Crops J. 5:31–61.

Grainger, A. 1990b. Modeling deforestation in the humid tropics. Pp. 51–67 in Deforestation or Development in the Third World?, Vol. III, M. Palo and G. Mery, eds. Bulletin No. 349. Helsinki: Finnish Forest Research Institute.

Grainger, A. 1990c. The Threatening Desert. Controlling Desertification. London: Earthscan Publications.

Grainger, A. 1990d. Modeling future carbon emissions from deforestation in the humid tropics. Pp. 105–119 in Tropical Forestry Response Options to Global Climate Change. Report No. 20P-2003. Washington, D.C.: Office of Policy Analysis, U.S. Environmental Protection Agency.

Grainger, A. 1991. The Tropical Rain Forests and Man. New York: Columbia University Press.

Grainger, A. In press. Population as concept and parameter in the modelling of tropical land use change. In Proceedings of the International Symposium on Population-Environment Dynamics, October 1–3, 1990, University of Michigan, Ann Arbor.

Hao, W. M., M. H. Liu, and P. J. Crutzen. 1990. Estimates of annual and regional releases of CO_2 and other trace gases to the atmosphere from fires in the tropics, based on the FAO statistics for the period 1975–1980. In Fire in the Tropical Biota, J. G. Goldammer, ed. Berlin: Springer-Verlag.

Harmon, M. E., W. Ferrell, and J. F. Franklin. 1990. Effect on carbon storage of conversion of old-growth forest to young forest. Science 247:699–702.

Hecht, S. B. 1982a. Agroforestry in the Amazon Basin: Practice, theory and limits of a promising land use. Pp. 331–372 in Amazonia: Agriculture and Land Use Research, S. B. Hecht, ed. Cali, Colombia: Centro Internacional de Agricultura Tropical.

Hecht, S. B. 1982b. Cattle Ranching in the Brazilian Amazon: Evaluation of a Development Strategy. Ph.D. dissertation. University of California, Berkeley.

Horne, R., and J. Gwalter. 1982. The recovery of rainforest overstorey following logging. I. Subtropical rainforest. Aust. Forestry Res. 13:29–44.

Houghton, J. T., G. J. Jenkins, and J. J. Ephraums, eds. 1990. Climatic Change: The IPCC Scientific Assessment. Cambridge, U.K.: Cambridge University Press.

Houghton, R. A. 1990a. The global effects of tropical deforestation. Environ. Sci. Technol. 24:414–422.

Houghton, R. A. 1990b. The future role of tropical forests in affecting the carbon dioxide concentration of the atmosphere. Ambio 19:204–209.

Houghton, R. A. 1991a. Releases of carbon to the atmosphere from degradation of forests in tropical Asia. Can. J. Forest Res. 21:132–142.

Houghton, R. A. 1991b. Tropical deforestation and atmospheric carbon dioxide. Climatic Change 19:99–118.

Houghton, R. A., J. E. Hobbie, J. M. Melillo, B. Moore, B. J. Peterson, G. R. Shaver, and G. M. Woodwell. 1983. Changes in the carbon content of terrestrial biota and soils between 1860 and 1980: A net release of CO_2 to the atmosphere. Ecol. Monogr. 53:235–262.

Houghton, R. A., R. D. Boone, J. M. Melillo, C. A. Palm, G. M. Woodwell, N. Myers, B. Moore, and D. L. Skole. 1985. Net flux of carbon dioxide from tropical forests in 1980. Nature 316:617–620.

Houghton, R. A., R. D. Boone, J. R. Fruci, J. E. Hobbie, J. M. Melillo, C. A. Palm, B. J. Peterson, G. R. Shaver, G. M. Woodwell, B. Moore, D. L. Skole, and N. Myers. 1987. The flux of carbon from terrestrial ecosystems to the atmosphere in 1980 due to changes in land use: Geographic distribution of the global flux. Tellus 39B:122–139.

Houghton, R. A., D. S. Lefkowitz, and D. L. Skole. 1991. Changes in the landscape of Latin America between 1850 and 1980. I. A progressive loss of forests. Forest Ecol. Management 38:143–172.

John, D. M. 1973. Accumulation and decay of litter and net production of forest in tropical West Africa. Oikos 24:430–435.

Jones, D. W., and R. V. O'Neill. In press. Land use with endogenous environmental degradation and conservation. Resources and Energy.

Juo, A., and R. Lal. 1977. The effect of fallow and continuous cultivation on the chemical and physical properties of an alfisol. Plant Soil 47:567–584.

Kartawinata, K., S. Adisoemarto, S. Riswan, and A. P. Vayda. 1981. The impact of man on a tropical forest in Indonesia. Ambio 10:115–119.

Lang, G. E., and D. H. Knight. 1979. Decay rates for tropical trees in Panama. Biotropica 11:316–317.

Lanly, J. P. 1982. Tropical Forest Resources. FAO Forestry Paper No. 30. Rome: Food and Agriculture Organization of the United Nations.

Lashof, D. A., and D. R. Ahuja. 1990. Relative contributions of greenhouse gas emissions to global warming. Nature 344:529–531.

Leite, L. L., and P. A. Furley. 1985. Land development in the Brazilian Amazon with particular reference to Rondônia and the Ouro Preto colonisation project. Pp. 119–140 in Change in the Amazon Basin. Volume II. The Frontier after a Decade of Colonization, R. Heming, ed. Manchester, United Kingdom: Manchester University Press.

Lerner, J., E. Matthews, and I. Fung. 1988. Methane emission from animals: A global high-resolution database. Global Biogeochem. Cycles 2:139–156.

Levine, J. S., W. R. Cofer, D. I. Sebacher, E. L. Winstead, S. Sebacher, and P. J. Boston. 1988. The effects of fire on biogenic soil emissions of nitric oxide and nitrous oxide. Global Biogeochem. Cycles 2:445–449.

Lugo, A. E. 1988. The future of the forest. Ecosystem rehabilitation in the tropics. Environment 30(7):16–20,41–45.

Lugo, A., and S. Brown. In press. Tropical forests as sinks of atmospheric carbon. Forest Ecol. Manage.

Lugo, A. E., M. J. Sanchez, and S. Brown. 1986. Land use and organic carbon content of some subtropical soils. Plant Soil 96:185–196.

Lugo, A. E., S. Brown, and J. Chapman. 1988. An analytical review of production rates and stemwood biomass of tropical forest plantations. Forest Ecol. Manage. 23:179–200.

Lugo, A. E., D. Wang, and F. H. Bormann. 1990a. A comparative analysis of biomass production in five tropical tree species. Forest Ecol. Manage. 31:153–166.

Lugo, A. E., E. Cuevas, and M. J. Sanchez. 1990b. Nutrients and mass in litter and top soil of ten tropical tree plantations. Plant Soil 125:263–280.

Luizão, F., P. Matson, G. Livingston, R. Luizão, and P. Vitousek. 1989. Nitrous oxide flux following tropical land clearing. Global Biogeochem. Cycles 3:281–285.

Marland, G., and T. Boden. 1989. Carbon dioxide release from fossil-fuel burning. Testimony presented at a hearing before the U.S. Senate Committee on Energy and Natural Resources, July 26, 1989.

Matthews, E., and I. Fung. 1987. Methane emission from natural wetlands: Global distribution, area, and environmental characteristics of sources. Global Biogeochem. Cycles 1:61–86.

Molofsky, J., E. S. Menges, C. A. S. Hall, T. V. Armentano, and K. A. Ault. 1984. The effects of land use alteration on tropical carbon exchange. Pp. 181–184 in The Biosphere: Problems and Solutions, T. N. Veziraglu, ed. Amsterdam: Elsevier.

Moore, B., R. D. Boone, J. E. Hobbie, R. A. Houghton, J. M. Melillo, B. J. Peterson, G. R. Shaver, C. J. Vorosmarty, and G. M. Woodwell. 1981. A simple model for analysis of the role of terrestrial ecosystems in the global carbon budget. Pp. 365–385 in Carbon Cycle Modelling, B. Bolin, ed. SCOPE 16. New York: Wiley.

Muzio, L. F., and J. C. Kramlich. 1988. An artifact in the measurement of N_2O from combustion sources. Geophys. Res. Lett. 15:1369–1372.

Myers, N. 1980. Conversion of Tropical Moist Forests. Washington, D.C.: National Academy Press.

Myers, N. 1984. The Primary Source. New York: W. W. Norton.

Myers, N. 1989. Deforestation Rates in Tropical Forests and Their Climate Implications. London: Friends of the Earth.

Nicholson, D. I. 1958. An analysis of logging damage in tropical rain forests, North Borneo. Malaysian Forester 231:235–245.

Olson, J. S., J. A. Watts, and L. J. Allison. 1983. Carbon in live vegetation of major world ecosystems. TR004. Washington, D.C.: U.S. Department of Energy.

Post, W. M., W. R. Emanuel, P. J. Zinke, and A. G. Stangenberger. 1982. Soil carbon pools and world life zones. Nature 298:156–159.

Post, W. M., T. S. Peng, W. R. Emanuel, A. W. King, V. H. Dale, and D. L. DeAngelis. 1990. The global carbon cycle. Amer. Sci. 78:310–326.

Ramanathan, V., L. Callis, R. Cess, J. Hansen, I. Isaksen, W. Kuhn, A. Lacis, F. Luther, J. Mahlman, R. Reck, and M. Schlesinger. 1987. Climate-chemical interactions and effects of changing atmospheric trace gases. Rev. Geophys. 25:1441–1482.

Ranjitsinh, M. K. 1979. Forest destruction in Asia and the South Pacific. Ambio 8:192–201.

Repetto, R. 1989. The Forest for the Trees? Government Policies and the Misuse of Forest Resources. Washington, D.C.: World Resources Institute.

Rodhe, H. 1990. A comparison of the contribution of various gases to the greenhouse effect. Science 248:1217–1219.

Saldarriaga, J. G., D. C. West, M. L. Tharp, and C. Uhl. 1988. Long-term chronosequence of forest succession in the Upper Rio Negro of Colombia and Venezuela. J. Ecol. 76:938–958.

Sanchez, P. A. 1976. Properties and Management of Soils in the Tropics. New York: Wiley.

Sanchez, P. A., and J. R. Benites. 1987. Low-input cropping for acid soils of the humid tropics. Science 238:1521–1527.

Sanchez, P. A., J. H. Villachica, and D. E. Bandy. 1983. Soil fertility dynamics after clearing a tropical rainforest in Peru. Soil Sci. Soc. Amer. J. 47:1171–1178.

Schlesinger, W. H. 1984. The world carbon pool in soil organic matter: A source of atmospheric CO_2. Pp. 111–124 in The Role of Terrestrial Vegetation in the Global Carbon Cycle: Measurement by Remote Sensing, G. M. Woodwell, ed. SCOPE 23. New York: Wiley.

Schlesinger, W. H. 1986. Changes in soil carbon storage and associated properties with disturbance and recovery. Pp. 194–220 in The Changing Carbon Cycle and Global Analysis, J. R. Trabalka and D. E. Reichle, eds. New York: Springer-Verlag.

Schlesinger, W. H. 1990. Evidence from chronosequence studies for low carbon-storage potential of soils. Nature 348:232–234.

Sedjo, R. A., and A. M. Solomon. 1989. Climate and forests. Pp. 105–119 in Greenhouse Warming: Abatement and Adaptation, N. J. Rosenberg, W. E. Easterling, P. R. Crosson, and J. Darmstadter, eds. Washington, D.C.: Resources for the Future.

Seiler, W., and R. Conrad. 1987. Contribution of tropical ecosystems to the global budgets of trace gases, especially CH_4, H_2, CO, and N_2O. Pp. 133–162 in Geophysiology of Amazonia, R. E. Dickinson, ed. New York: Wiley.

Seiler, W., and P. J. Crutzen. 1980. Estimates of gross and net fluxes of carbon between the biosphere and the atmosphere from biomass burning. Climatic Change 2:207–247.

Smith, R. F. 1970. The vegetation structure of a Puerto Rican rain forest before and after short-term gamma irradiation. Chapter D-3 in A Tropical Rain Forest, H. T. Odum and R. F. Pigeon, eds. Springfield, Va.: National Technical Information Service.

Southworth, F., V. H. Dale, and R. V. O'Neill. 1991. Contrasting patterns of land use in Rondônia, Brazil: Simulating the effects on carbon release. Int. Social Sci. J. 130:681–698.

Swift, M. J., O. W. Heal, and J. M. Anderson. 1979. Decomposition in Terrestrial Ecosystems. Berkeley: University of California Press.

Turner, B. L., R. Q. Hanham, and A. V. Portararo. 1977. Population pressure and agricultural intensity. Ann. Assoc. Amer. Geographers 67:384–396.

Uhl, C. 1987. Factors controlling succession following slash-and-burn agriculture in Amazonia. J. Ecol. 75:377–407.

Uhl, C., and I. C. G. Vieira. 1989. Ecological impacts of selective logging in the Brazilian Amazon: A case study from the Paragominas region of the state of Para. Biotropica 21:98–106.

Uhl, C., H. Clark, K. Clark, and P. Maquirino. 1982. Successional pattern associated with slash-and-burn agriculture in the Upper Rio Negro region of the Amazon Basin. Biotropica 14:249–254.

Uhl, C., R. Buschbacher, and E. A. S. Serrão. 1988. Abandoned pastures in eastern Amazonia. I. Patterns of plant succession. J. Ecol. 76:663–681.

Weaver, P. L. 1987. Structure and Dynamics in the Colorado Forest of the Luquillo Mountains of Puerto Rico. Ph.D. dissertation. Department of Botany and Plant Pathology, Michigan State University, East Lansing.

Weaver, P. L., and P. G. Murphy. 1990. Forest structure and productivity in Puerto Rico's Luquillo Mountains. Biotropica 22:69–82.

Whitmore, T. C., and J. N. M. Silva. 1990. Brazil rain forest timbers are mostly very dense. Commonwealth Forestry Rev. 69(1):87–90.

World Resources Institute. 1990. World Resources 1990–91. New York: Oxford University Press.

Young, A. 1989. Agroforestry for Soil Conservation. Wallingford, United Kingdom: CAB International.

Zinke, P. J., A. G. Stangenberger, W. M. Post, W. R. Emanuel, J. S. Olson. 1986. Worldwide Organic Soil Carbon and Nitrogen Data. ORNL/CDIC-18. Oak Ridge, Tenn.: Oak Ridge National Laboratory.

PART TWO

Country Profiles

The seven country profiles that constitute Part Two of this book are an integral part of the committee's report. They represent a portion of the data, observations, and insights that the committee amassed during the course of its study. Authors were selected based on broad recognition, by their scientific peers, of their authority and scientific knowledge of the deforestation and sustainable agriculture issues in the selected countries. The profiles on Brazil, Côte d'Ivoire, Indonesia, Malaysia, Mexico, the Philippines, and Zaire portray the pressures on natural resources that these countries face and ways they can be mitigated. They tell part of the story of what is happening in the humid tropics.

The profiles represent each of the three major humid tropic regions—Africa, Asia, and Latin America—and include discussions on land use and forest conversion, general causes and consequences of deforestation, sustainable land use alternatives, and policy implications. Discussions focusing on only 7 of the more than 60 countries lying within the humid tropics cannot and do not represent the status of science, agricultural and land use practices, and policy of all humid tropic countries. They do, however, illustrate the diversity of production systems, with their unique environmental, social, and market niches, that can be found in any given locale or region. These varied presentations reinforce the committee's three major findings concerning the potential to restore degraded lands, the range of appropriate land uses, and the capacity for general economic growth with real-world examples.

No single type of land use can simultaneously meet all the requirements for sustainability or fit the diverse socioeconomic and ecological conditions found throughout the humid tropics. The seven country profiles provide examples of many of the options within the land use continuum that the committee outlines in Part One. They also illustrate the committee's view that progress toward sustainability in the humid tropics depends not only on the availability of improved techniques of land use, but on the creation of a more favorable environment for their development, dissemination, and implementation.

Brazil

Emanuel Adilson Souza Serrão and
Alfredo Kingo Oyama Homma

Deforestation of the Brazilian Amazon, the largest tropical forest reserve on the planet, has attracted worldwide attention in recent years. The environmental disturbances have been claimed to be a result of agricultural developments over the past 3 decades. Because of the increasing rural and urban population demands for food and fiber and the need for environmental conservation and preservation, however, land in the Brazilian Amazon must be used on a sustainable basis. The search for a compromise between ecologic and population demands is a major challenge to those in governmental, nongovernmental, and private institutions. This profile addresses the questions of agricultural sustainability in the Brazilian humid tropics by analyzing the important present and potential land uses and by considering their sustainabilities and potential for improvement and expansion.

BASIS FOR SUSTAINABILITY ANALYSIS OF AMAZONIAN AGRICULTURE

Sustainability must be the basis for analysis and implementation of agricultural land use alternatives for the Brazilian Amazon, but

Emanuel Adilson Souza Serrão is a research agronomist and Alfredo Kingo Oyama Homma is a socioeconomist at the Centro de Pesquisa Agroflorestal da Amazônia Oriental (Center for Agroforestry Research of the Eastern Amazon), Empresa Brasileira de Pesquisa Agropecuária (Brazilian Enterprise for Agricultural Research), Belém, Brazil.

few analyses have provided insight (Alvim, 1989; Fearnside, 1983, 1986; Homma and Serrão, In preparation). The possibility of developing sustainable agriculture in the Amazon depends on its permanence in an area and on increasing land and labor productivity standards, thereby reducing the pressure for more deforestation. This concept of sustainability implies an equilibrium in time among agronomic and/or zootechnical, economic, ecologic, and social feasibility. Equilibrium is frequently fragile in Amazonian agricultural systems, and no agricultural land use system in the Amazon meets all four of these prerequisites for sustainability at highly satisfactory levels.

The land use systems analyzed here were selected because of their present and potential importance characterized by their scale of utilization (for example, total area used and number of farmers involved), the types of farmers that use each system, its economic importance, possibilities for future markets, environmental implications, and possibilities for agroindustries. Characterization also includes technological patterns (for example, land and labor use intensity, input utilization, adoption of technology, product processing, and management practices) and productivity patterns (for example, maintenance of productivity, productivity increase potential, and relationship between productivity and the environment).

More than enough land has already been deforested for agricultural development in the Amazon. From a technical point of view, by using only about 50 percent of the already deforested land and other less fragile ecosystems, such as well- and poorly drained savannahs and alluvial floodplains, it is possible to produce sufficient amounts of food and fiber to meet the demands of the region's population for the next decade at least. Future agricultural production in the Amazon will depend on higher levels of land use intensification with decreasing rates of deforestation (the decreasing deforestation brought about as a result of increasing national and international pressures for environmental conservation, increasing local environmental ethics, and increasing population density and, consequently, higher land prices). Productivity and sustainability must be the foundation for future agricultural development. In this scenario, agricultural technology will play the major role.

THE BRAZILIAN HUMID TROPICS

The Brazilian humid tropics encompasses the geographic area that has been named, for development purposes, the legal Amazon, an area of about 510 million ha, corresponding to 60 percent of Brazil's national territory.

Although there has been a significant increase in population density in the Amazon during the past 3 decades, only about 10 percent (16 million) of Brazil's population inhabits this immense region (Brazilian Institute of Geography and Statistics, 1991). This population is unevenly distributed throughout the region in densely populated nuclei separated by extensive, virtually uninhabited land.

The average population density in the Amazon is about 2.7 inhabitants per 100 ha. Presently, 61 percent of Brazil's population in the northern region lives in urban areas, and a significant portion of that population lives on the outskirts of Belém, Manaus, and other major cities. The region's population is expected to grow moderately in the next 2 decades, increasing from the present 16 million people (in 1990) to 26 million by 2010 (a 62 percent increase). This means that the Amazon population at the end of the first decade of the next century will be 13 percent of the country's population compared with the present 11.4 percent (Medici et al., 1990; Superintendency for the Development of the Amazon, 1991).

In general, per capita income in the Amazon region is very low, equivalent to US$1,271 (1991), which represents 51.5 percent of Brazil's per capita income (Superintendency for the Development of the Amazon, 1991).

The Environment

The Amazon hydrographic basin covers about 6 million km² and is considered the largest river network in the world. It is navigable along 20,000 km of waterways and has a total watershed area of about 7.3 million km². This network includes muddy-water rivers that originate in alluvial soil regions. The rivers deposit organic and inorganic sediments along their paths, forming floodplains locally called *várzeas*. These floodplains are rich in nutrients and organic matter and have a high potential for agricultural development.

The Amazonian climate is predominantly hot and humid and often presents conditions for high levels of biomass production. Relatively large amounts of solar radiation reach the earth's surface throughout the year. Average temperatures vary between 22° and 28°C, the daily variations being considerably higher than seasonal variations. Relative humidity tends to be high in most of the region, varying from about 65 to 90 percent. Total annual rainfall varies between 1,000 and over 3,000 mm. The rainy season is from December and January through May and June in most of the region, and a dry season occurs during the rest of the year.

The vegetation that covers the Amazon is related to climatic conditions, but rain forests are the predominant ecosystem. The main types of vegetation are dense upland forests, open upland forests, savannah-type vegetation that includes well- and poorly drained savannahs, and alluvial floodplain (*várzea*) vegetation (Nascimento and Homma, 1984). Dense upland forests, which have high levels of biomass and include the tallest tree species, occupy about 50 percent of the legal Amazon. Open forests, which have a considerably smaller biomass volume, shorter trees, and more palm species and lianas, occupy about 27 percent of the region. Well-drained savannah vegetation (*cerrado*) with different arboreal and herbaceous gradients occurs in extensive areas in the states of Amapá and Roraima and occurs less extensively in areas in other parts of the region, where the forest is interrupted.

About 80 percent of the legal Amazon (430 million ha) is upland, nonflooding area. The remaining 20 percent (70 million ha) is floodable area (Nascimento and Homma, 1984). Nascimento and Homma (1984) estimate that approximately 88 percent (450 million ha) of Amazonian soils are dystrophic (acidic and low in fertility) and that the remaining 12 percent (50 million ha) is eutrophic (less acidic and relatively high in fertility). Of the latter, 25 million ha is upland soils, and 25 million ha is floodable soils.

Macroecologic Units

At least one attempt (Nascimento and Homma, 1984) has been made to combine natural resources information by superimposing climate, soil, and vegetation maps to locate macroecologic units suitable for agricultural development, conservation, and preservation in the Amazon (Table 1). These macroecologic units and their distributions could be useful for making the first approximations of agroecological zoning in the Amazon.

AGRICULTURAL DEVELOPMENT

To evaluate agricultural sustainability in the Brazilian Amazon, it is important to examine agricultural development chronologically and from the physical and economic viewpoints.

Chronological Agricultural Development

The history of the development of the Amazon is pinpointed with ill-fated booms, badly oriented development projects, some partial successes, and ecologic and social mishaps (Norgaard, 1981).

TABLE 1 Macroecological Units of the Legal Amazon

Climate	Mapping Unit		Approximate Area (million ha)	Percentage of Total Legal Amazon Area
	Vegetation	Soil		
Afi	Dense forest	Upland dystrophic	60.0	11.66
		Floodplain eutrophic (*várzea*)	7.9	1.53
		Floodplain dystrophic	4.5	0.88
	Open forest	Upland dystrophic	13.0	2.53
		Floodplain eutrophic (*várzea*)	0.8	0.15
	Open native grassland	Upland dystrophic	0.2	0.04
		Floodplain eutrophic (*várzea*)	1.0	0.21
	Subtotal		87.4	17.00
Ami	Dense forest	Upland eutrophic	5.2	1.02
		Upland dystrophic	116.4	22.64
		Floodplain dystrophic (*várzea*)	11.3	2.19
		Floodplain dystrophic	13.7	2.66
	Open forest	Upland eutrophic	12.7	2.48
		Upland dystrophic	16.5	3.22
		Floodplain eutrophic (*várzea*)	2.3	0.45
	Savannah (*cerrado*)	Upland eutrophic	1.0	0.20
		Upland dystrophic	12.2	2.37
		Floodplain eutrophic (*várzea*)	0.5	0.10
	Open native grassland	Upland eutrophic	1.0	0.19
		Upland dystrophic	10.9	2.11
		Floodplain eutrophic (*várzea*)	4.0	0.77
		Floodplain eutrophic	3.1	0.60
	Subtotal		210.8	41.00

continued

TABLE 1 *Continued*

Climate	Mapping Unit Vegetation	Soil	Approximate Area (million ha)	Percentage of Total Legal Amazon Area
Awi	Dense forest	Upland eutrophic	1.5	0.30
		Upland dystrophic	27.7	5.38
		Floodplain eutrophic (*várzea*)	0.8	0.16
		Floodplain dystrophic	1.9	0.37
	Open forest	Upland eutrophic	5.6	1.09
		Upland dystrophic	80.8	15.72
		Floodplain eutrophic (*várzea*)	0.25	0.05
		Floodplain dystrophic	7.4	1.45
	Savannah (*cerrado*)	Upland eutrophic	3.4	0.67
		Upland dystrophic	66.0	12.83
		Floodplain dystrophic	5.1	1.00
	Open native grassland	Upland eutrophic	0.8	0.16
		Upland dystrophic	11.7	2.28
		Floodplain dystrophic	2.8	0.54
	Subtotal		215.75	42.00
Total			513.95	100.00

SOURCE: Adapted from Nascimento, C. N. B., and A. K. O. Homma. 1984. Amazônia: Meio Ambiente e Tecnologia Agrícola. Documento 27. Belém, Brazil: Brazilian Enterprise for Agricultural Research–Center for Agroforestry Research of the Eastern Amazon.

Even though mining and energy-producing projects have emerged as the main development thrusts in the Amazon, associated development activities, including agricultural activities, usually follow in their wake (Smith et al., In press-a,b). For this reason, some important historical aspects of agricultural development in the Amazon that will pave the way to a better understanding of the analysis of agricultural sustainability given later in this profile are presented here.

From the early seventeenth to the early twentieth centuries, agricultural development in the Amazon depended on extraction activities in existent forests. Even today, *extrativismo* (extractive land use) plays a very significant role in the regional economy, mainly because of the commercialization of timber, heart of palm, rubber, and Brazil nuts, among other forest products, in addition to hunting and fishing.

More modern agricultural and livestock development began to take place toward the end of the first quarter of the twentieth century along the relatively fertile *várzea* floodplains, not only because of the favorable conditions they offered for agricultural production but also because of favorable river transportation along the Amazon River network.

By the mid-1950s, the *várzea* development gave way to the upland *terra firme* development when road construction started crisscrossing the region. This phase was characterized by extensive agricultural development where forest slash-and-burn activity was the main feature. Road construction was then considered synonymous with progress and made the region attractive to immigrants. Cattle raising, shifting (slash-and-burn) subsistence agriculture, and timber exploration are now the dominant features of upland development (Homma and Serrão, In preparation).

Physical and Economic Agricultural Development

To analyze agricultural sustainability in the Brazilian humid tropics, it is important to have an idea of how and where agricultural development has taken place. More detailed descriptions are given in the literature (Homma, 1989; Homma and Serrão, In preparation; Nascimento and Homma, 1984; Serrão and Homma, In press).

From 1900 to 1953, extraction activities in the Amazon were greater than crop farming and cattle raising, contributing 50 percent of the agricultural gross national product (AGNP) in the region mainly because of the major influence of rubber extraction in the Amazon economy (Homma, 1989). After the mid-1940s, the decline of extraction began with the dissemination of jute cultivation along the Amazon *várzea* floodplains and with the expansion of black pepper agriculture in eastern Pará. From 1965 to 1971, for the first time, crop farming and cattle raising surpassed extraction activities.

The predominance of crop farming and cattle raising over extraction activities was observed in the 1970s and continues to the present. Most of those involved with extraction activities turned to crop farm-

---

Agricultural Development in the Brazilian Amazon

1616–1750	Agricultural activities were primarily the extraction of exotic herbs and medicinal plants as well as spices, especially cacao
1750–1822	Extraction activities and some small-scale expansion of shifting subsistence agriculture and cattle raising activities
1850–1912	Rubber extraction mostly displaced the then prevalent agricultural activities to meet international demand
1927	Henry Ford launched the first and largest private domesticated rubber plantation in Brazil, but the lack of agronomic sustainability led to the enterprise's failure; it was transferred to the Brazilian government in 1945
1932	Japanese immigrants introduced and expanded jute crop agriculture in the floodplains along the upper and mid-Amazon River
1933	Japanese immigrants introduced black pepper, an important source of revenue for the state of Pará
1939–1945	Rubber regained its importance as a strategic product as a result of the Washington Agreement signed in 1942, which guaranteed the supply of natural rubber to the Allied Forces (rubber tree plantations in southeastern Asia were controlled by the Japanese)
1953	Rubber production was greatly stimulated through several government development programs to meet the national rubber demand, but without success
1966	Operation Amazon gave ranchers incentives to raise cattle on pastureland that replaced forestland
1967	The Jari Agroforestry Project on the banks of the Jari River on the Amapá-Pará border was initiated; after a series of technical and political ups and downs, the project was sold to a consortium of Brazilian entrepreneurs in 1982
1970	The federal government launched aggressive development-through- colonization programs along recently built roads

ing and cattle raising, which was also the case with those who came with the migratory flux in that same period.

Shifting agriculture has become the major activity of a large number of small farmers. It is characterized by low levels of technology and low productivity, even though it is a reasonably good alternative for the partial recovery of soil fertility and for the recovery of weed-,

1970s	An important diversification process took place with the expansion and/or introduction of economically important crop production systems of black pepper, coffee, African oil palm, papaya, passion fruit, and melon, among others; this process continued into the 1980s with the expansion of citrus, coconut, Barbados cherry, cupuaçu, and other, less important crops
Early 1970s	Subsistence agriculture, which was initially carried out in the *várzea* floodplain areas, turned to the upland areas along the recently built roads and through the shifting agricultural systems
1976	Intensive cacao production began to be stimulated by the federal government through the Cacao Development Program
1980	The federal government set up the Grande Carajás Program in which the agricultural development component followed in the wake of the mineral exploration component
1987	Pressed by national and international ecologic movements and the autonomous rubber tappers movement, the federal government created the Extractive Allocation Project
1980s	The magnitude and intensity of deforestation and burning in the Amazon generated a great concern in national and international scientific communities and governments; this movement was stirred up in 1988 when rubber tapper leader Chico Mendes was assassinated because of land tenure conflicts
1989	The federal government conceived and created Our Nature Program; along with it, the Brazilian Environmental and Renewable Natural Resources Institute (IBAMA) was created in an attempt to, among other things, control deforestation and help to promote ecologically sustainable development in Brazil, particularly in the Amazon

pest-, and disease-infested areas, because of the accumulation of nutrients in the biomass during the various fallow periods imposed on cultivated tracts of land. However, this land use system has imposed substantial losses of forest resources and is subject to increasing socioeconomic instability when the population density increases.

Extensive cattle raising systems have been predominant in certain areas of the Amazon where natural grassland ecosystems (such as well- and poorly drained savannah grasslands and floodplain grasslands) are available and on the pasturelands that have replaced forests over the past 3 decades. Supported by tax incentive programs, this sector has been responsible for most of the deforestation in the Brazilian Amazon region (Browder, 1988).

The majority of the region's most important transformations in the primary (agricultural production) sector started in the 1960s with the expansion of the agricultural frontier, mostly as a result of tax incentive policies and the construction of important highways, which favored the development of colonization programs and the installation of large agricultural projects, the bulk involving cattle raising. Cattle raising expansion began in the mid-1960s because of the low utilization levels of labor, which was scarce at the time, and the abundance of land.

This most recent regional agricultural development phase is characterized by accelerated, large-scale, and aggressive exploration of natural resources. This replaces the humid tropical forests with land use systems with generally low ecologic and socioeconomic efficiencies (cattle raising projects and shifting agriculture) or large-scale predatory "industrial" extraction activities such as those for timber and heart of palm (*Euterpe oleracea*). Because of the environmental degradation that they cause, these land use systems have been severely criticized (Mahar, 1989).

During the past 3 decades, despite their still modest acreage in relation to shifting agriculture and cattle raising, perennial crop plants such as African oil palm (*Elaeis guineensis*), rubber (*Hevea* spp.), cacao (*Theobroma cacao*), Brazil nut (*Bertholletia excelsa*), guaraná (*Paullinia cupana*), and semiperennials such as black pepper (*Piper nigrum*) and, more recently, urucu (*Bixa orellana*) have become increasingly important. Special government financing programs such as the Cacao Development Program, PROBOR (the Natural Rubber Production Incentives Program), as well as a number of credit lines during the 1970s, give farmers incentives to expand these crops.

Today, there are different forms of agricultural production in the Amazon because of different environmental and basic infrastructural peculiarities. These range from extraction activities in remote areas

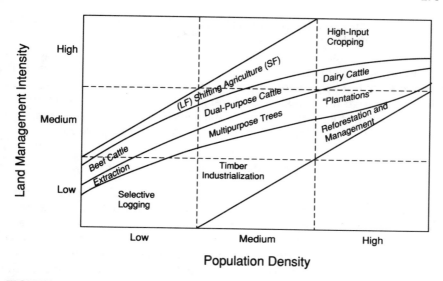

FIGURE 1   Effects of population density on land use in the Brazilian humid tropics. LF, Long fallow; SF, short fallow. Source: Adapted from Serrão, E. A. S., and J. M. Toledo. In press. Sustaining pasture-based production systems in humid tropics. In Development or Destruction: The Conversion of Tropical Forest to Pasture in Latin America, S. B. Hecht, ed. Boulder, Colo.: Westview.

with low population densities to extensive cattle raising, or from agricultural activities in recently opened frontier lands to those in long-occupied areas.

Land use intensification for forest product exploitation, traditional crop production, and cattle production has been influenced by population density and land prices (Figure 1). In areas with low population densities, where land prices are normally low, extraction activities, such as those for rubber, timber, and Brazil nuts, coexist with shifting agricultural systems with long fallow periods and extensive livestock activities (Serrão and Toledo, In press). In areas with medium population densities, land prices are higher, which brings about less extraction activity, shifting agricultural systems with shorter fallow periods, more intensive cattle production, and perennial cropping activities. In areas with high population densities, intensive annual and perennial cropping is expanded, subtracting from activities in areas previously devoted to extraction, shifting agriculture, and extensive cattle raising. Land prices become even higher and intensive agricultural practices are predominant. At this stage, more intensive integrated agricultural production (the agrisilvopastoral approach) begins to take place.

These contrasting situations of population and land use intensity form mosaics where areas have a virtual absence of development, intense spatial expansion, intense agricultural modernization, very intensive spatial expansion, and very high levels of modernization.

There are at least five distinct situations that characterize the present state of agricultural development in the Amazon (Figure 2).

AGRICULTURAL DEVELOPMENT IN NORTHEASTERN PARÁ

The northeastern part of the state of Pará was one of the first areas to be brought into upland agricultural production in the Amazon. After supporting rubber extraction activities by producing and supplying agricultural products to rubber-producing areas in the Amazon, this region went through a series of transformations and now produces about 90 percent of Brazil's black pepper; 50 percent of the national malva (*Urena lobata*) fiber; and most of the Hawaiian papaya, palm oil, passion fruit, oranges, and native fruits produced in the Brazilian Amazon region. This region also produces a significant amount of animal protein, from cattle and poultry.

With approximately 10 million ha (about 8.7 percent of the state's total area) and a population of about 2.5 million inhabitants (or 15 percent of the Amazon region's population), this region is the most densely populated area of the Amazon. About 0.5 million people live in rural areas, where small-scale shifting-agriculture farmers work the land alongside farming operations that use higher levels of technology (mechanization, fertilizers, improved crop management) and where social and physical infrastructures (roads, electricity, communication, health, and education) are satisfactory compared with those of other regional development poles. This region's development has been greatly influenced by the construction and operation of the Belém-Brasília Highway in the 1960s.

The northeastern part of the state of Pará has the most developed agroindustry in the Amazon region, mainly in relation to timber, African oil palm, jute (*Corchorus capsularis*), and malva fiber and meat processing. In Belém, extraction and agriculture of several products such as wood, Brazil nut, rubber, guaraná, native and exotic fruits, and other crops are industrialized.

In relative terms, and considering the Amazon as a whole, the northeastern part of the state of Pará is where agricultural development has the highest levels of sustainability because of its adaptation over time.

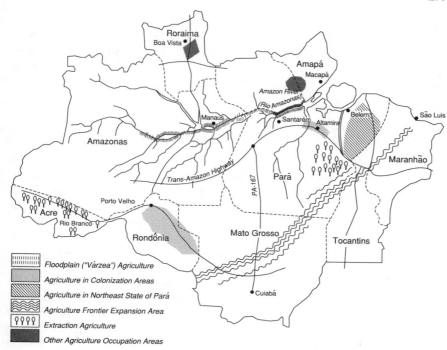

Floodplain ("Várzea") Agriculture

Agriculture in Colonization Areas

Agriculture in Northeast State of Pará

Agriculture Frontier Expansion Area

Extraction Agriculture

Other Agriculture Occupation Areas

FIGURE 2 Main agricultural development areas in the Amazon.   Source: Adapted from Nascimento, C. N. B., and A. K. O. Homma. 1984. Amazônia: Meio Ambiente e Tecnologia Agrícola. Documento 27. Belém, Brazil: Brazilian Enterprise for Agricultural Research–Center for Agroforestry Research of the Eastern Amazon.

AGRICULTURE IN VÁRZEA FLOODPLAINS

This type of agriculture has developed mainly along the margins of the Amazon and Solimões rivers on fertile *várzea* floodplain soils subjected to an annual flooding and receding water regimen. It was the first major agricultural development in the region, facilitated by river navigation, before the beginning of the road-building era in the 1960s. It has lost some if its importance over time, however, because of the decline in extraction activities (McGrath, 1991) and the increasing attraction of more dynamic areas in the region. There has been a strong tendency to migrate from the rural riverbank areas to main urban nuclei, resulting in almost stagnant agricultural development after 30 years of agricultural predominance by jute.

In addition to jute and malva fiber and subsistence food and fruit crops, beef and cow's milk (although limited somewhat by periodic flooding of the native floodplain grasslands) are also produced. There is also some timber, jute and malva fiber, rubber, and Brazil nut processing as well as good aquatic food sources, mainly fish. There is water buffalo raising potential in the floodplains and estuaries of the Amazon.

AGRICULTURE IN FRONTIER EXPANSION AREAS

At the outset of the 1970s, a dynamic period of agricultural development occurred primarily in the south of Pará, in the north of Mato Grosso, within Tocantins, and in the south of Maranhão. Road construction, tax incentives (where the Superintendency for the Development of the Amazon [SUDAM] has had a major role), and credit availability were the main driving forces for this development. In this development process, cattle ranches have been established. These are surrounded by small shifting agricultural plots cultivated by squatters, who also serve as labor for the cattle ranches.

Development in this area has been characterized by frequent land ownership conflicts in which religious groups and the government have played conflicting roles. In some areas, land conflicts are due to (1) invasion by squatters in areas already occupied by people who depend on the extraction of Brazil nuts and (2) large influxes of gold prospectors who, when they are unsuccessful in their search for ore, look for alternative livelihoods. The interconnection of the Belém-Brasília and Trans-Amazon highways, the construction of the Carajás-São Luís Railroad, and state roads such as the PA-150 made this region the point of entry of migratory fluxes from the northeastern part of Brazil. The implementation of the Carajás iron-processing plants and the discovery of gold in the Serra Pelada area, among other factors, induced the development of small farms and, consequently, the migratory flux to this particular region.

Large-scale cattle raising, which involves slash-and-burn destruction of the forest, has been severely criticized for its role in the region's deforestation. One of the reasons for land conflicts is the dichotomy of cattle raising, which demands large tracts of land for pasture establishment (to cover up for rapid pasture degradation) with low labor use, which then limits employment and becomes incompatible with the needs of small-scale farmers, who need to work outside their own plots to supplement their income.

Even though there has been development along important frontier highways, the infrastructures of frontier expansion areas are still deficient, particularly for small-scale farmers. Even so, many frontier areas in this region became municipalities in the 1980s. Large private colonization projects were also developed. The agricultural segments of these projects contemplate improved land use systems for coffee, cacao, black pepper, rubber, guaraná, and beef cattle.

Another agricultural development front is developing in western Maranhão. This region has Brazil's northeastern economic, social, and cultural characteristics and abundant labor force and roadways. The main agricultural activities are food crop production (mainly rice), cattle raising, and babassu palm (*Orbignya martiana*) extraction.

AGRICULTURE IN OFFICIAL COLONIZATION AREAS

Official colonization areas have been occupied mainly by farmers whose origins are in Brazil's northeastern and south-central regions and who were stimulated by the official colonization programs started in the early 1970s. While SUDAM played a major role in the agricultural development in frontier expansion areas, the Land Reform and Colonization Institute took the leading role in official colonization areas.

Two distinct regions were important in the context of official colonization. One was the region along the Trans-Amazon Highway, colonized mainly by landless northeastern Brazilians who left their region of origin because of socioeconomic constraints and prevailing severe droughts. Cacao, sugarcane, and food crop production were predominant agricultural activities. However, during the last 20 years of development, cattle raising also became important, causing the fusion of many agricultural lots owned by small-scale farmers.

Another colonization settlement was developed in different points in the former territory that is now the state of Rondônia. In this case, there was an intensive spontaneous and programed migratory flux of farmers from the northeast and south-central regions of Brazil who dedicated themselves to growing cacao, coffee, rubber, and food crops.

Agricultural lots have gone through significant amounts of fusion induced by a shortage of labor (displaced by gold mining activities), low cacao and coffee prices, and credit and tax incentives for cattle raising activities. Several milk-processing plants also operate in this region.

AREAS OF FOREST PRODUCT EXTRACTION

Areas where extraction of forest products is predominant are widespread in the Amazon and include different combinations of forest extraction and agricultural activities of various intensities. Some are very old, going back to the initial occupation of the region, and are now in a state of almost economic stagnation and population increase.

The most important area of extraction activity is in the state of Acre, where rubber tapping is the main activity for 55,000 gatherers who, in some measure, are also involved in complementary shifting agriculture and Brazil nut gathering.

Because of the expansion of the agricultural frontier, rubber tappers are able to maintain their activities with intensive support from national and international movements. This expansion pressured the Brazilian government to create, in 1987, the Settlement of Extractive Areas Project. This project established guidelines for the settlement of extractive reserve regions as a specific mode of agrarian reform in the Amazon region. That model was recently (1990) transformed into the Extractive Reserve. This initiative was an important factor in reducing the accelerated expansion of the agricultural frontier.

The rubber tapper's main drawback is their artificially maintained economic sustainability, which, because of the current weakness of their economic base, has been exogenously supported by the taxation of imported rubber. Their main strength is their successful organization.

After the assassination of rubber tapper Chico Mendes in December 1988, ample discussion has taken place in Brazil and elsewhere, bringing about an "extraction syndrome" that portrays the idea of extraction as the model for feasible development of the Amazon as a sustainable system. The emotional environment generally involved in the subject of extraction has been a limiting factor in discussing the matter technically and objectively.

DEFORESTATION FOR AGRICULTURAL DEVELOPMENT

Deforestation in the Brazilian Amazon region is closely connected to agricultural development, mainly with shifting agriculture, cattle raising, and logging activities. Because of this and because the extent, rate, causes, and consequences of deforestation have been a major concern worldwide, some highlights are stressed here.

Extent of Deforestation

A number of estimates of the extent of deforestation in the Amazon have been published previously (Brazilian Institute of Space Research, 1990; Fearnside, 1982, 1984; Mahar, 1989; Senado Federal, 1990). Some of those estimates and others publicized in leading national and international newspapers and magazines have overestimated the extent of deforestation and, in most cases, are associated with somewhat exaggerated and alarming trends in environmental degradation and its consequences.

The estimates of the Brazilian Institute of Space Research (Instituto de Pesquisas Espacias; INPE) are probably the most trustworthy. A Brazilian Senate committee's final report (Senado Federal, 1990), published in 1990 and reflecting INPE's estimates (Brazilian Institute of Space Research, 1990), indicated that until 1989, some 34 million ha of Amazon forest of various biomass gradients was deforested. This represents about 7 percent of the legal Amazon region and an area corresponding to seven Costa Ricas or to about the amount of cultivated land in Italy, England, and France. Table 2 gives the extent and rate of deforestation in the so-called Legal Amazon through 1990.

Rate of Deforestation

Even though the figures given above may not be considered alarming if the total Amazon forest area is taken into account, the speed with which deforestation has been taking place in the past 2 decades is disturbing.

The Brazilian Senate committee (Senado Federal, 1990) report shows that in only 11 years (from 1978 to 1989, when total deforestation reached 7 percent of the area of the legal Amazon), there was a rapid increase in deforestation (417 percent). This time frame coincides with the most active period of migration to the region. According to the report, the state of Rondônia suffered the most intensive deforestation (about 12 percent in 1989).

Since the creation of Our Nature Program (Programa Nossa Natureza) and the consequent advent of the Brazilian Institute of Environment (Instituto Brasileiro do Meio Ambiente e dos Recursos Naturais Renovaveis, IBAMA) in 1989, the trend has been in the direction of decelerating deforestation.

According to Alcântara (1991), deforestation was 2.1 million ha in 1989 and 1.4 million ha in 1990. Deforestation in 1991 was 1.11 million ha, according to the Brazilian Institute of Space Research. Besides ecologic conscientiousness and control of forest burning by government agencies—especially IBAMA—the economic crisis in Brazil explains the trend in deforestation. The exaggerated estimates for 1987, which indicated that 8 million ha was deforested, were probably due to the lack of experience during the first year of the INPE/IBDF (Brazilian Institute of Forest Development) (now IBAMA) agreement. In reality, 60 percent of the fires detected were the result of burning for pasture management in already existing pasturelands.

TABLE 2 Deforestation in the Legal Amazon Through 1990

State	Original Forest Area (km² [in thousands])	Deforested Area (km² [in thousands])				Deforested Area (percent of area originally in forest)				Rate of Deforestation (km² [in thousands]/year)		
		Jan. 1978	Apr. 1988	Aug. 1989	Aug. 1990	Jan. 1978	Apr. 1988	Aug. 1989	Aug. 1990	1978–1988	1988–1989	1989–1990
Acre	154	2.5	8.9	9.8	10.3	1.6	5.8	6.4	6.7	0.6	0.6	0.6
Amapá	132	0.2	0.8	1.0	1.3	0.1	0.6	0.8	1.0	0.1	0.2	0.3
Amazonas	1,561	1.7	17.3	19.3	19.8	0.1	1.1	1.2	1.3	1.6	1.3	0.5
Maranhão	155	63.9	90.8	92.3	93.4	41.2	58.5	59.5	60.2	2.7	1.4	1.1
Mato Grosso	585	20.0	71.5	79.6	83.6	3.4	12.2	13.6	14.3	5.1	6.0	4.0
Pará	1,218	56.3	129.5	137.3	142.2	4.6	10.6	11.3	11.7	7.3	5.8	4.9
Rondônia	224	4.2	29.6	31.4	33.1	1.9	13.2	14.0	14.8	2.3	1.4	1.7
Roraima	188	0.1	2.7	3.6	3.8	0.1	1.5	1.9	2.0	0.2	0.7	0.2
Tocantins/Goiás	58	3.2	21.6	22.3	22.9	5.4	37.0	38.3	39.3	1.7	0.7	0.6
Amazônia Legal	4,275	152.1	372.8	396.6	410.4	3.6	8.7	9.3	9.6	21.6	18.1	13.8
Hydroelectric, inundated forest	—	0.1	3.9	4.8	4.8	0.0	0.1	0.1	0.1	0.4	1.0	0.0
Deforestation, all sources	—	152.2	376.7	401.4	415.2	3.6	8.8	9.4	9.7	22.0	19.0	13.8

SOURCE: Comissão Interministerial para a Preparação da Conferência das Nações Unidas Sobre Meio Ambiente e Desenvolvimento. 1991. Subsídios Técnicos para a Elaboração do Relatório Nacional do Brasil para a CNUMAD. Brasília: Brazilian Institute for the Environment and Renewable Natural Resources.

In general, the importance of shifting subsistence agricultural activities in relation to deforestation in the Amazon region has been purposely overlooked for political and socioeconomic reasons. In 1985, the area in the northern region actually cultivated with short-cycle crops was estimated at about 1.35 million ha (Brazilian Institute of Geography and Statistics, 1991). However, despite the reduced individual lot sizes for shifting agriculture (between 10 and 50 ha), if one considers that there are more than 500,000 small-scale farmers who practice it in the Amazon, that each farmer cultivates an average of 2 ha for 2 consecutive years, and that these 2 ha are left to fallow for about 10 years, this activity is responsible for altering at least 10 million ha in a process of "silent deforestation" (Homma, 1989).

One implication for estimating the contribution to deforestation by different land use systems is the fact that farm plots devoted to annual crop farming are frequently sold or abandoned after only a few years of use, mainly because of rapidly declining yields. In general, they are then converted to pasturelands, increasing the area devoted to cattle raising. Therefore, some of the deforestation attributed to livestock development may have been caused by the spread of small-scale agriculture (Mahar, 1989).

Logging has been practiced in the Amazon for over 300 years (Rankin, 1985). For most of that time it was done manually and was restricted to relatively accessible, seasonally inundated forests. With the advent of road construction in the 1960s, interfluvial forests became more accessible to loggers. When this is combined with the depletion of native forests in southern Brazil and SUDAM's incentives for timber extraction operations, the result has been very large-scale logging activities in the region during the past decade (Uhl and Vieira, 1989). In 1978, 7.7 million m^3 of wood was harvested from the Amazon forest. In 1987, the harvest rose to 24.6 million m^3.

In 1987, the Amazon region contributed 55 percent of domestic timber production, in comparison with 24 percent in 1978 (IBGE, 1989). The advent of chainsaws in the 1970s resulted in technologically more efficient logging operations. This has resulted in a more than 30-fold increase in logging productivity over that from manual logging and has been a major factor contributing to logging intensity in the region. It is not clear how much deforestation can be attributed to logging because much of the timber extracted is a by-product of land clearing for other agricultural purposes (Mahar, 1989), mainly cattle raising and shifting agriculture.

Even though selective logging by itself results in the removal of only a few trees from the forest, the process causes considerable damage to the forest structure. In a selectively logged dense forest in the eastern part of the state of Pará, Uhl and Vieira (1989) found that although only 16 percent of the existing trees were harvested, 26 percent of the remaining trees were killed or damaged. On the basis of recent satellite imagery of disturbed forestlands, checking on the ground, and the number of sawmills (and their capacity to process timber), it is estimated that logging has accounted for about 10 percent of total deforestation in the state of Pará (Watrin and Rocha, In press).

These proximate causes (Mahar, 1989) of deforestation for agricultural development are consequences of government policies designed to open up the Amazon for human settlement and to encourage other types of economic activities.

Government policies and the consequent proximate causes of deforestation in the region do not reflect merely the regional needs for agricultural development, however. Most of the driving forces pushing deforestation in the Amazon result from a series of largely unseen causes nationwide, such as high population growth rates (more than 3 million people per year), high inflation, a socioeconomic environment in which land is a valuable reserve, unequal income distribution, lack of technological improvement in extra-Amazon areas, insufficient scientific knowledge of the region's natural resources, low levels of regional agricultural technology, external market growth for wood products, low education levels, high agricultural input costs, conflicting development and environmental policies, legislation inconsistent with the environmental conservation, weak law enforcement, and a large foreign debt.

The great problem, however, is the fact that the slash-and-burn practice is the cheapest alternative land preparation method for farmers. To use already deforested lands, mechanization and application of lime, fertilizers, and other modern inputs are required at an estimated cost of US$400/ha, in comparison with US$70/ha for the traditional slash-and-burn process.

Environmental Impacts of Deforestation

Deforestation for agricultural development in the Brazilian Amazon region has been closely connected with environmental disturbances, mainly climate change, loss of biodiversity, soil erosion, flooding, and the impact of smoke. Typical deforestation contributes to the increase in the atmospheric carbon dioxide concentration and, therefore, to the possible warming of the earth that may result from this increase.

To a large extent, agricultural development in forested areas of the Amazon has been based, for traditional and socioeconomic reasons, on slash-and-burn practices and pasture formation and management. Because of its intensity in the region—as many as 8 million ha were burned for agricultural purposes in the Brazilian Amazon in 1987, the highest annual incidence ever observed (Brazilian Institute of Space Research, 1990)—present and potential fire hazards have been a major concern. When the susceptibility to fire of four different dominant vegetation cover types in the eastern Amazon was studied, it was found that cattle pastures were the most fire-prone ecosystem; this was followed by selectively logged forests and second-growth (*capoeira*) vegetation. The primary forest is practically immune to fire (Uhl and Kauffman, 1990; Uhl et al., 1990a).

Despite its socioeconomic importance to agricultural development in the region (Falesi, 1976; Serrão, et al., 1979), fire has probably caused more damage than benefits in the process of agricultural development. In addition to destroying biomass, it contributes to losses in biodiversity (Uhl and Kauffman, 1990; Uhl et al., 1990a) and atmospheric pollution through the release of gases (principally carbon dioxide, methane, and nitrous oxide) that contribute to the greenhouse effect (Goldemberg, 1989; Salati, 1989, In press).

In general, estimates of the quantity of greenhouse gases released when forests are cleared are imprecise because of uncertainties regarding the extent of cleared areas, the amount of biomass per hectare, the amount of carbon in the biomass, and the conversion rates of carbon in biomass burning. Despite these uncertainties, Serrão (1990) estimates that during the past 20 years, conversion of forest to pasture consumed about 5.2 billion metric tons of forest biomass and caused a net increase in atmospheric carbon dioxide of about 2.4 billion metric tons. If carbon dioxide emissions from pasture management burning are added, it is possible that deforestation for pasture in the Amazon alone has contributed to up to 6 percent of carbon dioxide worldwide emissions.

Even though specific data are not available to quantify the local adverse effects of deforestation and burning for agricultural development in the Amazon, the local probable adverse effects are increases in temperature (20° to 50°C) and albedo (up to 100 percent) and decreases in evapotranspiration (30 to 50 percent), rainfall (20 to 30 percent), relative humidity (20 to 30 percent), and water infiltration (10 to 100 percent) (L. C. B. Molion, Instituto Nacional de Pesquisa da Amazônia, personal communication, 1990). The most relevant consequence of deforestation and burning at the local level is soil degradation, with soil loss rates of up to 300 metric tons/ha/year caused primarily by runoff (as a result of a 15 to 20 percent reduction in the interception of rainwater) carrying between 4,000 and 5,000 m³ of water (with soil) to streams and rivers (L. C. B. Molion, unpublished data).

The inability to predict the environmental impacts of deforestation by burning is partly because of a lack of understanding of the natural functions of the Amazon forest. Nepstad et al. (1991), for example, found that some Amazon forest trees have roots that extend to 12 m in depth and are therefore able to draw water from the soil throughout prolonged dry periods. The climatic aspect of the loss of these dry season functions is unknown.

MACROLIMITATIONS FOR SUSTAINABLE AGRICULTURAL DEVELOPMENT

Environmental and socioeconomic characteristics of the Brazilian Amazon region place important limitations on the existence, maintenance, or implementation of sustainable agricultural development. The present level of scientific knowledge and socioeconomic development precludes mid- and long-term generalizations. Therefore, the following are some exogenous and endogenous variables that influence agriculture sustainability in the Amazon but are not controlled by farmers.

Climate

Climatic factors are difficult to influence and almost impossible to control, despite their decisive influence on the types of crops that are planted and their dominant effect on almost all agricultural operations and biologic processes (Croxall and Smith, 1984).

The hot and humid climate reduces the efficiency of humans, animals, and land. Humans work less efficiently in hot climates (Kamarck, 1976). The hot and humid climate of the Amazon is frequently associated with high biotic pressures and acidic and infertile soils, conditions that are serious limiting factors for the sustainability of most crops in the region. In the humid tropics, unusually long dry spells determine agricultural sustainability. They have been occurring in the Amazon more frequently now than they did in the past.

Because of the Amazon's climatic characteristics, the most favorable environmental conditions for primary productivity are through photosynthesis by plants (Alvim, 1990). It is through photosynthesis that plants incorporate approximately 95 percent of their biomass components, namely, carbon (44 percent), oxygen (45 percent), and hydrogen (6 percent), from water and air, not from the soil. Chemical components from the soil make up only about 5 percent of the solid matter in the plant biomass. The total annual solar radiation reaching the Amazon is, for that reason, the greatest environmental factor that determines the primary productivity potential of the region.

Biotic Pressure

According to Goodland and Irwin (1977), the conversion of the humid tropical forest for agricultural production maximizes the return on a short-term basis, but this causes an invariable discontinuity of future production. This is because of high levels of soil leaching, organic matter decomposition, and biotic pressures.

Weeds, pests, and diseases are the most important limiting factors for increased production and productivity in the Brazilian humid tropics. Production losses because of biotic pressure have been estimated to be between 20 and 30 percent without including losses from storage (Croxall and Smith, 1984).

Despite the high economic importance of weeds as a limiting factor for sustainability in crop- and pasturelands, little is known about the extent to which they contribute to economic losses in the Amazon. However, a few million dollars is probably spent annually for weed control in crop- and pasturelands. Hundreds of weed species have been identified in croplands (Stolberg and de Souza, 1985) and cultivated pastures (Camarão et al., 1991; Dias Filho, 1990; Hecht, 1979) in the Amazon. This large number of weeds and their varied morphological features are limiting factors for their efficient control (Dias Filho, 1990). There is much yet to be learned about weed management and control in crop- and pasturelands in the Brazilian humid tropics.

Pests and diseases have been serious limiting factors for crop and pasture production in the Amazon. Some diseases are worth mentioning, such as rubber tree leaf blight caused by the fungus *Microcyclus ulei*, cacao witchbroom caused by the fungus *Crinipellis perniciosa*, black pepper fusarium caused by the fungus *Fusarium solani* f. sp. *piperis*, African oil palm fatal yellowing caused by a still-unknown agent, tomato bacterial wilt, and the *Phaseolus* bean mela caused by the fungus *Rhyzoctonia solani*. Insect pests such as pasture spittle bugs (mainly *Deois* species) and caterpillars and other short-cycle crop insects can cause severe damage and economic losses to crop- and pasturelands (Silva and Magalhães, 1980).

Soil-Related Limitations

About 70 percent of the existing land in the Brazilian humid tropics is appropriate for crop production, about 15 percent is appropriate for cultivated and native grasslands and forestry, and the remaining area has strong limitations for agricultural development and should be left as ecologic reserves (Silva et al., 1986).

Infrastructural deficiencies, price and market fluctuations, and the adoption of the same agricultural production practices that colonizers used on their original land explain why various agriculture-based products have failed on these relatively fertile lands. However, regions with low fertility and acidic soils have not been transformed into deserts, as some have foreseen (Goodland and Irwin, 1975). On the contrary, such regions have been very dynamic in terms of agricultural development.

Sociocultural Limitations

Agricultural sustainability in the Amazon is strongly influenced by sociocultural constraints (Homma and Serrão, In preparation). The low educational levels of most of the rural populations affects the dissemination of improved agricultural technologies because an inability to read and write increases the time and costs necessary for disseminating information.

Land, work, and capital have traditionally been considered the basic factors of productive agricultural systems. Land includes all natural resources, but soil and climate are the basic factors. Work includes labor and management. Capital is represented by funds for agricultural operations and infrastructure (Goedert, 1989). The failures of many agricultural development programs in the Amazon have been, among other factors, a result of inefficiency or neglect in the management of these programs, where misuse of government funds—for example, fiscal incentives or rural credit—has been a major limiting factor (E. B. Andrade, personal communication, 1991).

The solutions for small-scale farming in the Amazon are frequently complex. Basing his evaluation on scientific data, the technician tends to design a technology that saves land, inputs, or labor. However, the small-scale farmers's criteria for evaluating their own technologies are more complex and include factors such as the quantity and quality of certain agricultural products for consumption and sale, income, benefit per unit of work, and security offered by production systems in terms of reduced risk. These criteria are applied intuitively. For example, in a survey carried out in one colonization nucleus in the county of Altamira in the state of Pará (International Center for Tropical Agriculture, 1975), the farmers listed their limiting factors in the following order: health deficiency; lack of seeds, fertilizers, and transportation; low prices for their products; and the presence of pests and diseases. The project technicians, however, listed limiting factors in the following order: lack of transportation, low prices for products, pests and diseases, lack of seeds and fertilizers, and health problems.

Health factors undoubtedly affect agriculture-based colonization projects in the Amazon (Dias and de Castro, 1986). In an agricultural system whose efficiency depends on labor productivity, minimum health standards are needed. In frontier areas, high incidences of endemic diseases require that the health question be treated with proficiency.

In summary, farmers synthesize the human factor, but they are not the only humans involved. Even though they make the decisions for the agricultural operation, they are influenced by the willingness, intelligence, ability, and honesty of politicians, decision makers, consumers, and others. The capacities of farmers are limited not only because of their own limited abilities but also because of limited facilities and a limited work force.

Political Limitations

In general, development policies for the Brazilian Amazon region have shown low levels of efficacy in the internalization of income and labor, reinforcing the tendency to concentrate development activities within a few states, mainly Pará and Amazonas, and in the urban areas of state capitals. The penetration of capital into the field has determined the disarticulation of traditional activities in rural areas, stimulating large-scale rural-to-urban migration, which, in association with migratory fluxes, results in increasing social tensions regarding land ownership, swelling of populations in cities, and growing urban unemployment and underemployment. It has resulted in the deterioration of the population's quality of life (Homma and Serrão, In preparation).

This situation makes it clear that there is an "Amazonian cost of development"—that is, a set of difficulties for those who want to invest in developing the Amazon. It includes infrastructure deficiency, long distances, reduced stocks of technology, low labor and land productivities, limited access to capital, and other factors that aggregate more to regional than to national financial costs (Superintendency for the Development of the Amazon, 1986).

Agricultural development in the Amazon must be related to other sectors of the economy. The rural-to-urban migration that is under way does not correspond to significant changes in agricultural technology because of the deficient agrarian infrastructure and the search for a better life in the cities.

It seems that some rural activities begin to be implemented because of urban needs, for example, vegetable, fruit, and poultry production. Exportation of agricultural products, however, has been the driving force for improved agricultural production, with jute, malva, black pepper, papaya, oil palm, melon, and some extraction products (such as Brazil nuts and timber) being the main examples.

To date, technological evolution with a significant increase in agricultural productivity has been very limited. In general, an increase in production has been due to the expansion of the agricultural frontiers through land use systems with low levels of sustainability.

ENVIRONMENTAL BOTTLENECKS FOR SUSTAINABLE AGRICULTURAL DEVELOPMENT

Agricultural development in the Amazon has been faced with a number of environmental bottlenecks that have limited its bioeconomic sustainability. Along with the continental dimensions of the Brazilian Amazon, the complexity of the humid tropical ecosystems stands out, requiring that most of the technology be generated locally. This aspect and the region's socioeconomic environment limit the availability and the capacity of technology generation and transfer.

More specifically, environmental peculiarities, such as low fertility and high acidity of soils, favorable climatic conditions for the prevalence of pests and diseases, and aggressiveness of weed plants, are limitations for maintaining agricultural development with satisfactory levels of sustainability.

Even with the limited available knowledge and technology for agricultural development, the high costs of agricultural inputs as a result of a regional infrastructure have limited their utilization and, consequently, have impaired growth in production and productivity. As a result, traditional low-efficiency land use systems, despite their low productivity and high levels of environmental degradation, continue to be used because of their low costs and protectionist policies (Paiva, 1977).

The following are some general constraints under which agricultural development has taken place in the region and that limit sustainability.

- Insufficient knowledge of natural resources (climate, soil, fauna, flora, water resources);
- High biotic pressures (weeds, pests, and diseases);
- Low levels of sustainable production of annual food and fiber crops because of the reduced number of improved varieties and reduced knowledge of cultural practices;

- Low levels of sustainable production of perennial food and industrial crops because of a lack of improved varieties and reduced crop management knowledge;
- Low levels of sustainable production of pasturelands because of insufficient knowledge of forage species, pest and weed control, and pasture reclamation and management;
- Insufficient domestication of native plants with present and potential economic value for more intensive production;
- Reduced development of agroindustry of regional products, deficient transportation and storage, and distances to market;
- Difficulties in systematizing available research results and making them compatible with the agroecologic zoning of the region; and
- Reduced knowledge regarding reclamation of degraded lands and soil conservation.

There is a tendency to promote agroecologic and economic zoning of the Amazon as the panacea for preservation and conservation compatible with the needs for economic development. Conservationists tend to promote agroecologic and economic zoning in an attempt to limit economic activities as much as possible, while developmentalists see it as a guarantee for maintaining production activities. What must be realized is that 16 million people live in the Amazon and need to be fed and sheltered. They also have rights to health care, education, and a decent quality of life. Therefore, agroecologic and economic zoning makes sense only if it includes the participation of local communities. It should primarily consider the competitiveness of production costs and the ecologic implications involved, not just unilateral ecologic considerations. Agroecologic and economic zoning must be accompanied by strong technical assistance programs and a strong social infrastructure (Hirano et al., 1988).

PRESENT KNOWLEDGE BASE FOR AGRICULTURAL DEVELOPMENT

Knowledge about agriculture in the Amazon comes from research and experience gained regionally and from similar, extra-Amazon regions. Research has played a major role in the process of knowledge accumulation. Even though knowledge accumulation through research started as early as the 1930s, the greatest efforts began in the 1970s after which, among other events, the Brazilian Enterprise for Agricultural Research (EMBRAPA) and the Cooperative System of Agriculture Research (headed by EMBRAPA) were created. If agriculture-related publications can serve as an index of knowledge accumulation, from a total of about 1,400 publications produced up to 1985, about

Tropical forests in Brazil supply a variety of commercial products, including cashew nuts. Credit: James P. Blair © 1983 National Geographic Society.

1,200 were generated between 1970 and 1985 (Homma, 1989), a period that is strongly related to the beginning of economic development in the Amazon and the institution of EMBRAPA.

Recognizing the insufficiency of knowledge for sustainable agricultural development, the following sections summarize the present knowledge base for different areas.

Domestication of Nontimber Forest Extraction Products

Some significant advances have been accomplished in this area. Various native plant species that have been extracted from the forest have gone through a slow and difficult process of domestication (Homma, 1989). The available knowledge supports more intensive planting of rubber trees, Brazil nut, guaraná, cupuaçu (*Theobroma grandiflorum*), pupunha (*Guilielma gasipaes*), açaí (*Euterpe oleracea*), urucu (*Bixa orellana*), and malva (*Urena lobata*). As the region's population density increases and markets become available, presently and potentially valuable native forest plants will have to be domesticated.

Natural Resources—Climate, Soil, and Vegetation

A reasonable amount of knowledge about the natural resources of the Brazilian humid tropics, such as soil classification and potentialities, is available. Most of this information is still at a very reduced scale (1:2,500,000), however (Silva et al., 1986). A reasonable-approximation climatic classification supported by a network of small stations spread over the region is also available (Bastos et al., 1986). Also available are satisfactory vegetation classification and maps of the Amazon, which, along with edaphic and climatic information, allows for a reasonable approximation of agroecologic and economic zoning for more sustainable agricultural development (Nascimento and Homma, 1984; Silva et al., 1986).

Forest Exploration

Knowledge of forest exploration has gone in two directions. There is a search for valuable timber products by developing inventories of specific areas and extraction and sustainable management strategies (Superintendency for the Development of the Amazon, 1986; Yared, 1991). This is true also for medicinal forest products (Van den Berg, 1982). In the other direction, efforts have been made to domesticate tree species of high economic value, introduce exotic species, establish integrated systems involving agriculture and cattle raising, and select and test cellulose-producing plants.

Annual Food and Fiber Crops

Some knowledge has been gained for obtaining improved varieties of rice, beans, cassava, and maize, as well as for the development of cultural practices and of integrated systems with perennial crop plants. Rice growing in the *várzea* floodplains may be implemented because of a reasonable amount of field research and testing. Despite their decline in socioeconomic importance, jute and malva have been the most researched fiber-producing plants in the region (Da Silva, 1989a,b), with emphasis on the selection of more productive varieties, cropping systems, seed production, and decortication.

Perennial Crops

Some progress has been achieved in the selection and introduction of cultivars; cultural practices; pest and disease control; and processing of perennial crop plants such as rubber, black pepper, cacao, oil palm, coffee, guaraná, and native fruit trees (Alvim, 1989). For oil palm, one important achievement was the product resulting from crossing African oil palm with the native *caiaué* oil palm and the introduction of pollinating insects in the region.

Pastures and Animal Production

Significant progress has recently been achieved in the knowledge base of the environmental, technological, and socioeconomic interrelations involved in the process of pasture degradation, obtaining better-adapted forage plants, and reclamation of pastures formed after cutting and burning of forests (Dias Filho and Serrão, 1982; Serrão, 1986a; Serrão and Toledo, 1990; Serrão et al., 1979). Also, more recently, the knowledge base on the ecologic implications of pasture degradation and the ecologic and economic recuperation of degraded pasture ecosystems has increased (Buschbacher et al., 1988; Nepstad et al., 1990; Uhl and Kauffman, 1990; Uhl et al., 1988, 1990a,b).

A fair amount of knowledge on the potential and limitations of natural grassland ecosystems has also become available. If these grasslands are more efficiently utilized for cattle pasture (Serrão, 1986b) and other agricultural purposes, they can help to reduce the pressure on more forestlands.

Management techniques, genetic improvements in cattle herds, and sanitary measures have been developed for both cattle and water buffaloes. These allow for the design of production systems that are more efficient than traditional ones. The available stock of knowledge of water buffaloes is significant (da Costa et al., 1987; Lau, 1991; Moura Carvalho and Nascimento, 1986; Nascimento and Carvalho, In press).

Aquaculture

Although still rudimentary, the available knowledge on the fauna of Amazonian rivers has made it possible to develop simple, potentially sustainable fish production systems with native fishes such as tambaqui (*Colossoma* spp.), pirarucu (*Arapaima gigas*), and tucunaré (*Cichla ocellaris*), as well as exotic fishes such as tilapia (*Oreochromis niloticus*), in integrated systems with swine and water buffalo (Imbiriba, In press).

Agroindustrial Technology

Processing and industrialization of regional products have been given relatively high research priorities in the past 2 decades. Technology is becoming available, for example, for the processing of water buffalo milk (mainly for cheese making), tropical fruit nectar preservation, industrialization of black pepper by-products, powdered guaraná and açaí, cupuaçu chocolate, and cellulose from Amazonian wood species.

Basic Knowledge

Applied research and technology generation has been accompanied by some progress in basic research. Despite serious limitations in personnel, equipment, and infrastructure, knowledge has been obtained in the fields of botany, ecology, soil physics and chemistry, plant genetics and physiology (primarily rubber and cacao plants), plant pathology (mainly black pepper, cacao, and rubber plants), entomology, and climatology.

DIFFUSION AND UTILIZATION OF TECHNOLOGY

Diffusion of technology plays an important role in the utilization of knowledge and technology for agricultural development in the Brazilian humid tropics. Formal technical assistance and rural extension in the Brazilian humid tropics have been low in efficiency for supporting agricultural development. The reduced efficiency in the diffusion and adoption of technological improvements is still a major bottleneck in developing more sustainable agriculture in the Amazon.

Technology diffusion is apparent in the region in three main forms: (1) forms used by the Amazon Indians (for example, slash-and-burn planting of cassava and utilization of native plants); (2) imported forms, brought into the region by migrants, that tend to improve local technological standards (for example, Japanese immigrants introduced the jute fiber plant, black pepper, Hawaiian papayas, melon, and Barbados cherry and improved crop and soil management practices for those and other crops); and (3) forms developed by regional research institutions, which is still the weakest form. This low efficiency rating is associated with the still reduced stock of available technology, its feasibility level, and the fragile support provided by basic research. Nevertheless, the contribution of basic knowledge is important not only because it increases the frontier of knowledge that can be used in the future but also because it helps to form scientific judgments about the Amazon.

Because of the still relatively reduced dimension of agriculture in the Amazon, which functions by using the extremes of primitive and imported technologies, the market for technological improvements is small. Small-scale marketing of agricultural products in the region also limits the adoption of improved technologies. The adoption of developed technological practices may not result in success in terms of profitability, however, because of market deficiencies. For example, planting irrigated rice in some floodplain areas does not always result in improved standards of living for the farmers who adopt that technology.

The socioeconomic constraints, mainly in education and health, typically prevalent in the rural areas of the Brazilian Amazon region make agricultural technology a secondary priority. Owners of typical small- and medium-sized farms frequently have more important objectives than increasing land and labor productivity. In those cases, the social aspects of rural extension are more important than the technological aspects. This situation became more prevalent during the period of the New Republic (1984–1989), when technical assistance and extension focused almost exclusively on small farmers.

In a trend toward growing democratization, rural communities may be induced to take more responsibilities and play a more important role in the technology diffusion process.

AMAZONIAN AGRICULTURAL LAND USE SYSTEMS AND THEIR SUSTAINABILITIES

Agricultural development in the Amazon has taken place through the implementation of a number of agricultural production land use systems. The labor and technology utilization varies from very extensive to fairly intensive. This section evaluates the present states of sustainability of the most important agricultural land use systems, namely, extraction of forest products, upland shifting cultivation, *várzea* floodplain cropping, cattle raising, perennial crop plantation, and agrisilvopastoral systems (systems that combine crops, pastures, animals, and trees). An overview of these systems is given in Tables 3A, B, and C. The technological, socioeconomic, and ecologic sustainability parameters used in this analysis are listed in the sidebar entitled, "Parameters for Analyzing Sustainability of Land Use Systems."

Extraction of Nontimber Forest Products

Even though extraction activities are the oldest land use systems in the Amazon, only in the past decade have they become a subject of major interest for agronomists, ecologists, anthropologists, socioeconomists (Allegretti, 1987, 1990; Anderson, 1989, 1990; Fearnside, 1983, 1990; Homma, 1989; Peters et al., 1990) and even politicians, because of the national and international concern over the aggressive deforestation that has occurred over the past 25 years.

Economically important nontimber products that are extracted from forests include natural rubber (mainly from *Hevea brasiliensis*), nonelastic glues (waxes), fibers, oils, and food products (for example, fruits, heart of palm, and Brazil nuts).

In the Brazilian humid tropics, there are two types of extraction, namely, gathering extraction, in which the resource is extracted without any major damage to the plant, and destructive extraction, in which the extraction activity results in the destruction of the plant (Homma, 1989). Both forms of extraction can be sustainable if the extraction does not go beyond the species's regeneration capacity (Peters, 1990).

Unmanaged extraction has the tendency to be destructive in the long run. Because forests offer a fixed amount of products, the capacity to meet increasing demands for a particular product becomes limited, resulting in higher prices and replacement of the resource by domesticated or synthetic substitutes (Homma, 1989). Because of the fixed amount of a resource, expansion possibilities are limited and there is low land and labor productivity. Theoretically, extraction activities typically have a three-phase economic cycle: expansion, stagnation, and decline. Maintenance of extraction activities requires low population pressure, no synthetic substitutes or domestic products, special market conditions, and available stocks of forest products.

Plant domestication can make extraction activities unstable. When there is an adequate amount of extracted stock and domestication technology is not efficient, the extraction activity can compete; but when the extracted product is scarce, prices increase, stimulating domestication of the resource (Homma, 1989).

Synthetic resources also make extraction unstable, even though substitution is usually not perfect, such as for rubber, waxes, and lynalol. Forest food products are less vulnerable to competition from synthetic substitutes but are more vulnerable to domestication.

Frontier expansion and population growth also make extraction activities unstable. The survival of extraction depends on the maintenance of the primary forest. As forest areas become reduced, the cost for extraction in those areas increases. As a consequence, even with strict controls to avoid incorporation of these lands, the increase in the prices of agricultural lands tends to reduce even more the competitiveness of extraction.

In recent years, extraction of forest products has been suggested to be the model for sustainable development of the Amazon (Allegretti, 1987, 1990; Fearnside, 1990; Peters et al., 1990). A recent report (Peters et al., 1990) attempts to show the feasibility of extraction from the economic point of view. The authors concluded that 1 ha of standing primary forest near Iquitos, Peru, can yield US$6,820 annu-

TABLE 3A Land Use Systems in the Brazilian Humid Tropics: Producers, Products, and Technological Intensity

System and Region[a]	Type[b]; Number of Producers or Farmers	Products Explored	Technological Intensity	
			Knowledge-Dependent[c]	Capital-Dependent
Nontimber extraction; Acre, Amapá, Rondônia, Pará	Small; 70,000	Rubber, Brazil nut, heart of palm, oil, fruits	Very low	Very low / low
Timber extraction; Pará, Rondônia, Mato Grasso	Medium/large; 25,000	Timber	Low	High
Upland shifting agriculture; Amazon region	Small; 400,000	Beans, cassava, malva, rice, maize, fruits, cotton	Medium	Low
Traditional *várzea* floodplain crop production; Amazonas, Pará	Small; 50,000	Jute, cassava, maize, beans, fruits	Medium	Very low
Upland perennial and semi-perennial crop production; Pará, Rondônia, Mato Grasso	Medium/large; 20,000	Oil palm, rubber, cocoa, guraná, Brazil nut, black pepper, coffee, urucu, coconut, citrus, cupuaçu, Barbados cherry	Medium/high	Medium/high/very high
Wood plantation production; Pará, Amapá	Large; very few	Timber, cellulose	High	Very high

	Size[b]	Products	Knowledge[c]	Value
Agroforestry systems (Nippo-Brazilian type); Pará	Medium; 500	Multiple annual and perennial crops	High	High/very high
Cattle production on first-cycle forest-replacing open pastures; Amazon region	Large; 5,000 (?)	Beef cattle	Low/medium	High/very high
Cattle production on second-cycle forest-replacing open pastures; Amazon (Pará, Mato Grasso, Tocantins)	Large; 1,000–1,500[d]	Beef cattle	Medium/high	High/very high
Cattle production on forest-replacing, pasture-based agrisilvopastoral systems; Pará	Medium/large; 100–200	Beef and dairy cattle, timber, crops	High	High
Cattle production on native *várzea* floodplain grassland; Pará, Amazonas, Amapá	Small/medium/large; at least 2,000	Beef and dairy cattle	Low	Low
Cattle production on native, well-drained savannah grassland; Amapá, Roraima, Rondônia	Medium/large; 3,500	Beef cattle	Low	Medium
Cattle production on native, poorly drained savannah grassland; Pará, Mato Grasso, Maranhão	Medium/large; at least 1,000	Beef cattle	Low	Medium

[a]Other regions may have a particular land use system, but on a smaller scale.
[b]In relation to landholding size.
[c]Knowledge of natural resources, species, and technical practices.
[d]These farmers were included in the number of first-cycle pasture farmers.

TABLE 3B   Land Use Systems in the Brazilian Humid Tropics: Productivity and Present Sustainability

System and Region[a]	Productivity per[b]		
	Area	Capital	Person
Nontimber extraction; Acre, Amapá, Rondônia, Pará	Very low	Medium (?)	Very low
Timber extraction; Pará, Rondônia, Mato Grosso	Very low	Very high	Medium (?)
Upland shifting agriculture; Amazon region	Medium	High	High
Traditional *várzea* floodplain crop production; Amazonas, Pará	Medium/ high	High	High
Upland perennial and semi-perennial crop production; Pará, Rondônia, Mato Grosso	High	Low/ medium	Medium
Wood plantation production; Pará, Amapá	High	Medium	Medium
Agroforestry systems (Nippo-Brazilian type); Pará	High	Medium	Medium
Cattle production on first-cycle forest-replacing open pastures; Amazon region	Low	Low	Medium/ high
Cattle production on second-cycle forest-replacing open pastures; Amazon (Pará, Mato Grasso, Tocantins)	Medium	Medium	Medium/ high
Cattle production on forest-replacing, pasture-based agrisilvopastoral systems; Pará	Medium/ high	High	High
Cattle production on native *várzea* floodplain grassland; Pará, Amazonas, Amapá	Medium	Medium/ high	Medium/ high
Cattle production on native, well-drained savannah grassland; Amapá, Roraima, Rondônia	Low	Low/medium	Medium/ high
Cattle production on native, poorly drained savannah grassland; Pará, Mato Grasso, Maranhão	Low	Medium/ high	Medium/ high

NOTE: Question marks indicate uncertain or unknown information; NA, not applicable.

[a]Other regions may have a particular land use system, but on a smaller scale.
[b]Wealth generated over time.

Present Sustainability Level

Agronomic	Zootechnical	Ecological	Economic	Social	Cultural
Very high	NA	Very high	Low	Low	Very high
Low	NA	Low (?)	Low	Low	?
Low/ medium	NA	Low	Low/ medium	Low/ medium	Medium
Medium	NA	Medium	Low/ medium	Low/ medium	Medium
Low/ medium	NA	Low/ medium	Low/ medium	Medium	?
Medium	NA	High	Medium	Low/ medium	?
Medium/ high	NA	Medium/ high	Medium/ high	Medium	Medium
Low	Medium/ high	Low	Low/ medium	Low	?
Medium	Medium/ high	Low/ medium	Low/ medium	Low/ medium	?
Medium/ high	Medium	Medium/ high	Medium	Medium	?
High	Low/ medium	Medium/ high	Medium	Medium	Medium
Medium/ high	Medium	Medium/ high	Medium	Medium	?
Medium/ high	Medium	Medium/ high	Medium/ high	Low	Medium

TABLE 3C   Land Use Systems in the Brazilian Humid Tropics:
Potential for Increasing Sustainability

System and Region[a]	Potential for Increasing Sustainability			
	Agronomic	Zootechnical	Ecological	Economic
Nontimber extraction; Acre, Amapá, Rondônia, Pará	Medium	NA	Low	Low/ medium
Timber extraction; Pará, Rondônia, Mato Grosso	Medium	NA	Medium (?)	Medium
Upland shifting agriculture; Amazon region	Medium/ high	NA	Medium	Medium
Traditional *várzea* floodplain crop production; Amazonas, Pará	Medium/ high	NA	Low/ medium	Medium
Upland perennial and semi-perennial crop production; Pará, Rondônia, Mato Grosso	Low/ medium	NA	Medium	Medium/ high
Wood plantation production; Pará, Amapá	Medium	NA	Medium	Low
Agroforestry systems (Nippo-Brazilian type); Pará	Medium/ high	NA	Medium/ high	Medium/ high
Cattle production on first-cycle forest-replacing open pastures; Amazon region	Medium	Medium	Medium	Medium
Cattle production on second-cycle forest-replacing open pastures; Amazon (Pará, Mato Grasso, Tocantins)	Medium	Medium	Medium	Medium
Cattle production on forest-replacing, pasture-based agrisilvopastoral systems; Pará	Medium	Medium	Medium	Medium
Cattle production on native *várzea* floodplain grassland; Pará, Amazonas, Amapá	Medium	Medium	Medium	Medium
Cattle production on native, well-drained savannah grassland; Amapá, Roraima, Rondônia	Medium/ high	Medium	Low/ medium	Medium
Cattle production on native, poorly drained savannah grassland; Pará, Mato Grasso, Maranhão	Medium	Medium	Medium	Medium

NOTE: Question marks indicate uncertain or unknown information; NA, not applicable; gradients G_1 and G_2 correspond to the well-drained savannah grassland ecosystem.

[a]Other regions may have a particular land use system, but on a smaller scale.
[b]In relation to socioeconomic and ecological practices.

Social	Cultural	Potential for Expansion[b]	Research Needs
Medium (?)	Low	Low	Exploration and management techniques; enrichment; integration with agroforestry; marketing
Medium	?	Very high	Exploration and management techniques; enrichment
Medium	Medium (?)	Low/ medium	Organic matter management; improved crop varieties; integration with agroforestry
Medium	?	High	Water control for crop production; selection of adapted crop varieties; non-water-polluting intensive crop systems
Medium/ high	?	Medium	Disease control; agroforestry; domestication of high-value native perennial crop plants
Low/ medium	?	High	Domestication of high-value timber and cellulose-producing trees; integration into agrosilvopastoral systems
Medium/ high	Medium	High	Domestication of high-value tree, food, and forage crops; development of alternative systems
Medium	?	Low	Should not be stimulated now
Medium	?	Medium/ high	Selection of grasses and legumes for open pastures; selection of forages, crops, and trees for agrosilvopastoral systems
Medium	?	Medium	Selection of forages, crops, trees, and animals for integrated systems; designing, testing, and implementing alternative agrosilvopastoral systems
Medium	Low	Low/ medium	Selection of forages for pasture establishment on adjacent upland areas; integration of native and cultivated pastures
Low/ medium	?	Medium/ high	Fire and grazing management; selection of forages, crops, and trees for integrated systems
Medium	Low/ medium	Low/ medium	Mineral supplementation; selection of forage grasses and legumes for pasture establishment in gradients G_1 and G_2

Parameters for Analyzing Sustainability of Land Use Systems

Technological Parameters
    Demand for technical assistance
    Demand for mechanization
    Demand for fertilizers, lime, herbicides, insecticides, fungicides
    Demand for quality seed
    Demand for equipment
    Incidence of pests and diseases
    Management intensity
    Weed control
    Possibility of combination with other systems
    Production fluctuation
    Resilience to attacks of pests and diseases
    Need for organic fertilization
    Labor need
    Need for a high level of specialization
    Soil conservation practices
    Harvesting ease
    Establishment ease
    Stability
    Productivity
Ecological Parameters
    Level of environmental degradation
    Receptiveness from ecological community (national, international)
    Degradation of fauna and flora
    Loss of biodiversity
    Cause of water pollution (streams, rivers)
    Extent of deforestation needed
    Extent of burning needed
    Long-term implication in relation to the ecology
    Current judgment of producer in relation to ecology
    Present extent of environmental degradation because of use
    Support from environmental institutions
    Possibility of being used in degraded lands
    Effect on climate change
    Effect on greenhouse gases
    Potential for improving environmental conditions
Economic Parameters
    Subject to price fluctuations
    Need for intermediaries for commercialization
    Trustworthy policies for the sector
    Need for credit
    Problems of overproduction
    Competitiveness with other activities (production systems)

Cost of labor needed

Cost of modern inputs (for example, mechanization, seeds, fertilizer, and pest control)

Ease of acquiring modern inputs

Extension services (easy, difficult)

Research support need

Physical infrastructure (for example, roads and transport)

State or national price policies

Ease of product commercialization

Local, regional, national, and international markets

Environmental protection pressures

Future scenarios for the Amazon (for example, price liberation)

Level of technology

Dysfunction between producing what, how, and for whom

Social Parameters

Labor offer (for example, planting, weeding, harvesting, and industrialization)

Labor intensive by nature (for example, extractivism)

Level of education required for farmer or labor

Length of tradition required

Immigrants from other regions

Mutirão practices

Level of income required

Allowable social infrastructure (for example, school, health centers, and social clubs)

Interaction among producers (for example, Japanese and rubber tappers)

Strong political participation (lobbying capabilities)

Also serving as labor for other agricultural activities (for example, small farmers also serving as labor for weeding pastures in large neighboring cattle ranches)

Mobilization

Equitability

Cultural Parameters

Dependence on cultural tradition (for example, farmers from Bahia for cacao and from São Paulo for coffee)

Cultural background versus adoption of technology

Fear of being a pioneer (wait for others)

Extension service's familiarity with local ecological and socioeconomic environment

Parochialism

Mixture of farmers' origins

Strength of political leadership

Access to local, regional, and national news

Access to newspapers and magazines

Length of time dedicated to agricultural activity

Knowledge of day-to-day life in the Amazon

ally, at present values. However, such an analysis is of a static nature and does not take into account the above-mentioned factors that affect the stability of extraction.

Extraction activities are agronomically and ecologically sustainable. However, their economic and social sustainabilities are restricted to the short term. In most cases extraction activities are associated with the acquisition of food products from agricultural activities. For example, the autonomous rubber tappers of Acre integrate shifting agriculture with cattle raising activities.

Extractive reserves have the advantage of being entirely open to management options. They also cause minimal micro- and macro-environmental damage (Fearnside, 1983, 1990).

SUSTAINABILITY OF NONTIMBER RESOURCE EXTRACTION

Within the scenario of nontimber extraction activities, what can be done to promote a more realistic and sustainable use of extractive reserves? Many of the inherent problems of extraction systems in the Amazon may be solved, as long as extraction is not seen as a panacea. These systems have marginal economic viabilities, and because they lack strong economic and social structures, they can be, and frequently are, replaced by other agricultural land use systems, such as shifting agriculture and cattle raising (Anderson, 1989).

Therefore, if extractive reserves are to function, they must evolve. To be successful, in addition to simple extraction practices, they must incorporate other land use systems that would ideally intensify production per unit area with a minimal reduction in their ecologic sustainabilities.

According to Anderson (1989), in the Amazon humid tropics, agroforestry systems represent the best alternative to conciliate these demands (see below). Maintenance of a forestlike canopy that is typical of those systems maintains ecologic sustainability, while other activities under the canopy increase production in economic terms. The rate of this increase is related to the management intensity of natural resources.

Anderson (1989) analyzed three real-world commercial land use systems with increasing management intensities, namely, extraction of forest products, extensive agroforestry, and intensive agroforestry. Each system has weak and strong points. Extraction requires minimum input but produces minimum returns. Intensive agroforestry gives high levels of return, but costs of labor, input, and capital are also very high. Even though extensive agroforestry seems to be able to combine the best features of the two extremes of land use intensity,

TABLE 4   Comparison of Three Land Use Strategies in the Brazilian Amazon Region

Factor	Extraction of Forest Products	Extensive Agroforestry	Intensive Agroforestry
Area utilized per household (ha)	372	36	28
Annual labor requirements			
Person-days per holding	199	661	2,477
(percent family labor)	(100)	(92.2)	(23.3)
Person-days per hectare	0.53	18.36	88.46
Hired labor costs per holding ($)	0	134.05	4,939.63
Hired labor costs per hectare ($)	0	3.72	176.42
Material costs ($)			
Fertilizers, pesticides	0	0	13,490.02
Utensils, machinery	87.65	51.77	1,738.24
Material costs per holding	87.65	51.77	15,228.26
Material costs per hectare	0.24	1.44	543.87
Gross return ($)			
Per holding	960.00	2,733.45	29,667.39
Per hectare	2.58	75.93	1,059.55
Net return ($)			
Per holding	872.35	2,547.63	9,499.50
Per hectare	2.35	70.77	339.27
Per person-day of family labor	4.38	4.18	16.46

SOURCE: Anderson, A. B. 1989. Estratégias de uso da terra para reservas extrativistas da Amazônia. Pará Desenvolvimento 25:30–37.

it is only feasible under highly specific ecologic conditions (Table 4). Perhaps the best strategy for extractive reserves is a combination of the three systems.

According to Anderson (1989), one scheme to accomplish integration might involve the utilization of swidden plots (plots where the vegetative cover has been burned) as sites for agroforestry systems since, in most areas where extraction activities occur, swidden plots are abandoned after a few years of cultivation. Instead of being abandoned, such plots could be used to establish plantations of perennial tree crops.

As in other swidden-fallow agroforestry systems in the Amazon (Denevan and Padoch, 1987; Posey, 1983), the degree of intervention could increase from the center of the plot, with intensively maintained plantations giving way to manipulated forest fallow. Along this management gradient, depending on the stage of land use intensiveness in the extractive reserve, a wide range of plant products and game resources could be exploited. The local market must be able to absorb the resulting products, however. In this way, higher levels of overall sustainability of the integrated system would be secured (Anderson, 1989).

RESEARCH NEEDS

To increase the sustainability of extraction activities, there must be a search for the alternative land use models. It seems most logical to follow the agroforestry approach, since extraction per se is a land use system with low levels of socioeconomic sustainability. Research efforts and policies should consequently be aimed at transforming extractive reserves into viable enterprises. The selection of high-value, low-input, easy-to-establish annual and perennial crops and trees for extractive reserve enrichment should be the most important goal of research.

Extraction of Timber Products

Timber extraction—a subsystem of extraction of forest products—has had accelerated growth during the past 2 decades because of wood scarcity in the extra-Amazon regions of Brazil and in southeastern Asia and because of the increased value of some regional wood species such as mahogany and cerejeira *(Amburana acreana)* (Yared, 1991).

About 50 percent of Brazil's native forest timber is extracted from the northern region; 85 percent of that is extracted from the state of Pará.

Even though timber extraction may be seen as a threat to the region's forest resources, timber is second in economic value only to mineral products in the export market. In 1988, for example, the states of Pará and Amapá exported about 500 m^3 of wood worth US$150 million (Associaçáo das Indústrias de Madeiras dos Estados do Pará e Amapá, 1989). It also contributes significantly to regional employment. Each sawmill employs an average of 34 workers and each veneer and plywood plant employs about 300 workers, contributing to the employment of about 125,000 people in the Brazilian Amazon region in 1989 (this does not include indirect employment) (Yared, 1991).

The only source of timber for the wood industry in the Amazon is native forest. Timber comes from selective logging operations or from deforestation for other purposes (for example, for cattle pasture establishment and shifting agriculture). In areas with high timber extraction pressures, selective logging is characterized by destructive management practices that include incursions into logged forests at intervals too short to allow sufficient time for the biologic regeneration of the forest, resulting in genetic erosion of important species (Yared, 1991). In addition, selective logging is frequently the first step toward the occupation of the logged forest by other land use systems, mainly cattle pastures.

A more recent development is the link between logging and ranching (Uhl et al., In preparation). This link arose because of the high costs involved in reclaiming first-cycle degraded pastures in the Amazon. (First-cycle pastures are those formed after slashing and burning of the primary forest vegetation.) The present cost of pasture reformation is about US$250/ha (Mattos et al., In press), which is too costly because of the high interest on credit and the lack of tax incentives. Therefore, ranchers selectively log their remaining forest segments to finance the formation of second-cycle pastures. (Second-cycle pastures are reformed degraded first-cycle pastures.) The forest now plays a critical role in sustaining cattle-raising activities, which creates pressures for additional deforestation.

Because of logging's important role in the regional ranching economy and in the accumulation of wealth by a new entrepreneurial class, Uhl et al. (1991) evaluated its social and environmental impacts. They concluded that the impacts have been substantial. Even though employment is considerable, those employed in the logging sector spend most of their wages satisfying their basic needs, with little prospect for improving their lives or those of their children.

Logging results in substantial damage to the forest (Uhl et al., 1991). Canopies are opened by 30 percent or more, and 25 trees are damaged for each tree that is harvested. These open conditions favor the growth of vine species, which frequently dominate logged sites for many years.

Economically, technologically, and environmentally, natural forest management for timber extraction has been deficient (Uhl et al., 1991; Yared, 1991). However, there are possibilities for improvement. Technologies developed by the research and development institutions in the region, such as EMBRAPA and SUDAM, are gradually becoming available. For example, in the polycyclic system (Yared, 1991), timber extraction is planned in such a way as to minimize irreversible damage to the forest. Experiences with large-scale operations of this system show that it is possible to log about 40 m³ of wood per ha at a cost of about US$10/m³, including transportation to distances of up to 100 km. Since the price of logged timber varies between US$9.50/m³ (light wood) and US$17.5/m³ (heavy, dark wood, the type that contributes to 90 percent of total extracted volume), extraction by this system is profitable (Yared, 1991).

Even though the actual and potential environmental effects of logging are considerable (Uhl et al., 1991), research results show that logged forests in the Amazon have satisfactory resilience (Yared, 1991). Although the opening of the forest canopy after selective logging favors the growth of a larger number of trees with low economic value, the regeneration of presently and potentially valuable trees is adequate, allowing for new harvests in the future. On the basis of the polycyclic method of sustained timber production systems (de Graaf and Poels, 1990), simulation studies show that an adequate volume of wood is expected 30 years after logging (Silva, 1989) and that the expected volume can be doubled or even tripled if appropriate silvicultural treatments are carried out during and after logging. In this system, for a continuous annual supply of wood (as logs) of about 30 million m^3 (demand in 1987 was 24.6 million m^3) and considering harvest cycles of 30 years and average extraction of 40 m^3/ha, it would be necessary to immobilize an area of about 22 million ha, which represents almost 10 percent of the total dense forest area of the Amazon. With this system, timber production presumably would not require additional deforestation.

SUSTAINABILITY OF TIMBER EXTRACTION

Use of a sustainable management system for timber extraction is far from being realistic. There are serious restrictions to the proposed sustainable native timber extraction management system for adoption on a commercial scale (Pearce, 1990). There are biologic restrictions because of low humid tropical forest growth rates, resulting in unfeasible time spans between harvests, and there are economic restrictions because of high-interest bank loans, management is costly, returns on capital investments are long term, and minimum-sized forests are too large to rotate. This ties up capital in an inflationary economy with high rates of interest. Therefore, sawmills prefer to buy wood from occasional independent suppliers.

Forest timber resources are abundant and cheap in the Brazilian humid tropics. Therefore, there is little incentive on the part of the industry to engage in constructive management (Uhl et al., 1991). Management will only begin to make sense if or when forest timber resources become scarce. Then, timber industries will be able to manage timber forest resources for sustainable yields and still possibly make profits. Although this is not occurring at present, sustainable timber exploration in the Amazon may be possible in the future.

According to Uhl et al. (1991), government policies that encourage sustainable management for timber exploration should be designed to make timber resources artificially scarce. This could be done by allowing logging only in designated areas of state forests and prohibiting sawmill owners from relocating their operations. In turn, each sawmill could be given a license to log a specified area of forest adequate for supplying the mill indefinitely, if it were properly managed. In the meantime, enforceable guidelines should be developed. These guidelines should specify how logging and management operations should be conducted.

RESEARCH NEEDS

Research should concentrate on the search for feasible sustainable extraction (methods that will result in the minimum wastage of timber and other nontimber forest resources) of native forest timber products and on the domestication of presently and potentially important high-value timber-producing trees.

Shifting Agriculture in Upland Areas

Shifting (slash-and-burn) agriculture is still probably the most important land use system in the region; it still accounts for at least 80 percent of the region's total food production. It is also important because of the number of people who depend on it directly and indirectly. Yet, despite its importance to the regional macroeconomy, its feasibility has declined with the declining process of agricultural frontier expansion because of deforestation restrictions, increasing consolidation of already existing poles of development, and increasing demographic density and the consequent increasing food demand and land prices (see Figure 1). Under these conditions, long fallow periods—the prime condition necessary for maintaining the agronomic sustainability of the system—are not as feasible as before, and in the long run, shifting agriculture will be replaced naturally by more intensive land use systems.

From the socioeconomic point of view in Brazil, and particularly in the Amazon, annual subsistence crops (mainly cassava, beans, malva, rice, and maize) are connected with those small-scale farmers who have lower standards of living (Kitamura, 1982). Higher standards of living are necessary for increasing the sustainability of shifting agriculture. Nakajima's (1970) classification of the agricultural properties of small farms can be used to illustrate this point (Figure 3): on the basis of the rate of production by the family and the rate of

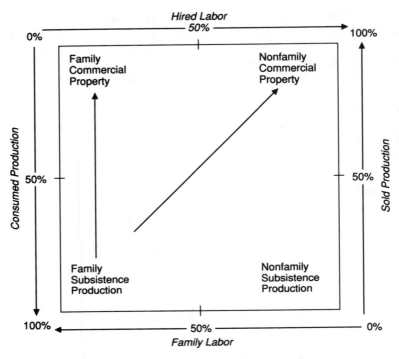

FIGURE 3 Possible forms of production in relation to labor utilization and production destination in a typical small-farm (including shifting agriculture) enterprise. Source: Adapted from Nakajima, C. 1970. Subsistence and commercial family farms: Some theoretical models of subjective equilibrium. Pp. 165–185 in Subsistence Agriculture and Economic Development, C. R. Wharton, ed. Chicago: Aldine Publishing.

participation of family labor, Nakajima classified properties as those dedicated exclusively to subsistence production and those dedicated exclusively to commercial production. In the Brazilian humid tropics, the first situation is rarely found, except in indigenous communities. On the other hand, very few shifting-agriculture farmers are dedicated exclusively to production commercialization.

Improvement in socioeconomic sustainability is possible for commercial family or nonfamily properties. However, limiting factors such as the prevailing inadequate infrastructural and technological conditions impose severe constraints on improvement efforts. Therefore, although favoring equity in income distribution among those who practice it, shifting agriculture offers few possibilities for socioeconomic improvements (Alves, 1988; Alvim, 1989; Homma and Serrão, In preparation).

An evaluation of small farms in the eastern Amazon (Burger and Kitamura, 1987) suggests that external factors such as population pressure, integration of a market economy, and cultural and technological influences are disrupting small-farm production systems, causing their degradation in three dimensions—namely, ecologic degradation as a consequence of shorter fallow periods, resulting in low, unstable, and undiversified production; economic degradation caused by unfavorable price relations for basic food products that are controlled by the government and that prevent agricultural modernization (Alvim, 1989); and human resource degradation as a result of insufficient work force replacement because of low levels of nutrition and formal and informal education as well as the loss of skilled labor to urban areas.

SUSTAINABILITY OF SHIFTING AGRICULTURE IN UPLAND AREAS

From the biologic point of view, annual crops such as rice, maize, cassava, beans, and sugarcane demand substantial quantities of soil nutrients for satisfactory yields (Goodland and Irwin, 1975), but Amazon upland soils are generally dystrophic, and the environment is favorable for pests and diseases that affect cultivated plants. Improved adapted varieties and cultural practices that include minimum amounts of agricultural inputs (mainly fertilizers and pesticides) are needed to improve agronomic sustainability.

Although some technological improvements may be achieved, however, incorporation of technology by small-scale food crop farmers has been practically nil. According to Pastore (1977), ignorance, impotence, and lack of interest are the main factors limiting the use of new technological developments by Brazilian small-scale farmers. First, farmers are unaware of the available new technologies. Second, even though they have a reasonable knowledge of new technologies, they cannot adopt them because of cultural and socioeconomic restrictions. Third, although they are aware of and are able to adopt new agricultural techniques, small-scale farmers prefer to take other courses of action.

Despite its low sustainability levels and the tendency that it will disappear in the remote future because of population pressures and other factors (see Figure 1), shifting agriculture will continue to be an important agricultural land use system in the Amazon. Therefore, it is necessary to raise the socioeconomic standards of farmers who practice it. An increase in the level of their income from agricultural activities may be accomplished by encouraging them to use improved technologies with as few inputs as possible and by making appropriate credit available.

Reductions in the cycle of shifting agriculture would also considerably reduce ecologic disturbances. For example, by cropping 2 ha for 3 years instead of 2 years, silent deforestation (as discussed above) would be reduced by about 30 percent. Annual food crop production models, such as the Yurimagua model (Nicholaides et al., 1985; Sanchez et al., 1982), which involves intensive land use, including fertilizers, need to be implemented in the Brazilian humid tropics, as long as they are adjusted to the socioeconomic environment of the region (Fearnside, 1987).

RESEARCH NEEDS

Research support should be directed toward a gradual transformation of shifting agriculture into more sustainable agroforestry and even agropastoral systems, thus preventing farmers who practice shifting agriculture from being displaced from their lands. Research should focus on the development of annual and perennial crop varieties and their integrated utilization in agroforestry systems to improve the sustainability of upland agriculture by small farmers in the Brazilian humid tropics.

Várzea Floodplain Agriculture

Várzea floodplain agricultural systems, which have mainly been developed along the floodable margins of the Amazon River and its tributaries with their muddy, sediment-rich waters, can also be considered systems of shifting agriculture because they have some common features such as slash-and-burn practices, growth of predominantly annual food crops, and small-scale farmers with similar socioeconomic situations.

There are differences, however. Floodplain vegetation is less heterogeneous and includes large tracts of herbaceous, mostly grassy vegetation. Floodplain soils are more fertile than upland soils. Shifting cycles are considerably shorter in floodplains than they are in uplands because of higher soil fertility. Floodplains are subject to an annual flooding and receding cycle, with its consequent flooding risks. Agricultural activities complement subsistence fishing activities in the floodplain system; jute and malva as fiber are important products of floodplain agriculture.

Typically, agricultural practices consist first of selecting areas of the floodplain with the least probability of being totally flooded during the high-water season. Then, the arboreal and herbaceous vegetation is cleared and burned during the dry season, and crops are planted in the beginning of the rainy season and harvested before the onset of the following dry season. Soil fertility conditions allow these same operations to be carried out for years on the same patch of land.

On average, if atypical floodings are not a limiting factor and minimal cultural management is practiced, yields can be considerably higher than those in the standard upland shifting agricultural system.

SUSTAINABILITY OF FLOODPLAIN AGRICULTURE

The possibility of agronomic sustainability of floodplain food crop agriculture is certainly higher than that in uplands, mainly because of more favorable soil conditions. However, weed invasion, pests, and diseases and the risks of flooding are serious constraints to agronomic sustainability.

Socioeconomic sustainability, though, is lower than that in the upland shifting agricultural system because of deficient basic infrastructure conditions (education, health, transportation) in the floodplain areas. In particular, commercialization of agricultural products is deficient because river transportation from the interior to the commercial centers is slow and generally precarious. To counterbalance this situation, however, floodplain farmers can get most of their dietary animal protein needs from fish.

At the present levels of demographic density and low technological intensity, the ecologic sustainability of the floodplain agricultural system is satisfactory because the extent and intensity of clearing and burning are relatively low.

It has been emphasized that the Amazon's *várzea* floodplains should be used as an alternative to intensive agricultural production (mainly annual food crops) in forested areas, thus reducing the pressure of silent deforestation brought about by the shifting agricultural system in upland regions (Lima, 1956; Nascimento and Homma, 1984). To date, this possibility has been explored mostly on paper and in conferences and debates within political and scientific communities. This certainly can and must be achieved with technological improvements involving better crop cultivars for appropriate production systems under either controlled or uncontrolled water conditions and an appropriate socioeconomic environment for development of this system.

Intensive agricultural production in the floodplains would involve intensive pest and disease control. Therefore, precautions should be taken to avoid agrotoxic water pollution in streams and lakes. This type of water pollution could cause serious, unpredictable environmental consequences (Goulding, 1980).

RESEARCH NEEDS

If the development described above is to take place, research must concentrate on the development of production systems with minimum inputs and with the least possible damage to the aquatic ecosystem of the floodplains.

Cattle Raising on Pastures that Have Replaced Forests

A major agricultural development in the Brazilian humid tropics has been the turning of rain forests into pastures to raise cattle. This was a result of the road construction developments that began in the mid-1960s. This type of land use system has been seriously questioned in view of its agronomical-zootechnical, socioeconomic, and, principally, ecologic implications (Browder, 1988). It has been blamed for being the main cause of environmental degradation and for being infeasible biologically and socioeconomically (Fearnside, 1983, 1990; Hecht, 1983; Hecht et al., 1988). It is defended, however, as being an adequate activity for opening frontiers for development and making good use of the available land and labor force (Falesi, 1976; Montoro Filho et al., 1989).

SUSTAINABILITY OF CATTLE RAISING

Analyses that contemplate more recent, improved pasture-based cattle raising developments point toward the possibility of increasing levels of sustainability (Serrão, 1991; Serrão and Toledo, 1990, In press). The economic and ecologic sustainability of the cattle raising activities that have replaced forests in the Amazon depends to a large extent on the sustainability of the pastures. In general, it is agreed that zootechnical (animal component) sustainability is much less limiting than agronomic (pasture) sustainability is. Beef cattle (mainly zebu) breeds are well adapted to the Brazilian humid tropics, where parasites and diseases are less limiting to beef cattle than are other environmental conditions in the country (Serrão, 1991).

In general, during the first 3 to 4 years after the first-cycle pasture formation by cutting and burning forest biomass and then sowing grass seeds, primary pasture production is relatively high, supporting stocking rates of up to two 300-kg (live weight) head of cattle per ha. After that period, a gradual but fairly rapid decline in productivity takes place. This is accompanied by weed encroachment and results in an advanced stage of degradation that occurs between 7 and 10 years after pasture establishment. It is estimated that, to date, at least 50 percent (about 10 million ha) of the total first-cycle pastures formed in the past 25 years have reached advanced stages of degradation (Serrão, 1990, 1991). At this stage, the carrying capacity cannot exceed 0.3 head of cattle (100 kg [live weight]) per ha. The average carrying capacity of first-cycle pastures during their life cycle is about 0.7 head per ha (Mattos et al., In press), which is considered too low for improved pasture standards.

In their average 6- to 7-year productive life, first-cycle pastures have produced as much as 250 to 300 kg of beef. This level of productivity is very low, especially when it is compared with those of other agricultural products, such as cassava, rice, maize, beans, cacao, and Brazil nuts, in terms of protein and energy production as well as monetary value per unit area (Mattos et al., In press).

These problems, which have resulted in low levels of sustainability, were typical of cattle raising activities in the 1960s and 1970s. The 1980s was the beginning of a new and more sustainable cattle raising trend in forested areas. The knowledge obtained from research in the late 1970s and early 1980s made it evident that first-cycle pasture degradation is caused by an interrelation of environmental, technological, and socioeconomic constraints. Environmental constraints included low soil fertility, with phosphorus being the main limiting factor; high biotic pressures, principally of insects (spittle bugs, for the most part) and weed aggressiveness; and water stress. Technological constraints included low adaptability of pioneer forage grasses (mainly guinea grass, *Brachiaria decumbens,* and *Hyparrhenia rufa*), poor pasture establishment and management, nonutilization of forage legumes, and fertilization. Socioeconomic constraints included unfavorable input/product ratios, inadequate development policies, land speculation, and deficient governmental and nongovernmental technical support. Beginning in the early 1980s, however, progressive ranchers began to adopt technological innovations in the search for higher levels of sustainability in their operations. Thus, a significant proportion of first-cycle pastures that were formed from the use of better-adapted forages such as *B. humidicola, B. brizantha* cultivar Marandu, and *Andropogon gayanus* cultivar Planaltina had considerably higher levels of agronomic sustainability than those formed in the 1960s and 1970s.

Higher land use intensification in cattle development areas in the Amazon was induced by considerable reductions in tax incentives and subsidies for cattle in the past decade, the increased area of degradation of first-cycle pastures, increasing pressures for environmental preservation, the increased availability of scientific knowledge and technologies for pasture production, the decreased availability of forest areas in already established ranching projects, increasing population density in already established development poles, and consequent increases in land prices (see Figure 1).

With land use intensification, much degraded first-cycle pastureland has been converted to second-cycle pastures. In this second-cycle pasture generation, more modern agricultural technologies are being used. These technologies include mechanization for preparation and seeding of degraded pasturelands, soil fertilization, better forage grasses, higher-quality forage seeds, and improved pasture management. Official data are not available, but Serrão (1991) estimated that at least 10 percent of the total degraded first-cycle pastures formed to date have been reclaimed and converted to second-cycle pastures. Despite the recent improvements in pasture sustainability, socioeconomic, environmental, and agronomic constraints are still pending for the expansion of second-cycle pastures. One aspect is the high cost involved with transforming degraded pastures to second-cycle pastures. High-interest governmental and private bank credit has induced the logging and ranching link (Mattos et al., In press). This link is one more driving force toward deforestation. This constraint may be minimized by the utilization of cash crops (such as maize, rice, and beans) in association with forage grasses and legumes in the process of second-cycle pasture establishment. Returns from growing cash crops can considerably reduce the cost of pasture establishment (Veiga, 1986), minimize the need for the logging and ranching link, and add more to the subsistence food supply in the region.

Second-cycle pastures will continue to be monoculture open pastures with low levels of biomass accumulation; however, is it correct to keep searching for higher levels of sustainability for cattle raising in the humid tropics on the basis of the traditional pasture systems (open monoculture pastures) used in the region? It is known that the monoculture—whether domesticated, naturalized, or exotic—that has replaced the humid tropical forest without taking into account its environmental (climatic, edaphic, and biotic) adversities and its great biodiversity has had serious agronomic sustainability limitations. This is the case, for example, for rubber, cacao, black pepper, and more recently, African oil palm. In the case of pastures, it is probable that the dissemination of spittle bugs (the most economically significant pasture insect pest) has been the result of extensive deforestation to form monoculture pastures of *Brachiaria decumbens* in the early 1970s, *B. humidicola,* and other, less important *Brachiaria* species.

In view of this environmental and socioeconomic scenario, there should be a search for alternative models of pasture-based cattle raising systems that can be agronomically, ecologically, and socioeconomically more sustainable than those in use. Within that context are the agrisilvopastoral systems. These systems are defined by King and Chandler (1978) as agricultural production systems in which arboreal and nonarboreal crops are grown simultaneously or sequentially in planned association with annual food crops and/or pastures. They have recently claimed the attention of research and commercial agricultural operations.

By this integrated approach, high levels of sustainability are expected as follows:

- Agronomically—reduction of risks caused by pests and diseases and improved cycling and, consequently, better utilization of nutrients;
- Economically—different sources of income;
- Socially—production of different products, more direct and indirect employment opportunities, higher levels of labor specialization; and
- Ecologically—higher levels of biomass accumulation, improvement in the hydrological balance, improvement in soil conservation, and improved environmental conditions for micro- and macroflora and -fauna (Serrão and Toledo, In press).

It is expected that the pasture-based integrated approach will be significantly implemented during the 1990s in the process of reclamation of already degraded pasturelands and that this approach will be a common practice in the first decade of the next century (Serrão, 1991).

With technological intensification and the consequent improvement in the sustainability of forest-replacing pastures, complemented by more efficient utilization of the native grassland ecosystem (see below), productivity from cattle raising operations in the Amazon can be doubled or tripled. Therefore, from the technical point of view, no more than 50 percent of the area already used for cattle raising is actually necessary to meet the regional demand for beef, milk, and other agricultural products at least through the 1990s. If this is correct, and given the relatively favorable resilience of degraded pasture ecosystems (Buschbacher et al., 1988; Uhl el al., 1988, 1990b), a considerable amount of already degraded pastureland can be reclaimed or regenerated toward forest formation and biomass accumulation (Nepstad et al., 1990, 1991).

RESEARCH NEEDS

Although there has been some progress in increasing the sustainability of cattle raising operations on forest-replacing pastures in the Brazilian humid tropics, from a technological point of view, insufficient adapted forage germplasm is probably the most important constraint to continued progress. The main priority of applied research should be to correct this problem by developing adapted cultivars of grasses and legumes. This should be combined with additional applied research efforts for designing and implementing integrated agrisilvo-pastoral systems (Serrão and Toledo, 1990, In press; Veiga and Serrão, 1990). Applied research is also necessary to develop a means of restoring forest biomass in degraded pasturelands, especially through the strategic introduction of high-value timber and fruit trees to provide some economic return from the regeneration process.

More sustainable future development of cattle raising on forest-replacing pasture systems should be based on high-knowledge and low-input land use systems. Basic research is essential for this and studies should be concentrated on the ecology of the weed community in regional pastures, the biotic and abiotic mechanisms of forest regeneration in degraded pasture, the phosphorus cycling mechanism in pasture ecosystems, and the microbiology of soil organisms in pastures, especially in relation to *Rhizobium* species and mycorrhizae.

Cattle Raising on Native Grassland Ecosystems

Before the advent of pasture development in forested areas in the 1960s, cattle raising in the Brazilian Amazon was carried out almost exclusively on native grassland ecosystems with varied botanical, hydrological, edaphic, and productivity characteristics (Serrão, 1986b). After the more-negative-than-positive results of cattle raising on forest-replacing pastures and the need to minimize the pressure of cattle raising on new segments of forested areas, the emphasis is on the importance of native grasslands. Native grasslands can complement more sustainable and more intensive pasture development in already explored forested areas.

Nascimento and Homma (1984) and Serrão (1986b) estimate that there are between 50 and 75 million ha of land in the Brazilian humid tropics with varying gradients of herbaceous and arboreal vegetation and with varying grazing potentials. Serrão (1991) estimates that these lands carry about 6 million head of cattle but could potentially carry 30 million head. Economically, the most important ecosystems are well-drained *cerrado*-type savannah grasslands with varying herbaceous and arboreal gradients, poorly drained *cerrado*-type savannah grasslands with varying flooding gradients, and *várzea* floodplain grasslands (Serrão, 1986b).

WELL-DRAINED SAVANNAH GRASSLANDS (WDSG)

WDSG correspond to the typical *cerrado* grassland. WDSG have little edaphic and floristic variation, are found in smaller patches where the forest's vegetation is interrupted, and have varying gradients of herbaceous and arboreal strata.

The herbaceous stratum is of major interest for animal production. It is mainly made up of grasses of the genera *Andropogon, Eragrostis, Trachypogon, Paspalum*, and *Mesosetum* and, on a much smaller scale, of legumes of the genera *Stylosanthes, Desmodium, Zornia*, and *Centrosema* (Coradin, 1978; Eden, 1964; Serrão and Simão Neto, 1975).

One of the main limitations of WDSG for cattle production is its low forage productivity. Available data (Brazilian Enterprise for Agricultural Development, 1980, 1990) indicate that primary production of WDSG herbaceous extracts rarely exceeds 5 metric tons of dry matter per ha. Consequently, the carrying capacity varies from 4 to 10 ha per animal unit (AU) (1 AU equals 450 kg live weight), which is very low. The low nutritive value of the available forage is the main limitation of WDSG. Even under the most favorable conditions, during the rainy season, available forage, protein, phosphorus, and dry matter digestibility of the grasses in WDSG are below standard critical levels for beef production (Brazilian Enterprise for Agricultural Research, 1990; National Research Council, 1976; Serrão and Falesi, 1977).

Serrão and Falesi (1977) suggest that the low productivity and quality of WDSG are related to the low levels of soil fertility in the ecosystem and the high rate and speed of lignification of the available grasses in the herbaceous stratum. These constraints are accentuated during the dry season, when the contributions of native legumes are probably insignificant because of their sparse presence in the ecosystem. The use of fire to burn WDSG toward the end of the dry season helps to alleviate the low-quality constraint for at least the first 2 or 3 months of the following growing season (Serrão, 1986b). Despite its economic and ecologic importance, research on the burning of WDSG has been neglected.

Cattle raising productivity in the WDSG of the Brazilian humid tropics can be increased by more intensive utilization of the natural ecosystem per se and by supplemental feeding of cattle on nearby improved cultivated pastures. These types of pastures provide higher production and quality potentials, have a positive effect on increasing the carrying capacity of the land, and reduce the problem of low quality in the system as a whole (Serrão, 1986b; Serrão and Falesi, 1977). Selection of adapted improved grasses such as *Brachiaria humidicola*, *B. decumbens*, *B. brizantha* cultivar Marandu, and *Andropogon gayanus* cultivar Planaltina as well as research on pasture fertilization have contributed to increased WDSG productivity (Brazilian Enterprise for Agricultural Research, 1980; Serrão, 1986b).

Despite their inherent low productivity, WDSG have relatively high levels of ecologic and agronomic sustainability because of their resilience after burning disturbances, the very low soil fertility conditions, and the relatively harsh climatic conditions that prevail in the ecosystem. To date, however, socioeconomic sustainability has been marginal.

Applied research must be prioritized for the selection of adapted and more productive forage germplasm, pasture establishment and management, mineral supplementation, and fire management in the native savannah. Basic research should concentrate on physical and biologic characterization and on water stress pressures in WDSG.

CATTLE RAISING ON ALLUVIAL FLOODPLAIN (VÁRZEA) GRASSLANDS (FPG)

FPG ecosystems are found mainly in association with "white" muddy-water rivers. The Amazon River is the main contributor to their formation, as are other tributaries whose waters are rich in the organic and mineral sediments deposited annually on the floodplains when river waters recede (Sioli, 1951a,b).

Prototype FPG (Figure 4) have mainly been developed along the lower and mid-Amazon River regions. They are also found, on a smaller scale, on Marajó Island and in the state of Amapá. The predominant soils are fertile alluvial inceptisols, which generally support a herbaceous vegetation with high productivity and quality potential. "Amphibian" grasses, that float when the water is high and thrive on the *restingas* (the highest part of the *várzea* ecosystem) in the dry season after the water recedes, are dominant (Brazilian Enterprise for Agricultural Research, 1990). The amphibian grasses *Echinochloa polystachya, Hymenachne amplexicaulis, Leersia hexandra, Luziola spruceana, Paspalum fasciculatum, Oryza* species, and *Paspalum repens* are the most important from the standpoint of animal production (Brazilian Enter-

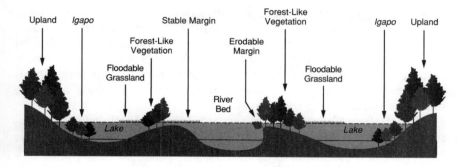

----- *Maximum Flood Water Level*

——— *Minimum Flood Water Level*

FIGURE 4 Profile of a typical *várzea* floodplain grassland (FPG) ecosystem of the lower and mid-Amazon River region. "Igapo" is the inundated parts of riverine woodlands. Sources: Sioli, H. 1951a. Sobre a sedimentação na várzea do baixo Amazonas. Pp. 42–66 in Boletim Técnico 24. Belém, Brazil: Instituto Agronomico do Norte; Serrão, E. A. S. 1986b. Pastagens nativas do trópico úmido brasileiro. Conhecimentos atuais. Pp. 183–205 in Simpósio do Trópico Úmido I, Vol. V. Anais. Belém, Brazil: Brazilian Enterprise for Agricultural Research–Center for Agroforestry Research of the Eastern Amazon.

prise for Agricultural Research, 1990; Serrão, 1986b; Serrão and Falesi, 1977; Serrão and Simão Neto, 1975).

In addition to being the main source of feed for cattle, the importance of FPG has increased as interest has increased in raising water buffaloes because of their proved higher efficiency in utilizing floodplain grasslands (da Costa et al., 1987; Nascimento and Moura Carvalho, In press).

FPG produce relatively high levels of forage, up to 20 metric tons or more of forage dry matter per ha, depending on the flooding gradient (Camarão et al., 1991; Serrão, 1986b). The forage quality of FPG is considerably higher than that of WDSG and is similar or superior to that of upland sown pastures. Daily live weight gains of between 400 and 600 g for cattle and water buffaloes are fairly common, mainly during the dry season (September through February), when grazing conditions are adequate (Camarão et al., 1991; da Costa et al., 1987; Serrão, 1986b).

The agronomic sustainability of FPG is high because of the favorable edaphic and hydrologic conditions of *várzea* and *várzea*-like ecosystems. Forage production potential is higher in the dry season, when adjacent upland native (savannah-type) and cultivated pastures have less available forage and are lower in quality. Utilization of FPG during the flooding season (March through August) is difficult, resulting in poor animal performance and the frequent loss of animals, mainly cattle, since water buffaloes are better able to thrive under partial flooding conditions.

The high-productivity (dry season)/low-productivity (flood season) fluctuations of FPG affect their economic sustainability because animals are ready for market only when they are 48 to 54 months old. Results of recent research (da Costa et al., 1987; Serrão et al., In preparation) and from commercial operations indicate that the integration of improved upland pastures of *Brachiaria* species, mainly *B. humidicola* (for grazing in the wet season), with adjacent FPG (which are grazed in the dry season) can considerably increase production and the economic sustainability of cattle raising activities in FPG. These integrated systems reduce the age at which cattle are ready for market by as much as 40 percent (da Costa et al., 1987; Serrão et al., In preparation).

Cattle raising on FPG has the potential for more intensive production with a more favorable socioeconomic environment. Owners of small- and medium-sized farms are the main practitioners of this activity, but the main constraint on sustainability in agricultural development in the floodplains of Brazil's humid tropics is the lack of a better socioeconomic environment for the farmers.

Research is needed to obtain higher levels of technical sustainability for cattle raising in FPG. Research should concentrate on more efficient means of managing FPG per se and on the selection of better-adapted and more-productive forages for pasture establishment and utilization in upland areas adjacent to FPGs.

CATTLE RAISING ON POORLY DRAINED SAVANNAH GRASSLANDS (PDSG)

PDSG are drainage-deficient native grasslands typical of the eastern part of Marajó Island in the state of Pará (Figure 5). A typical PDSG ecosystem is frequently associated with FPG when the PDSG is in its more humid gradient. (In Figure 5, gradients G_1 and G_2 correspond to the WDSG ecosystem, and gradient G_3 is similar to the FPG ecosystem [Serrão, 1986b].) Inceptisols (mainly groundwater laterites), entisols (mostly groundwater podzolic soils and quartz sands), and oxisols (latosols) are the predominant soils. Herbaceous, grassy vegetation is predominant in the ecosystem. Grasses of the genera *Axonopus, Andropogon, Trachypogon, Eragrostis, Eleusine, Paspalum,* and

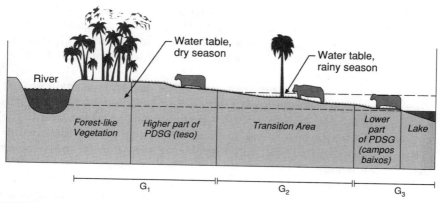

FIGURE 5 Profile of a typical poorly drained savannah grassland (PDSG) ecosystem on the Island of Marajó, state of Pará. Gradient G_1 corresponds to the well-drained savannah grassland ecosystem; G_2 is the transition area from G_1 to G_3; and G_3 corresponds to the floodplain grassland ecosystem (Serrão, 1986). Sources: Organization of American States and Instituto do Desenvolvimento Economico e Social do Pará. 1974. Marajó: Um Estudo para Seu Desenvolvimento. Washington, D.C.: Organization of American States; Serrão, E. A. S. 1986b. Pastagens nativas do trópico úmido brasileiro. Conhecimentos atuais. Pp. 183–205 in Simpósio do Trópico Úmido I, Vol. V. Anais. Belém, Brazil: Brazilian Enterprise for Agricultural Research–Center for Agroforestry Research of the Eastern Amazon.

Panicum are the main components in gradients G_1 and G_2, while those of the genera *Eriochloa, Echinochloa, Hymenachne, Leersia, Luziola,* and *Oryza* tend to dominate in gradient G_3.

Various gradients of PDSG occupy about 2 million ha (Organization of American States and Instituto do Desenvolvimento Economico e Social do Pará, 1974) of the eastern portion of Marajó Island, where cattle raising has been the main activity for the past 300 years (Teixeira, 1953). More than 1 million head of cattle and water buffalo are grazed on PDSG, mostly in cow-calf operations. PDSG are intermediate between WDSG and FPG for cattle production. Productivity is generally low. The annual primary productivities of gradients G_1 and G_2 (Figure 5) are rarely higher than 6 metric tons of dry matter per ha, and their carrying capacities vary from 3 to 5 ha/AU (Brazilian Enterprise for Agricultural Research, 1980; Organization of American States and Instituto do Desenvolvimento Economico e Social do Pará, 1974; Teixeira Neto and Serrão, 1984). Although the forage quality of PDSG is slightly higher than that of WDSG, it is intrinsically low, resulting in relatively low animal performance (Serrão, 1986b).

As in WDSG, low levels of productivity and quality of PDSG are associated with low levels of soil fertility, although, because of higher soil moisture levels during most of the year in gradients G_1 and G_2, pasture productivity and quality in PDSG tend to be somewhat higher than in WDSG (Serrão, 1986b).

PDSG on Marajó Island are subjected to strong seasonal climatic fluctuations. This results in corresponding seasonal forage and animal production fluctuations that, in turn, considerably extend the age at which cattle are ready for market. Therefore, cattle are finished on improved upland forest-replacing pastures on lands other than on the Island.

Despite the above-mentioned floristic, edaphic, hydrological, and management limitations, PDSG have good potential for extensive cattle raising activities. The resilience of PDSG in light of edaphic, climatic, and management constraints is high, resulting in relatively high agronomic and ecologic sustainabilities.

Typically, cattle raising on PDSG is carried out by a few employees and their families on large ranches owned by individual proprietors. The employees generally have low socioeconomic standards of living, which renders low levels of socioeconomic sustainability to the system.

Because of ecologic limitations on Marajó Island, cattle raising on PDSG has reached its limit for expansion. However, research results (Brazilian Enterprise for Agricultural Development, 1980; Marques et al., 1980; Teixeira Neto and Serrão, 1984) indicate that there is room for sustainable increased production by intensifying the utilization of PDSG or, as with WDSG, by replacing patches of native savannahs in gradients G_1 and G_2 with more productive improved pastures to qualitatively and quantitatively supplement the native pasture.

Additional research is necessary to promote more sustainable use of PDSG. Basic research is needed to generate knowledge on the ecology and ecophysiology of the native grassland for its sustainable use. Applied research efforts should concentrate on the selection of adapted and more productive pasture grasses and legumes, mainly for gradients G_1 and G_2 (see Figure 5), mineral supplementation, and native savannah grassland management.

Perennial Crop Agriculture

Perennial crop farming has been considered an ideal model for agriculture in the Brazilian humid tropics as a means of minimizing local environmental disturbances and maintaining the ecologic equilibrium in the region (Alvim, 1978).

Ecologically, perennial crops—as well as forest and agroforestry plantations—are the closest to natural forests in their efficiency in protecting the soil from erosion, leaching, and compaction (Alvim, 1989). In addition, in comparison with short-cycle crops, perennial crops have lower demand for soil nutrients, because of their efficient soil nutrient recycling mechanisms, and higher tolerance to high acidity and aluminum toxicity, which are common limitations of about 80 percent of Amazonian soils (Nicholaides et al., 1985).

SUSTAINABILITY OF PERENNIAL CROP AGRICULTURE

The potential of perennial crops in the agricultural development of the humid tropics has been underestimated or neglected. Although there are ecologic and agronomic reasons for being optimistic, there are important considerations limiting economic sustainability, since for most of the important perennial crop products, there is limited market potential, which is a constraint for large-scale plantations.

Although perennial crops are recognized as having fairly high levels of agronomic sustainability, high biotic pressure caused by the variety of pests and diseases these crops are plagued by is probably the most limiting factor in the Brazilian humid tropics (Morais, 1988). Leaf blight disease (caused by the fungus *Microcyclus ulei*, which attacked rubber tree plantations in the 1930s) continues to be a major limiting factor of rubber tree plantations today. Fusariose, or dry rot (caused by the fungus *Fusarium solani* f. sp. *piperis*), has caused serious agronomic and economic problems to the black pepper industry for many years. Witchbroom disease (caused by the fungus *Crinipellis perniciosa*), which affects cacao; and, more recently, the fatal yellowing disease of African oil palm (caused by an unknown pathogen) have been serious threats to the agronomic and economic sustainabilities of important perennial crops.

The social sustainability of perennial crop agriculture may be high (Alvim, 1989; Fearnside, 1983). These crops are appropriate to both small and large operations and are labor intensive, generating high levels of employment in small areas. However, profits are marginal (Flohrschütz, 1983) and cannot finance the infrastructural adaptation and economic and ecologic changes necessary for prolonged sustainability of the land use system.

A major limitation to expanding perennial crop plantations in the Amazon is the market dimension. Regional experiences have shown rapid market saturation for products such as black pepper and urucu (*Bixa orellana*). This market saturation creates serious economic sustainability problems for those land use systems. Use of only a small fraction of the Amazon for perennial crop production may saturate national and international markets. For example, 200,000 ha of rubber tree plantations would be enough to make Brazil self-sufficient in natural rubber, 160,000 ha of cacao plantations would be enough for the Amazon region to contribute 50 percent of the Brazilian cacao production, and 10,000 ha of guaraná is sufficient to saturate national and international markets. Growth of the black pepper market is subject to the rate of population growth. These considerations also apply to Brazil nuts, coffee, and African oil palm.

Present and potential national and international timber markets seem to be unlimited. Therefore, timber production in reforestation projects should be emphasized and stimulated, whether directly in homogeneous plantations or indirectly in integrated agroforestry and silvopastoral (pasture, animal, and tree) systems.

In addition to the presently economically important perennial plants, there are many others in the forest that also are or may be important as fruit, medicinal, timber, fiber, and oil products. These products need to be domesticated for future plantation or agroforestry land use systems. Association of perennial crops with other plants with shorter cycles, and even pastures, should reduce the biologic risks and make the system more accommodating to market fluctuations.

RESEARCH NEEDS

Research will be the basis for more sustainable perennial crop systems. Economically important diseases of the present high-value perennial crops must be the priority of applied and basic research. Emphasis should also be given to research of the domestication of potential high-value perennial crops and to the definition of production systems.

Agroforestry

Agroforestry systems (AFSs) have recently been examined as land use systems that will use land resources in the Brazilian humid tropics more sustainably. They should gradually replace or be associated with present extensive low-sustainability land use systems such as open monoculture pasture-based cattle raising systems, upland shifting agricultural systems, and extractive forest reserves. Possible combinations of AFSs are presented in Figure 6. The reasons for this emphasis of AFSs are as follows.

• AFSs may increase the productive capacity of certain agricultural lands that have had reduced productive capacity because of mismanagement that resulted in compaction and loss of fertility.

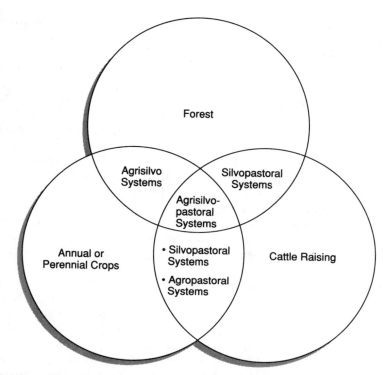

FIGURE 6   Possible combinations involving annual and perennial crops with trees and cattle raising.   Source: Homma, A. K. O., and E. A. S. Serrão.   In preparation.   Será Possível a Agricultura Autosustentada na Amazônia?

• AFSs allow the growth of combinations of species with different demands for energy, resulting in the more efficient use of solar energy because of the vertical stratification of associated plants. If the association includes leguminous plants, soil fertility can also be increased.

• In AFSs, crop diversification reduces biologic risks and is more adaptable to market fluctuations. The introduction of a tree component in annual or perennial cropping systems or in cattle-raising systems may favor the replacement of unsustainable slash-and-burn agricultural systems.

AFSs present peculiarities in relation to market, technological practices, farm administration, and management.   For example, the rubber tree–cacao systems recommended by research institutions result in yield reductions, in relation to the single-crop system, of about 75 percent for rubber and 50 percent for cacao.   From the market point of view, between 100,000 to 120,000 ha of rubber plantation in production is needed today to neutralize rubber imports, while the market for cacao is fairly restricted.

Anderson et al. (1985) described and analyzed a commercial AFS with relatively high levels of sustainability that is being developed by riverbank dwellers. This system is based on the extraction of forest products with and without management and is being developed in a periodically inundated *várzea* floodplain of the Amazon River estuary, in the vicinity of Belém, where it is difficult to use conventional agricultural practices. The main activities in the system include hunting, fishing, raising of small domestic animals, and harvesting of fruits, heart of palm, wood, organic fertilizer, ornamental plants, latex, fibers, oil-bearing seeds, and medicinals. These products are sold in the Belém farmer's open market. This is an example of a semiextractive agroforestry system in which a proportion of the economically valuable trees in the system are domesticated or semidomesticated.

An important example of sustainable agroforestry agriculture is one developed by Japanese immigrants and their offspring (Nippo-Brazilian farmers) who have farmed remote forest regions of the Amazon Basin since the late 1920s (Subler and Uhl, 1990). In the mid-1950s black pepper fusariose became the most serious constraint to sustainability of black pepper production, the main activity of those farmers at the time. In the early 1970s these farmers had to diversify their agricultural systems.

Nippo-Brazilian farmers have replaced most of their black pepper agriculture with diverse agroforestry arrangements. Farmers rely on intensive cultivation, producing a diversity of high-value cash crops through mixed cropping of perennial plants. These plants include a wide variety of perennial trees (such as cacao, rubber, cupuaçu [*Theobroma grandiflorum*], graviola [*Annona muricata*], papaya, avocado, mango, and Brazil nut) and palms (such as açai [*Euterpe oleracea*], coconut, oil palm, peach palm), shrubs and vines (pineapple, Barbados cherry [*Malpighia glabra*], banana, coffee, passion fruit, black pepper, and urucu), and annuals (such as cotton, cowpea beans, pumpkin, cassava, melon, pepper, cucumber, cabbage) (Subler and Uhl, 1990).

Most farms are operated by single families, and the average size is between 100 and 150 ha. On average, however, each farm cultivates only about 20 ha (Flohrschütz et al., 1983). The rest of the area is generally in secondary forest regeneration, following pepper field abandonment or previous slash-and-burn activity, or is undisturbed forest. Figure 7 shows a typical Nippo-Brazilian agroforestry farm in Tomé-Açu.

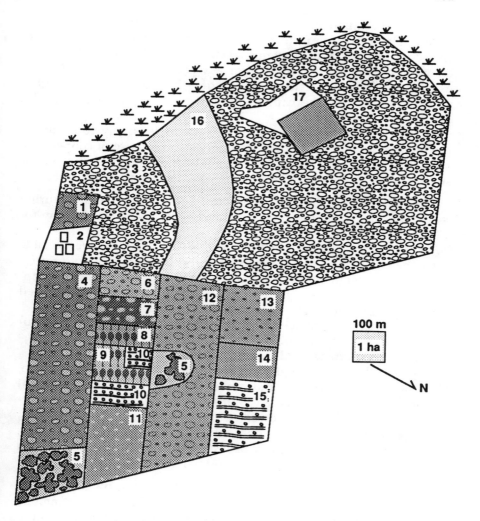

FIGURE 7 Land use on a representative Nippo-Brazilian farm in Tomé-Açu. 1, cacao, erythrina; 2, household area; 3, coconut, citrus, mangosteen, graviola; 4, cacao, erythrina, andiroba, Brazil nut; 5, secondary forest regeneration; 6, cacao, vanilla, palheteira, freijó; 7, cacao, paricá,; 8, rubber trees; 9, rubber trees, black pepper, cacao; 10, rubber trees, passion fruit; 11, black pepper, cacao; 12, cacao, banana, *Cecropia* sp.; 13, black pepper, cupuaçu; 14, black pepper; 15, passion fruit, cupuaçu; 16, pasture grasses; 17, black pepper, clearing. Source: Subler, S., and C. Uhl. 1990. Japanese agroforestry in Amazonia: A case study in Tomé-Açu, Brazil. Pp. 152–166 in Alternatives to Deforestation: Steps Toward Sustainable Use of the Amazon Rain Forest, A. B. Anderson, ed. New York: Columbia University Press.

Nippo-Brazilian AFSs (NBAFSs) rely on fairly heavy inputs of chemical and organic fertilizers, although the amounts tend to decrease as the trees in the systems reach maturity. There is also a high labor requirement. A typical farm with about 20 ha in cultivation uses approximately six to eight full-time laborers, which, together with inputs, also make capital investments high (Subler and Uhl, 1990).

The basis for the success of those systems is largely constant experimentation with innovative techniques and the use of cooperative marketing systems. From an overall analysis of these systems, Subler and Uhl (1990) came to the following conclusions about NBAFSs:

• NBAFSs are conservative of forest and soil resources, requiring relatively small-scale forest clearing and maintaining soil fertility for a long time.

• The long-term sustainability of NBAFSs may be questionable since there is a trend toward increasing fertilizer and energy prices.

• Even though transportation is a limiting factor to the development of NBAFSs in remote frontier areas, they may be largely used with the increasing road network in the region.

• Rather than displacing rural inhabitants, NBAFSs use local human resources, but their high labor requirements make them vulnerable to labor shortages and increasing labor costs.

• Even though the high prices received for crops such as cacao, black pepper, passion fruit, and rubber make up for the heavy capital investments required by NBAFSs, market saturation may be a limiting factor for large-scale adoption of the system.

• Some form of institutional support through training, credit, and community services seems to be necessary to encourage the adoption of NBAFSs by Brazilian small-scale farmers.

In the case of silvopastoral systems, as trees grow taller, integrated management difficulties become more evident. For example, fire outbreaks cannot be overlooked, since fire may be a major limitation for arboreal vegetation. According to Veiga and Serrão (1990), the success of integration depends mainly on the equilibrium of the interaction among the animal, tree, and pasture components. The competition for light, water, and nutrients between tree and pasture must be well understood.

Silvopastoral systems are in their initial stages of development in the Amazon. Most of those land use systems are concentrated in the eastern state of Pará on small- and medium-sized properties, where Veiga and Serrão (1990) found associations of rubber, coconut, African oil palm, cashew, urucu, pine, mango, and Brazil nut trees with strata of grasses and legumes for cattle grazing. They observed that the main management and sustainability limitations of the varied integrated system are related to pasture production and persistence—the pasture is overgrazed in most cases and maintenance management is deficient (for example, insufficient weed control). Under those conditions, since the available forage in the system tends to be overestimated, extra buffer pasture areas should complement the integrated system for more flexible grazing management.

Promising silvopastoral system combinations are being tested and evaluated by EMBRAPA researchers in Paragominas in the eastern state of Pará (Veiga and Serrão, 1990). Two native timber-producing trees, namely, *paricá* (*Schizolobium amazonicum*) and *tatajuba* (*Bagassa guianensis*), and one exotic tree species (*Eucalyptus teriticornis*) are each associated individually with three forage grasses (*Brachiaria brizantha* cv. Marandu, *B. humidicola*, and *B. dictyoneura*). Five years after establishment and 3 years under grazing management, the combination of *paricá* × *B. brizantha*, for example, is showing satisfactory levels of agronomic and ecologic sustainability.

Undoubtedly, AFSs rank high in terms of sustainability among the agricultural land use systems used in the Brazilian humid tropics, and there is a probability of expansion in the near future. The probability is so high that EMBRAPA's agricultural research centers in the Amazon have recently been changed into agroforestry research centers.

Although they rank high in sustainability, AFSs cannot be considered a panacea for the Amazon. Their expansion will depend on the market for the products involved, labor use intensity, and most important, their economic profitability. Monocultures of cupuaçu, Barbados cherry, and black pepper have higher profitabilities than do some arboreal associations because of the present market demand characteristics of the region. Therefore, appropriate market conditions need to be developed to ensure the expansion of AFSs.

Research priorities for developing more sustainable AFSs should include the domestication and introduction of high-value, multipurpose native and exotic trees and food and forage crops for the development and management of integrated systems of crops, pastures, animals, and trees.

LAND USE INTENSITY, RESEARCH, AND TECHNOLOGY: THE KEY FOR SUSTAINABILITY

The low sustainability of agricultural development in the frontier expansion process has been an important cause of high rates of deforestation and the consequent negative environmental and socioeco-

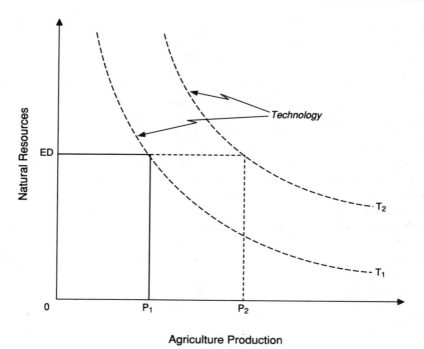

Agriculture Production

FIGURE 8   Exchange relations between agricultural production and natural resource disturbances affected by technological development. ED, environmental disturbances; T_1, inappropriate technology; T_2, more appropriate technology; P_1, agricultural production with technology T_1; P_2, agricultural production with technology T_2. Source: E. B. Andrade, personal communication, 1990.

nomic implications. A major reason for this is the fact that, in the past 30 years, the most important political decisions regarding regional agricultural development have largely bypassed scientific and technological considerations.

Because of society's demand for food and fiber and deforestation restrictions in the Brazilian Amazon, more production must be realized mostly from already deforested lands. This implies increasing land and labor productivities, which can only be achieved with land use intensification. This, in turn, can only be achieved with the strong support of science and technology, but the levels of technology used for the most important agricultural land use systems that replace forests have typically been low.

Figure 8 illustrates the importance of technology for agricultural production in relation to the conservation of natural resources. Logically, for each degree of agricultural development there is a corresponding degree of environmental degradation. In the Amazon, use of inappropriate technologies has resulted in low levels of agricultural products with high levels of environmental degradation. However, scientific and technological developments can propitiate increases in agricultural production with more appropriate technologies at the same (or even lower) level of environmental degradation. The low technological level of agricultural production in the Amazon indicates a high potential for improvement.

From these considerations and considering the insufficiency of the available knowledge basis, the search for sustainability will depend to a large extent on research development. Research should be directed mainly toward increasing the productivities of already deforested areas to guarantee a local supply of food and fiber and the export of products that are exclusive to the Brazilian Amazon region and toward reducing the pressure on new forest frontiers. Research should also be directed toward supporting the conservation and preservation of natural resources.

To accomplish those more general goals that integrate the needs of society with the conservation of natural resources, future agricultural development should be built fundamentally on the diversity that characterizes the humid tropical ecosystem and should mirror as much as possible its complexity (National Research Council, 1991). Therefore, research should focus on the following:

- Increasing basic knowledge of Amazonian natural ecosystems;
- Surveying, classifying, and analyzing presently and potentially successful agricultural land use and land resource management systems;
- Developing and promoting principles and components of land management that sustain land resources under the constraints of humid tropical ecosystems;
- Reclaiming degraded ecosystems for intensive agricultural production and regeneration of the ecosystem; and
- Promoting the agroecologic zoning of the Brazilian humid tropics.

Basic research on the following topics is immediately relevant for increasing the sustainability of Amazonian agricultural systems:

- Nutrient, water, and biomass cycling in forest ecosystems that have been disturbed by agriculture as well as those that are undisturbed;
- Climatic, edaphic, and biologic disturbances caused by deforestation and fire utilization for agricultural development purposes;

• Evaluation of biotic and abiotic factors that influence degradation and regeneration of forest ecosystems disturbed by agriculture; and

• Survey, classification, and analysis of presently and potentially important agricultural land use systems.

Applied research should focus on the continuous search for alternative sustainable agricultural production systems and on improving the sustainability of important systems already in use. Applied research priorities for the most important agricultural land use systems in the Brazilian Amazon are given in Table 3. In addition, applied research for fish production systems should focus on domestication of economically important freshwater fish; controlled native fish reproduction and management; and development of integrated systems that include fish, crop, and cattle production.

Institutional Capacity

More than ever, research is fundamental for agricultural development in the Amazon. The present agricultural production limitations and the need for natural resource conservation demand a research agenda that requires an enormous institutional effort.

Figure 9 lists the research institutions that are directly and indirectly involved with agricultural research and natural resources conservation in the Amazon. Paradoxically, those institutions have been practically stagnant during the past decade from the standpoint of infrastructure, personnel (quantitatively and qualitatively), and financial situation. In addition, intense politicization and lack of stimuli (for example, low salaries) within research institutions have reduced the research impetus. It is difficult to foresee any short-term improvement in institutionalized agricultural research in Brazil as a whole and in the Amazon in particular.

A FUTURE SCENARIO

Throughout the history of the Amazon, economic features have reflected its dependence on more developed nations. During the "drogas do sertão" phase (extraction of cacao, medicinal and aromatic plants, and plant and animal oils), it depended on Portugal, and during the rubber cycle it depended on rubber-importing countries. Starting in the 1970s, national and international capitals directed the occupancy of the Amazon, extrapolating the dimension of occupied area to include future economic possibilities.

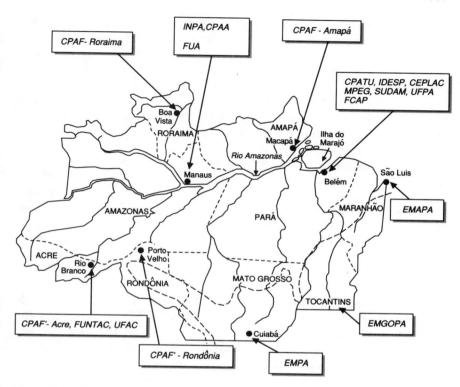

FIGURE 9   Research institutions directly and indirectly involved with agricultural development in the Amazon.   Key to acronyms: CPAF-Roraima, Centro de Pesquisa Agroflorestal de Roraima; INPA, Instituto Nacional de Pesquisas da Amazônia; CPAA, Centro de Pesquisa Agroflorestal da Amazônia Ocidental; FUA, Fundação Universidade do Amazonas; CPAF-Amapá, Centro de Pesquisa Agroflorestal do Amapá; CPATU, Centro de Pesquisa Agroflorestal da Amazônia Oriental; IDESP, Instituto de Desenvolvimento Econômico Social do Pará; CEPLAC, Comissão Executiva do Plano da Lavoura Cacaueira; MPEG, Museu Paraense Emilio Goeldi; SUDAM, Superintendência do Desenvoluimento da Amazônia; UFPA, Universidade Federal do Pará; FCAP, Faculdade de Ciências Agrarias do Pará; EMAPA, Empresa Maranhense de Pesquisa Agropecuária; EMGOPA, Empresa Goiana de Pesquisa Agropecuária; EMPA, Empresa Matogrossense de Pesquisa Agropecuária; CPAF-Rondônia, Centro de Pesquisa Agroflorestal de Rondônia; CPAF-Acre, Centro de Pesquisa Agroflorestal do Acre; FUNTAC, Fundaçãe Tecnologia do Acre; UFAC, Universidade Federal do Acre.

The greater concern with the environment that started in the 1980s as a result of the alarming rates of deforestation will direct the future economic development of the region. The future scenario of development in the Amazon is therefore discussed at the national and international levels, with the environmental question being the backdrop. Other variables, such as the Acre-Pacific Highway through Peru, minimization or cancellation of support to agricultural activities, and road construction restrictions, will also direct the level of human occupation of the Amazon.

Environmental aggression should be reduced considerably in the future. However, the growth of pockets of poverty cannot be eliminated if environmental policy is directed exclusively toward zero deforestation. Small-scale farmers will probably be the main victims, rural to urban migration will be enforced, and unemployment and underemployment will be stimulated if more ample development policies are not implemented.

One probable consequence of environment-oriented policies will be increasing land value, which will likely induce utilization of more capital-intensive technologies in already deforested lands. Agricultural activities will be restricted to meet the regional demands for products that are not exclusively Amazonian and the external demand for Amazon-exclusive products that are competitive with products from other regions.

Despite criticism, native timber extraction will probably grow in intensity to meet growing national and international market demands. Contradictions about its sustainability will probably induce silvicultural development in already deforested areas of the Amazon. In that direction, the FLORAM (Forest Environment) megasilviculture project (Universidade de São Paulo, 1990) is being proposed. Besides economics, the project is also intended to study atmospheric carbon fixation. The Forest Poles Project for the Eastern Amazon is another example; it aims to forest 1 million ha of land along the Carajás-Itaquí Highway at a cost of US$1.2 billion.

Extraction activities, and specifically extractive rubber tapping (in this case, even with external support that is now under way), should gradually decline in importance. Some extractors will move toward agroforestry.

Other activities with low levels of sustainability such as traditional shifting agriculture will not be able to be maintained in the long run because of increasing population density in addition to deforestation restrictions.

What will happen to the regional development of science and technology? Research activities in the Amazon are stagnant, and the future is cloudy. The conservation, preservation, and rational utilization of many natural resources will largely depend on the future generation of knowledge and technology.

The tendency to reduce environmental disturbances is due more to economic and/or legal impediments that are created rather than to environmental ethics or consciousness. Day-to-day regional life includes high demographic densities, urbanization, the need for more employment, low income, and low quality of life. If poverty, unemployment, underemployment, and the lack of a basic infrastructure persist, conservation and preservation intentions will gradually lose the support of the population.

EXPANSION POTENTIAL OF PRESENT LAND USE SYSTEMS

Extraregional forces will likely direct the pace of production activities in the Amazon. With the label of environmental cause, a set of measures to discourage production activities, except for agroforestry and extraction activities, are being launched. Some have proposed that extraction activities should be the land use system for about 25 percent of the Brazilian Amazon region.

On the other hand, a set of intraregional forces reacts to the impropriety of agricultural systems from the point of view of macroeconomics in relation to the region's inhabitants. This presupposes that agricultural activities must supply the local population's needs for food, generate employment, guarantee better living standards, and promote the region's development.

Within the not-so-remote future, it is probable that the extractive reserve syndrome will be weakened when realistic and impartial evaluations are made. The conclusion will likely be that it is not easy to propose simple solutions for the Amazon.

Environmentally oriented proposals have not been accompanied by reasonable development alternatives. Consequently, they may induce rural as well as urban socioeconomic adversities such as unemployment, which is already high in the region. This stagnation scenario might favor extraction activities and even become their justification. In that scenario, production activities considered to be harmful to the environment will continue in the search for new adaptations to the prevalent biosocioeconomic environment.

The closing of the agricultural frontier will make land more expensive, which will induce the use of more capital-intensive technologies. Small farmers will find it difficult to maintain their activities because of restrictions on deforestation and burning, the basic ingredient of shifting agriculture. Unless other alternatives are offered, deforestation reduction of 500,000 ha/year may cause serious adversities to small-scale farmers in the Amazon.

Várzea floodplain agriculture will probably remain stagnant. If political measures are taken to increase the food supply to the main urban nuclei, food production along the floodplain may be stimulated. Because of the favorable conditions for raising water buffalo in the *várzeas*, it may be even more strongly stimulated than it was previously.

Although environmental restrictions tend to be reinforced, the survival strategy of farmers will prevail. The emergence of new, alternative products exclusive to the Brazilian Amazon region are always possible, whether they supply regional needs or are exported. With strict environmental controls, the prices of these products will increase. This will, in turn, stimulate more intensive production, resulting in the displacement of small farmers. As long as they do not have external market competition, export products, because they are exclusive, will have a good chance for sustainable production.

The possibility for developing an "Amazonian agriculture" cannot be discarded. This may be the positive side of the exaggerated interest in extraction activities. Agricultural development based on domesticated natural resources, such as medicinal plants, toxic plant products, native fruits, oils, and heart of palm, may have ample markets in the future. The beginning of that trend seems to be under way. The success of these new alternatives will depend on the research capacity for plant domestication and market dimension.

The local society will likely react to environmental policies that come from outside the region. In that sense, a more progressive vision for the Amazon cannot be overlooked. It may be that the production sector will demand regional access to the Pacific and more investments in rural areas in terms of social infrastructure, besides tax incentives, subsidies, and export taxes, with all of these demands being under environmentally oriented premises. The maintenance of uneconomic extraction systems by the state—with a social crisis dilemma—may be the result of society's acceptance of more progressive measures.

These facts may create a new equilibrium in the sustainability of the production system as a whole. The international capitalistic system itself will favor these actions because of its implicit interest in the timber and mineral markets. The growth of timber extraction is inevitable because of the increasing internal and external demand for wood products. Under the assumption of a not-yet-proved sustainability, timber extraction will probably continue for the next few decades and will probably be the last extraction activity in the Amazon. The need for maintaining biodiversity and the slow vegetative growth cycles of forest timber resources will restrict timber extraction to some selected areas.

Increasing prices of timber products will induce production on timber plantations, the only alternative to meet future demands because of population increases. Future plantations will also be needed to meet the future demands of the paper and cellulose industries. Ecologically, these plantations will be justified as a means of absorbing atmospheric carbon.

Integrated systems to increase agronomic and ecologic sustainabilities will be stimulated even if economic sustainability is marginal. Within this context, agrisilvopastoral systems are included. Intelligent, appropriate combinations will be proposed. Their implementation will largely be limited by market dimension, management, and the availability of technology.

Other activities will probably be implemented. Fish production—whether through cultivation of native and exotic fish under controlled conditions or through the replenishing of rivers and lakes—and domestication of high-value native wildlife will be developed.

With the present technological standards of agriculture in the Amazon, the possibilities for high levels of agronomic and ecologic sustainability are reduced. Socioeconomic limitations for sustainable agriculture are also important barriers, since agronomic and ecologic sustainability is generally economically infeasible.

To maintain productivity gains, maintenance of sustainability requires continuous investments in research. Environmental constraints will always be a challenge to research in the search for agricultural sustainability in the humid tropics.

In the long run, the comparative advantages of abundance of natural resources and unqualified labor will be abandoned. It is probable that increasing technological advances and labor qualification will be the main supports of future agricultural activities.

Despite these limitations, there are ample possibilities for increasing agricultural sustainability in the Brazilian humid tropics without having to incorporate new segments of forest and within global perspectives of sustainability. Continuous technological development within the farmer's capacity to accompany technical progress is indispensable to implementing production systems that are more compatible with agronomic and ecologic sustainability. Economic viability must be within short- and long-term horizons, preferably without any protectionist measures.

Economic profitability is a key factor for agricultural sustainability in the Amazon. Rural poverty will not allow high ecologic sustainability.

Even in the case of cattle raising activities, the adoption of fewer ecosystem-degrading processes will depend on higher values of cattle-related products. However, an awakening of society's awareness and the formation of a new ethic in relation to profitability, which includes environmental costs, are necessary.

From this analysis of traditional and presently developing land use systems in the Brazilian humid tropics, it is clear that some land use systems are more appropriate for implementation. Because these have demonstrated moderate to high levels of sustainability and high expansion potential for mid- and long-term agricultural development, and on the basis of their favorable present and potential sustainability features, priority for expansion and research support should be given to the following land use systems:

- Nippo-Brazilian-type agroforestry,
- Integrated pasture-based (agrisilvopastoral) systems,
- Native forest timber extraction with sustainable management,
- Reforestation for timber and cellulose production, and
- *Várzea* floodplain agriculture.

Technological and educational deficiencies are the main factors limiting farmers in their attempts to practice agriculture that allows higher levels of sustainability in the Amazon. Research is not the panacea for meeting high levels of agricultural sustainability as defined here. The reduced success of most agricultural enterprises in the Amazon is not so much due to the productive potential of the land as it is due to deficient social, economic, and infrastructural conditions; lack of stable and coherent agricultural policies; and fluctuations in the prices of agricultural products. More investments are needed in the rural environment to improve quality of life, thus avoiding (or minimizing) a rural exodus and continuous migration to new areas.

REFERENCES

Alcântara, E. 1991. A ciência afasta o perigo do desastre global. Rev. Veja, São Paulo 24(41):78–84.

Allegretti, M. H. 1987. Reservas Extrativistas: Una Proposta de Desenvolvimento da Floresta Amazônica. Curitiba, Brasil: Instituto de Estudos Amazônicos.

Allegretti, M. H. 1990. Extractive reserves: An alternative for reconciling development and environmental conservation in the Amazon. Pp. 252–274 in Alternatives to Deforestation: Steps Toward Sustainable Use of the Amazon Rain Forest, A. B. Anderson, ed. New York: Columbia University Press.

Alves, E. 1988. Pobreza Rural no Brasil: Desafios da Extensão e da Pesquisa.

Brasilia: Companhia de Desenvolvimento do Vale do Rio São Francisco.

Alvim, P. T. 1978. Floresta Amazônica: Equilíbrio entre utilização e conservação. Ciência Cultura 30(1):9–16.

Alvim, P. T. 1989. Tecnologias apropriadas para a agricultura nos trópicos úmidos. Agrotrópica 1(1):5–26.

Alvim, P. T. 1990. Agricultura apropriada para uso contínuo dos solos na região Amazônica. Espaço, Ambiente Planejamento 2(11):1–71.

Anderson, A. B. 1989. Estratégias de uso da terra para reservas extrativistas da Amazônia. Pará Desenvolvimento 25:30–37.

Anderson, A. B. 1990. Extraction and forest management by rural inhabitants in the Amazon estuary. Pp. 65–85 in Alternatives to Deforestation: Steps Toward Sustainable Use of the Amazon Rain Forest, A. B. Anderson, ed. New York: Columbia University Press.

Anderson, A. B., A. Gely, J. Strudwick, G. L. Sobel, and M. G. C. Pinto. 1985. Um sistema agroflorestal na várzea do estuário Amazônico (Ilha das Onças, Município de Barcarena, Estado do Pará). Acta Amazon. Manaus 15(Suppl.):195–224.

Associação das Indústrias de Madeiras dos Estados do Pará e Amapá. 1989. Comércio Exterior: Produtos Exportados Pelo Estado do Pará. Fonte, Brasil: Carteira de Comércio Exterior, Banco do Brasil.

Bastos, et al. 1986. O estado atual de conhecimentos de clima da Amazônia brasileira com finalidade agricola. Pp. 19–36 in Simpósio do Trópico Úmido I, Vol. VI, Anais. Belém, Brazil: Brazilian Enterprise for Agricultural Research–Center for Agricultural Research of the Humid Tropics.

Brazilian Enterprise for Agricultural Research. 1980. Centro de Pesquisa Agropecuária do Trópico Úmido, Belém, Projeto Melhoramento de Pastagem da Amazônia (PROPASTO). Relatório Técnico 1976/79. Belém, Brazil: Center for Agricultural Research of the Humid Tropics.

Brazilian Enterprise for Agricultural Research. 1990. Relatório Técnico Anual do Centro de Pesquisa Agropecuária do Trópico Úmido. Belém, Brazil: Center for Agricultural Research of the Eastern Amazon.

Brazilian Institute of Geography and Statistics. 1981. Anuário Estatístico do Brasil. Rio de Janeiro: Brazilian Institute of Geography and Statistics.

Brazilian Institute of Geography and Statistics. 1991. Anuário Estatístico do Brasil. Rio de Janeiro: Brazilian Institute of Geography and Statistics.

Brazilian Institute of Space Research. 1990. Avaliação da Alteração da Cobertura Florestal na Amazônia Legal Utilizando Sensoriamento Remoto Orbital. São Paulo: Brazilian Institute of Space Research.

Browder, J. O. 1988. The social costs of rainforest destruction: A critique and economic analysis of the "hamburger debate." Interciencia 13:115–120.

Burger, D., and P. Kitamura. 1987. Importância e viabilidade de uma pequena agricultura sustentada na Amazônia oriental. Tübinger Geog. Studien 95:447–461.

Buschbacher, R., C. Uhl, and E. A. S. Serrão. 1988. Abandoned pasture in eastern Amazônia. II. Nutrient stocks in the soil and vegetation. J. Ecol. 76:682–699.

Camarão, A. P., and E. A. S. Serrão. In press. Produtividade e qualidade de pastagens de várzeas inundáveis.

Camarão, A. P. C., M. Simão Neto, E. A. S. Serrão, I. A. Rodrigues, and C. Lascano. 1991. Identificação e composiças química de espécies invasoras de pastagens cultivadas consumidas por bovinos em Paragominas, Pará. Boletim de Pesquisa 104. Belém, Brasil: Empresa Brasileira de Pesquisa Agropecuária.

Comissão Interministerial para a Preparação da Conferência das Nações Unidas Sobre Meio Ambiente e Desenvolvimento. 1991. Subsidios Técnicos para a Elaboração do Relatório Nacional do Brasil para a CNUMAD. Brasilia: Brazilian Institute for the Environment and Renewable Natural Resources.

Coradin, L. 1978. The Grasses of the Natural Savannahs of the Territory of Roraima, Brazil. Master's thesis. Herbert H. Lehman College of the City University of New York, New York.

Croxall, H. E., and L. P. Smith. 1984. The Fight for Food; Factors Limiting Agricultural Production. London: George Allen.

da Costa, N. A., J. B. Lourenco Junior, A. P. Camarão, J. R. F. Margues, and S. Dutra. 1987. Produção de carne de bubalinós em sistema integrado de pastagem nativa de terra inundável e cultivada de terra firme. Boletim de Pesquisa No. 86. Belém, Brasil: Empresa Brasileira de Pesquisa Agropecuária.

Da Silva, J. F. 1989a. Malva. Informações Básicas para Seu Cultivo. Documento 7. Belém, Brazil: Brazilian Enterprise for Agricultural Research–Unidade de Execução de Pesquisa de Âmbito Estodual.

Da Silva, J. F. 1989b. Juta. Informações Básicas para Seu Cultivo. Documento 8. Belém, Brazil: Brazilian Enterprise for Agricultural Research–Unidade de Execução de Pesquisa de Âmbito Estodual.

de Graaf, N. R., and R. L. H. Poels. 1990. The Celos management system: A polycyclic method for sustained timber production in South American rainforest. Pp. 116–127 in Alternatives to Deforestation: Steps Toward Sustainable Use of the Amazon Rain Forest, A. B. Anderson, ed. New York: Columbia University Press.

Denevan, W. M., and C. Padoch. 1987. Swidden-fallow agroforestry in the Peruvian Amazon. Adv. Econ. Bot. 5:1–7.

Dias, G. L. D., and M. D. de Castro. 1986. A Colonização Oficial no Brasil: Erros e Acertos na Fronteira Agrícola. São Paulo: Instituto de Pesquisas Economicas, Universidade de São Paulo.

Dias Filho, M. B. 1990. Plantas Invasoras em Pastagens Cultivadas da Amazônia: Estratégias de Manejo e Controle. Documento 42. Belém, Brazil: Brazilian Enterprise for Agricultural Research–Center for Agroforestry Research of the Eastern Amazon.

Dias Filho, M. B., and E. A. S. Serrão. 1982. Recuperação, Melhoramento e Manejo de Pastagens na Região de Paragominas, Pará; Resultados de Pesquisa e Algumas Informações Práticas. Documento 5. Belém, Brazil: Brazilian Enterprise for Agricultural Research–Center for Agricultural Research of the Humid Tropics.

Eden, M. J. 1964. The Savannah Ecosystem—Northern Rupununi, British Guiana. McGill University Savanna Research Project. Report 1. Savannah Research Report Series. Montreal: McGill University.

Falesi, I. C. 1976. Ecossistema de Pastagem Cultivada na Amazônia Brasileira. Boletim Técnico No. 1. Belém, Brazil: Brazilian Enterprise for Agricultural Research–Center for Agricultural Research of the Humid Tropics.

Fearnside, P. 1982. Desmatamento na Amazônia: Com que intensidade vem ocorrendo? Acta Amazôn. 10:579–590.

Fearnside, P. 1984. A floresta vai acabar? Ciência Hoje 2(10):43–52.

Fearnside, P. M. 1983. Development alternatives in the Brazilian Amazon: An ecological evaluation. Interciencia 8(2):65–78.

Fearnside, P. M. 1986. Human Carrying Capacity of the Brazilian Rainforest. New York: Columbia University Press.

Fearnside, P. M. 1987. Rethinking continuous cultivation in Amazônia. BioScience 37:209–214.

Fearnside, P. M. 1990. Predominant land uses in Brazilian Amazon. Pp. 233–251 in Alternatives to Deforestation: Steps Toward Sustainable Use of the Amazon Rain Forest, A. B. Anderson, ed. New York: Columbia University Press.

Flohrschütz, G. H. H. 1983. Análise Econômica de Estabelecimentos Rurais no Município de Tomé-Acu, Pará. Um Estudo de Caso. Documento 19. Belém, Brazil: Brazilian Enterprise for Agricultural Research–Center for Agricultural Research of the Humid Tropics.

Goedert, W. J. 1989. Região dos cerrados: Potencial agrícola e política para seu desenvolvimento agrícola. Pesq. Agrope. Bras. Brasília 24(1):1–17.

Goldemberg, J. 1989. Amazônia and the greenhouse effect. Pp. 13–17 in Amazônia: Facts, Problems and Solutions, Vol. I. São Paulo: Universidade de São Paulo.

Goodland, R. J., and H. Irwin. 1975. A Selva Amazônica: Do Inferno Verde ao Deserto Vermelho? São Paulo: Itatiaia.

Goodland, R. J., and H. Irwin. 1977. O cerrado e a floresta amazônica. Pp. 9–37 in Seminário Regional de Desenvolvimento Integrado 1, Vol. 2. Manaus and Belém, Brazil: Superintendency for the Development of the Amazon.

Goulding, M. 1980. The Fishes and the Forest. Berkeley: University of California Press.

Hecht, S. B. 1979. Leguminosas espontâneas en praderas Amazônicas cultivadas esu potencial forragero. Pp. 71–78 in Producción de Pastos en Suelos Ácidos de los Trópicos, P. A. Sanchez and L. E. Tergas, eds. Cali, Colombia: International Center for Tropical Agriculture.

Hecht, S. B. 1983. Cattle ranching in eastern Amazon: Environmental and social implications. Pp. 155–188 in The Dilemma of Amazonian Development, E. F. Moran, ed. Boulder, Colo.: Westview.

Hecht, S. B., R. B. Norgaard, and G. Possio. 1988. The economics of cattle ranching in eastern Amazônia. Interciencia 13(5):233–240.

Hirano, C., F. C. S. Amaral, F. Palmieri, J. O. I. Larach, and Souza Neto.

1988. Delineamento Macro-ecológico do Brasil. Rio de Janeiro: Serviço Nacional de Levantamento e Conservação de Solos.

Homma, A. K. O. 1989. A Extração de Recursos Naturais Renováveis: O Caso do Extrativismo Vegetal na Amazônia. Ph.D. dissertation. Universidade Federal de Viçosa, Viçosa, Brazil.

Homma, A. K. O., and E. A. S. Serrão. In preparation. Será Possível a Agricultura Autosustentada na Amazônia?

Imbiriba, E. P. In press. Produção e manejo de alevinos de pirarucu, *Arapaima gigas* (Cuvier). Belém, Brazil: Brazilian Enterprise for Agricultural Research–Center for Agroforestry Research of the Eastern Amazon.

International Center for Tropical Agriculture. 1975. Informe Anual 1974. Cali, Colombia: International Center for Tropical Agriculture.

Kamarck, A. M. 1976. The Tropics and Economic Development. Baltimore: Johns Hopkins University.

King, K. F. S., and M. T. Chandler. 1978. The Wasted Lands: The Programme of Work of International Council for Research in Agroforestry. Nairobi, Kenya: International Council for Agroforestry Research.

Kitamura, P. C. 1982. Agricultura Migratória na Amazônia: Um Sistema de Produção Viável. Documento 12. Belém, Brazil: Brazilian Enterprise for Agricultural Research–Center for Agroforestry Research of the Eastern Amazon.

Lau, H. D. 1991. Manual de Práticas Sanitárias para Bubalinos Jovens. Circ. Tec. 60. Belém, Brazil: Brazilian Enterprise for Agricultural Research–Center for Agroforestry Research of the Eastern Amazon.

Lima, R. R. 1956. A Agricultura nas Várzeas do Estuário do Amazonas. Belém, Brasil: Instituto Agronômico do Norte.

Mahar, D. J. 1989. Government Policies and Deforestation in Brazil's Amazon Region. Washington, D.C.: World Bank.

Marques, J. R. F., J. F. Teixeira Neto, and E. A. S. Serrão. 1980. Melhoramento de Pastagens na Ilha de Marajó: Resultados e Informações Práticas. Miscelânea 6. Belém, Brazil: Brazilian Enterprise for Agricultural Research–Center for Agricultural Research of the Humid Tropics.

Mattos, M. M., C. Uhl, and D. A. Gonçalves. In press. Perspectivas econômicas e ecológicas da pecuária na Amazônia Oriental na década de 90. Paragominas como estudo de caso. Pará Desenvolvimento.

McGrath, D. G. 1991. Varzeiros, Geleiros, and Resource Management in the Lower Amazon Floodplain. Belém, Brasil: Núcleo de Altos Estudos Amazônicos–Universidade Federal do Pará.

Medici, A. C., H. A. Moura, L. A. P. Oliveira, M. M. Moreira, and T. F. Santos. 1990. Déficits Sociais na Amazônia. Belém, Brazil: Superintendency for the Development of the Amazon.

Montoro Filho, A. F., A. E. Comune, and F. H. de Melo. 1989. A Amazônia e a Economia Brasileira—A Integração Econômica, os Desafios e as Oportunidades de Crescimento. São Paulo: Associação dos Empresários da Amazônia.

Morais, F. I. D. 1988. O cultivo do cacaueiro na Amazônia brasileira. Pp. 41–55 in Faculdade de Ciências Agrárias do Pará, Departamento de Solos, Simpósio Sobre Produtividade Agroflorestal da Amazônia: Problemas e Perspectivas. Programa e Resumos. Belém, Brasil: Faculdade de Ciencias Agrarias do Pará.

Moura Carvalho, L. O. D., and C. N. B. Nascimento. 1986. Tecnologia de criação de búfalos no Trópico Úmido brasileiro. Pp. 239–249 in Simpósio do Trópico Úmido, Vol. V. Anais. Belém, Brazil: Brazilian Enterprise for Agricultural Research–Center for Agricultural Research of the Humid Tropics.

Nakajima, C. 1970. Subsistence and commercial family farms: Some theoretical models of subjective equilibrium. Pp. 165–185 in Subsistence Agriculture and Economic Development, C. R. Wharton, ed. Chicago: Aldine Publishing.

Nascimento, C. N. B., and A. K. O. Homma. 1984. Amazônia: Meio Ambiente e Tecnologia Agrícola. Documento 27. Belém, Brazil: Brazilian Enterprise for Agricultural Research–Center for Agroforestry Research of the Eastern Amazon.

Nascimento, C. N. B., and L. O. D. Moura Carvalho. In press. Criaçao de Búfalos: Alimentação, Manejo, Melhoramento e Instalaçoés. Belém, Brazil: Brazilian Enterprise for Agricultural Research–Center for Agroforestry Research of the Eastern Amazon.

National Research Council. 1976. Nutrient Requirements of Beef Cattle, 5th ed. Washington, D.C.: National Academy of Sciences.

National Research Council. 1991. Toward Sustainability: A Plan for Collaborative Research on Agriculture and Natural Resource Management. Washington, D.C.: National Academy Press.

Nepstad, D., C. Uhl, and E. A. S. Serrão. 1990. Surmounting barriers to forest regeneration in abandoned, highly degraded pastures: A case study from Paragominas, Pará, Brasil. Pp. 215–229 in Alternatives to Deforestation: Steps Toward Sustainable Use of the Amazon Rain Forest, A. B. Anderson, ed. New York: Columbia University Press.

Nepstad, D. C., C. Uhl, and E. A. S. Serrão. 1991. Recuperation of a degraded Amazonian landscape: Forest recovery and agricultural restoration. Ambio 20:248–255.

Nicholaides, J. J., III, D. E. Bandy, P. A. Sanchez, J. R. Benitez, J. H. Villachica, A. J. Coutu, and C. S. Valverde. 1985. Agriculture alternative for the Amazon Basin. BioScience 35:279–285.

Norgaard, R. B. 1981. Significado do potencial para produzir arroz com irrigação controlada na várzea Amazônica. Rev. Econ. Rural 19(2):287–313.

Organization of American States and Instituto do Desenvolvimento Economico e Social do Pará. 1974. Marajó: Um Estudo para Seu Desenvolvimento. Washington, D.C.: Organization of American States.

Paiva, R. M. 1977. Modernização agrícola e processo de desenvolvimento econômico: problema dos países em desenvolvimento. Pp. 37–86 in Ensaios sobre Política Agrícola Brasileira, A. Veiga, ed. São Paulo: Secretaria de Agricultura.

Pastore, J. 1977. Agricultura de subsistência e opções tecnológicas. Estudos Econ. 7(3):9–18.

Pearce, D. 1990. Recuperação ecológica para conservação das florestas: A perspectiva da economia ambiental. Trabalho apresentado no Seminário "Recuperação Ecológica para Conservação das Florestas," Promovido pelo. Instituto Brasileiro do Meio Ambiente e Recursos Naturais Renováveis, Overseas Development Administration, and Imperial Chemical Industries, Brasília. Mimeograph.

Peters, C. M. 1990. Population ecology and management of forest fruit trees in Peruvian Amazon. Pp. 86–98 in Alternatives to Deforestation: Steps Toward Sustainable Use of the Amazon Rain Forest, A. B. Anderson, ed. New York: Columbia University Press.

Peters, C. M., A. H. Gentry, and R. O. Mendelsohn. 1990. Valuation of an Amazonian rain forest. Nature 339:655–656.

Posey, D. A. 1983. Indigenous knowledge and development: An ideological bridge to the future. Ciência Cultura 35:877–894.

Rankin, J. M. 1985. Forestry in the Brazilian Amazon. Pp. 369–392 in Amazônia, G. T. Prance and T. E. Lovejoy, eds. Oxford: Pergamon.

Salati, E. 1989. Soil, water and climate of Amazônia. An overview. Pp. 265–319 in Amazônia: Facts, Problems and Solutions, Vol. I. São Paulo: Universidade de São Paulo.

Salati, E. In press. Possible climatological changes. In Development or Destruction: The Conversion of Tropical Forest and Pasture in Latin America, T. E. Downing, S. B. Hecht, H. A. Pearson, and C. Garcia-Downing, eds. Boulder, Colo.: Westview.

Sanchez, P. A., D. E. Bandy, J. H. Villachica, and J. J. Nicholaides III. 1982. Amazon basin soils: Management for continuous crop production. Science 216:821–827.

Senado Federal. 1990. CPI [Senate Committee Inquiry] Hiléia Amazônica. Relatório Final. Brasília: Senado Federal.

Serrão, E. A. S. 1986a. Pastagem em área de floresta no trópico umido brasileiro. Conhecimentos atuais. Pp. 147–174 in Simpósio do Trópico Úmido I, Vol. V. Anais. Belém, Brazil: Brazilian Enterprise for Agricultural Research–Center for Agroforestry Research of the Eastern Amazon.

Serrão, E. A. S. 1986b. Pastagens nativas do trópico umido brasileiro. Conhecimentos atuais. Pp. 183–205 in Simpósio do Trópico Úmido I, Vol. V. Anais. Belém, Brazil: Brazilian Enterprise for Agricultural Research–Center for Agroforestry Research of the Eastern Amazon.

Serrão, E. A. S. 1990. Pasture development and carbon emission/accumulation in the Amazon (topics for discussion). Pp. 210–222 in Tropical Forestry Response Options to Global Climate Change. São Paulo Conference Proceedings. Washington, D.C.: U.S. Environmental Protection Agency.

Serrão, E. A. S. 1991. Pastagem e pecuária. Pp. 85–137 in O Futuro Econômico da Amazônia. Revista do Partido do Movimento Democrático Brasileiro. Brasília: Senado Federal.

Serrão, E. A. S., and I. C. Falesi. 1977. Pastagem do trópico úmido brasileiro.

In Simpósio Sobre Manejo de Pastagens, 4. Piracicaba, Brasil: Escola Superior de Agricultura Louis de Queiroz.

Serrão, E. A. S., and A. K. O. Homma. In press. A Questão da Sustentabilidade da Pecuária Substituindo Florestas na Amazônia: A Influénca de Variáveis Agronómicas, Biológicas e Socioeconomicas. Documento Elaborado a Pedido do Banco Mundial. Washington, D.C.: World Bank.

Serrão, E. A. S., and M. Simão Neto. 1975. The adaptation of forages in the Amazon region. Pp. 31–52 in Tropical Forages in Livestock Production Systems. Special Publication 24. Madison, Wis.: American Society of Agronomy.

Serrão, E. A. S., and J. M. Toledo. 1990. The search for sustainability in Amazonian pastures. Pp. 195–214 in Alternatives to Deforestation: Steps Toward Sustainable Use of the Amazon Rain Forest, A. B. Anderson, ed. New York: Columbia University Press.

Serrão, E. A. S., and J. M. Toledo. In press. Sustaining pasture-based production systems in the humid tropics. In Development or Destruction: The Conversion of Tropical Forest to Pasture in Latin America, S. B. Hecht, ed. Boulder, Colo.: Westview.

Serrão, E. A. S., A. P. Camarão, and J. A. Rodrigues Filho. In preparation. Sistema Integrado de Pastagens Nativas de Terra Inundável e da Terra Firme na Engorda de Bovinos. Belém, Brazil: Brazilian Enterprise for Agricultural Research–Center for Agroforestry Research of the Eastern Amazon.

Serrão, E. A. S., I. C. Falesi, J. B. Veiga, and J. F. Teixeira Neto. 1979. Productivity of cultivated pastures on low fertility soils in the Amazon of Brazil. Pp. 195–225 in Pasture Production in Acid Soils of the Tropics, P. A. Sanchez and L. E. Tergas, eds. Cali, Colombia: International Center for Tropical Agriculture.

Silva, A. B., and B. P. Magalhães. 1980. Insetos Nocivos de Pastagens no Estado do Pará. Boletim de Pesquisa No. 8. Belém, Brazil: Brazilian Enterprise for Agricultural Research–Center for Agroforestry Research of the Eastern Amazon.

Silva, B. N. R., et al. 1986. Zoneamento agrossilvopastoril da Amazônia: Estado atual do conhecimento. Pp. 225–240 in Simpósio do Trópico Úmido I, Vol. VI. Anais. Belém, Brazil: Brazilian Enterprise for Agricultural Research–Center for Agricultural Research for the Humid Tropics.

Silva, J. N. M. 1989. The Behaviour of Tropical Rain Forest of the Brazilian Amazon after Logging. Ph.D. thesis. Oxford University, Oxford, England.

Sioli, H. 1951a. Alguns resultados dos problemas da limunologia Amazônica. Pp. 3–41 in Boletim Técnico 24. Belém, Brasil: Instituto Agronomico do Norte.

Sioli, H. 1951b. Sobre a sedimentação na várzea do baixo Amazonas. Pp. 42–66 in Boletim Técnico 24. Belém, Brasil: Instituto Agronomico do Norte.

Smith, N. J. H., P. T. Alvim, E. A. S. Serrão, A. K. O. Homma, and I. C. Falesi. In press-a. Environment and Sustainable Development in Amazônia.

Smith, N. J. H., P. T. Alvim, E. A. S. Serrão, A. K. O. Homma, and I. C. Falesia. In press-b. Amazônia. In Critical Environmental Zones in Global Environmental Change, J. Kasperson and R. Kasperson, eds. Tokyo: United Nations University Press.

Stolberg, A. V., and V. S. F. de Souza. 1985. Catálogo de ervas daninhas da Amazônia. Brazilian Enterprise for Agricultural Research–Center for Agroforestry Research of the Eastern Amazon, Belém, Brazil. Mimeograph.

Subler, S., and C. Uhl. 1990. Japanese agroforestry in Amazônia: A case study in Tomé-Açu, Brazil. Pp. 152–166 in Alternatives to Deforestation: Steps Toward Sustainable Use of the Amazon Rain Forest, A. B. Anderson, ed. New York: Columbia University Press.

Superintendency for the Development of the Amazon. 1986. I Plano de Desenvolvimento da Amazônia, Nova República 1986/1989. Belém, Brazil: Superintendency for the Development of the Amazon.

Superintendency for the Development of the Amazon. 1991. Cenários da Amazônia. Ciência Hoje 13(78):52–61.

Teixeira, J. F. 1953. O Arquipélago do Marajó. Rio de Janeiro: Brazilian Institute of Geography and Statistics.

Teixeira Neto, J. F., and E. A. S. Serrão. 1984. Produtividade Estacional, Melhoramento e Manejo de Pastagem na Ilha de Marajó. Comunicado Técnico 51. Belém, Brazil: Brazilian Enterprise for Agricultural Research–Center for Agricultural Research for the Humid Tropics.

Toledo, J. M., and E. A. S. Serrão. 1982. Pasture and animal production in Amazônia. Pp. 282–309 in Amazônia: Agriculture and Land Use Research, S. B. Hecht, ed. Cali, Colombia: International Center for Tropical Agriculture.

Uhl, C., and J. B. Kauffman. 1990. Deforestation, fire susceptibility, and potential tree responses to fire in eastern Amazon. Ecology 71:437–449.

Uhl, C., and I. C. G. Vieira. 1989. Ecological impacts of selective logging in the Brazilian Amazon: A case from the Paragominas region of the state of Pará. Biotropica 21(2):98–106.

Uhl, C., R. I. Buschbacher, and E. A. S. Serrão. 1988. Abandoned pasture in eastern Amazônia. I. Patterns of plant succession. J. Ecol. 76:663–681.

Uhl, C., J. B. Kauffman, and E. D. Silva. 1990a. Os caminhos do fogo na Amazônia. Ciência Hoje 11(65):24–32.

Uhl, C., D. Nepstad, R. Buschbacher, K. Clark, B. Kauffman, and S. Subler. 1990b. Studies of ecosystem response to natural and anthropogenic disturbances provide guidelines for designing sustainable land-use systems in Amazônia. Pp. 24–42 in Alternative to Deforestation: Steps Toward Sustainable Use of the Amazon Rain Forest, A. B. Anderson, ed. New York: Columbia University Press.

Uhl, C., A. Veríssimo, M. M. Mattos, Z. Brandino, and I. C. G. Vieira. 1991. Social, economic, and ecological consequences of selective logging in an Amazon frontier: The case of Tailandia. Forest Ecol. Manag. 46:243–273.

Uhl, C., A. Veríssimo, M. M. Mattos, P. Barreto, and R. Tarifa. In preparation. Aging of the Amazon frontier: Opportunities for genuine development.

Universidade de São Paulo. 1990. Projecto Floram: Uma plataforma. Estudos Avançados, São Paulo 4(9):7–280.

Van den Berg, M. E. 1982. Plants Medicinais na Amazônia; Contribuição ao Seu Conhecimento Sistemático. Belém, Brasil: Conselho Nacional de Desenvolvimento Científico e Tecnológico and Programa do Trópico Úmido.

Veiga, J. B. 1986. Associação de culturas de subsistência com forrageiras na renovação de pastagens degradadas em área de floresta. Pp. 175–181 in Simpósio do Trópico Úmido I, Vol. V. Anais. Belém, Brazil: Brazilian Enterprise for Agricultural Research–Center for Agroforestry Research of the Eastern Amazon.

Veiga, J. B., and E. A. S. Serrão. 1990. Sistemas silvopastoris e produção animal nos trópicos úmidos: A experiência da Amazônia brasileira. Pp. 37–68 in Pastagens. Piracicaba, Brasil: Sociedade Brasileira de Zootecnia.

Watrin, O. S., and A. M. A. Rocha. In press. Levantamento da Vegetação Natural e do Uso da Terra no Município de Paragominas (PA) Utilizando Imagens TM/LANDSAT. Boletin de Pesquisa 124. Belém, Brazil: Brazilian Enterprise for Agricultural Research–Center for Agricultural Research for the Humid Tropics.

Yared, J. A. G. 1991. Exploraçãuo florestal. Pp. 141–159 in O Futuro Econômíco da Amazônia. Revísta do Partido do Movimento Democrático Brasileiro 16. Brasília: Senado Federal.

Côte d'Ivoire

Simeon K. Ehui

Côte d'Ivoire is located in western Africa on the Gulf of Guinea (Atlantic Ocean) between Liberia and Ghana. It covers an area of 322,463 km^2. With the exception of a relief zone in the western region, where the altitude reaches above 1,300 m, the land rises gradually from the coast to the north and does not exceed 800 m (Persson, 1977). The country has three main types of vegetation. The southern part of the country consists of closed, humid forests (humid evergreen and semideciduous forests), and then, toward the north, there is a transition zone (forest-savannah mosaic). The transition zone turns into open country in the north, with vast woodlands or savannah (Figure 1).

The most important timber species in the humid evergreen forests are *Tieghemella heckelii* (makoré), *Tarrietia utilis* (niangon), and *Mansonia altissima* (bété), which require annual rainfall of 1,600 mm. *Celtis* species are an important part of the dominant layer in the humid semideciduous forests, which require annual rainfall of 1,350 to 1,600 mm. The most important timber species exclusive to this zone is *Triplochiton scleroxylon* (samba). In the dry season the trees of the upper layer shed their leaves. The forest-savannah mosaic is found north of the moist semideciduous forest and is a transition between

Simeon K. Ehui is a senior economist with the International Livestock Center for Africa, Addis Ababa, Ethiopia.

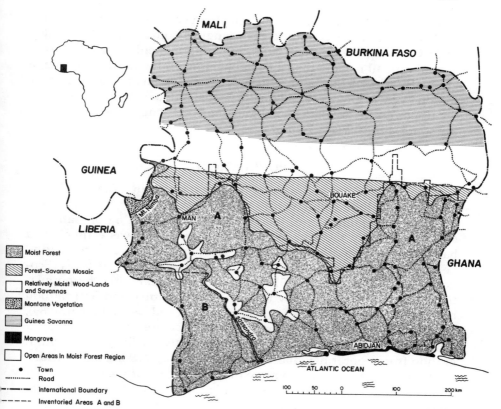

FIGURE 1   Côte d'Ivoire and its forests.   Source: Adapted from Persson, R.
1977.   Forest resources of Africa.   Part II.   Regional Analysis Research Notes
No. 22.   Stockholm, Sweden: Royal College of Forestry.

moist semideciduous forests and the savannah woodlands in the north,
which are deciduous and require annual rainfall of 1,000 mm.   They
are characterized by *Isoberlinia doka, Uapaca togoensis,* and *Anogeissus
leiocarpa.*   Gallery forests are also found along rivers.   Other vegetation types in the country include the humid highland mountain forests, found in the mountains in the western part of the country, and
mangroves, found along the Atlantic coast.   There are areas of littoral
savannah in the humid evergreen forest zone (Persson, 1977).

In the coastal region, the climate is tropical, with two dry and
two rainy seasons each year.   Dry seasons are from December to
April and from August to September; rainy seasons are from May to

July and from October to November. Temperatures generally remain fairly constant throughout the year, ranging from about 22°C at night to 33°C during the day, and humidity is permanently high. Average annual rainfall is more than 1,800 mm. Toward the north, however, it gradually diminishes, and seasonal variations change to one rainy season (May to October) and one dry season (November to April). In the upper north, the climate exhibits more extreme variations than in the south, but it is less humid.

POPULATION

The average annual population growth rate in Côte d'Ivoire is one of the highest in the world (3.6 percent). In 1960, the population was about 3.8 million, and the 1975 census recorded a population of 6.67 million. By the end of 1985, the population was estimated to have risen to more than 10 million. The population was estimated to be 13.02 million as of mid-1991, more than tripling in 3 decades (Economic Intelligence Unit, 1991). Projections indicate that the population will reach 18 million by the end of the century and 39.3 million by 2025 (International Bank for Reconstruction and Development, 1989), which is equivalent to an average annual increase of 3.6 percent.

The high population growth rate is partly attributable to immigration from poorer neighboring countries (mainly Mali and Burkina Faso). Immigrants make up more than 20 percent of the total population of Côte d'Ivoire. Other contributing factors are the high fertility rate (7.4 births per woman) and improvements in the health of Ivoirians. Life expectancy at birth rose from 44 years in 1965 to 52 years in 1987. Although the crude birth rate changed little over this period (52 per 1,000 population in 1986), the crude death rate fell from 22 to 15 per 1,000 population (Economic Intelligence Unit, 1991; International Bank for Reconstruction and Development, 1989). An increasing proportion of the population lives in urban areas. For example, in 1960 the urban population as a proportion of the total population was estimated to be 19.3 percent; it had increased to 32.2 percent by 1975, and by 1990, it was estimated to be about 46.6 percent. Between 1960 and 1990, the urban population grew at an annual average rate of 7.2 percent, whereas the rural population grew at only 2.7 percent (World Resources Institute, 1990).

Despite the rapid population growth, Côte d'Ivoire still appears to have a relatively low population density (39 inhabitants per km^2 in 1990). However, when taking into account only the usable land (that is, total land area less nonarable land, including inland water bodies, wasteland, built-up areas, parks and reserves, and 50 percent of the

TABLE 1 Agricultural Population Densities in Forest and
Savannah Zones in Côte d'Ivoire, 1965–1989[a]

Year	Forest Zone Population (1,000s)	Density (no./km²)	Savannah Zone Population (1,000s)	Density (no./km²)	Nationwide Population (1,000s)	Density (no./km²)
1965	2,030	14.9	1,378.0	12.0	3,408	13.8
1975	3,003	22.0	1,443.0	12.6	4,446	17.7
1985	3,932	28.9	1,518.0	13.2	5,450	21.7
1989	5,303[b]	38.9[b]	1,551.0[b]	13.5[b]	6,854	27.3

[a]The total usable land area in the country is 251,120 km², including 136,249 km² for the forest zone and 114,871 km² for the savannah zone.
[b]Values are estimates.

SOURCES: Modified from Durufles, G., P. Bourgerol, B. Lesluyes, J. C. Martin, and M. Pascay. 1986. Desequilibres Structurels et Programmes d'Ajustement en Côte d'Ivoire. Paris, France: Mission d'Evaluation, Ministère de la Cooperation. Data for the agricultural population in 1989 are from Food and Agriculture Organization. 1989. 1989 Production Yearbook. Rome, Italy: Food and Agriculture Organization of the United Nations.

reserved forestlands), the population density increases to 50 inhabitants per km².

A good indicator of the rate at which forestlands are being used is the agricultural population density, which is defined as the ratio of agricultural population divided by the total area of usable land. Table 1 presents the agricultural population densities over a 24-year period (1965–1989) for the forest and savannah zones. The data indicate that agricultural population densities have increased over time nationwide and that they are higher in the forest zone than they are in the savannah zone (Figure 2). By 1989, the forest and savannah zones had densities of 38.9 and 13.5 inhabitants per km², respectively.

FOREST RESOURCES

The status of forest resources in Côte d'Ivoire is difficult to describe because data on the extent and condition of tropical forest areas are widely scattered and frequently inaccurate (U.S. Office of Technology Assessment, 1984). Accuracy is further impaired by the lack of standard definitions and classifications of forest types. Table 2 presents the status of tropical forests in Côte d'Ivoire in the 1980s and their evolution since 1900. The Food and Agriculture Organization and United Nations Environment Program (1981) indicate that

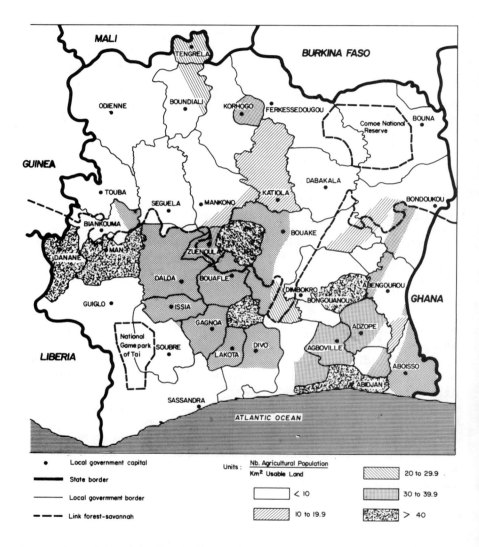

FIGURE 2 Agricultural population density in Côte d'Ivoire in 1985. The numbers next to the symbols are in agricultural population per square kilometer of usable land. Source: Adapted from Durufles, G., P. Bourgerol, B. Lesluyes, J. C. Martin, and M. Pascay. 1986. Desequilibres Structurels et Programmes d'Ajustement en Côte d'Ivoire. Paris, France: Mission d'Evaluation, Ministère de la Cooperation.

TABLE 2   Evolution of Tropical Forest Endowments in Côte d'Ivoire and Rates of Deforestation from 1900 to 1990

Year or Period	Forest Cover, Dense Humid Tropical Forest (millions of ha)	Average Annual Area Deforested (ha/year)	Average Annual Rate of Deforestation (as percentage of forested area)
1900	14.50	—	—
1955	11.80	—	—
1965	8.98	—	—
1973	6.20	—	—
1980	3.99	—	—
1985	2.55	—	—
1990	1.55[a]	—	—
1956–1965	—	280,000	2.37
1966–1973	—	350,000	3.90
1974–1980	—	315,000	5.80
1981–1985	—	290,000	7.26
1986–1990	—	200,000a	7.84[a]

NOTE: Information that is not available is denoted by a dash.

[a]Estimated.

SOURCE: Modified from Food and Agriculture Organization and United Nations Environment Program. 1981. Pp. 124–125 in Tropical Forest Resources Assessment Project (in the Framework of GEMS). Forest Resources of Tropical Africa. Part II. Country Briefs. Rome, Italy: Food and Agriculture Organization of the United Nations.

total forest cover at the beginning of the colonial period (1900) was on the order of 15 million ha. In 1990, forest cover was estimated to be 1.55 million ha.

To appreciate the rapid rate of forest clearing in Côte d'Ivoire, it is useful to compare the country's rate of deforestation with that of Indonesia, the world's leading producer of tropical logs from 1973 to 1983. Table 3 shows that the annual level of deforestation has been about half that of Indonesia, a poorer country with 6 times the area of Côte d'Ivoire and a population that is 16 times greater than that of Côte d'Ivoire. However, the estimated annual rate of deforestation in Côte d'Ivoire (7.26 percent) after 1980 was more than 12 times that of Indonesia (0.5 percent).

Given the current trend in deforestation rates, it is estimated that in 10 to 20 years, natural forests will not satisfy the local demand for logs in Côte d'Ivoire. Furthermore, it is estimated that Côte d'Ivoire, which until 1983 was the most prolific exporter of logs in Africa, will become a net importer by the end of the century (Bertrand, 1983).

TABLE 3   Deforestation in Indonesia Versus that in Côte d'Ivoire

Parameter	Indonesia	Côte d'Ivoire
Population in 1985 (millions)	162.000	10.1
Area (ha) deforested annually (1981–1985)	600,000	290,000
Annual deforestation rate (deforestation annually as percentage of forested area)	0.5	7.26
Per capita income in 1985 (US$)	530	660
Area (thousand km²)	1,919	322

SOURCE: Adapted from Gillis, M. 1988. West Africa: Resource management policies and the tropical forest. Pp. 299–351 in Public Policies and the Misuse of Forest Resources, R. Repetto and M. Gillis, eds. New York: Cambridge University Press.

This is not surprising since, solely on the basis of the commercial benefits of tropical forests, Ehui and Hertel (1989) showed that the optimal steady-state forest stock in Côte d'Ivoire exceeds what is considered to be needed to meet current levels for social discount rates less than 8 percent. (A social discount rate, measured in percent, expresses the preference of a society as a whole for present rather than future returns.) Only when the social discount rate reaches the relatively high value of 9 percent does some further deforestation appear to be socially optimal. The optimal steady-state forest stock decreases in direct proportion to higher social discount rates because future forest stocks are valued less than present well-being, thus there is the motivation to clear the forest faster. The critical value of forests increases when one takes into account the noncommercial benefits of tropical forests, for example, the preservation of genetic diversity and climatic benefits. Thus, it is likely, even on strictly commercial grounds, that Côte d'Ivoire has already excessively depleted its forest resources.

Today, the main forestry policy question facing the government of Côte d'Ivoire is how to manage effectively what is left of the original 15 million ha of tropical rain forest, which has been reduced to less than 2 million ha (Ehui and Hertel, 1989; Spears, 1986). Current government policy objectives, as defined in the 1976–1980 and 1981–1985 5-year plans, include preservation and protection of the forest stock (Borreau, 1984). A first step toward those objectives was the creation in 1978 of a permanent forestry domain of 4.7 million ha and a rural forestry domain of 731,750 ha that is reserved for agriculture. However, because of continual encroachment of uncontrolled shifting cultivation onto forestlands, it has become difficult, if not impossible, to achieve the forest protection objective (Bourreau, 1984). As a result, the officially preserved forest area has continuously been reduced to keep pace with the remaining forest stock.

DOMESTIC ECONOMY

Côte d'Ivoire is essentially an agricultural country, relying on its two principal cash crops—cacao and coffee—for almost 50 percent of its export revenues (Economic Intelligence Unit, 1991). In the first 2 decades following independence (in 1960), Côte d'Ivoire's gross domestic product (GDP) grew by 7.5 percent annually, which ranked among the highest in Africa and among the top 15 in the world (Michel and Noël, 1984). In 1965, Côte d'Ivoire had a per capita GDP of about US$169. By 1980, it had risen to about US$1,150, ranking second among developing countries in sub-Saharan Africa. Apart from a brief respite in 1985–1986 because of excellent harvests and improved agricultural exports, a severe slowdown has occurred since 1980, and in 1987 the per capita GDP was estimated to be only US$690, a decline of 40 percent from its 1980 level. From a peak level of US$1,170 in 1980, the per capita gross national product (GNP) declined to US$740 in 1987. During the period from 1980 to 1987, Côte d'Ivoire experienced net negative growth of –3.0 percent/year (Table 4).

There are several reasons for the slowdown in Côte d'Ivoire's

TABLE 4 Average Annual Change in Growth and Structure of Production in Côte d'Ivoire, 1965–1987

				Percent Change		
Parameter	1965	1980	1987	1965–1973	1973–1980	1980–1987
Per capita GNP (US$)	N.A.	1,170	740	4.5	1.2	–3.0
GDP (millions of US$)	760	8,482	7,650	8.6	4.7	2.2
Population (millions)	4.5	8.3	11.1	4.1	4.3	4.2
Distribution of GDP (percent)						
Agriculture	47	33	36	4.9	3.3	1.6
Industry	19	20	25	12.5	11.7	–2.4
Services	33	47	39	11.2	3.6	4.2

NOTE: N.A., not available; GNP, gross national product; GDP, gross domestic product.

SOURCES: Compiled from Food and Agriculture Organization. Various issues. Agricultural Production Yearbook. Rome, Italy: Food and Agriculture Organization of the United Nations; International Bank for Reconstruction and Development. 1989. Sub-Saharan Africa: From Crisis to Sustainable Growth, a Long-Term Perspective Study. Washington, D.C.: World Bank.

economy: (1) a dramatic adverse shift in the country's terms of trade in the early 1980s mainly because of the continuing slump in commodity prices and (except in 1985–1986) the depreciation of the dollar against the CFA (Communauté Financière Africaine) franc (in 1991, US$1 = CFA franc 275); (2) a serious drought during 1982–1984 that affected both agricultural production and hydroelectricity generation, thereby reducing power supplies to industry; and (3) the high cost of servicing the debt incurred to finance ambitious investment projects launched during the boom years of the late 1970s.

The total external public debt at the end of 1989 totaled US$15.4 billion, representing about 182 percent of the country's total GNP. In 1970 total public debt was only US$255 million, 19 percent of GNP. By 1980 it had risen to US$4.3 billion, equivalent to 44 percent of the country's GNP. Interest payment on the public debt in 1989 was estimated at US$517 million. The total debt service ratio (measured as a proportion of exports of goods and services) during the same period (1989) was estimated to be about 41 percent. In 1980 it was estimated to be 24 percent of the exports of goods and services. It was swollen in 1980 by the increase in the value of the U.S. dollar, in which more than 40 percent of the country's debt is denominated. In 1970 the debt service ratio was only 7.1 percent (Economic Intelligence Unit, 1991; International Bank for Reconstruction and Development, 1989).

AGRICULTURE

The overall performance of Côte d'Ivoire's economy springs from its agriculture. With a consistent annual growth rate of 5 percent, Côte d'Ivoire achieved the highest agricultural growth rate in sub-Saharan Africa during the first 2 decades after its independence in 1960 (Lee, 1983). Despite an apparent decline of its share in the GDP (Table 4), agriculture still remains the pillar of the country's economy. It contributes about 33 percent of the GDP, provides between 50 and 75 percent of the nation's total export earnings, and employs an estimated 79 percent of the labor force, of which 13 percent are immigrants (Economic Intelligence Unit, 1991). Table 5 presents details of the structure of merchandise import and export trade in Côte d'Ivoire during 1965, 1980, and 1987.

Export Crops

Export of agricultural products was the primary source for agricultural growth. Agricultural products account for more than 75 per-

TABLE 5 Structure of Merchandise Imports and Exports in Côte d'Ivoire, 1965, 1980, and 1987 (Percent Share)

Merchandise	1965		1980		1987	
	Imports	Exports	Imports	Exports	Imports	Exports
Food	18	—	17	—	19	—
Fuel	6	2	17	6	15	4
Other primary commodities	3	93	3	84	4	86
Machinery and transport equipment	28	1	28	2	28	2
Other manufactures	46	4	35	7	35	7

SOURCE: Compiled from International Bank for Reconstruction and Development. 1989. Sub-Saharan Africa: From Crisis to Sustainable Growth, a Long-Term Perspective Study. Washington, D.C.: World Bank.

cent of export earnings. The major agricultural exports are coffee, of which Côte d'Ivoire is the world's fifth largest producer; cacao, of which it became the world's largest producer in 1977–1978, surpassing Brazil and Ghana; and cotton. Together, these three commodities account for more than 60 percent of the area under cultivation, 50 percent of export earnings, and 75 percent of total cash earnings from agricultural activities. Cacao production has expanded rapidly, rising from 140,000 metric tons in 1965 to 388,000 and 543,000 metric tons in 1980 and 1987, respectively. The average annual rate of growth is estimated to be about 6 percent (Table 6).

Coffee production followed a different pattern. As a result of the producer price parity (by which farmers receive the same price for a product regardless of whether it is good or substandard) for cacao and coffee, which has been in place since the mid-1970s, production of coffee has been falling steadily. Coffee is more difficult to produce than cacao, and it is also taxed more heavily. Output fell from 210,000 metric tons in 1980 to an estimated 163,000 metric tons in 1987. Production of cotton rose from 2,000 metric tons in 1965 to 39,000 metric tons in 1980 and 68,000 metric tons in 1987. As a result, Côte d'Ivoire is now Africa's third largest cotton producer, after Egypt and Sudan (Economic Intelligence Unit, 1991).

Another important export commodity is timber, which accounted for almost 7 percent of export earnings in 1988, but forest resources have been greatly depleted and timber exports have been falling. The forestry industry was traditionally the country's third main export earner. The total area of timber harvested for export was esti-

TABLE 6   Volume, Percentage of Total Merchandise Export Value, and Growth in Volume of Major Agricultural Exports in Côte d'Ivoire, 1965–1987

Parameter and Year or Period	Cacao	Coffee	Cotton
Volume (1,000s of metric tons)			
1965	140	186	2
1980	388	210	39
1987	543	163	68
Percentage of total merchandise export value			
1965	17.6	36.6	0.2
1980	29.9	22.0	2.2
1987	36.6	13.7	2.5
Volume growth (percent)			
1965–1973	4.6	1.7	26.3
1973–1980	6.1	−1.1	14.1
1980–1987	6.2	−2.9	10.7

SOURCE: Compiled from International Bank for Reconstruction and Development. 1989. Sub-Saharan Africa: From Crisis to Sustainable Growth, a Long-Term Perspective Study. Washington, D.C.: World Bank.

mated to have fallen from 15.6 million ha at the beginning of the century to only 1 million ha in 1987.

Food Crops

The principal food crops in Côte d'Ivoire are cassava, yams, cocoyam (taro), maize, rice millet, sorghum, and plantains. The country is self-sufficient in manioc (cassava), yams, bananas (plantains), and maize. Table 7 presents estimates of the compositions of Ivoirian diets. Yams are the most consumed commodity, followed by bananas (plantain) and manioc (cassava). The principal grain that is produced and consumed is rice; it has become a staple for much of the urban population and is also popular in rural areas because of its ease of preparation and storage. Although rice production has risen steadily, it has not increased rapidly enough to keep pace with per capita consumption. The result is that Côte d'Ivoire meets more than half of its current rice needs through imports (Figure 3) (Trueblood and Horenstein, 1986). Overall, Côte d'Ivoire's agricultural sector has performed well relative to those sectors throughout the rest of sub-Saharan Africa. Figures 4 and 5 present per capita food and agricultural production, respectively, in Côte d'Ivoire and sub-Saharan

TABLE 7 Composition of Ivoirian Diets, 1980

Region	Annual per Capita Consumption (kg)							
	Wheat	Rice	Maize	Other Cereals	Yams	Manioc (Cassava)	Bananas (Plantains)	Cocoyams (Taro)
Rural area average								
Forest	11.6	39.9	25.4	6.3	189.2	155.4	162.4	34.0
Savannah	11.4	40.0	17.8	*a*	155.5	186.4	226.4	47.0
Urban area average	12.0	39.8	39.8	20.8	266.3	84.6	16.8	4.3
Forest	40.2	74.2	17.5	7.9	73.1	81.3	90.2	12.9
Savannah	31.2	48.4	21.2	8.0	103.8	96.5	154.9	29.5
Bouake	25.5	60.6	41.7	23.2	102.7	44.3	25.0	9.8
Abidjan	39.0	90.5	14.5	6.5	120.4	76.9	45.4	8.0
National average	49.5	91.6	10.5	4.9	38.4	79.5	62.9	3.2
	22.9	53.5	22.6	7.2	144.6	124.7	131.3	25.1

*a*Insignificant.

SOURCES: Adapted from Côte d'Ivoire, National Development Plan, 1981–1985. Abidjan; Trueblood, M. A., and N. R. Horenstein. 1986. The Ivory Coast: An Export Market Profile. Foreign Agricultural Economic Report No. 223. Washington, D.C.: Economic Research Service, U.S. Department of Agriculture.

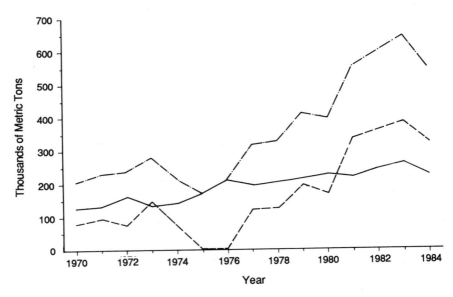

FIGURE 3   Milled rice production (———), imports (– – – –), and consumption (– · – · –) in Côte d'Ivoire. Source: Adapted from Trueblood, M. A., and N. R. Horenstein. 1986. The Ivory Coast: An Export Market Profile. Foreign Agricultural Economic Report No. 223. Washington, D.C.: Economic Research Service, U.S. Department of Agriculture.

Africa. Although sub-Saharan Africa has received much publicity for its recent famines (for example, the famine caused by drought in Ethopia from 1984 to 1986) and declining per capita food production, per capita food production in Côte d'Ivoire has actually increased considerably over time; agricultural production (which includes non-food crops) has generally increased as well, albeit with more fluctuation (Trueblood and Horenstein, 1986).

Sources of Agricultural Growth

The factors responsible for Côte d'Ivoire's general economic performance can be credited to a carefully implemented agricultural policy. Since independence in 1960, agriculture in Côte d'Ivoire has been promoted by planning, research, and investment aided by significant inflows of foreign labor and capital and (on average) by relatively high world prices for Ivoirian exports such as coffee and cacao. The government has lent strong support to the agricultural sector and, in particular, to the numerous smallholders through its programs of guaranteed producer price, input subsidies, and agricultural extension services (Trueblood and Horenstein, 1986).

Most cacao and coffee production is in the hands of smallholders who employ foreign labor. They sell their crops to the state marketing agency at prices that are fixed by the government. By means of a stabilization fund, the government has been able to sustain the development of agricultural exports by providing producers with minimum guaranteed prices, despite the sharp fluctuations in world market prices. At times, however, these producer prices were far below world market prices (most notably during the boom years of 1975 to 1977), thus enabling the government to exact surpluses from the producers and use the proceeds to invest in other sectors of the economy, as well as subsidize inputs to farmers. The stabilization fund has been able to make transfers to public enterprise budgets and to pay

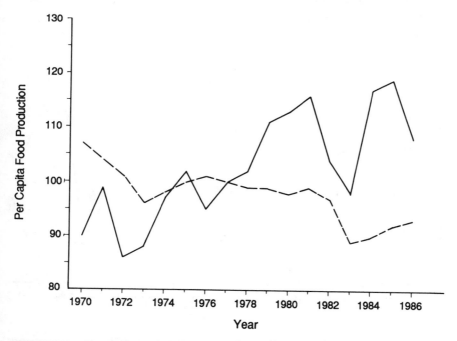

FIGURE 4 Per capita food production in Côte d'Ivoire (——) and sub-Saharan Africa (– – –), 1970–1986. Source: U.S. Department of Agriculture, Economic Research Service. 1988. World Indices of Agricultural and Food Production 1977–86. Statistical Bulletin No. 759. Washington, D.C.: U.S. Government Printing Office.

for food production development projects, as in the case of rice in northern Côte d'Ivoire (Gbetibouo and Delgado, 1984).

Although government revenues have been generated mainly through predatory price policies that exact surpluses from farm exports (coffee and cacao in particular), farmers in Côte d'Ivoire have received prices that, on average, have assured them incomes higher than those of farmers in the sub-Saharan region (den Tuinder, 1978; Gbetibouo and Delgado, 1984). World prices were depressed during most of the 1980s, and the government was unable either to exact surpluses from the export crop sector or to maintain the real purchasing power of the planters (Economic Intelligence Unit, 1991). The surpluses generated by the stabilization fund during the boom years were not enough to support producer prices, which were halved in 1989.

Another factor that has contributed to agricultural growth is the expansion in the agricultural land frontier (which arises solely from deforestation). Table 8 presents estimates of agricultural land utilization for cash and food crops and their growth rates between 1960 and 1984.

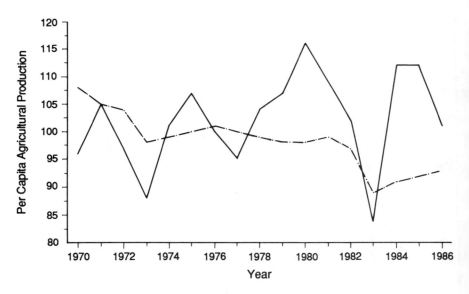

FIGURE 5 Per capita agricultural production in Côte d'Ivoire (————) and sub-Saharan Africa (– · – · –), 1970–1986. Source: U.S. Department of Agriculture, Economic Research Service. 1988. World Indices of Agricultural and Food Production 1977–86. Statistical Bulletin No. 759. Washington, D.C.: U.S. Government Printing Office.

TABLE 8   Agricultural Land Utilization for Cash and Food Crops in Côte d'Ivoire, 1960–1984 (in Thousands of Hectares)

Agricultural Land Utilization	1960	1970	1980	1984	Annual Growth Rate (percent), 1960–1984
Cash crops[a]	1,022 (56)	1,592 (56)	2,827 (67)	3,025 (64)	5.3
Food crops[b]	817 (44)	1,112 (44)	1,419 (33)	1,698 (36)	3
Total cropped area	1,839 (100)	2,714 (100)	4,246 (100)	4,723 (100)	4.3

NOTE: Numbers in parentheses are the percent shares of the total cropped area.

[a]Cash crops include coffee, cacao, oil palm, coconut, rubber, and cotton.
[b]Food crops include rice, maize, millet, sorghum, yams, cassava, and groundnuts.

SOURCE: Compiled from International Bank for Reconstruction and Development. 1985. Côte d'Ivoire Agricultural Sector Statistical Annex 7. Washington, D.C.: World Bank.

CAUSES OF DEFORESTATION

The causes of deforestation in Côte d'Ivoire are varied but can be categorized as principal (direct) and underlying (indirect).

Principal Causes

The conversion and use of forestlands for agriculture and logging activities are the principal causes of deforestation in Côte d'Ivoire. Use of forest for fuelwood and clearing forests for cattle grazing are also causative factors, but to a lesser extent.

AGRICULTURE

Increased agricultural production has been a result of expansion of the land area devoted to agricultural uses. With huge untapped reserves of arable land, economic growth was fueled by the rapid extension of the land frontier (Lee, 1983). The expansion, however, has often been onto marginal soils and sloping uplands that cannot support permanent cropping as do the temperate areas, where agricultural production has increased in recent decades mainly through the more intensive use of already cleared land (Ehui and Hertel, 1992a). Table 9 summarizes changes in cropland area in the forest regions of Côte d'Ivoire in 1965 and 1985. During this 20-year period, untouched primary forests were reduced by about 66 percent, whereas the area under cultivation more than doubled (Spears, 1986).

TABLE 9   Vegetative Cover in the Forest
Regions of Côte d'Ivoire, 1965 and 1985
(in Millions of Hectares)

Vegetative Cover	1965	1985
Untouched primary forest	9.0	3.0
Tree crop	1.1	2.8
Food crop	0.5	0.9
Forest fallow (secondary forest and bush)	3.1	7.0
Total	13.7	13.7

SOURCE: Adapted from Spears, J. 1986. Key forest policy issues for the coming decade. Côte d'Ivoire forestry subsector discussion paper. World Bank, Washington, D.C. May. Photocopy.

LOGGING

It is unclear the extent to which selective logging has contributed to deforestation; however, it is known that the use of heavy equipment for the extraction of timber causes substantial secondary tree losses. Deforestation and land degradation occur with the removal of best-tree species, and harvested trees fall against and destroy other trees. Figure 6 depicts the level of timber production and exports from 1965 to 1983.

The building of roads and passages to reach logging sites is another direct cause of deforestation. Not only are forests destroyed to make room for the roads, the roads and passages then provide access to previously undisturbed areas. For example, a road program funded by the African Development Bank has led to the construction of a major highway along the Atlantic coast (Economic Intelligence Unit, 1991). This road provided access to formerly undisturbed coastal forests and mangroves, and since 1988 an inrush of immigrants has lead to massive destruction of the coastal forests.

FUELWOOD

Fuelwood, which constitutes the most important source of energy in Côte d'Ivoire, accounts for nearly 53 percent of all wood extracted in the country. However, deforestation caused by fuelwood extraction from the humid forest zone is limited compared with that from the savannah zone, where vegetation is characterized by open woodlands.

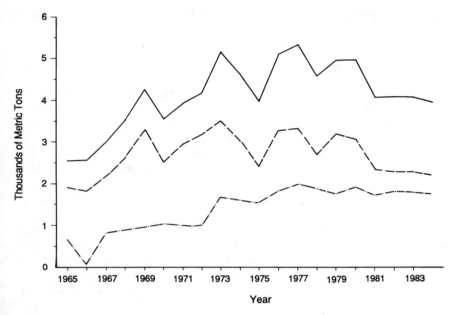

FIGURE 6   Total production (———), exports (– – – –), and domestic use of timber (– · – · –) in Côte d'Ivoire, 1965–1984.   Source: Adapted from Bourreau, C.   1984.   Plan Quinquennal (1986–1990).   Bilan Diagnostic Première Partie: Les Foréts et la Production Forestrière.   Abidjan, Côte d'Ivoire: Direction des Eaux et Foréts, Ministère de l'Agriculture.

CATTLE GRAZING

Grazing is rare in the forest zone of Côte d'Ivoire, as it is in most of the humid tropical areas of Africa.   This is primarily because of the occurrence of tsetse flies, which carry trypanosomiasis (sleeping sickness), and because of the topographic limitations of the forest cover, that is, the high tree density and a highly developed root network that prevents the use of animals.   Unlike the Amazon region, where one of the main causes of deforestation has been conversion of forests to pastures by livestock ranchers, livestock plays a limited role in deforestation in Côte d'Ivoire.

Underlying Causes of Deforestation

Some of the underlying causes of deforestation are the result of the combined effects of the spread of shifting cultivation, which, in turn, is caused by population pressures, unclearly defined land tenure regimes (property arrangements), and government agricultural and forestry policies.

SHIFTING CULTIVATION

The major impetus for the increases in agricultural lands in Côte d'Ivoire is shifting (slash-and-burn) cultivation. It is an extensive system of food crop production in which natural forests, secondary forests, or open woodlands are felled and burned. Theoretically, the cleared area is cultivated for a few years (usually 1 to 3 years), after which the land is abandoned and allowed to return to forest or bush fallow. The process is repeated after a period of time that ranges between 4 and 20 years. It is necessary to practice shifting cultivation in the tropics because of the low nutrient content of many tropical soils. Most of the nutrients are in living plants, and the nutrients are made available when an area is cleared and burned; the resultant nutrient-rich ash fertilizes the soil (Persson, 1975). The system, however, operates effectively only when there is sufficient land to allow a long fallow period so that soil productivity, which is exhausted during the short cropping cycle, can be restored.

Today, because of increasing populations, fallow periods are being reduced and smallholders are compelled to clear more forests or to exploit the more fragile, marginal lands that cannot support an increasingly large population. Considerable deforestation occurs because of the movement of shifting cultivators into areas opened up by logging. It is estimated that for each 5 m³ of logs harvested in Côte d'Ivoire, 1 ha of forest is converted into cropland by subsequent cultivators (Myers, 1980).

LAND TENURE REGIMES

Excessive clearing of forestlands also occurs because of the open-access nature of forest resources in Côte d'Ivoire. The open-access nature of management can best be expressed by popular sayings of Ivoirians: "[L]and belongs to whoever cultivates it" or to ". . . whoever uses it and values it" (Bertrand, 1983). What is happening is, in effect, the result of government policies that attempt to supplant local tenure regimes. By ignoring the distinction between common property and open access, the government has failed to offer legal mechanisms for protecting communal land rights. Instead, attempts are often made to convert common properties into government lands and private properties, even though the public sector's capacity to manage the forest resources and the legal infrastructure needed to enforce private tenure are poorly developed. As a result, people gather what they need from the forest and freely exploit forest resources, in spite of the fact that current legislation forbids unauthorized clearing of state-owned forests (Southgate et al., 1990).

Close examination of the Ivoirian land tenure regime structure indicates that there is a juxtaposition of informal, customary laws and formal government legislation. Customary laws regulate the traditional land use patterns, which are based on group or communal ownership. The government legislation (which was initially inherited from the colonial power, France, and has slowly been reformed) distinguishes among three forms of land tenure.

State Ownership The first and most important is forestland under state ownership. Commonly called reserved forests, these are vast areas of forestlands surveyed to be protected from illegal encroachment. The people who settle in reserved forests clear them and illegally take valuable forest products, apparently because the laws are not well enforced or because the forests are not well policed. As a result, 100,000 people per year have spontaneously migrated into and settled in the forest zone for the past 20 years. In reality, enforcement of laws is particularly difficult, if not impossible, because peasants obey customary laws, which sometimes run counter to the spirit and provision of state forestland ownership laws. In a society like Côte d'Ivoire's, in which the institutions that govern the use of resources overlap, enforcement must deal with several institutional structures. The weakening of traditional property arrangements without the provision of a viable institutional alternative diminishes the incentives for forest dwellers to conserve natural resources (Bromley and Cernea, 1989).

Collective Ownership The second form of forestland tenure is communal, or collective, ownership. Under this category, local communities (villages) are recognized as the owners of the forestlands, but the government and others may manage them. Group ownership constitutes the most common form of land ownership in Côte d'Ivoire. Land is viewed as belonging to a common ancestor, and any member of the extended family can use it when it becomes vacant, but it cannot strictly be sold or transferred to someone outside the family. Although individual cultivators have control over the crops they produce, the group (or the extended family) has the power to decide on the use of a particular area of land.

The problem is that the communal landholdings are poorly delineated. They cannot be distinguished unambiguously, nor can communal landholdings be distinguished from state holdings. Because of this lack of clearly defined property rights, the economies of the people who live in the forests of Côte d'Ivoire are largely geared to the extensive use of land. Peasants sometimes view the forests as an

Women sell pineapples, maize, coconuts, yams, and other local produce at a market in Côte d'Ivoire. Credit: James P. Blair © 1983 National Geographic Society.

obstacle to the development of their plantations and fields. As a result, open-access types of exploitative behavior arise. Under the open-access system, no individual or group of individuals wants to incur the costs required to protect and maintain forest resources. On the contrary, individual forest users have every incentive to clear forestlands as soon as possible because they have no guarantee that whatever they leave untouched will be available in the near future.

Private Individual Ownership The third and last form of land tenure is private individual ownership. This form of ownership is the least developed, however, because few individuals own forestland outright (Bertrand, 1983; Food and Agriculture Organization and United Nations Environment Program, 1981).

GOVERNMENT POLICIES

The final underlying cause of deforestation in Côte d'Ivoire discussed here is government agricultural and forestry policies. One example is the marketing policy for the major export crops in Côte d'Ivoire (notably, coffee, cacao, cotton, and palm oil). The prices of these commodities are regulated by a marketing board, the Caisse de Stabilization et de Soutien des Prix des Produits Agricoles (CSSPA; Agricultural Product Price Support and Stabilization Fund). The board guarantees a fixed price to planters throughout the crop year and, at times, for several consecutive seasons. Prices are set on a cost-plus basis and are lower than international prices, thus enabling the government to generate surpluses. As long as producer prices are low, farmers will not be able to afford to use intensive means of production. It is therefore more profitable to cultivate extensively at the expense of forests. It appears, however, that cash crop farmers have found it more profitable to cultivate extensively than to intensify their cultivation practices. Also, despite the creation of the Agricultural Development Bank in 1968, farmers still face severe capital constraints. Many smallholders are unable to gather sufficient funds for investment because the cost of credit is very high. Also, the titling problems exacerbated by unclearly defined property rights place small-scale farmers at a distinct disadvantage in negotiating with banks and government entities for credit.

Some forestry-based policy instruments have also contributed to the rapid rate of deforestation in Côte d'Ivoire. The fiscal policy in the forestry sector distinguishes among four types of royalties and license fees (Gillis, 1988): (1) a timber royalty, (2) a concession license, (3) a public work fee, and (4) an annual area charge.

Timber royalty rates (imposed on harvested volumes rather than on a per tree basis) were set in 1966 and have remained unchanged. Despite some differentiation in the royalty schedule according to tree species, timber royalties are judged to be too low relative to free on-board (FOB) log export values to have serious implications for lower rates of deforestation. The cost of the concession license is only US$0.25/ha, and public work fees amount to US$0.79 and US$0.40/ha on the richer and poorer stands, respectively. Both are one-time levies. The annual area charge is levied at the rate of US$0.05/ha/year. These charges are estimated to be too low to have notable effects on forest-clearing decisions. The very low fees that are charged for the right to clear forests encourage the exploitation of marginal stands by providing a large profit margin while offering little incentive for more intensive exploitation of more valuable stands because expanding the area of harvest is less costly than intensifying cultivation. If these fees had been increased substantially by 1970, the nation might have experienced a somewhat lower rate of deforestation than actually occurred in the 1970s and 1980s (Gillis, 1988).

EFFECTS OF DEFORESTATION

The conversion of forestlands to other uses produces a broad range of effects, including (1) changes in climate and microclimate, (2) erosion of biodiversity, (3) long-term decline of agricultural productivity and income, and (4) forest damage associated with the loss of timber production potential. Together, these effects constitute a serious threat to agricultural sustainability in Côte d'Ivoire.

Climate and Microclimate

Scientists are concerned that tropical deforestation might affect climate on a global scale by increasing the levels of carbon dioxide (CO_2) in the atmosphere (Sedjo, 1983). This is because a significant portion of the world's carbon is locked in the wood of the tropical forests. Some of the carbon that is stored in forest soils is also released as the land is converted from forestland to cropland. Climatologists are engaged in a continuing debate, however, regarding the global effects of deforestation in this regard. One analysis suggests that the amount of CO_2 released by the clearing and burning of wood from dense tropical forests may be roughly equivalent to the amount of CO_2 released by fossil fuel combustion (Woodwell, 1978). Concerns about the concentrations of CO_2 in the atmosphere arise from the hypothesis that rising atmospheric CO_2 concentrations will cause a greenhouse effect, with disruptions of the world's agricultural productivity in the twenty-first century (U.S. Department of State, 1980).

There is little science-based information on the effect of deforestation on the microclimate in Côte d'Ivoire. Spears (1986) measured the bioclimatic impact of different vegetative covers and showed that different vegetative covers result in distinctly different transpiration and energy exchange characteristics. In the past 2 decades, rainfall levels have generally decreased and the soils of forest regions have become progressively drier, particularly in the south-central part of Côte d'Ivoire. However, it is necessary to interpret carefully the climatic and ecologic data obtained over long time periods. The lower rainfall of the past 2 decades could represent downswings in rainfall that are part of the 30-year rainfall cycles of the region. Such trends are apparent from the rainfall records of Côte d'Ivoire and other countries in the region.

Ghuman and Lal (1988) reported experimental results of a study done in a region of Nigeria in which the climate is similar to that in Côte d'Ivoire. The study quantified the magnitude and trends in alterations of the soil, hydrology, microclimate, and biotic environments resulting from the conversion of a tropical rain forest to different land use systems and agricultural practices. The rainfall results showed that the amount of rainfall under the forest canopy was about 12 percent less than that in cleared areas. The amount of solar radiation received in the cleared area was 25 times greater than that received under forests. On average, soil and air temperatures and evaporation rates were lower in areas under forest cover than they were in cleared areas. Relative humidities (which inversely correspond to variations in air temperature) were higher in forest areas than in cleared ones.

Biodiversity

There are no empirical data on the extent of erosion of biodiversity because of deforestation in Côte d'Ivoire. However, forests are known to contain a wide variety of plant and animal species, many of which have not been examined by scientists. For example, they contain the gene pools of parent species from which many agricultural crops were originally bred and, therefore, may be needed for future breeding efforts if crops are devastated by new diseases or other catastrophes. Some of these species may be critically important for pest and disease resistance in agricultural crops. For example, because of a smaller gene pool, it will be harder to counteract a weakness such as reduced disease resistance in varieties of plants and animals used for economic production. Other species have important potential as pharmaceutical agents, some of which are known only to people indigenous to the forests. The erosion of the genetic base as a result of deforestation will make it increasingly difficult to maintain economic production from biologic resources.

Agricultural Productivity

After forests are cleared from the land, the soil's physical and chemical properties undergo significant changes, leading to nutrient losses, accelerated rates of soil erosion, and declining yields (Lal, 1981; Seubert et al., 1977). Forests protect the soil by regulating stream flows (thereby minimizing soil erosion), modulating seasonal flooding, and preventing the silting of dams and canals. Forests help to accelerate the formation of topsoil, create favorable soil structures, and store nutrients. Using data from Côte d'Ivoire, Ehui and Hertel (1989, 1992a) showed that part of the agricultural growth in Côte d'Ivoire has been accomplished at the expense of the natural resource base and is therefore unsustainable. In particular, they showed that deforestation contributes positively to crop yields, but that increases in the cumulative amount of deforested lands cause yields to fall. This study thus confirms soil scientists's hypotheses that crop yields increase immediately after deforestation because of the nutrient content of the ash that is present after burning. However, yields decline over time because of the loss of the soil productivity as a result of movement of cropping activity onto marginal lands, removal of organic matter, and erosion. This affects the overall productivity and sustainability of the agricultural sector. Ehui and Hertel (1989, 1992a) also showed that aggregate yields are somewhat insensitive to deforestation in the same year, but are sensitive to the cumulative amount of deforestation over several years. A 10 percent increase in cumulative deforested land results in a 26.9 percent decline in aggregate yields.

In a follow-up study, Ehui and Hertel (1992b) conducted simulation studies that measured the value of conserving marginal forestlands in Côte d'Ivoire by taking into account the short- and long-term impacts of deforestation on agricultural productivity. Examination of the impacts of deforestation and cumulative deforested lands on food crop revenues indicated that forest conservation results in net benefit to agriculture. For example, with a one-time 20 percent decrease in the rate of deforestation, the net value of food crop revenues rose by US$21.3 million. This translated into approximately US$507/ha of forest saved. Ehui and Hertel (1992b) concluded that at current rates of deforestation, Côte d'Ivoire has been forgoing long-term agricultural revenues in pursuit of short-term gains.

Forest Damage and Timber Production Potential

The lack of proper forest management, which leads to excessive logging and agroconversion, also leads to losses in earnings from the timber industry. Using the average timber export tax rate as the opportunity costs of unmanaged forestlands and annual deforestation of 300,000 ha, Bertrand (1983) estimated that the annual cost of deforestation is between US$69 million and US$295 million. Bertrand also estimated that lost FOB earnings range between US$80 million and US$200 million. These are important losses because the forestry-based sector plays a larger role in Côte d'Ivoire's economy than it does in any other African country (Gillis, 1988). The contribution of the forestry-based sector in the decade prior to 1981 was consistently about 6 percent of GDP, the highest in Africa. The value of wood extracted from forests rose from nearly US$600 million in 1977 to US$900 million in 1980, by which time the value of log and wood product exports reached US$562 million, or about 11 percent of total export earnings, down from the peak of 35 percent in 1973 (Gillis, 1988). In the decade prior to 1981, the Ivoirian forestry-based sector was a fairly strong source of tax revenues, providing, in all years except 1973, an annual average of 6 percent of government revenues, which was greater than the average for all other African countries except Liberia. The decline in the nation's forest export taxes and fees in relation to total government revenues has primarily been due to a reduction in total exports of higher value logs (for example, sappelli, sipo [*Entandrophragma utile*] and samba [*Triplochiton scleroxylon*], which are used to make furniture and to build houses). By 1978, lower value species constituted more than 50 percent of Ivoirian timber exports, the richer stands of sappelli and sipo trees having been largely depleted.

AGRICULTURAL INTERVENTIONS AND SUSTAINABILITY

About 60 percent of Côte d'Ivoire's population lives in the forests. Spontaneous settlement and migration into the forest zone has averaged 100,000 people a year for the past 20 years. Instead of forcing people out of the forests, the best hope for slowing deforestation is to provide the people already there with the means of intensifying agricultural productivity and to combine sound agricultural and forestry policies to slow future migration into forest zones (Spears, 1986). Interventions to increase agricultural productivity can be divided into two categories: technological interventions and policy interventions.

Technological Interventions

Shifting (slash-and-burn) cultivation is still the dominant land use system in vast areas of Côte d'Ivoire. This traditional food crop production system, which is based solely on the restorative properties of woody species, has sustained agricultural production on uplands in many parts of the tropics for many generations. The system involves partial clearing of the forest or bush fallow. The cropping period is marked by a random spatial arrangement of crops and "regrowth" of woody perennials. Long fallow periods (10 to 20 years) are necessary to allow regeneration of soil productivity and weed suppression. However, the annual population growth rate of 4 percent increases the need for food, which, in turn, increases the need for land, causing increased deforestation and shorter fallow periods (2 or 3 years), which reduces the productive capacity of the land, decreases crop yields, and increases the opportunity for weed and pest infestation.

In the forests of Côte d'Ivoire, as in many other parts of the humid tropics, the maintenance of soil fertility constitutes the major constraint to increased agricultural sustainability. One of the basic characteristics of soils in the humid tropical lowlands of Africa is the susceptibility of the soils to degradation and the tendency for soil productivity to decline rapidly with repeated cultivation (Carr, 1989; Lal, 1986). The greatest challenge to research and extension staff is to maintain soil fertility in a sustainable manner. Farmers need sustainable land use systems that allow them to achieve the necessary levels of production while conserving the resources on which that production depends, thereby permitting the maintenance of productivity. According to Lal (1986), sustainable land use management technologies should include the following:

• Preservation of the delicate ecologic balance, namely, that among vegetation, climate, and soil;
• Maintenance of a regular, adequate supply of organic matter on the soil surface;
• Enhancement of soil fauna activity and soil turnover by natural process;
• Maintenance of the physical condition of the soil so that it is suitable for the land use;
• Replenishment of the nutrients removed by plants and animals;
• Creation of a desirable nutrient balance and soil reaction;
• Prevention of the buildup of pests and undesirable plants;
• Adaptation of a natural nutrient recycling mechanism to avoid nutrient losses from leaching; and
• Preservation of ecologic diversity.

All of these requirements are met in the traditional shifting cultivation systems that allow short cropping periods followed by long fallow periods. The scarcity of arable land because of increasing population pressures, however, has drastically shortened fallow periods, making a change or an adaptation of technology inevitable. Recent studies of farming systems have been done at national agricultural research centers, such as country-based research centers and universities in sub-Saharan Africa, and at international agricultural research centers, such as the International Institute of Tropical Agriculture (IITA), Ibadan, Nigeria; the International Livestock Center for Africa (ILCA), Addis Ababa, Ethiopia; and the International Center for Research on Agroforestry (ICRAF), Nairobi, Kenya. Based on the developments of this research, a survey of the literature indicates that five basic technologies are used to restore soil fertility in annual mixed (livestock and crops) cropping systems (Carr, 1989).

ORGANIC MATTER

This technology is based on the importation of organic matter from outside the system. It usually relies on wood ash as a soil enhancer. Because of the high levels of mineralization of organic matter in the humid tropics, the system requires heavy and frequent application of organic matter and does not provide a viable technology for field-scale crop production when only human labor is used.

MULCHES AND COVER CROPS

Mulch cover is an essential ingredient of conservation farming. Without an adequate amount of mulch, the soil structure deteriorates rapidly and crop yields decline. Mulch can be procured from crop residues, a cover crop, or a combination of both of these. In a crop residue system, substantial crop residue mulch is regularly added to the soil surface. It has proved to be beneficial for a wide range of soils and agroecologic environments in the tropics. The main benefits include better soil and water conservation, improved soil moisture and temperature regimes, amelioration of soil structure, favorable soil turnover through enhanced biotic activity of soil fauna, and protection of the soil from intense rains and desiccation. Because of amelioration of the soil structure and the effect of mulch on weed suppression, mulching is generally beneficial to crop growth (Lal, 1986).

When crop residue is inadequate, a practical means of procuring mulch is by the incorporation of an appropriate cover crop or the use of a planted fallow in the rotation. Research results have shown that in addition to providing mulch residues, planted fallows are more effective in restoring soil physical and nutritional properties than long bush fallow (Lal, 1986:77–81). Organic matter can be built up and soil structure can be improved, even on eroded and degraded lands, by growing appropriate planted fallow for 2 to 3 years (Wilson et al., 1982, 1986).

Despite the potential benefits that can be derived from the use of crop residues or herbaceous cover crops, their use has never gained popular acceptance in the humid tropics (Wilson et al., 1986), perhaps because farmers may be averse to using green manure crops that occupy the land during the rainy season without providing a direct return or because the herbaceous crops do not survive the dry period before the cropping season in areas with low total annual rainfall.

INORGANIC FERTILIZERS

Appropriate fertilizer regimes have been developed. These regimes enhance crop growth but do not cause soil acidification or toxicity problems. For example, experiments conducted at the IITA (Ibadan, Nigeria) have shown that low-level application of lime and inorganic fertilizer results in lower rates of degradation of acidic soils (which are predominant in the tropical humid forests) and reduced acidity and toxicity, permitting significantly improved yields for crops such as maize. Other work shows, however, that if fertilizer is the only input, yields decline over time. In addition, lime and other, related fertilizers are not always available. Another problem related to the use of lime is that many soil nutrients can be lost through leaching because they are released as a result of changes in soil acidity (International Institute of Tropical Agriculture, 1990). In addition, fertilizers cannot readily be found because of high prices and difficulties in transporting them to the areas where they are needed.

AGROFORESTRY

For many generations, farmers have exploited the potential of trees and shrubs for soil fertility regeneration and weed suppression in traditional slash-and-burn agricultural systems. The effectiveness of the role of trees and shrubs depends not only on the compositions of the woody species and soil characteristics but also on the length of the fallow periods (Nye and Greenland, 1960). Work at international research centers, such as the IITA, ILCA, and ICRAF, over the past 2 decades has demonstrated that replacing traditional species with trees that are both leguminous and tolerant of frequent pollarding can help slow down soil degradation. This led to the development of and research on alley cropping systems (Kang et al., 1981).

Alley Cropping   Alley cropping is an agroforestry system in which crops are grown in alleys formed by hedgerows of trees and shrubs, preferably legumes (Figure 7). The hedgerows are cut back at the time of planting of food crops and are periodically pruned during cropping to prevent shading and to reduce competition with the associated food crops. The hedgerows are allowed to grow freely to cover the land when there are no crops (Kang et al., 1981). The major advantage of alley cropping over the traditional shifting and bush

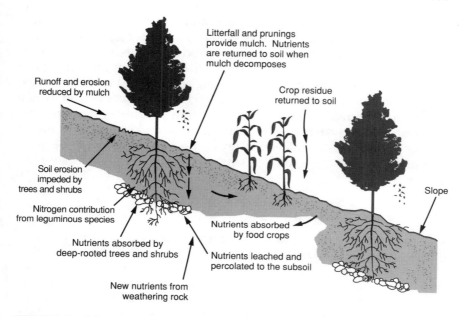

Litterfall and prunings provide mulch. Nutrients are returned to soil when mulch decomposes

Runoff and erosion reduced by mulch

Crop residue returned to soil

Soil erosion impeded by trees and shrubs

Slope

Nitrogen contribution from leguminous species

Nutrients absorbed by food crops

Nutrients absorbed by deep-rooted trees and shrubs

Nutrients leached and percolated to the subsoil

New nutrients from weathering rock

FIGURE 7   Schematic representation showing the benefits of nutrient cycling and erosion control in an alley cropping system. Source: Kang, B. T., A. C. B. M. van der Kruijs, and D. C. Cooper. 1989. Alley cropping for food production. Pp. 16-26 in Alley Farming in the Humid and Sub-humid Tropics, B. T. Kang and L. Reynolds, eds. Ottawa, Canada: International Development Research Center.

fallow system is that the cropping and fallow phases can take place concurrently on the same land, thus allowing farmers to crop the land for an extended period of time without returning to a fallow period.

The ILCA has extended the concept of alley cropping to include livestock by using a portion of the hedgerow's foliage for animal feed (the alley farming method) (Kang et al., 1990). Use of woody legumes provides rich mulch and green manure to maintain soil fertility, enhance crop production, and provide protein-rich fodder for livestock. On sloping lands, planting of hedgerows along the contours greatly reduces soil erosion. Alley cropping or farming is a potentially beneficial technology, but despite the improved basic knowledge about this technology, it is still in the development phase in the humid tropics. Additional technical and economic analysis is required.

Recently, Ehui et al. (1990) conducted an economic analysis of the effect of soil erosion on alley cropping and on no-till and bush fallow systems. They concluded that, in general, when access to new forest-lands is costless in terms of foregone production because the land is fallow, slight decreases in yields from erosion will not detract significantly from the profit obtained by using traditional bush fallow systems with long fallow periods. However, in those cases in which land values increase because of population pressures, farmers who use bush fallow systems have incurred costs by keeping land out of production (that is, in fallow). Alley cropping was shown to be more profitable during the growing season, despite its higher labor requirement.

CONSERVATION TILLAGE

Studies at the IITA and elsewhere have shown the advantage of conservation tillage, an approach to soil surface management that emphasizes use and improvement of natural resources rather than exploitation and mining for quick economic return. Conservation tillage is defined as any system that leaves at least 30 percent of the previous crop residue on the surface after planting (Lal et al., 1990:207). When it is successfully applied, conservation tillage may maintain soil fertility and control erosion. The various types of conservation tillage include minimum tillage, chisel plowing, plow-plant, ridge tillage, and no-tillage.

In the humid tropics, no-till farming, which involves seeding through a crop residue mulch or on unplowed soil, has several advantages. One is the conservation of soil and water. Other advantages are the lowering of the maximum soil temperature and the maintenance of higher levels of organic matter in the soil. Experimental data from Ibadan, Nigeria (a subhumid zone), indicate that conservation tillage can be extremely effective in controlling soil erosion. For example, mean soil erosion rates for areas with slopes of up to 15 percent were estimated to be 0.1 and 9.4 metric tons/ha for no-till and plowed systems, respectively. Ehui et al. (1990) showed that, in areas with increasing population pressures, the no-till system is more profitable than the traditional bush fallow systems. The alley cropping system with 4 m of space between hedgerows is more profitable than the no-till system.

Policy Interventions

Government intervention is required when there are market failures. Some causes of market failure are the lack of clearly defined or secure property rights, variable external market pressures, inappropriate timber taxation, and a short-sighted plan that pursues quick profits at the expense of long-term, sustainable benefits (Panayotou, 1983). These causative factors characterize the economy of Côte d'Ivoire and emphasize the fact that policy reforms that address fundamental issues are needed.

SECURE PROPERTY RIGHTS

The pressure for shorter fallow periods, spurred by population growth, requires investments in land improvements to retain soil fertility and investments of capital to expedite the preparation of land for farming and to increase productivity. The incentive to undertake such investments is based in part on secure future access to that land. Inappropriate land tenure regimes or the lack of a secure means of land ownership forces farmers to take actions—encroachment onto marginal lands, deforestation, and cultivation of steep slopes—that help them only in the short term. The main effect of insecure land tenure is the land operators's uncertainty about their ability to benefit from any investments they might make to improve and sustain the productive capacities of their farms (Feder and Noronha, 1987). Francis (1987) noted that community-controlled rotations of land parcels discouraged the adoption of alley farming in southeastern Nigeria. Survey results by Lawry and Stienbarger (1991) showed that most farmers who practice alley cropping obtained their land through divided inheritances, which allows them full control over their land.

Ownership security reinforces both investment incentives and the availability of investment capital. Availability of credit from institutional sources in particular frequently depends on the borrower's ownership security because unsecured loans are more risky for institutional lenders and less likely to be granted. In Côte d'Ivoire, proper titling of rural land areas is necessary to provide sufficient land tenure security for the people because the rights to most forest areas belong to the government. There are only a few individuals with property rights in Ivoirian forest areas. A unified, state-controlled system of rural land registration is one way of enhancing ownership security.

Goodland (1991) proposed that, in addition to being secure, land holdings should be of a size that can sustainably support families and provide them with a reasonable standard of living. Adequate parcel size promotes agricultural intensification and conservation of soils and forestland.

Promotion of sustainable use of forest lands can be achieved by granting long-term forest concessions to timber exploiters. Long-term concessions increase the forest exploiters' land tenure security and promote the efficiency of resource use. Such concessions should be revoked, however, and the concessionaires fined if the land is used in an unsustainable manner. Implementation of this policy would require that the government properly monitor logging activities of the concessionaires.

FISCAL POLICIES

Earlier in this profile it was noted that one of the causes of deforestation is that timber license fees and royalties are, collectively, too low to encourage sustainable management of forest resources. Ehui and Hertel (1989, 1992a) showed that, although deforestation in Côte d'Ivoire increases aggregate yields in the short term, it has long-term deleterious effects on productivity. Depletion of forest resources is associated with external factors, which have not been properly accounted for. (An "external factor" being the resultant effect when the action of one individual or farm has a positive or negative effect on other individuals or farms that are not parties to the activity but, as a consequence, incur the costs or enjoy the benefits.) For example, loggers and shifting cultivators receive the full benefits from extraction of timber and slash-and-burn land preparation, respectively, but they incur only some of the costs; the rest of the costs are incurred by downstream farmers—and by the society at large—in the forms of flooding, siltation, and erosion.

Theoretically, the preferred policy for controlling excessive deforestation would be taxation. A proper level of taxation on forest exploiters would reduce the level of deforestation to a point at which the marginal social costs of deforestation would be equal to the marginal benefits. (Marginal social costs are defined as the direct costs of clearing the forest plus the associated opportunity or user costs.) Because the forest stock is fixed, any unit cleared or consumed is unavailable for use in the future. Consequently, current deforestation comes at the expense of future benefits from forest endowment, resulting in opportunity or user costs.

CREDIT, PRICE POLICIES, AND MARKETS

Other reasons for the excessive rate of deforestation in Côte d'Ivoire include capital constraints faced by the farmers combined with the often highly imperfect and distorted capital markets and relatively low producer prices. Often, it is cash funds for consumption and investment—not land—that is the scarcest resource for farmers. Capital constraints prevent the optimal use of resources. It is at this point that affordable credit is needed. In many rural areas, institutional credit either is not available or is too costly. The result is that many farmers are unable to put their land to its best use, even if they have the knowledge and motivation to do so. The lack of credit is also exacerbated by the low prices, relative to world market prices, that farmers receive for their products. One solution to excessive deforestation is to intensify agricultural productivity, thus negating the need to deforest more land. Intensification occurs through the use of improved inputs and extension services and when farmers are encouraged to mechanize their farming operations and apply pesticides. Without adequate prices and credit farmers will not be able to acquire these inputs.

The proper role of markets in sustainable soil management needs to be outlined as well. In studying agricultural mechanization and the evolution of farming systems in sub-Saharan African, Pingali et al. (1987) showed that for a given population density, an improvement in access to markets causes further intensification of the farming system (in this case, use of the plow). Their survey results support the hypothesis that, with poor access to markets, extensive forms of farming such as forest fallow and bush fallow are usually practiced.

SUMMARY

Côte d'Ivoire achieved the highest agricultural growth rate (5 percent) in sub-Saharan Africa during the first 2 decades after independence in 1960 (den Tuinder, 1978; Lee, 1983). This growth rate was driven primarily by increases in the area under cultivation (Lee, 1983; Spears, 1986), which arose solely from deforestation (see Table 2). As a result, agricultural expansion has often involved movement onto poorer soils and sloping uplands that cannot support permanent cropping (Bourreau, 1984) and is therefore unsustainable; this has mitigated rural poverty.

Planners must implement an agricultural system that can feed an increasing population without irreparably damaging the natural resource base on which agricultural production depends. Today, with an annual population growth rate of close to 4 percent, the main forestry policy question facing the government of Côte d'Ivoire is how to effectively manage what is left of the tropical rain forest.

TABLE 10 Forest Loss Scenarios in Côte d'Ivoire, 1990–2029

Scenario and Time Period	Forest Cover at Beginning of Decade (millions of ha)	Average Loss (millions of ha/year)	Total Loss for Decade (millions of ha)	Forest Cover at End of Decade (millions of ha)	Percent Loss for the Decade
Baseline scenario					
1990–1999	1.55	0.08	0.8	0.15	52
2000–2009	0.75	0.06	0.6	0.15	80
2010–2029	0.15	0.05	0.15	0	100
Worst-case scenario					
1990–1999	1.55	0.20	1.55	0	100
2000–2009	0	0	0	0	0
2010–2029	0	0	0	0	0
Best-case scenario					
1990–1999	1.55	0.05	0.55	1.05	32
2000–2009	1.05	0.02	0.20	0.85	19
2010–2029	0.85	0.01	0.10	0.75	12

Three Deforestation Scenarios

Table 10 presents the expected patterns of deforestation over the next 30 years using three scenarios: a base-case scenario (scenario A), a worst-case scenario (scenario B), and a best-case scenario (scenario C).

In the base-case scenario, it is assumed that there will be some reformation of government policy toward forest resource management but no real high-level political commitment. Because forest resources have decreased to such a large extent, the rate of deforestation in this scenario will, in the 1990s, decline to about 80,000 ha/year. The rate will decline to 60,000 ha/year from 2000 to 2009 and to 50,000 ha/year from 2010 to 2029 before the forests are depleted of their resources.

The worst-case scenario is based on a laissez faire policy, in which the government will, as in the past, have no overall land use policy. Price and fiscal policies will be unchanged, and there will be no effort to intensify agriculture. In this scenario, the rate of deforestation is hypothesized to be at least the same as that during the previous decade (that is, almost 200,000 ha/year). At this rate, there will be no remaining highland forest by the end of 2000. This hypothesis is based on the assumption that there will be no population growth control, that the population will continue to grow at an average rate of 3.6 percent per year, and that the major source of food and agricultural growth for the country will be through the expansion of the agricultural land frontier into presently forested areas rather than through land-saving technologies. Also, projecting the current slump in prices for Côte d'Ivoire's major export crops (cacao and coffee) and the increasing debt burden and unemployment rate in the cities, farmers and loggers will be encouraged, in an effort to increase foreign exchange earnings, to cut the remaining tracts of natural forests.

In the best-case scenario, the rate of deforestation is expected to average 50,000 ha/year between 1990 and 1999, 20,000 ha/year between 2000 and 2009, and 10,000 ha/year between 2010 and 2029. With these levels of deforestation there will be 1.05 million ha of forest remaining by 2009 and 0.85 million ha of forest remaining by 2029. This scenario is based on the assumption that policy and technology options listed below (see also, Spears [1986]) will be supported by the government, with high-level political commitment.

Technology options lie in the direction of sustainable and economically efficient agricultural practices—that is, practices that can maintain protective organic mulches on the soil surface by maximizing biomass production (organic residue production) while minimizing the negative competitive effects on the crops or animals produced. Policy options lie in the direction of reformation of land tenure rights and taxation and fees for timber extraction.

TECHNOLOGY OPTIONS

Technology options include the following:

• Use of organic manure and inorganic fertilizers;
• Use of mulches and cover crop systems;
• Intensification of agricultural production in humid forest zones through the use of tree-based technologies—such as alley cropping—that can reduce dependence on bush fallowing;
• Development of intensive food crop production in lowland areas;
• Conservation tillage; and
• Creation of a buffer zone of intensive agricultural perennials (coffee, cacao, oil palm, and rubber) around or adjacent to the most imminently threatened forest areas.

POLICY OPTIONS

Policy options include the following:

• Continue public awareness, mass education about and moral persuasion against deforestation;
• Incorporate environmental conservation curricula in schools, including intensive forestry and agroforestry education, training, and research, with special emphasis on topics such as tree breeding and genetic improvement in order to increase productivity and shorten plantation rotations;
• Establish a mechanism for defining proper land tenure regimes (for example, a unified, state-controlled system of land titling);
• Improve timber pricing and fiscal policies (for example, sales of permits for the extraction of forest products and strict monitoring of current extraction and transportation procedures);
• Raise timber extraction taxes to reflect the true price of forest resources and to help fund reforestation;
• Institute subsidies, investment tax credits, and other incentives for reforestation by private and government agencies;
• Support large-scale government and private investments in reforestation;
• Improve agricultural pricing and credit policies; and

• Prepare a land use plan for forest zones, demarcating areas suited to perennial agricultural tree crops, food crops, and forestry and setting up a more effective government mechanism for land use allocation in forest zones.

In Côte d'Ivoire, most forestlands are owned by the government, and prices for extraction of forest resources are fixed far below what is necessary to make sustainable practices cost-effective and to stimulate capital formation for replanting operations. With the costs of deforestation externalized (for example, the impact of deforestation on the future productivity of the land), forestland pricing policy needs a thorough revamping if forest regeneration is to be boosted and excessive deforestation reduced.

Illegal encroachments of forests because of unclearly defined property rights have become increasingly common, and the multiple activities that follow encroachment (for example, cattle grazing and shifting cultivation) intensify the deleterious effects of deforestation. Policies regarding land titling must, therefore, also be revamped.

Among the agricultural technology options, alley cropping appears to be the most promising. Even though alley cropping has proved to be agronomically and economically more viable than alternative land use systems, its successful adoption depends on the prevailing policy environment. Without sound economic policies that support agriculture—such as investment in infrastructure, proper incentives to farmers, adequate supplies of production inputs, effective marketing, and credit facilities—it will be difficult to achieve increased agricultural productivity through new land use technologies.

REFERENCES

Bertrand, A. 1983. La déforestation en zone de forêt en Côte d'Ivoire. Rev. Bois Trop. 220:3–17.

Bourreau, C. 1984. Plan Quinquennal (1986–1990). Bilan Diagnostic Première Partie: Les Forêts et la Production Forestrière. Abidjan, Côte d'Ivoire: Direction des Eaux et Forêts, Ministère de l'Agriculture.

Bromley, D., and M. M. Cernea. 1989. The Management of Common Property Natural Resources: Some Conceptual and Operational Fallacies. Washington, D.C.: World Bank.

Carr, S. J. 1989. Technology for Small Scale Farmers in Sub-Saharan Africa: Experience with Food Crop Production in Five Major Ecological Zones. World Bank Technical Paper No. 109. Washington, D.C.: World Bank.

den Tuinder, B. A. 1978. Ivory Coast: The Challenge of Success. Baltimore: Johns Hopkins University Press.

Durufles, G., P. Bourgerol, B. Lesluyes, J. C. Martin, and M. Pascay. 1986.

Desequilibres Structurels et Programmes d'Ajustement en Côte d'Ivoire. Paris, France: Mission d'Evaluation, Ministère de la Cooperation.

Economic Intelligence Unit. 1991. Côte d'Ivoire Country Profile: Annual Survey of Political and Economic Background. London: Business International Limited.

Ehui, S. K., and T. W. Hertel. 1989. Deforestation and agricultural productivity in the Côte d'Ivoire. Amer. J. Agric. Econ. 71:703–711.

Ehui, S. K., and T. W. Hertel. 1992a. Testing the impact of deforestation on aggregate agricultural productivity. Agric. Ecosystems Envir. 38:205–218.

Ehui, S. K., and T. W. Hertel. 1992b. Measuring the value of marginal forest conservation in Côte d'Ivoire. Draft paper. Addis Ababa, Ethiopia: International Livestock Center for Africa.

Ehui, S. K., B. T. Kang, and D. S. C. Spencer. 1990. Economic analysis of soil erosion effects in alley cropping, no-till and bush fallow systems in south western Nigeria. Agric. Syst. 34:349–368.

Feder, G., and R. Noronha. 1987. Land systems and agricultural development in sub-Saharan Africa. World Bank Res. Observer 2(2):143–169.

Food and Agriculture Organization and United Nations Environment Program. 1981. Tropical Forest Resources Assessment Project (in the Framework of GEMS). Forest Resources of Tropical Africa. Part II. Country Briefs. Rome, Italy: Food and Agriculture Organization of the United Nations.

Francis, P. A. 1987. Land tenure systems and agricultural innovation: The case of alley farming in Nigeria. Land Use Policy (July):305–319.

Gbetibouo, M., and C. L. Delgado. 1984. Lessons and constraints for export crop-led growth. Pp. 115–147 in the Political Economy of Ivory Coast, I. W. Zartman and C. Delgado, eds. New York: Praeger.

Ghuman, B. S., and R. Lal. 1988. Effects of deforestation on soil properties and micro-climate of a high rain forest in southern Nigeria. Pp. 225–244 in The Geophysiology of Amazônia: Vegetation and Climate Interactions, R. E. Dickinson, ed. New York: Wiley, for the United Nations University.

Gillis, M. 1988. West Africa: Resource management policies and the tropical forest. Pp. 299–351 in Public Policies and the Misuse of Forest Resources, R. Repetto and M. Gillis, eds. New York: Cambridge University Press.

Goodland, R. 1991. Tropical Deforestation: Solutions, Ethics and Religion. Environment Working Paper No. 43. Washington, D.C.: World Bank.

International Bank for Reconstruction and Development (IBRD). 1985. Côte d'Ivoire Agricultural Sector Data Base Statistical Annex 7. Washington, D.C.: World Bank.

IBRD. 1989. Sub-Saharan Africa: From Crisis to Sustainable Growth, A Long-Term Perspective Study. Washington, D.C.: The World Bank.

International Institute of Tropical Agriculture. 1990. IITA Annual Report 1989/1990. Ibadan, Nigeria: International Institute of Tropical Agriculture.

Kang, B. T., G. F. Wilson, and L. Spiken. 1981. Alley cropping maize with leucaena in southern Nigeria. Plant Soil 63:165–179.

Kang, B. T., A. C. B. M. van der Kruijs, and D. C. Cooper. 1989. Alley cropping for food production. Pp. 16–26 in Alley Farming in the Humid and Sub-humid Tropics, B. T. Kang and L. Reynolds, eds. Ottawa, Canada: International Development Research Center.

Kang, B. T., L. Reynolds, and A. N. Atta-Krah. 1990. Alley farming. Adv. Agron. 43:315–359.

Lal, R. 1981. Clearing a tropical forest. II. Effects on crop performance. Field Crops Res. 4:345–354.

Lal, R. 1986. Soil surface management in the tropics for intensive land use in higher and sustained production. Adv. Agron. 5:1–109.

Lal, R., O. J. Eckert, N. R. Fausey, and W. M. Edwards. 1990. Conservation tillage in sustainable agriculture. Pp. 203–225 in Sustainable Agricultural Systems, C. A. Edwards, R. Lal, P. Madden, R. H. Miller, and G. House, eds. Ankeny, Iowa: Soil and Water Conservation Society.

Lawry, S. W., and D. M. Stienbarger. 1991. Tenure and Alley Farming in the Humid Zone of West Africa. Final Report of Research in Cameroon, Nigeria and Togo. Land Tenure Center Research Paper No. 105. Madison: University of Wisconsin.

Lee, E. 1983. Export led rural development: The Ivory Coast. Pp. 99–127 in Agrarian Policies and Rural Poverty in Africa, D. Ghia and S. Radwan, eds. Geneva: International Labor Office.

Michel, G., and M. Noël. 1984. Short-Term Responses to Trade and Incentive Policies in the Ivory Coast: Comparative Static Simulations in a Computable General Equilibrium Model. World Bank Staff Working Paper No. 647. Washington, D.C.: World Bank.

Myers, N. 1980. Conversion rates in tropical moist forests: Review of a recent survey. Pp. 48–66 in Proceedings of International Symposium on Tropical Forest, Utilization and Conservation, Ecological, Socio-Political and Economic Problems and Potentials, F. Mergen, ed. New Haven, Conn.: University of Connecticut.

Nye, P. H., and D. J. Greenland. 1960. The Soil Under Shifting Cultivation. Commonwealth Bureaus of Soils Technical Communication No. 51. Harpenden, England: Farnham Royal.

Panayotou, T. 1983. Renewable resource management for agricultural and rural development: Research and policy issues. Agricultural Development Council, Bangkok. Photocopy.

Persson, R. 1975. Forest Resources of Africa. Part I. Regional Analysis. Research Notes No. 22. Stockholm, Sweden: Royal College of Forestry.

Persson, R. 1977. Forest Resources of Africa. Part II. Regional Analysis Research Notes No. 22. Stockholm, Sweden: Royal College of Forestry.

Pingali, P., Y. Bigot, and H. P. Binswanger. 1987. Agricultural Mechanization and the Evolution of Farming Systems in Sub-Saharan Africa. Baltimore: John Hopkins University Press.

Sedjo, R. A. 1983. How serious is tropical deforestation? Are the world's tropical forests being rapidly deforested? Only in places say the authors. J. Forestry 81:792–794.

Seubert, C. E., P. A. Sanchez, and C. Valverde. 1977. Effects of land clearing methods on soil properties of an Ultisol and crop performance in the Amazon jungle of Peru. Trop. Agric. (Trin.) 54:307–321.

Southgate, D., J. Sanders, and S. Ehui. 1990. Resource degradation in Africa and Latin America: Population pressure, policies, and property arrangements. Amer. J. Agric. Econ. 72:1259–1263.

Spears, J. 1986. Key forest policy issues for the coming decade. Côte d'Ivoire forestry subsector discussion paper. World Bank, Washington, D.C. May. Photocopy.

Trueblood, M. A., and N. R. Horenstein. 1986. The Ivory Coast: An Export Market Profile. Foreign Agricultural Economic Report No. 223. Washington, D.C.: Economic Research Service, U.S. Department of Agriculture.

U.S. Department of Agriculture, Economic Research Service. 1988. World Indices of Agricultural and Food Production 1977–86. Statistical Bulletin No. 759. Washington, D.C.: U.S. Government Printing Office.

U.S. Department of State. 1980. The World's Tropical Forests: A Policy, Strategy and Program for the United States. U.S. Interagency Task Force on Tropical Forests. A Report of the President. Washington, D.C.: U.S. Government Printing Office.

U.S. Office of Technology Assessment. 1984. Technology to Sustain Tropical Forest Resources. Washington, D.C.: U.S. Government Printing Office.

Wilson, G. F., R. Lal, and B. N. Okigbo. 1982. Effect of cover crops on soil structure and on yield of subsequent arable crops grown under strip tillage on an eroded Alfisol. Soil Tillage Res. 2:233–250.

Wilson, G. F., B. T. Kang, and K. Mulongoy. 1986. Alley cropping: Trees as sources of green-manure and mulch in the tropics. Biol. Agric. Horticult. 3:251–267.

Woodwell, G. M. 1978. Biotic Interactions with Atmospheric Carbon Dioxide, Forests, Soil Humus. Oxford: Pergamon.

World Resources Institute. 1990. World Resources 1990–91. New York: Oxford University Press.

Indonesia

Junus Kartasubrata

Indonesia is the world's largest archipelago, consisting of some 13,700 islands. It is physiologically, biologically, and culturally one of the most diverse countries in the world. Some 70 percent of Indonesia is sea, while its land area is greater than 195 million ha. Massive mountain ranges containing a large number of volcanic formations run through the islands of Sumatra, Java, and the Lesser Sunda and also extend throughout the islands of Sulawesi and Irian Jaya. The highlands consist of broad alluvial plains.

DESCRIPTION OF THE COUNTRY AND ITS TROPICAL FORESTS

Indonesia is part of the Malesian botanical region, which is characterized by a large number of endemic species, a rich flora, and a complex vegetation structure. The Malesian rain forests are the richest in the world in terms of number of species (Whitmore, 1984). One of their most important features is the abundance of trees in the family Dipterocarpaceae.

Junus Kartasubrata is the general editor for Plant Resources of South-East Asia, Bogor, Indonesia.

Population

Indonesia is a country of villages, with 67,949 villages spread over 3,542 subdistricts within 246 regencies in 27 provinces. Indonesia is the fifth most populous country in the world, with over 184 million people (World Resources Institute, 1992). The population is unevenly distributed. Approximately 100 million people, 61 percent of the population, are concentrated on the island of Java, which accounts for only 6.7 percent of the total land area of Indonesia.

The island of Java, which has rich volcanic soils and high agricultural productivity, is one of the most populous regions in the world (population density, 768 people/km^2). The islands of Kalimantan and Irian Jaya, on the other hand, which together account for 50 percent of the country's land area, have population densities of 14 and 3 people/km^2, respectively. Urban populations are also higher in Java and Bali. Thirty percent of the population of Java is concentrated in cities, compared with 20 percent in the Outer Islands.

Indonesia's population increased at an average annual rate of 2.3 percent from 1965 to 1986. The growth rate decreased to about 2.15 percent in the 1980s. The annual growth rate varies markedly among the provinces, for example, 3.1 percent for Sumatra and 1.8 percent for Java in 1985, with the other regions having growth rates between those for Sumatra and Java (Asian Development Bank, 1989).

Urban populations have also been increasing considerably faster than rural populations, reflecting the country's industrialization. In 1971, for example, of the total population, the urban population was 17 percent in 1983 it had increased to 26 percent, and in 1993 it is expected to reach 32 percent, that is, 61 million of 193 million people (Asian Development Bank, 1989).

Demographic policies have focused on controlling population growth through family planning and regional population distribution. The government's target of annual population growth for REPELITA V (Rencana Pembangunan Lima Tahun), Indonesia's Fifth Five Year Development Plan (1989–1990 to 1993–1994), is 1.9 percent (Government of Indonesia/National Development Planning Agency, 1989). Even so, Indonesia's population is expected to increase substantially, to about 193 million people by 1993 (Government of Indonesia/National Development Planning Agency, 1989) and to 307 million people by 2030 (Government of Indonesia/Ministry of Forestry and Food and Agriculture Organization of the United Nations, 1990).

The uneven population distribution between the islands of Java and Bali and the Outer Islands is perceived as a major problem. Therefore, transmigration programs that resettle people from one region to another have been a priority of the Indonesian government. Migrants from Java and Bali are resettled in the provinces of Sumatra, Kalimantan, Sulawesi, Maluku, and Irian Jaya. According to government records, during the first 4 years of the REPELITA IV plan (1984–1985 to 1988–1989), 504,941 families were relocated; the target for the REPELITA V plan is 750,000 families (Government of Indonesia/Department of Information, 1989).

Indonesia's work force amounts to 74.5 million people, or 42 percent of the total population, with 61 percent in Java and 39 percent in the Outer Islands (Government of Indonesia/Department of Information, 1989). In 1985, the proportion of the work force employed in various sectors was as follows: 54.7 percent in the agricultural sector (compared with 64.2 percent in 1971); 15.0 percent in the commercial sector; 13.3 percent in public services; 9.3 percent in industry; 3.3 percent in construction; 3.1 percent in transportation and communication; 0.7 percent in mining; 0.4 percent in finance and insurance; 0.1 percent in electricity, gas, and water; and 0.1 percent in other sectors.

In 1985 the work force increased at an annual rate of 4 percent. During the REPELITA V plan, the work force is expected to increase at an average annual rate of 3.0 percent, with 2.2 percent in Java and 4.2 percent in the Outer Islands (Government of Indonesia/Department of Information, 1989).

Agriculture

Indonesian statistics on food crop production distinguish between production of wet paddy rice, dryland rice, and secondary crops, such as maize, cassava, sweet potatoes, peanuts, and soybeans. The agricultural survey of 1985 provided annual statistics for food crop production (Table 1).

Milled rice is a staple food in Indonesia. Milled rice production more than tripled in 40 years (1950 to 1987); consequently, rice imports have decreased, whereas the per capita supply of rice has almost doubled. Production and imports of milled rice from 1950 to 1987 are given in Table 2. A detailed account of milled rice production and imports from 1981 to 1987 has been compiled by Sadikin (1990) and is presented in Table 3. In 1985, Indonesia became self-sufficient in rice production. This balanced situation has mostly been maintained.

Agricultural (including forestry) product exports include rubber, tea, coffee, oil palm, tobacco, white and black pepper, and timber mainly as plywood. Exports totaled about 4 million metric tons in 1988 (Biro Pusat Statistik [Central Bureau of Statistics], 1988).

TABLE 1   Production of Food Crops in Indonesia, 1985

Food Crop	Area Harvested (ha)	Total Production (metric tons)
Wet paddy rice	8,755.721	37,027.443[a]
Dryland rice	1,146.572	2,005.502[a]
Maize	2,439.966	4,329.503
Cassava	1,291.845	14,057.027
Sweet potatoes	256.086	2,161.493
Peanuts	510.037	527.852
Soybeans	896.220	869.718

[a]Unmilled rice.

SOURCE: Summarized from Biro Pusat Statistik (Central Bureau of Statics). 1989. Input-Output Table 1985. Jakarta: Biro Pusat Statistik.

TABLE 2   Average Production and Imports of Milled Rice, 1950–1987

Period	Metric Tons (in millions) Production	Imports	Per Capita Supply (kg)
1950–1960	7.26	0.56	86.09
1961–1970	9.79	0.59	91.20
1971–1980	15.93	1.31	120.80
1981–1987	25.12	0.37	148.57

SOURCE: Biro Pusat Statistik (Central Bureau of Statistics). 1988. Statistik Indonesia. Statistical Year Book of Indonesia 1988. Jakarta: Biro Pusat Statistik.

TABLE 3   Production and Import of Milled Rice, 1981–1987 (in Thousands of Metric Tons)

Year	Production	Import
1981	22,236	530
1982	23,007	300
1983	24,006	1,160
1984	25,932	380
1985	26,542	0
1986	27,014	0
1987	27,253	0.05

SOURCE: Sadikin, S. W. 1990. The diffusion of agricultural research knowledge and advances in rice production in Indonesia. Pp. 106–123 in Sharing Innovation. Global Perspectives on Food, Agriculture, and Rural Development. Washington, D.C.: Smithsonian Books.

Forest Resources

The forests of Indonesia can be classified into the following 10 aggregations on the basis of the characteristics of their vegetation (Government of Indonesia/Ministry of Forestry and Food and Agriculture Organization of the United Nations, 1990):

- Coastal forests on beaches and dunes;
- Tidal forests, including mangrove, nipa, and other coastal palms;
- Heath forests associated with sandy, infertile soils;
- Peat forests associated with organic soils with peat layers at least 50 cm deep;
- Swamp forests seasonally inundated by fresh water;
- Evergreen forests, including moist primary lowland, riparian, and dry deciduous forests;
- Forests on rocks that contain basic (pH more than 7) minerals (for example, hornblend, augite, biotite, and plagiolass);
- Mountain forests (at elevations above 2,000 m);
- Bamboo forests; and
- Savannah forests.

DESIGNATED FORESTLANDS

Records from the Tata Guna Hutan Kesepakatan (TGHK; Forest Land Use by Consensus) inventory indicate that areas designated as forestlands cover 144.0 million ha, about 74 percent of the total land area of Indonesia. They are subdivided into the following four forest classes: conservation forests (18.8 million ha), protection forests (30.3 million ha), production forests (64.4 million ha), and conversion forests, including some unclassified forestlands (30.5 million ha) (Government of Indonesia/Ministry of Forestry and Food and Agriculture Organization of the United Nations, 1990). These functional classes are not demarcated on the ground, and forestlands have been used for other purposes, for example, human settlements as a result of transmigration, mining, and agricultural perennial crops.

Forestland on Java (about 3 million ha) is legally declared as such and is referred to as "gazetted" (set-aside) forestland and is demarcated in the field. Most of the TGHK forestland outside Java is in the process of becoming legal forestland (pregazetted—that is, preset-aside—forestland). Of the 144.0 million ha comprising the four forest classes, only 109 million ha has forest cover at present. This constitutes 9 to 10 percent of the world's total area of closed tropical forests (Government of Indonesia/Ministry of Forestry and Food and Agriculture Organization of the United Nations, 1990). The distributions

TABLE 4   Distribution of Forest Classes among Various Indonesian Islands (in Thousands of Hectares)

Island	Permanent Forest	Production Forest	Total for Permanent and Production Forests	Conversion Forest[a]	Total
Sumatra	10,777	14,399	25,176	5,032	30,208
Java and Madura	999	2,014	3,013	0	3,013
Kalimantan	11,025	25,650	36,675	8,293	44,968
Sulawesi	5,274	6,018	11,292	1,587	12,879
Bali/Lesser Sunda	2,016	1,349	3,365	3,008	6,373
Maluku Island	1,991	3,106	5,097	436	5,533
Irian Jaya	16,960	11,856	28,816	11,775	40,591
Total	49,042	64,392	113,434	30,131	143,565
Percent[a]	34.2	44.9	79.0	21.0	100.0

NOTE: Percent totals may not add to 100 because of rounding.

[a]Conversion forests are forests on government lands that can be converted to other uses, such as agriculture, industry, and settlements, after the removal of timber or with the approval of the government.

SOURCE: From Statistik Kehutanan Indonesia, 1982/1983 Department of Forestry, Jakarta, 1984. In Government of Indonesia/International Institute of Environment and Development. 1985. A Review of Policies Affecting the Sustainable Development of Forest Lands in Indonesia. Jakarta: Government of Indonesia.

of TGHK forests among various islands of Indonesia are given in Table 4.

PRODUCTION FORESTS

Major timber products from forests used for tree production (production forests) outside Java are mainly members of the family Dipterocarpaceae and include the genera *Shorea, Hopea, Dipterocarpus, Dryobalanops, Anisoptera, Parashorea,* and *Vatica.*

Satellite imagery, aerial photographs, and terrestrial inventories indicate that of the area designated as production forests, only 39,200 million ha (60.90 percent) is productive. The remaining 25,200 million ha (39.10 percent) is no longer productive (Prastowo, 1991). The TGHK area of permanent-production forests is 33.9 million ha, of which 21.0 million ha (52.0 percent) is productive. The TGHK area of limited-production forests is 30.5 million ha, of which 18.2 million ha (48.0 percent) is productive (Prastowo, 1991).

According to various surveys, potential production in limited- and permanent-production forests is as follows. In Java and Madura, the production forest extends to 1.9 million ha, consisting of tree plantations of, for example, the genera *Pinus, Agathis, Swietenia, Dalbergia,* and *Altingia,* with an average production potential of 908.773 m³/ year from a harvested area of 50,549 ha/year. Of the 66.6 million-ha concession area (forestlands leased to private companies for 20 years for logging and replanting) in the Outer Islands, 56.3 million ha is productive forest and is located in production and conversion forests (a conversion forest is forest on land that can be used for other purposes, for example, agriculture, settlements, or industry). The average production potential of a stand of a commercial species with diameters of ≥50 cm is more than 90 m³/ha for species consisting mostly of the dipterocarp family but including members of the genera *Agathis* and *Gonystylus,* among others. The largest standing volumes are in the provinces of Kalimantan Timur (1,751 million m³), Kalimantan Tengah (764 million m³), Irian Jaya (661 million m³), Kalimantan Barat (476 million m³), and Riau (365 million m³).

Ecologic Characteristics and Issues

Indonesia is outstandingly rich in plants and animals. Only 1.3 percent of the earth's land surface is occupied by Indonesia; yet 10 percent of the world's plant species, 12 percent of the world's mammal species, 16 percent of the world's reptile and amphibian species, and 17 percent of the world's bird species can be found in Indonesia (Government of Indonesia/Ministry of Forestry and Food and Agriculture Organization of the United Nations, 1990). Therefore, Indonesia has a great responsibility to maintain the biodiversity found in that country. For that purpose Indonesia has promulgated laws and regulations pertaining to the protection of nature (these are discussed in greater detail later in this profile) and has earmarked 341 locations (a total of 13 million ha) as conservation forests or protected areas. Nevertheless, many species in Indonesia are already threatened with extinction: 126 birds, 63 mammals, and 21 reptiles.

BIOGEOGRAPHICAL DIVERSITY

Indonesia also has a famed diversity of ecosystems—from the ice fields of Irian Jaya to a wide variety of humid lowland forests, from deep lakes to shallow swamps, from coral reefs to mangrove forests. Indonesia also has valuable genetic resources.

Indonesia is not a uniform country, as demonstrated by the 416 land systems identified in the Regional Physical Program for Trans-migration report (1990). This biogeographical diversity is reflected in its biologic resources. For example, the Sulawesi-Maluku-Lesser Sunda area, known as the "wallacea area" (named for the nineteenth century British biologist Alfred Wallace), is biologically complex. It is characterized by animals that are neither particularly Asian nor particularly Australian but, rather, commonly unique to a single is-land. There is much concern about the degraded ecologic conditions resulting from shifting (slash-and-burn) cultivation and forest clear-ing in mountainous areas for use as agricultural land—conditions such as the formation of large areas of alang-alang (*Imperata cylindrica*) fields in the Outer Islands and accelerated soil erosion in the upland areas of Java. These concerns are described in detail below.

The Alang-Alang Problem   Alang-alang is a notorious weed found in the humid tropics. It is known as lalang in Malaysia and as blady grass in Australia. Alang-alang is a climax plant community that spreads rapidly after burning of the land, maintaining its dominance in the ecosystem. About 15 million ha (8 percent of Indonesia's land area) is classified as alang-alang fields. Although Irian Jaya contains more alang-alang than the other provinces do, the Sulawesi, Sumatra Utara, Kalimantan Selatan, and Timor Timur regions are most criti-cally affected by alang-alang vegetation.

Soil Erosion   The problem of soil erosion has attracted public atten-tion since the middle of the nineteenth century, when there was heavy flooding of some rivers in Java and the emergence of critically de-graded lands (Utomo, 1989). It was assumed that the floods were caused by excessive clearing of forested areas for the establishment of large agricultural estates in upland areas, thus critically degrading the land. Sukartiko (1988) reported on the alarming erosion rates of soils in the watershed areas of some rivers in Java and Sumatra. They varied from 1.28 mm/year in the Asahan watershed in Sumatra to 8.0 mm/year in the Cisanggarung watershed in Java. Erosion has also caused sedimentation in reservoirs and irrigation systems and a subsequent loss of their water-holding capacities.

Economic Activity

The following data were derived from a joint report of the Gov-ernment of Indonesia/Ministry of Forestry and the Food and Agri-culture Organization (FAO) of the United Nations (1990) relating to the situation and outlook for forestry in Indonesia.

Indonesia's gross domestic product (GDP) in 1987 amounted to

114.5 trillion rupiah (Rp) (US$69.4 billion). From 1965 to 1980, Indonesia's GDP grew at an average annual rate of 7.9 percent (in U.S. dollars). From 1980 to 1986, annual GDP growth averaged 3.4 percent (World Bank, 1989). Indonesia's economy was actually in recession in 1982, with the GDP declining 2.2 percent (Government of Indonesia/Department of Information, 1989). Further declines because of declines in oil prices were observed in 1985 and 1986.

Indonesia is still an agricultural country, despite the sharp decline in the contribution of the agricultural sector to the country's GDP. In 1961, the agricultural sector contributed 47 percent of the GDP, but its contribution declined to 26 percent in 1986. As an oil-exporting country, oil has been one of Indonesia's main sources of foreign exchange. The mining and the oil and gas sectors increased their contributions to GDP from 12.3 percent in 1973 to 19 percent in 1983; this declined to 13.5 percent in 1986.

The various regions of Indonesia have developed at different rates. The fastest-growing area has been the island of Bali, with a GDP annual growth rate of 13.3 percent from 1980 to 1986, while the Riau archipelago has a recessionary economy, with a negative annual growth rate of –7.4 percent.

Further industrialization is a national goal for the REPELITA V plan. The annual growth target for the manufacturing sector is 8.5 percent, while that for the agricultural sector is 3.6 percent. Another goal is to further diversify the manufacturing sector away from oil. Although the target for the oil and gas sectors is an annual increase of 4.2 percent, the target for the non-oil and gas sectors is 10 percent annually.

Indonesia's average per capita income in 1988 was US$440. Indonesia ranked close to last in the lower-middle income country groupings, behind the Philippines and Papua New Guinea (World Bank, 1989). Among 120 reporting countries, however, Indonesia had the eighth fastest rate of growth in income per capita from 1965 to 1986. The target for per capita GDP growth for the REPELITA V plan is approximately 3.1 percent per annum.

Total domestic investment amounted to 20 trillion rupiah in 1986, which was 20.7 percent of GDP. Private investment contributed 48 percent of total investment in 1985–1986 and 57 percent in 1988–1989. Annual fixed investment grew considerably (11.7 percent) from 1971 through 1981, but registered negative growth (–0.5 percent) from 1981 to 1988 because of the contraction of public investment. Investment growth recovered considerably in 1988 (Government of Indonesia/ Ministry of Forestry and Food and Agriculture Organization of the United Nations, 1990).

Economic Importance of Forestry

During the last 25 to 30 years there has been rapid change in the forestry sector in Indonesia. During the early 1960s timber production was confined mostly to teakwood in Java and a limited number of valuable wood species in the more accessible natural forests in the Outer Islands. Since then, most forestry activities have moved from Java to the Outer Islands.

TIMBER PRODUCTION AND DEVELOPMENT OF PRIMARY WOOD-BASED INDUSTRIES

During the past 30 years annual log production increased from about 2 million to 36 million m^3, originating mostly (96 percent) from the natural forests. This has resulted in an increase in the number of processing units, mostly sawmills and plywood mills, and in the volume of manufactured wood-based products (Government of Indonesia/Ministry of Forestry and Food and Agriculture Organization of the United Nations, 1990).

Prastowo (1991) reported on the development of log and lumber production from 1969–1970 to 1988–1989 (Table 5). The development of wood processing industries, in particular sawmills and plywood mills, is described in Table 6. In 1973 there were 14 sawmill units with a rated capacity of 200,000 m^3/year. This total grew to 364 units in 1988 with a capacity of 11,400,000 m^3, a growth of 26 times in the total number of units and 57 times in capacity. Plywood mills increased from 2 units in 1973 to 114 units in 1988 (57-fold growth), and the capacity went from 28 m^3 in 1973 to 9,013,000 m^3 in 1988 (321-fold growth).

DEVELOPMENT OF SECONDARY WOOD-BASED INDUSTRIES

The development of primary industries (sawmills and plywood mills) was considered satisfactory up to the end of the REPELITA IV plan. Secondary wood-based industries, such as pulp and paper, furniture, and other woodworking industries, are now on the agenda for development. The objective is to obtain more added value and to expand employment opportunities.

The production level for furniture and other woodworking industries in 1986–1987 was 1,494,178 m^3. This increased to 1,904,231 m^3 in 1988–1989 (Prastowo, 1991). Faster development of secondary wood-based industries is anticipated in the years to come, as was experienced with the plywood industry.

TABLE 5   Development of Log and Lumber Production
(in Thousands of Cubic Meters)

Year[a]	Logs	Lumber
1969–1970	6,206	177
1970–1971	10,899	1,164
1971–1972	13,706	998
1972–1973	17,717	1,037
1973–1974	26,297	1,350
1974–1975	21,752	1,819
1975–1976	16,296	2,400
1976–1977	21,428	3,000
1977–1978	22,939	3,500
1978–1979	26,256	1,512
1979–1980	24,557	1,637
1980–1981	23,995	1,793
1981–1982	14,024	2,659
1982–1983	13,377	3,686
1983–1984	15,209	2,711
1984–1985	15,958	2,119
1985–1986	14,551	2,643
1986–1987	19,758	7,442
1988–1989	27,566	9,750

[a]Data for 1987–1988 were not included in the original source.

SOURCE: Prastowo, H. 1991. The system of production forest management in the future. In Homecoming Day Alumni VIII/1991. Faculty of Forestry, Bogor Agricultural University, Bogor, Indonesia.

The growth of the pulp and paper industry is also promising. In 1979–1980 the production level was 220,000 metric tons, which increased to 600,000 metric tons in 1986–1987. At the beginning of 1990 there were 43 pulp and paper mills, with an annual capacity of 1 million metric tons of pulp and 1.7 million metric tons of paper (Prastowo, 1991). Indonesia is ambitiously trying to become one of the world's largest pulp and paper producers. To achieve this goal, the government has embarked on the large-scale development of forest industrial plantations, which are expected to become the main source of raw materials for the pulp and paper industry.

Contribution of Forestry to the National Economy

Forestry, together with downstream forest-based industries, has become an important sector in the Indonesian economy, even without considering the various nonmarket benefits arising from forests and

TABLE 6   Development of Sawmills and Plywood Mills in Indonesia, 1973–1988

Year	Sawmill		Plywood	
	Number of Units	Capacity (1,000 m^3)	Number of Units	Capacity (1,000 m^3)
1973	14	200	2	28
1974	31	300	5	103
1975	54	400	8	305
1976	65	1,500	14	405
1977	81	1,800	17	535
1978	121	3,200	19	799
1979	145	4,100	21	1,809
1980	188	5,500	29	1,989
1981	239	7,100	40	2,601
1982	257	7,600	61	3,292
1983	286	8,500	79	4,477
1984	294	8,700	95	5,327
1985	297	9,600	101	6,228
1986	331	10,500	111	6,500
1987	364	11,400	114	8,130
1988	364	11,400	114	9,013

SOURCE: Prastowo, H. 1991. The system of production forest management in the future. In Homecoming Day Alumni VIII/1991. Faculty of Forestry, Bogor Agricultural University, Bogor, Indonesia.

forest activities. In 1987 forestry contributed 1.2 percent to the Indonesian GDP, and the forest-based industries contributed another 1.5 percent, bringing the total to 2.7 percent. That same year, agriculture and fishing contributed 25.5 percent, oil contributed 14 percent, and non-oil manufacturing contributed another 13.9 percent to the GDP.

Forestry has been particularly important for foreign exchange earnings. In 1987 forestry and the forest industries led to export revenues of US$2.7 billion, or 16 percent of the value of Indonesia's total exports. In the same year, agriculture and fisheries contributed 19 percent of the value of Indonesia's total exports, non-oil manufacturing contributed 15 percent, and petroleum and gas contributed 41 percent.

Among the various forestry-based industries, the plywood industry is the most important one, making up 52 percent of the total contribution of the forest industry to Indonesia's GDP. Sawn wood and other wood products contributed 37 percent, and pulp and paper contributed 11 percent.

As a result of limitations on log exports and a later total ban on log exports, Indonesia's sawmill and plywood industries grew dramatically from 1980 to 1987. Exporters were able to penetrate world markets, and Indonesia is now a dominant exporter, accounting for 48 percent of the world's plywood market and 17 percent of the nonconiferous sawn wood market.

Other products, such as rattan, are also important sources of foreign exchange. Furthermore, large numbers of forest dwellers and rural people eke out livelihoods and earn cash incomes by extracting products from the forests.

The benefits arising from the environmental functions of forests are also important. These functions include regulation of river water flow, which prevents floods in the wet season and water shortages in the dry season; control of soil erosion; curtailing extreme temperatures and reducing wind velocities in and around forests (producing a more favorable microclimate); and oxygen production and carbon dioxide utilization, which mitigate the effects of the greenhouse effect. However, there is no adequate mechanism to quantify these functions (Government of Indonesia/Ministry of Forestry and Food and Agriculture Organization of the United Nations, 1990).

HISTORICAL ASPECTS AND CAUSES OF DEFORESTATION

In this profile, the term *deforestation* means the removal or destruction of all or most of the trees of a forest such that reproduction is impossible except by artificial means. Deforestation is also used to refer to the loss of natural forest cover. In Indonesia, deforestation includes conversion of forestlands into estate crops (large tracts of land [200 to 300 ha] on which crops such as tea, rubber, coconut, oil palm, and cacao are cultivated), as well as clearing of forestlands for settled agriculture (farming of the same piece of land without fallow periods); shifting cultivation; and such things as human settlements, infrastructure, and mining. Indiscriminate and excessive logging may also cause deforestation.

Reforestation in Indonesia means the planting of trees on bare forestlands, that is, land designated by law as permanent forest. *Regreening* means the planting of trees on private land.

Rates of Deforestation

Average rates of deforestation in Indonesia (by island) were computed by using observations of forest cover from various assessments carried out between 1950 and 1984. The rates of deforestation (Table 7)

TABLE 7   Average Deforestation Rates in Indonesia, by Island

Province or Island	Period	Loss of Forest Cover (percent/year)
Sumatra	1950–1984	−1.30
Kalimantan	1950–1982	−0.49
Sulawesi	1950–1982	−0.82
Lesser Sunda	1950–1982	−0.35
Maluku	1950–1982	−0.44
Irian Jaya	1950–1982	−0.50
Outer islands except Timor Timur	1950–1982	−0.71

NOTE: The compounded rate of deforestation between 1950 and 1982 given here is based on regression analysis by using different sources of information for different years.

SOURCE: Government of Indonesia/Ministry of Forestry and Food and Agriculture Organization of the United Nations. 1990. Situation and Outlook of the Forestry Sector in Indonesia. Jakarta: Government of Indonesia.

measure the average percent decline in area under forest cover. For Indonesia the average was an annual decline of 0.71 percent.

The total annual rate of deforestation was estimated to be about 300,000 ha in the early 1970s and about 600,000 ha in the early 1980s. Using the estimates of smallholder conversion, shifting cultivation, development projects, poor logging practices, and losses caused by fire, the World Bank (1989) estimated a deforestation rate of between 700,000 and 1,200,000 ha in 1989, or an average of 1.2 percent per year (Table 8) (Government of Indonesia/Ministry of Forestry and Food and Agriculture Organization of the United Nations, 1990). The estimated area of closed forests (forests in which the tree canopies completely cover the land) was 109 million ha in 1990 (Government of Indonesia/Ministry of Forestry and Food and Agriculture Organization of the United Nations, 1990).

Population Pressure and Demand for Agricultural Land

In principle, deforestation can be seen to be a result of demand for agricultural land, depending on a variety of factors. In a developing country such as Indonesia, population pressure is one of those factors. Other factors may be also important. In communities where there is industrial development and a nonsubsistence economy (an economy in which not only basic needs but also nonbasic needs such as a higher standard of living, education, and recreation are fulfilled), demand for agricultural land is lower because there are sources of income other than those from farming activities. In a market economy, food can readily be imported and exchanged for other goods produced in the country.

Furthermore, economic development brings about changes in the structure of demand, away from food commodities and toward industrial goods. When the manufacturing sector grows faster than the agricultural sector, there is increasing urbanization. Thus, higher per capita income is likely another explanation for the decline in deforestation trends. In developed countries, for example, deforestation has stopped, and in many cases the forestland base is increasing.

Income disparities also play an important role in deforestation. Thus, if increases in per capita income are not evenly distributed, the pressure on forestland from the rural poor and land-hungry farming communities may continue (Government of Indonesia/Ministry of Forestry and Food and Agriculture Organization of the United Nations, 1990). Gains in agricultural productivity, if coupled with economic development, may reduce the demand for agricultural land by releasing farm labor to move to other sectors of the economy and contribute further to urbanization. In this case, a smaller amount of land is required to produce the same amount of food, and deforestation pressures are reduced. If sectors of the population do not have means of economic survival other than working the land, pressures on forestlands continue independent of gains in food production.

A classic example of deforestation brought about by population pressures and demand for agricultural land is that of the islands of Java and Bali. Deforestation in Java and Bali started some 300 to 400 years ago. At the end of the nineteenth century, forestland was pushed

TABLE 8   Sources of Deforestation (in Thousands of Hectares per Year)

Source	Best Estimate	Range
Smallholder conversion	500	350–650
Development projects	200	200–300
Logging practices	80	80–150
Fire loss	70	70–100
Total	950	700–1,200

SOURCE: Government of Indonesia/Ministry of Forestry and Food and Agriculture Organization of the United Nations. 1990. Situation and Outlook of the Forestry Sector in Indonesia. Jakarta: Government of Indonesia.

back to the summits and higher slopes of the mountains to provide land for agriculture. The lower hill areas of the northern parts of central and east Java were unaffected, however. Since the seventeenth century, the United Dutch India Company maintained teak forests to provide timber for their merchant fleet and for use as merchandise in their Asian-European trade. Using the domain principle, which was part of the Dutch Agrarian Law of 1870, the Dutch Indian government declared that the remaining forested area was classified as forestland, demarcating it with boundary poles in the field. Today, Java has about 3 million ha of forestland, which is about 22 percent of the island's land area.

In the meantime, pressure on forestlands continues to increase with increases in population density (on average, 768 inhabitants per km^2 at present). As a consequence, large areas of forestland are used for agriculture. According to Perum Perhutani (State Forest Corporation), the total area of critically degraded forestland in Java is estimated to be 230,000 ha. In addition to other efforts through social forestry programs, serious efforts are being made to reforest the critical forestlands and to regreen degraded agricultural lands.

In the lower parts of the island of Java, in particular, which have sufficient water supplies, wet paddy rice fields, a productive and sustainable form of agriculture, have been constructed. Rice production has increased manyfold in the past 20 years because of improved rice cultivation technology, including the use of high-yield varieties, fertilizers, and insecticides; support by soft loan credits (money lent on favorable terms from government banks) for operational costs as well as for seeds, fertilizers, and insecticides; and a well-organized extension network. However, this increase in the productivity of wet paddy rice fields cannot prevent landless farmers from looking for more land to farm, even on steep slopes. To prevent further degradation of the natural resources in Java, two strategies are used by the Indonesian government: soil conservation in the uplands and transmigration of needy farmers to the Outer Islands.

Logging in Natural Forests

Increased exploitation of natural forests in the islands outside Java was stimulated by the enactment of laws on foreign capital (Law No. 1, 1967) and domestic capital investment (Law No. 6, 1968). Through these laws, the government of Indonesia opened the possibility of forest exploitation to foreign as well as domestic private companies by providing incentives such as tax exemptions. Forest concessions are granted under a right for forest exploitation (Hak Pengusahaan

Hutan [HPH]) after the application for concession is approved in a Forestry Agreement contract, in which the rights and obligations of the HPH holders are stipulated. For example, companies are required to pay license fees and royalties and are obliged to adhere to proper and sustainable forest operations.

Within the HPH system, based on the actual conditions and needs of the forests, the forests are managed under a combination of the following three systems:

- Tebang Pilih Indonesia (TPI), Indonesian selective cutting system;
- Tebang Habis dengan Permudaan Alam (THPA), clear-cutting with natural regeneration; and
- Tebang Habis dengan Permudaan Buatan (THPB), clear-cutting with artificial regeneration.

In practice, however, the TPI system is mostly practiced in the management of natural forests by HPH holders. The TPI system assumes a 70-year rotation, and harvesting is carried out on 35-year cutting cycles. Trees must have a minimum diameter of 50 cm, measured over the bark, before they can be cut. In the cutting area, at least 25 trees with diameters of more than 20 cm must be kept for regeneration purposes. If there are fewer than 25 remaining mother trees (trees for seed production), enrichment planting (planting of additional seedlings) must then be carried out.

Other provisions that must be observed for sustainable forest management include determination of the annual allowable cut by the Ministry of Forestry—in consideration of the existing standing stock—and prescription in a forest management plan of pre- and postfelling inventories as well as postlogging silvicultural treatments and tending of regeneration and advance growth.

There were serious lapses, however, in the implementation of the TPI system. Several evaluations carried out over the past 4 to 5 years indicated that in general the production forests are managed inadequately and improperly (Government of Indonesia/Ministry of Forestry and Food and Agriculture Organization of the United Nations, 1990). Logged-over stands are frequently damaged, sometimes by up to 60 percent. Moreover, many license holders select only the most valuable trees ("creaming"), and exceed the allowable annual cutting area, so that the whole concession area is logged over after 20 years instead of the prescribed 35 years.

As a consequence, degradation of the growing stock in many concession areas has taken place. In addition, ill-designed skid and logging roads have contributed to the acceleration of erosion rates. The same logging roads are also frequently used by migrants to gain access to land for shifting cultivation. Therefore, logging operations in natural forests can directly or indirectly cause environmental degradation and, in some cases, outright deforestation.

In 1991, the TPI system was replaced with the Indonesian selective cutting and planting system (the TPTI system), which places greater emphasis on forest regeneration. The effectiveness of the TPTI system has not yet been evaluated because of its recent implementation.

Shifting Cultivation

Shifting or slash-and-burn cultivation, in general, is regarded by some as a menace to the environment, a harmful practice that causes widespread deforestation and erosion. Others view shifting cultivation as the benign and productive use of poor soils by those who live under poor socioeconomic conditions.

Because of the increasing numbers of the rural population who have no secure access to land, many people have become shifting cultivators. These landless people do not practice a form of shifting cultivation based on cultural heritage, nor do they have any local community or legal system that provides them with the ability to use sustainable (perpetually productive and ecologically sound) agricultural practices. As a result, their shifting cultivation activities are detrimental to forestlands. The problem is further exacerbated when these "transitional" shifting cultivators work for an urban-based entrepreneurial system that employs them to carry out shifting cultivation. These transitional shifting cultivators have access to chain saws and outboard motors, which they use to cut primary forests to produce surplus products for nearby markets. After 2 to 6 years of shifting cultivation, these areas are often degraded into alang-alang grasslands. A cropping phase that is too long and a fallow period that is too short result in rapidly declining crop yields, loss of soil nutrients, and soil erosion. Greater population pressure has also stimulated spontaneous migrant cultivators who convert (primary) forestland to land on which destructive forms of shifting cultivation is practiced.

According to estimates of the Ministry of Forestry of the government of Indonesia, 10 percent of the total forestland in Kalimantan is degraded because of shifting cultivation. The areas of the forest under shifting cultivation and the total number of households that practice shifting cultivation on islands outside Java (except Irian Jaya) are given in Table 9. Forest losses resulting from shifting cultivation are estimated to be between 300,000 and 500,000 ha annually (Govern-

TABLE 9   Forest Area Under Shifting Cultivation and Shifting Cultivation Households

| Island | Total Forest-land Area (1,000 ha) | Shifting Cultiva-tion Area (1,000 ha) | Percentage of Forestland Under Shifting Cultivation | Households Practicing Shifting Cultivation | |
				Number (1,000s)	Percent
Sumatra	30,208	924	3.1	262	4.9
Kalimantan	44,968	4,477	10.0	228	17.1
Sulawesi	12,879	1,352	10.5	244	12.7
Lesser Sunda	5,547	568	10.2	251	23.0
Total	93,601	7,321	7.8	985	9.2

SOURCE: Government of Indonesia/International Institute of Environment and Development. 1985. A Review of Policies Affecting the Sustainable Development of Forest Lands in Indonesia. Jakarta: Government of Indonesia.

ment of Indonesia/Ministry of Forestry and Food and Agriculture Organization of the United Nations, 1990; World Bank, 1989).

Transmigration Program

Since 1969, some 613,700 families have transmigrated to islands other than Java (Sumatra, Kalimantan, Sulawesi, and Irian Jaya). Each family receives 2 ha of dryland (or 2.5 to 3.0 ha of land in wetland reclamation areas in Kalimantan and Sumatra, where conditions are less favorable) under the sponsorship of the Indonesian government. Most of this land originated from forestland. By the year 2000, an estimated 10 million families are expected to be transmigrated and settled on these islands. This means that 20 million ha of predominantly forested land will likely be transformed into agricultural land by the end of the century. Parts of the 30.5 million-ha conversion forest could be used for this purpose.

Table 10 shows the average annual amount of forestland that was released to the transmigration program during REPELITA III and REPELITA IV (1979–1980 to 1988–1989).

In theory, the transmigration program should result in systematic sustained and productive development of the land in the underpopulated Outer Islands. In the first 2 to 4 years, the most common use of the opened land areas has been for continuous cultivation of traditional food crops, predominantly upland rice. Most of the soils of the newly opened upland are not fit for that purpose. The soils in the area consist

TABLE 10 Annual Average of Forestland Released to Transmigration Program During REPELITAS III and IV (in Hectares)

Island or Province	Limited-Production Forest	Permanent-Production Forest	Conversion Forest	Unclassified Forest	Protection Forest	Total
Sumatra	24,223	42,712	116,354	101,143	8,655	293,087
Lesser Sunda						
Barat	273	173	2,940	—	90	3,476
Timor	—	87	1,578	250	15	1,930
Kalimantan	23,090	59,385	86,501	63,542	3,040	235,558
Sulawesi	8,485	10,710	17,350	34,485	3,285	74,315
Maluku	3,835	1,597	18,889	—	—	24,321
Irian Jaya	6,118	14,044	51,534	688	1,157	73,541
Total	66,024	128,708	295,146	200,108	16,242	706,228

SOURCE: Government of Indonesia/Ministry of Forestry and Food and Agriculture Organization of the United Nations. 1990. Situation and Outlook of the Forestry Sector in Indonesia. Jakarta: Government of Indonesia.

of Latosols (Oxisols) and red-yellow podzols (Alfisols, Ultisols), which are moderately to highly acidic (pH 4 to 5). Drainage is unusually poor, the mineral and organic content is low, the erosion rate is high, phosphorus-fixing capacity is high, and the aluminum concentration and levels of aluminum saturation are high (Kaul, 1990).

As a consequence, sustainable production of food crops, including rice, is not attainable without heavy inputs and proper soil and crop management. Maize, cassava, various legumes, amaranthus, chiles, eggplant, coffee, and some minor spice plants are grown in these continuous-cultivation cropping systems. Yields are generally very low because of weather conditions and the high incidence of pests. The lack of resources for mechanization or draft power and the high incidence of weeds (mainly alang-alang) have made the cropping systems nonsustainable in many transmigration areas (Kaul, 1990). Those constraints may lead to encroachment onto forested lands, which are generally preferred for use in these cropping systems.

Kaul (1990) also asserts that serious problems have arisen in wetland reclamation areas cleared for transmigration schemes in Sumatra and Kalimantan. Settlers are allocated about 2.75 ha of land, of which 0.25 ha is for home gardens, 0.75 ha is for dryland crops, and 1.75 ha is for tidally irrigated rice. They lay artificial drainage channels and remove commercially valuable tree species in an attempt to force the existing ecosystems to convert to irrigated rice fields, in some cases in association with coconut plantations.

The originally planned double rice cropping (two rice crops in one year) has been achieved in few locations because of the paucity of water during the dry season. The rapid deterioration of these ecosystems is traceable to the heavily eutrophic peat soils. In comparison, economic and sustainable yields of sago palm (*Metroxylon* sp.) have been obtained in permanently inundated swamp forests. According to Kaul (1990), coupling of irrigated rice to the transmigration program, particularly in swamplands, has been a mistake of the transmigration policy.

Tree Crop Development

Tree crop development has been carried out mainly through the Nucleus Agriculture Estates Program (NES/PIR). These programs are organized by the Directorate General of Estate Crops of the Ministry of Agriculture and through private and government agricultural estates.

The area of estate crops (rubber, oil palm, coconut, cacao, and other tree crops) in Sumatra, Java, Lower Sunda, Kalimantan, Maluku,

TABLE 11 Production of Tree Crops in Indonesia, 1988 (in Metric Tons)

Crop	Sumatra	Java	Kalimantan	Sulawesi	Others	Total
Rubber	896,926	98,548	191,549	1,616	650	1,189,289
Coconut	560,119	508,642	102,719	533,025	330,393	2,034,898
Palm oil	1,600,732	30,925	48,477	5,851	5,015	1,691,000
Coffee	275,592	67,526	5,160	15,326	30,778	394,382
Tea	28,494	115,367	—	—	—	143,861
Pepper	35,767	293	5,042	298	—	41,400
Clove	13,549	19,582	1,173	14,426	12,641	61,371
Cacao	18,997	16,114	441	10,959	7,333	53,844
Nutmeg	17,274	1,502	—	7,130	4,008	29,914
Cassia vera	17,274	700	16	—	10	18,000
Vanilla	22	343	—	3	1,180	1,548
Cashew nut	65	13,085	33	12,976	840	26,999
Percent	60.8	15.4	6.3	10.6	6.9	—

SOURCE: Kaul, A. 1990. Indonesian Farming Systems: Types and Issues. Unpublished manuscript.

and Irian Jaya established up to fiscal year 1987 was 11,572,337 ha. The area needed for rubber, oil palm, and coconut alone was 7,140,040 ha up to 1988.

Based on the development of all tree crops from 1984 to 1989, it is estimated that increases of 300,000 to 400,000 ha/year could be expected in future. (For tree crop production in 1988, see Table 11.)

Although only a small area of forest has so far been used for estate crop development, it is becoming increasingly difficult to earmark new lands, except forestlands, for estate crop development. Priority should be given to the development or upgrading of idle degraded land instead of the conversion of more forestland.

Fires

Fire is a great destroyer of forests. It leads to increased soil erosion, lowering of water quality, an erratic water supply, loss of species, less biodiversity, and the loss of genetic resources. Forest fires are more common in Java than they are in the Outer Islands, but fire control is better in Java. In most years, fires in Sumatra and Kalimantan are set by farmers to clear land that is neither marked as forestland nor actively protected. The enormous fire in Kalimantan in 1982–1983 was the result of a combination of climatic and biotic factors. These fires are unnatural in humid tropical forests and are stimulated by the drying that occurs because of shifting cultivation, cattle grazing activities, and forest plantations. They are also aggravated by smoldering fires in the arid peat soils and the coal layers in the subsoil. Webster (1984) reported that the great fire in Kalimantan in 1982–1983 destroyed plant and animal life over an area of 2,925,000 ha.

In October 1991, a fire also raged in parts of Kalimantan and Sumatra ignited by the same forces that ignited the one in 1982–1983—in particular, the long dry season of 1991. Tentative data indicate that 5,400 ha of industrial forests in Lampung, 7,600 ha of forestland in central Kalimantan, 7,000 ha in southern Sumatra, and 90,000 ha in eastern Kalimantan were damaged by fire (Kusumah et al., 1991).

PROGRAMS FOR SUSTAINABLE LAND USE DEVELOPMENT

In its broad and specific sense, sustainability has been discussed intensively in Indonesia over the past 5 years. At the broad conceptual level, it has been said that a sustainable society is one that satisfies its needs without jeopardizing the prospects of future generations. Sustainable land use development is geared to the attainment of these societal needs on a perpetual basis, that is, with due consideration of environmental conservation. Within the past 40 years, society has been warned of potential environmental collapse if economic development proceeds without considering the impacts of that development on the environment.

Deforestation, in the sense that it removes natural forest cover for other development purposes such as agriculture, human settlements, and infrastructure, is a logical process of development and can be justified if it is implemented in an orderly manner until forest areas considered sufficient to maintain an ecologic balance in watershed areas are obtained. This could be realized through a policy of designating permanent forestlands, which should then be managed on a sustainable basis.

Legislation and Policies on the Management of Forest Resources

The constitution of the Republic of Indonesia of 1945 (Article 33) states that land and water resources should be administered by the state and used for the greatest possible prosperity of the Indonesian people. The provisions in that article express the need for sustainable management of forest resources.

The basic principles for forest administration and forest management are laid down in a law (No. 5, 1967) concerning the basic provisions on forestry. The essence of the policy in that law (Article 9) states that, "The administration of forests has the objective to get maximum multipurpose and sustainable benefits, directly or indirectly, in the context of developing a just and prosperous Indonesian society based on Pancasila." The law also prescribes ways to make forestry plans and implement activities in forest utilization and protection. These activities are prescribed in more detail by the following government regulations: No. 22 (1967), concerning royalties and license fees for forest utilization; No. 21 (1970), concerning forest utilization and forest product harvesting rights; No. 33 (1970), concerning forest planning; and No. 28 (1985), concerning forest protection.

In the field of land tenure, a law (No. 1, 1960) concerning the basic principles of agrarian affairs was enacted. The law regulates the tenure rights of individuals as well as legal bodies. Because the land tenure and forestry laws overlap, compromises must be achieved on a local level. The law on land tenure as well as the forestry law recognize, in principle, the right to (forest) land tenure by local communities, provided that it is actually being practiced in the field and is not deemed to be contrary to the interests of the state.

In the field of nature conservation, the following laws have been enacted: the Law on Wild Animals, 1931; the Law on Natural Reserve and Wildlife Refuge, 1939; the Law on Hunting in Java and Madura of 1940. Other laws and regulations that cover broader areas have been issued: Law No. 4 (1982), concerning basic provisions of environmental management; Government Regulation No. 9 (1986), concerning environmental impact analyses; Law No. 5 (1990), concerning the conservation of biologic natural resources.

Government policies regarding the management of natural resources and environmental conservation for REPELITA V are stipulated in directives from the National Consultative Assembly (Majelis Permusyawaratan Rakyat) of 1988. Some of the points closely related to deforestation and ecologic sustainability are maintained in the following statements.

1. The natural resources of the country—whether they are on land, in the sea, or in the air; whether they are minerals, flora, or fauna; and including genetic resources—should be managed and used for the greatest possible benefit of the community. At the same time, the environment should always be preserved to produce the greatest possible advantage for development and public welfare for both present and future generations.

2. The exploitation of natural resources should be continued, by appropriate means, so that damage to the environment is minimal and the quality and conservation of resources and the environment can be assured. In this way, development can proceed unhampered.

3. Rehabilitation of degraded natural resources calls for a concerted approach to the problems of river basins. In this context, rehabilitation of forests and critical land areas; soil conservation; and rehabilitation of rivers, lakes, swamps, marshlands, and coral reefs should be intensified, while the function of river basins needs to be reinstated. To control the emergence of poor-quality forests and critical lands, measures should be taken to halt damage to forests and to improve the control of forests, dryland cultivation, and shifting cultivation. Reforestation activities should be increased to improve the productivity of forestlands and to save forest areas. Public participation in these activities should be encouraged.

These policies are translated into various development programs aimed at achieving environmental stability and sustainability and pertain to the ecologic, economic, as well as social aspects of development programs. These programs are carried out by the Indonesian government as well as by nongovernment organizations, including private companies, cooperatives, and self-help organizations. The programs can be placed into the following four broad categories: (1) conservation of forest ecosystems, (2) stricter control on logging operations in natural forests, (3) reforestation and regreening programs, and (4) rationalization of shifting cultivation in which the respective activities that are part of the shifting cultivation system are related to or mutually supportive of each other. For example, rationalization of shifting cultivation includes planting of industrial type crops to replace natural fallow vegetation, which supports the reforestation program, enhancing ecologic stability and sustainability and at the same time providing raw materials for wood-based industries. It also increases the incomes of the indigenous people involved in the program.

In addition to these programs, which are geared to the better use of forest resources and increases in agricultural productivity by extension of dryland agricultural areas, much has also been done in intensification of wet paddy rice agriculture to step up rice production, in Java in particular.

Designation of Permanent Forests

Principles for the designation of permanent forests were stipulated in an FAO paper in 1952 (Food and Agriculture Organization of the United Nations, 1952) in Basjarudan (1978), as follows:

• Each country must designate certain areas as forest area.
• The designation of forest areas must be done prudently, in accordance with the social economic policy of the country, with due consideration to other forms of land use.
• Forest areas must be protected against damage by humans or other agents, such as fire, pests, and diseases.
• Priority must be given to the protective function of forests; other functions can be defined.
• In the harvesting of forests, the best method of exploitation should be applied so that maximum yields can be obtained from the forest; harvesting should be carried out in an economic and efficient manner under a sustained yield principle.
• To facilitate the application of proper forest management principles, the status of the forest area must be classified as such; this must be followed by a demarcation of boundaries on a map and in the field.

Government Regulation No. 33, 1970, concerning forest planning described the steps required to prepare areas as permanent forestlands. After being given a legal designation as permanent forestland, the forests are classified according to their respective functions, that is, protection, production, and conservation (including wildlife refuges and national parks) forests. For forestlands in Java, this procedure has been followed since the 1890s. Approximately 3 million ha is currently classified as permanent forestland and work is continuing, in particular in a program that establishes settlements on disputed forestlands.

The designation and classification of permanent forestlands outside Java started in the 1980s through the forest use planning by consensus (TGHK) procedure after large-scale forest operations in concessions areas had begun 2 to 3 years earlier. The TGHK procedure is solely a desk exercise in which the boundaries of forestlands to be designated are drawn on maps after a consensus has been reached among the concerned government agencies. This procedure must be followed by work in the field, including negotiations with local communities, and placement of clear boundary markers, as has been done in Java. The work must be done consistently and intensively, and the work will take several decades to complete because of the large area involved (140 million ha).

Conservation of Forest Ecosystems

The government's policy for conserving forest ecosystems is based on the desire to promote the cultural and economic development of the Indonesian people in harmony with their natural environment. The policy states that all forms of natural life and examples of all ecosystems within Indonesia—in particular, air, water, soil, plants, and animals—must be protected for the benefit of future generations.

The main conservation policies can be summarized as follows:

• Nature reserves must be used rationally and wisely without jeopardizing their functions.
• Natural resources and the living environment should be managed wisely to provide maximum benefit for the people.
• Appropriate technology should be used to sustain the high quality of natural resources and the natural environment.
• Rehabilitation of damaged forests, degraded soils, and the water supply should be improved through integrated watershed and regional management approaches.
• Important marine and coastal habitats should be conserved.

The policy objectives of REPELITA V emphasize the proper utilization of natural resources as well as the need to:

• Further develop the ecotourism industry to increase foreign exchange earnings and initiate employment opportunities;
• Improve management of terrestrial and marine conservation areas;

- Increase the people's participation in conservation efforts;
- Increase the preservation of animal and plant species; and
- Control threats to forestry and forest security.

To conserve genetic resources, viable examples of all distinct ecosystems and species must be protected within a system of reserves. The types of protected natural reserves are as follows:

- Wildlife sanctuaries—medium-sized area, 200–1,600 km^2; relatively undisturbed, stable habitats of moderate to high conservation importance;
- National parks—medium- to large-sized areas, 500–7,000 km^2; relatively undisturbed areas of outstanding natural value with high conservation importance, high recreational potential because of easy access for visitors;
- Strict nature reserves—50–1,300 km^2; undisturbed fragile habitats of high conservation importance, unique natural sites, or homes of particular species;
- Hunting parks—medium- or large-sized area of natural or seminatural habitats with relatively easy access for hunters, with large populations of legal game species, for example, pigs, deer, and feral buffalos; of low conservation importance;
- Protection forests—medium- to large-sized areas of natural or planted forestlands on steep, high, extremely erodible lands that have high levels of rainfall thus making forest cover important to protect water catchment areas and prevent landslides and erosion;
- Natural recreation parks and grand forest parks—generally some disturbed areas designated for high-intensity use and limited ex situ genetic conservation; and
- Marine reserves—large-sized areas, 1,000–5,000 km^2.

Human settlements, food crop agriculture, and commercial logging are prohibited in all of the protected areas, but activities such as recreational camping and mineral exploration are permitted in wildlife reserves, and hunting is permitted in protection forests.

As of August 1990, there were 336 classified conservation areas with an area of 16.02 million ha (Government of Indonesia/Ministry of Forestry and Food and Agriculture Organization of the United Nationals, 1990).

Stricter Control on Logging Operations in Natural Forests

To induce more orderly forest operations, corrective measures are prescribed and stricter control on the implementation of the operations are exercised by the Provincial Forestry Service of the Ministry of Forestry. The TPI method is improved with the Tebang Pilih dan Tanam Indonesia (TPTI; Indonesian Selection Felling and Planting) system. TPTI is a silvicultural system that regulates tree felling and regeneration in natural production forests. The objective of the TPTI system is to utilize the forest and, at the same time, to qualitatively and quantitatively increase the value of the forest in the logged-over area for the next rotation period to ensure sufficient and perpetual production of raw material for the wood-based industries and to improve the protective value of production forests, for example, control of the water regime, minimization of soil erosion, and induction of the beneficial effects on micro- and macroclimates.

The silvicultural treatment consists of the following activities:

- Regulating the compositions of tree species in forest stands, which will be more beneficial from an ecologic as well as economic point of view;
- Developing an optimum stand density to produce more logs than in the previous rotation period;
- Enhancing the beneficial functions of the forest in soil and water conservation; and
- Boosting the protective functions of the forest.

To ensure strict and complete implementation of the TPTI system and to impose efficient and just disciplinary measures, the concession holders are classified as companies that have (1) not yet implemented the TPTI system; (2) implemented the TPTI system, but not correctly and completely, according to the rules; (3) correctly and completely implemented the TPTI system.

Penalties for failing to implement the TPTI system completely and correctly consist of, for example, reducing the annual production target or determining the annual production target without approving the annual working plans. For concessionaires that have approved annual working plans but fail to implement the TPTI system, the forest operation will be stopped if necessary. For companies that have implemented the TPTI system actively and strictly but have no wood-based industries or no stock relationships with wood-based industries in which the stocks (that is, shares) are partly or entirely owned by the concession holder, the concession certificate may be withdrawn. Companies that implement the TPTI system correctly and completely are eligible for an award from the government and to be named as a model company; the concession period may also be extended.

To implement the TPTI system correctly and completely, a climate of law and order must be created in the field. This means that the companies must be equipped with a clear working plan, must have a proper organization for forest development, must be supported by qualified personnel sufficiently trained in forestry, and must have financial support sufficient for an effective operation. Extension and supervision on the proper implementation of the TPTI system by well-trained and experienced forestry personnel is necessary to provide information and the necessary correction of the activities carried out by the concessionaires. Penalties must be imposed for every deviation in the implementation of the TPTI system in the field. These steps to ensure the continuity of forest production have already produced some satisfactory results.

Reforestation and Regreening

Reforestation activities have a relatively long history and tradition in Indonesia. Teakwood (*Tectona grandis*) was first planted in Java in 1880, and by the end of 1988 teakwood plantations covered about 0.88 million ha. *Pinus merkusii*, a pine indigenous to Sumatra, has been planted in Sumatra and Java since 1916. Large-scale plantations began in 1935, and in 1975 these were extended to Kalimantan, Sulawesi, and Bali. At the end of 1988, there were about 600,000 ha of pine plantations compared with about 134,000 ha of natural pine forests in Sumatra.

From the 1920s to the 1940s, other, mostly long-rotation, high-value timber species were planted in trial plots and pilot plantations on the islands of Java, Sumatra, Sulawesi, and Lesser Sunda. These species included mahogany (*Swietenia macrophylla*), rosewood (*Dalbergia latifolia*), New Giomea Kauri (*Agathis loranthifolia*), rasamala (*Altingia excelsa*), and black wattle (*Acacia decurrens*).

Since the 1950s, increasing population pressure in Java and parts of the Outer Islands have led to increased clearing of forests for cultivation and fuelwood, resulting in land degradation and soil erosion problems. This led, in the 1970s, to a program to establish fuelwood plantations on nearly all islands. Fast-growing species were mostly planted, including Kaliandra (*Calliandra* species), akasia (*Acacia auriculiformis*), kayuputih (*Melaleuca leucadendra*), lamtoro gung (*Leucaena leucocephala*), sengon (*Paraserianthes* [*Albizia*] *falcataria*), and turi (*Sesbania* species).

In 1980, the Indonesian government established Dana Jaminan Reboasasi (the Reforestation Guarantee Deposit Fund) to encourage establishment of forest plantations in timber concession areas. Concessionaires were required to contribute to the fund US$4/m^3 of logs and US$0.50/m^3 of harvested chipwood. Upon proper fulfillment of their regeneration and reforestation obligations, the concessionaires could claim reimbursement of their expenses from the fund. However, the fund did not generate interest among concessionaires to increase their reforestation efforts for two reasons: (1) the actual costs of reforestation were much higher than the level of reimbursement provided, and (2) the 20-year concession period did not provide sufficient tenure to justify investment in reforestation.

This situation, coupled with forecasts of a timber supply deficit in Indonesia from the year 2000 on, prompted the government to launch the Timber Estates Development Program in 1984. This program aimed to establish 4.4 million ha of new industrial plantation forests (Hutan Tanaman Industri [HTI]) for a total of about 6 million ha of such forests by the year 2000. The reforestation fund/fee for log harvests was increased to US$7/m^3 beginning July 1, 1990. The fund/fee was redesignated Dana Reboasasi (Reforestation Fund) to clearly reflect its purposes.

The major roles of forest plantations in the continued development of Indonesia can be summarized as follows:

- To increasingly take the pressure off natural forests;
- To meet the timber supply deficit from natural forests that is anticipated to occur within the next 5 to 10 years;
- To rehabilitate watersheds that have been extensively degraded by increasing population pressure, particularly in Java, Sumatra, and Lesser Sunda (in terms of the land area and population involved, this is a much greater issue than the development of timber estates); and
- To provide socioeconomic benefits; plantations can provide, in addition to soil and water protection, a wide range of wood and nontimber products—fuelwood, wood poles, wood posts, food, fodder, medicinal plants, and essential oils—for local communities, either for their own consumption or to generate income, employment, or both.

The targets for the year 2000 are very ambitious: about 6 million ha of industrial timber estates, about 13 million ha of critical watersheds to be regreened (in principle, on private land), and about 7 million ha of critical watershed to be reforested (on government forestlands). Of these targets, about 1.44 million ha of timber plantations had been established by the end of 1988, of which 1.36 million ha was on Java island, and since 1984, only about 69,000 ha had been established under the HTI program (the target was 1.5 million ha). About 5.8 million and 1.2 million ha of critical watersheds were regreened and reforested, respectively, by the end of 1988. Only 57 percent of reforestation plantations are estimated to be successful, whereas the survival of regreening plantations is reported to vary between 6 and 71 percent (Government of Indonesia/Ministry of Forestry and Food and Agriculture Organization of the United Nations, 1990).

Rationalization of Shifting Cultivation

Programs and projects that addressed the problem of shifting cultivation were started in the early 1970s. Descriptions of the various programs implemented by different agencies are given below. In this profile of Indonesia, *rationalization of shifting cultivation* means minimization of the adverse effects of shifting cultivation by introducing perennial crops (timber and other products—for example, fruit and bark—that can be used or sold at market) to replace the fallow natural vegetation (which is only slashed and then burned in the next cropping period), and better soil conservation techniques (agroforestry technologies) supported by intensive extension and training.

MINISTRY OF FORESTRY

The Ministry of Forestry has three programs that either directly or indirectly have an impact on shifting cultivation: (1) a program to control shifting cultivation, (2) a village development program, and (3) a social forestry program.

From its inception in 1971 until 1981 the development activities of the Forestry Department, which was then a part of the Ministry of Agriculture, in addressing issues affecting shifting cultivation were oriented toward resettlement (the ex situ approach). This resettlement program was generally considered to be unsuccessful. In some locations, resettled shifting cultivators moved back to their former places of residence. Beginning in 1981 the emphasis was changed to nonresettlement (the in situ approach). The program had three types of activities: provision of work for wage laborers in reforestation and industrial forest plantation programs, sedentary subsistence dry farming, and flooded rice farming. More people can be involved in in situ-type programs. By 1986, in situ-type programs included about 1,900 households involved in wet paddy rice agriculture, some 10,600 households involved in sedentary subsistence dry farming activities, and some 24,900 households involved in land rehabilitation activities (reforestation) or in the development of export crops.

The related training programs for cadres of people involved in sedentary agriculture also included cadres of people from the programs of other agencies, for example, the Nucleus Agricultural Estate

(NES) of the Ministry of Agriculture, the Resettlement Program of the Ministry of Transmigration, and the Sedentary Agriculture Program of the Ministry of Forestry.

The Social Forestry Program in Java began in 1984. It was developed from similar programs that started in the 1970s. The primary objective of this program is to induce sustainable forest management through successful forest plantations and to induce forest protection with the participation of local communities by providing them better incentives in the use of forestlands (agroforestry technologies) and forest products. By 1987, some 10,000 farming households that used 10,000 ha of forestland were involved. The Social Forestry Program, which is partly financed by the Ford Foundation, intends to rehabilitate and develop 270,000 ha of degraded forestland.

The Social Forestry Program in the Outer Islands, which began in 1986, has five approaches for involving people in forestry activities:

- Participatory forestry—Members of local communities are recruited as forest exploitation workers by state forest corporations.
- Community forestry—Patches of forestland are cultivated and exploited by local communities.
- Village forestry—Existing farming methods are continued, but farmers receive assistance in the form of training and inputs.
- Farmers forestry—This is like village forestry, but the activities are undertaken by individual farmers or small enterprises.
- Tree farming—Farmers grow tree crops on their own lands for use as timber, firewood, or charcoal.

Village Development Programs have not yet developed. In 1988, it was decided to include HPH holders in these programs under the name Timber Concession Holders Village Development Program (HPH Bina Desa). Implementation is to be coordinated by the Masyarakat Perhutanan Indonesia (Indonesian Forestry Association). The Ministry of Forestry will train farmers and will develop demonstration plots.

MINISTRY OF AGRICULTURE

In 1979, Perusahaan Inti Rakyat Perkebunan (the Nucleus Agriculture Estates Program, commonly known as NES/PIR projects) was established with the support of the World Bank. Outside Java, NES/PIR projects are often developed on forestlands. This program is organized by the Directorate General of Estate Crops. Its aim is to integrate people living in villages near agricultural estates into the activities of those estates. NES/PIR projects distribute land to villagers, offer them technical assistance in establishing estate crops, and subsequently buy their produce. The people living on land designated for NES/PIR activities either are integrated into the project or are resettled. As participants of the program, they are given 2 ha of tree crop land and 0.5 to 1 ha for a house lot and home garden. This allotment is known as the smallholder component of the NES/PIR program and is owned by the individual participants.

Smallholder allotments constitute about 80 percent of total land under the control of NES/PIR projects, with the remaining 20 percent being the estates of private or state companies. Companies are obliged to provide the overall infrastructure for the project. They must provide technical assistance to smallholders and buy their produce. Some 75,000 households are engaged in NES/PIR projects, of which some 15,000 households (20 percent) are supposedly former shifting cultivators.

Another program, the Rehabilitation and Expansion of Export Crops, began in 1979 under the Directorate General of Estate Crops. The program's main activity is to provide credit to farmers to improve the quality of their smallholder plantations. Special funds are set aside in the Bank Rakyat Indonesia. From its inception, this program has emphasized six specific cash crop commodities: rubber, coconut, coffee, tea, cacao, and pepper. (The World Bank supports rubber and coconut plantations in eight provinces.) The program supports farmers who are already engaged in planting these cash crops. Outside of Java farmers receive support to plant between 1 and 2 ha of land, while on Java, program support is limited to only 1 ha. It can be assumed that many of the people included under this program are shifting cultivators. The question is whether these people have given up shifting cultivation or whether the activities associated with this program are in addition to shifting cultivation activities.

MINISTRY OF TRANSMIGRATION

Shifting cultivators and local participants of the resettlement program are integrated into the overall transmigration program through the Allocation Scheme for people living in transmigration areas. In conjunction with these activities, the Ministry of Transmigration and the Ministry of Forestry have begun cooperative actions to remove people from the forest and to resettle them in transmigration project sites. These people include shifting cultivators as well as other residents of the forests. According to official data from the Directorate General of Reforestation and Land Rehabilitation of the Ministry of Forestry, until 1986 there were an estimated 33,000 households of shifting cultivators integrated into the transmigration program. The figure was an estimated 34,540 households until 1988.

A related activity for resettling shifting cultivators was based on a cooperative decision between the Ministries of Agriculture and Forestry to control shifting cultivation by using the NES/PIR program. This later became the joint NES-Transmigration Program, known by the acronym PIRTRANS.

Another new concept is known as *parallel transmigration*. It is envisioned to be a long-term program to familiarize shifting cultivators with more sedentary methods of agricultural production. Under this program it is not necessary to move people from their original settlements. This concept is based on the idea of integrating the transmigration program into the overall regional development plan of a province or region.

MINISTRY OF SOCIAL AFFAIRS

In 1971, the Ministry of Social Affairs started the Social Welfare Development for Isolated Societies program. The target was people who live in "isolated societies," which the Ministry for Social Affairs defined on the basis of four criteria: (1) people who live in small bands, mostly without a sedentary settlement and in isolation from the modern world; (2) people who are only loosely governed by the central government administration and who are primarily governed by traditional political organizations (for example, tribal communities, which have very limited or no contact with the government or the mainstream population); (3) people who still hold animistic or traditional beliefs; and (4) people whose main source of living is hunting and gathering or shifting cultivation.

This program has three main objectives: (1) to raise the standard of living of the target groups through the development of sedentary and productive sources of living and to integrate these people into the regional and provincial market economies, (2) to introduce government administration at the village level, and (3) to develop stable communities that have ecologically sound production systems.

Until 1987 a total of 13,440 households were resettled and placed under the administrative responsibility of the respective local governments. As of 1987 there were 11,520 households still under the administration of the program.

MINISTRY OF HOME AFFAIRS

Under the Directorate of Settlement and Village Infrastructure of the Directorate General of Village Development, known by its acronym BANGDES, the Ministry of Home Affairs has had its own resettlement program called the Village Resettlement Project (Proyek Pemukiman Kembali Penduduk Desa). The main objective was to resettle people from scattered and isolated villages to more easily accessible locations that conform to the standard criteria set by the Indonesian government. The standard criteria includes, for example, an administration unit of not more than 3,000 people in one village (desa), the existence of a village government (chief, secretary, security, and welfare), compulsory elementary school attendance by children, provision of health services, and an agricultural extension program.

Besides the largely administrative nature of this program, people and communities are chosen for resettlement for a variety of reasons, for example, people who are nonsedentary because they practice shifting cultivation, people who live in protected forests or degraded watersheds, and people who are affected by natural disasters or who are moved for their own or for national security. From 1972–1973 to 1984–1985, BANGDES reportedly resettled 11,570 households of shifting cultivator. Because of budgetary constraints since 1986, however, BANGDES no longer undertakes direct implementation and financing of any resettlement programs.

Program Results

The total number of households that practice shifting cultivation and were involved in the different programs can be summarized as follows: Ministry of Forestry Program, 37,000; NES/PIR projects, 15,000; Ministry of Transmigration, 34,540; Ministry of Social Affairs, 24,960; and Ministry of Home Affairs, 123,470 (Government of Indonesia/Ministry of Forestry and Food and Agriculture Organization of the United Nations). Even though a substantial number of the approximately 123,500 participating households are no longer involved in the various programs, the total number of households involved indicates a magnitude that should be compared with the targets for the REPELITA IV and REPELITA V plans. The aim of the REPELITA IV plan was to include 500,000 families. The targets were roughly the same for the REPELITA V plan. Although the current emphasis is on in situ development rather than resettlement (ex situ programs), considerable and concerted efforts are required to achieve the targets of the REPELITA V plan. Significant changes in strategy and approach are also needed (Government of Indonesia/Ministry of Forestry and Food and Agriculture Organization of the United Nations, 1990).

Intensification of Wet Paddy Rice Agriculture

This section is derived in large part from a report by Sadikin (1990). During the 2 decades after Indonesia's independence in 1945, significant efforts were made to increase food and agricultural production. But the absence of ingredients for development rendered many projects ineffective. Ingredients for development include infrastructural improvements and development program support, for example, political support; use of high-yielding plant varieties, fertilizers, and insecticides (if necessary); well-maintained irrigation systems; improved communications among groups of farmers; improved transport facilities; provision of credit; and reasonable market prices.

In the late 1950s, the campaign to achieve self-sufficiency in rice production through the use of improved Indonesian varieties and the intensification of production met with only limited success because the security of the public and national security as a result of public unrest in the main rice-producing centers, such as West Java, South Sulawesi, and East Java, were poor, and irrigation systems and transportation infrastructures were dilapidated.

Plans for the expansion of agricultural land and rice production areas into the tidal swamps of Kalimantan and into the upland rainfed environments in Sumatra, Kalimantan, and Sulawesi depended on the use of heavy equipment. The poor infrastructure caused the transport, maintenance, and repair of the equipment to be difficult and costly.

Rice imports, which, on average, were less than 300,000 metric tons/year from 1950 to 1955, rose to an average of 810,000 metric tons/year from 1956 to 1960 and exceeded 1 million metric tons/year in the 1960s. In the late 1970s and 1980, Indonesia imported the most rice of any country in the world, with imports being as high as 2 million metric tons/year (Table 12).

An encouraging sign in rice production emerged in 1963–1964, when students at the Bogor Agricultural University, using a demonstration area of 50 ha, showed that rice production could be nearly tripled if the recommended packages of technologies for use with improved Indonesian varieties were properly used. An important lesson emerged: improved production depends on a secure supply of agricultural inputs and on face-to-face communications with farmers. The experiment led to the creation in 1965–1966 of a mass guidance program Bimbingan Masal (BIMAS; Mass Guidance program) to increase rice production by encouraging and enabling farmers to take full advantage of technological innovations. In 1966 and 1967, rice

TABLE 12   Imports of Milled Rice, 1971–1987

Year	Metric Tons (in thousands)	Year	Metric Tons (in thousands)
1971	490	1980	2,030
1972	730	1981	530
1973	1,660	1982	300
1974	1,070	1983	1,160
1975	670	1984	380
1976	1,280	1985	0
1977	1,960	1986	0
1978	1,850	1987	0.05
1979	1,950		

SOURCE: Sadikin, S. W. 1990. The diffusion of agricultural research knowledge and advances in rice production in Indonesia. Pp. 106–123 in Sharing Innovation. Global Perspectives on Food, Agriculture, and Rural Development. Washington, D.C.: Smithsonian Institution Press.

yields at adaptive trials and on farmer's plots that were planted with the modern varieties of the International Rice Research Institute (IRRI; Los Baños, Philippines) were found to be impressive in comparison with the yields of the popular improved Indonesian varieties.

After this first success, the government mobilized considerable resources to secure a sufficient supply of fertilizers and pesticides to support a national campaign of introducing the modern varieties and set an ambitious target of planting 150,000 ha of rice in 1968. Within 5 years, the areas planted with modern varieties increased to over 3 million ha. After 1972 farmers also planted modern Indonesian varieties, which have cooking and taste qualities favored by Indonesians and produce fewer green, chalky grains when they are planted in the rainy season. In 1989 the area planted with the Indonesian and the IRRI modern varieties was 7.78 million ha, or 85 percent of the total area of harvested rice in Indonesia.

Because the program expanded too rapidly, shortcomings could not be avoided, such as in the application of the recommended packages of technology as well as in the management of the supply of farm inputs and the recovery of production credits. Nevertheless, aggregate rice production increased faster than the population. As a result of general increases in incomes, however, per capita rice consumption also increased substantially, with the effect that rice imports continued to increase.

Another serious problem emerged in the form of an insect infestation involving the brown planthopper. An outbreak in 1974–1975 destroyed lands planted in the popular high-yielding Indonesian rice varieties PELITA I and II, affecting an area of 240,000 ha. Indonesians learned to live in peaceful coexistence with the brown planthopper, and the high-yielding PELITA I and II varieties yielded another stream of benefits. Rice production jumped in 1978 from 15.8 million to 17.5 million metric tons of milled rice. Although a modest drought intervened in 1979, production again increased in 1980 and 1981 to unprecedented levels of 20.1 million and 22.2 million metric tons, respectively.

As other environmental obstacles were removed, the diffusion of technological innovations to farmers gradually and substantially accelerated. The number of extension personnel and specialists with competence to help the 18 million farm households in Indonesia grew rapidly. Improved irrigation and drainage facilities provided a more secure base for ensuring yield and production stability. There was a growing awareness and understanding among policy makers, legislators, and development professionals at the national, provincial, and district levels about the way to solve problems in the agricultural sector. As a result, rice production rose sharply, reaching a production level of 25.9 million metric tons of milled rice in 1984 (see also Table 3). This progress in production capabilities, along with the presence of government-held reserves of 2 million metric tons at the end of 1984, allowed the government to halt rice imports and was an historic turning point in Indonesia's quest for self-sufficiency in its staple food commodity. Significant efforts are now being made to maintain this level of food security and to diversify food production and consumption.

DISCUSSION AND FINDINGS

The estimated annual rate of deforestation in Indonesia has increased from 300,000 to more than 1,000,000 ha in the past 20 years. The average rate of deforestation between the 1950s and the early 1980s was 0.7 percent. This increased to about 1.2 percent annually between 1982 and 1990 (Government of Indonesia/Ministry of Forestry and Food and Agriculture Organization of the United Nations, 1990).

Forest Potential and Causes of Deforestation

The main causes of deforestation have been identified as population pressure and demand for agricultural land, logging in natural forests, shifting cultivation, transmigration programs, smallholder tree

crop development, and fires. Population pressure, shifting cultivation, and fires are social-economic (and natural) causes of deforestation, whereas the other causes—logging, transmigration, and smallholder
tree crop development—constitute pressures resulting from development activities.

In 1980, 64 percent of Indonesia's population was concentrated in
the islands of Java and Bali. This skewed population distribution has
both a positive and a negative effect on Indonesia's development. It
has centralized development and service activities in Java and Bali at
the expense of these activities in the Outer Islands. On the other
hand, because the population was concentrated in Java and Bali, this
allowed conservation of the immense natural resources in the Outer
Islands. And although many countries have nearly exhausted their
forest resources, Indonesia has significant areas of natural forest remaining. Indonesia has 144 million ha of set-aside and pre-set-aside
forestlands, providing a very high forestland-to-total land ratio (74
percent). Considering the land use changes and deforestation during
the past several years, the area of forested land in 1990 was estimated
to be about 109 million ha (Government of Indonesia/Ministry of
Forestry and Food and Agriculture Organization of the United Nations, 1990). More than about 60 million ha is leased out to private
and state-owned logging concessions, which form the core of the forest industry sector of Indonesia. These industries are major contribu-
tors to Indonesia's economic growth.

The substantial achievements in the forest industry sector, however, have aroused concern about the sustainability of forest management. Because it is aware of the dangers of overexploitation of forestlands, Indonesia has embarked on an intensive plan of developing
forest plantations and rehabilitating critical lands in watersheds through
reforestation and greening programs and through the improvement
of logging operations in natural forests. The government has also
encouraged rural households to raise fuelwood and light construction wood in their home gardens and farms to supply household
energy and timber needs, so that the natural forests will not be over-
burdened.

Because of the great population pressure, Indonesia has embarked
on a family planning (birth control) program since the 1950s, with
the set target that the growth rate of the population in Indonesia will
decline. However, even with the targeted reduction in growth rates,
from an annual rate of 2.34 percent in 1980 to 1.0 percent by 2011 and
beyond, Indonesia's present population (184 million in 1991 [World
Resources Institute, 1992]) will almost double by 2050.

Although the family planning and transmigration programs to

relax population pressures in densely populated areas are considered to be relatively successful, they are not expected to be able to alleviate the increased demand for food and, hence, the demand for agricultural land if the soil productivity of the agricultural sector is not adequately increased in the near future and if other sectors such as industry and trade are not effectively developed to offer alternative employment opportunities.

One of the constraints in the transmigration program is that some of the new settlers eventually revert to shifting cultivation. This is usually caused by the inability to sustain food production on the 1 to 2 ha of land provided by the program. Low productivity rates are also a direct result of low soil fertility and insufficient water supplies. Options for promoting more sustainable agriculture within shifting cultivation communities include the various agroforestry systems and practices. From the conservation point of view, these alternative systems are far superior to traditional shifting cultivation. The growing of perennial crops to cover fallow areas will also discourage alang-alang formation.

In the Outer Islands, agricultural and other development programs organized by different government agencies, such as transmigration smallholder tree crop development programs, have been identified as affecting deforestation. The same programs, however, have been aimed at controlling shifting cultivation to offer more sustainable agricultural systems. Avoiding the use of the designated forestlands and training shifting cultivators to become sedentary agriculturalists are pivotal parts of the program. The challenge is to better integrate the activities to achieve better results.

Intensive agriculture has been practiced for more than a century in Java and Bali. Because of the intensified productivity of wetland rice paddies, population growth has been accompanied by a steady increase in rice production. These achievements in agricultural productivity have helped to reduce rates of deforestation. The problem now is how to maintain this situation, considering the high rate of population growth, and at same time gearing to diversify the types of food crops, which may bring about a diversification of food production systems.

ASSESSMENT OF FOREST LOSS IN THE NEAR FUTURE

The government of Indonesia's Ministry of Forestry, in conjunction with the Food and Agriculture Organization of the United Nations (1990), has developed a model for investigating the causes underlying deforestation and projecting forest cover. An increase of 100 kg per hectare per crop of wet paddy rice (treated as a proxy for average agricultural productivity) increases forest cover by 4.6 percent. An increase in income per capita of 1,000 rupiah increases forest cover by 0.015 percent. An increase in population density of one person per square kilometer decreases forest cover by 0.8 percent. In addition, unexplained factors (for example, the cumulative amount of logging and other roads opened) contribute to an average decline in forest cover of 3.7 percent per year.

To project deforestation rates, assumptions regarding, for example, population growth rates, economic development and investment policies, foreign assistance policies, and agricultural and forest industry policies are necessary. These three scenarios reflect the following assumptions:

• Baseline scenario, which assumes government programs use the same strategies from the 1980s and at about the same rate as they did in the 1980s.
• Worst-case scenario, which assumes a least-favorable yet still plausible combination of factors that could cause the rate of deforestation to increase more than baseline scenario estimates.
• Best-case scenario, which assumes a most-favorable yet still plausible combination of factors that could cause the rate of deforestation to increase less than baseline scenario estimates.

The estimated annual deforestation rates of the World Bank (1989) and Government of Indonesia/Ministry of Forestry and Food and Agriculture Organization of the United Nations (1990) presented in Table 8 are used to estimate future deforestation rates under the three scenarios. For the baseline scenario, the best estimate of the rate of deforestation is taken; for the worst-case scenario, the minimum estimate is taken; and for the best-case scenario, the maximum estimate is taken. Table 13 provides forest losses for different time periods under the three scenarios.

Findings

1. Annual deforestation rates in Indonesia were 0.7 percent in the 30 years between the 1950s and early 1980s, increasing to 1.2 percent within the past decade. It could increase to an estimated 1.5 percent by 2030 if efficient measures are not taken to control the causes of deforestation effectively.

2. Removal of forest cover for development purposes cannot be avoided. Forested lands in lower areas (0 to 250 m above sea level) that are fit for agriculture and other related activities could be ear-

TABLE 13   Analysis of Forest Loss Estimates in Indonesia, 1990–2029 (in Millions of Hectares)

Scenario and Time Period	1990 Forest Cover	Average Loss per Year	Total Loss for Decade	Total Forest Remaining at End of Decade	Percent Loss for Decade
Baseline					
1990–1999	108.6	0.9	9	99.6	8.3
2000–2009	99.6	0.9	9	90.6	9.0
2010–2029	90.6	0.9	9	81.6	9.9
Worst case					
1990–1999	108.6	1.2	12	96.6	11.0
2000–2009	96.6	1.2	12	84.6	12.4
2010–2029	84.6	1.2	12	72.6	14.6
Best case					
1990–1999	108.6	0.7	7	101.6	6.4
2000–2009	101.6	0.7	7	94.6	6.9
2010–2029	94.6	0.7	7	87.6	7.4

marked for development purposes. The Ministry of Forestry has classified some forested lands (about 30 million ha) as conversion forests that are to be used for purposes other than forestry after the timber stands have been removed. Coordination between government and private development agencies is necessary so that forestland classified as permanent forest (or candidate permanent forest) will not be used for other development purposes.

3. After implementation of the Constitution of 1945, the primary authority for forestry administration was Law No. 5 (1967), Basic Provisions on Forestry. In the execution of the law, however, in particular, forestland use, there are provisions that are thought to be in conflict with those in the Basic Law on Agrarian Matters of 1960, for example, land tenure aspects of forested lands. Legislation on forestland and general land use should be adjusted to allow for better land use—including forestland use—and land tenure arrangements.

4. The first steps to designate permanent forestlands outside Java have been done through the forestland use by consensus (TGHK) approach. Because this method is limited to desk exercises, field operations are necessary—that is, surveys should be followed by demarcation of the designated forestlands with easily recognizable boundary markers. In this way, misunderstandings between the Indonesian government and local communities and governments or private development agencies could be reduced to a minimum. Special attention should be paid to the existing tenure rights of local communi-

ties. The many conflicts concerning the existing TGHK boundaries urgently need solutions. The stewardship certificate system could be studied in this respect. The stewardship certificate system of the Philippines, for example, states that occupants of forestlands can use the forest and the land for 25 years (usufruct rights) but they must use agroforestry practices prescribed by the Ministry of Forestry and must maintain the existing forests.

5. The designation and demarcation of forest ecosystems to conserve biodiversity and genetic resources should be given special attention. Between 1979 and 1984 over 10 million ha of reserves was added to the existing conservation forests, but the rate of setting aside forestlands has fallen since then. As a target, about 18 million ha of conservation forests is envisaged by TGHK. Another disturbing aspect is the incompleteness of the conservation forest system across the seven major biogeographic zones in Indonesia.

6. Provisions regarding the implementation of logging and other forest operations in the concession areas, in particular, regarding forest regeneration as prescribed in the Indonesian selective cutting system (TPI) and, later, in the Indonesia selective cutting and planting system (TPTI), are not adequately observed in general, so that the reality of forest operations is far from an ideal sustainable forest management system. To overcome this problem, stricter controls in the implementation of forest operations by concession holders should be exercised, and stiff penalties should be imposed on those operations that deviate from the regulations, in particular, those that deviate from the annual allowable cut and the allowable harvesting area. On the other hand, a possible extension of the 20-year concession period (which does not stimulate sustainable forest operations), for example, to 35 years (the same as the silvicultural rotation period) or on a variable basis with periodic performance reviews of logging and other forest operations, could be considered as alternatives.

7. Shifting (slash-and-burn) cultivation is considered a severe land use problem, causing deforestation and the formation of extensive areas of alang-alang grasslands and other unproductive lands, in particular, as a result of shortened fallow periods and influxes of migrants. Rationalization of shifting cultivation should take into account, for example, land use, cultural, land tenure, and other socioeconomic factors related to the issue. Decisions should be made in consultation with the affected communities. For rationalized shifting cultivation, better sites should be allocated. Proper extension and provision of credits can act as positive incentives and can upgrade land use practices to a more sustainable level. An integrated approach to rationalizing shifting cultivation, which has a greater

chance of success, should include education, health services, and the provision of other community services. Also important are common policies and strategies among the agencies whose programs partly or entirely involve shifting cultivation—that is, rationalization of shifting cultivation should be implemented as a concentrated effort of a general local community development program.

8. To alleviate the impacts of deforestation in terms of declining forested lands or a worsening of ecologic conditions, reforestation (on forestland) and regreening (on private land) programs have been initiated. Although the concepts of the programs are commendable, because of inadequate planning and execution the present rates of success of reforestation and regreening are low. A well-designed plan for reforestation and regreening must address seed availability, seedling production, proper site selection and preparation, and above all, continued care and management after the establishment of plantations. A national plan for reforestation and regreening should also involve the public, private, and community sectors. The program should be supported by well-coordinated research.

9. The role of industrial plantations is, in principle, to supplement natural forest resources and to improve ecologic conditions, in particular, in those areas where degraded forestlands have been selected. Industrial forest plantations, including agroforestry systems, can also provide valuable services to local communities by providing employment and, in some cases, better housing, education, and health care as well as agricultural extension services and loan credits.

10. Forest development programs were, in principle, designed to generate awareness of conservation issues by the public and private sectors as well as communities. The roles of nongovernment organization (NGOs) can be substantial in this respect. There are hundreds of NGOs that have shown interest in conservation issues; however, the lack of coordination and resources prevent them from being well-functioning organizations. NGOs and other community groups should be involved in training and education on a community level. NGOs with major extension plans should be given funding and personnel training priority.

11. To maintain self-sufficiency in food production diversification in agriculture, production and consumption of agricultural crops must be encouraged.

In addition to their economic importance, the forests of Indonesia are also considered a gigantic carbon sink. The perpetuation of Indonesia's forest cover is therefore necessary for long-term global survival (Government of Indonesia/Ministry of Forestry and Food and Agri-

culture Organization of the United Nations, 1990). The commitment
of Indonesia to sustainable development of its tropical forests is am-
plified in a statement by President Suharto (Government of Indone-
sia/Ministry of Forestry and Food and Agriculture Organization of
the United Nations, 1990):

> Our tropical forests are the lungs of the world. Their degradation
> brings disaster not only to our nation, but also to other nations and
> inhabitants of the earth. We must manage our forests under sus-
> tainable development for our next generations in particular, and for
> all mankind in general.

REFERENCES

Asian Development Bank. 1989. Operational Strategy 1989. Study for Indo-
    nesia. Manila, Philippines: Asian Development Bank.
Basjarudan, H. 1978. Forest policy and legislation. Lecture notes. Forestry
    Faculty, Bogor Agricultural University, Bogor, Indonesia.
Biro Pusat Statistik (Central Bureau of Statistics). 1988. Statistik Indonesia.
    Statistical Year Book of Indonesia 1988. Jakarta: Biro Pusat Statistik.
Biro Pusat Statistik (Central Bureau of Statistics). 1989. Input-Output Table
    1985. Jakarta: Biro Pusat Statistik.
Food and Agriculture Organization. 1952. Principles of Forest Policy. Rome,
    Italy: Food and Agriculture Organization of the United Nations.
Government of Indonesia/Department of Information. 1989. Indonesia 1989.
    An Official Handbook. Jakarta: Government of Indonesia.
Government of Indonesia/International Institute of Environment and Devel-
    opment. 1985. A Review of Policies Affecting the Sustainable Develop-
    ment of Forest Lands in Indonesia. Jakarta: Government of Indonesia.
Government of Indonesia/Ministry of Forestry and Food and Agriculture
    Organization of the United Nations. 1990. Situation and Outlook of the
    Forestry Sector in Indonesia. Jakarta: Government of Indonesia.
Government of Indonesia/National Development Planning Agency. 1989.
    REPELITA V: Indonesia's Fifth Five Year Development Plan. Basic Data
    and Main Targets. Jakarta: Government of Indonesia.
Kaul, A. 1990. Indonesian Farming Systems: Types and Issues. Unpub-
    lished manuscript.
Kusumah, B., D. S. Irawanto, G. Aji, and I. Qodar. 1991. Indonesian forest:
    How are you? Tempo Mag. XXI(35):23–24.
Prastowo, H. 1991. The System of Production Forest Management in the
    Future. Homecoming Day Alumni VIII/1991 Faculty of Forestry. Bogor
    Agricultural University, Bogor, Indonesia.
Regional Physical Program for Transmigration. 1990. The Land Resources
    of Indonesia: A National Review. Direktorat Bina Program, Direktorat
    Jendral Penyiapan Pemukiman, Departemen Transmigrasi, Jakarta.
Sadikin, S. W. 1990. The diffusion of agricultural research knowledge and

advances in rice production in Indonesia. Pp. 106–123 in Sharing Innovation. Global Perspectives on Food, Agriculture, and Rural Development. Washington, D.C.: Smithsonian Institution Press.

Sukartiko, B. 1988. Soil conservation program and watershed management in Indonesia. Paper presented at the Regional Workshop on Ecodevelopment Process for Degraded Land Resources in Southeast Asia, Bogor, Indonesia, August 23–25, 1988 (Man and Biosphere, Indonesia–United Nations Educational, Scientific, and Cultural Organization, South-East Asia).

Utomo, W. H. 1989. Konservasi Tanah di Indonesia (Soil Conservation in Indonesia). Jakarta: C. V. Rajawali.

Webster, B. 1984. Devastated forest offers a rare view of rebirth. New York Times, April 24, 1984.

Whitmore, T. C. 1984. Tropical Rain Forests of the Far East. Oxford: Oxford University Press.

World Bank. 1989. Indonesia Strategy for Growth and Structural Change. Report No. 7758-IND. Washington, D.C.: World Bank.

World Resources Institute. 1992. The 1992 Information Please Environmental Almanac. Boston: Houghton Mifflin.

Malaysia

Jeffrey R. Vincent and Yusuf Hadi

The tandem of commercial logging and shifting cultivation has been blamed as the leading cause of deforestation in the humid tropics (Lanly, 1982; Myers, 1978). (The term *deforestation* is used here in the strict sense favored by Lanly [1982]: conversion of forests to a nonforest land use. Thus, logging of a primary—that is, virgin or old-growth—forest is not regarded as deforestation unless the logging is so intensive that tree cover is essentially eliminated.) In several countries, however, other agricultural activities are more responsible. Peninsular Malaysia provides a notable example. In the late 1800s, the peninsula was virtually completely forested. Today, natural forests cover less than half of their original extent. Forests have been converted primarily to agricultural use, but not by a process of shifting cultivation. Shifting cultivation affected less than 0.1 percent of the peninsula's land area in 1966 (Wong, 1971) and the late 1980s (Rambo, 1988).

Instead, tree crops represent the principal agricultural land use in Peninsular Malaysia. Rubber, oil palm, coconut, and cacao accounted for 83 percent of the area devoted to agriculture in 1988 (Ministry of

Jeffrey R. Vincent is an institute associate at the Harvard Institute for International Development, Harvard University, Cambridge, Massachusetts; Yusuf Hadi is dean of Fakulti Perhutanan (Faculty of Forestry) at the Universiti Pertanian Malaysia, Selangor, Malaysia.

Agriculture [Malaysia], 1991). Unlike shifting cultivation, the opening of new areas for tree crops has not been driven by the need to replace abandoned, exhausted lands. New plantations represent net additions to an essentially permanently productive agricultural land base.

This profile analyzes the role played by tree crops in the conversion of forests in Peninsular Malaysia during the past century. It addresses four broad questions: (1) Are tree crop plantations a sustainable land use? (2) Are tree crop plantations economically feasible? (3) How have policies affected the expansion of tree crop plantations? (4) What are the environmental impacts of conversion of natural forests to tree crop plantations? In addition, deforestation projection rates up to the year 2030 are provided. It also highlights policy implications and identifies principal research needs.

Under Malaysia's federal constitution, individual states retain substantial autonomy over land development and forestry policies. Policies are coordinated more among the states of Peninsular Malaysia than between Peninsular Malaysia and either Sabah or Sarawak (Vincent, 1988). Because of this autonomy and because there are profound differences among the three regions in demography (Peninsular Malaysia had 82 percent of the nation's population in 1990), economic activity (Peninsular Malaysia is more industrialized and accounted for 84 percent of Malaysia's gross domestic product [GDP] in 1987), and agricultural activity (74 percent of Malaysian land in agricultural use was in Peninsular Malaysia in 1990), this profile focuses only on Peninsular Malaysia.

DESCRIPTION OF PENINSULAR MALAYSIA AND ITS FORESTS

Malaysia is a federation of 13 states. Eleven of the states comprise Peninsular Malaysia, which was the British colony of Malaya until it became independent in 1957. The other two states, Sabah and Sarawak, share the island of Borneo with Brunei and Kalimantan (part of Indonesia).

Topography, Climate, and Soils

Peninsular Malaysia is located entirely within the equatorial zone. It covers 13.2 million ha (40 percent of Malaysia's land area). Aiken et al. (1982) and Tija (1988) have summarized the peninsula's physical and climatic characteristics. Figure 1 shows how climate and topography vary within the peninsula. No part of the peninsula is more than about 150 km from the sea. The interior of the northern

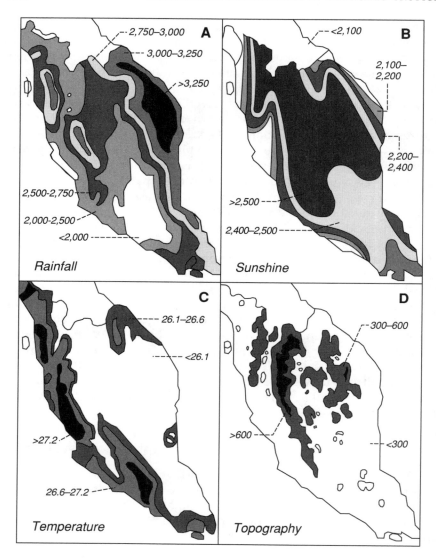

FIGURE 1   Climate and topography of Peninsular Malaysia.   A.   Mean annual rainfall in millimeters: <2,000, 2,000–2,500, 2,500–2,750, 2,750–3,000, 3,000–3,250, and >3,250.   B.   Mean annual bright sunshine in hours: <2,100, 2,100–2,200, 2,200–2,400, 2,400–2,500, and >2,500.   C.   Mean annual temperature in °C: <26.1, 26.1–26.6, 26.6–27.2, and >27.2.   D.   Elevation in meters: <300, 300–600, and >600.   Source: Tija, H. D.   1988.   The physical setting.   Pp. 1–19 in Key Environments: Malaysia, Earl of Cranbrook, ed.   Oxford, U.K.:   Pergamon.   Reprinted with permission from the publisher, © 1988 by Pergamon Press Ltd.

two-thirds contains mountain ranges that run approximately north-south. The highest peak, Gunung Tahan, is 2,188 m in elevation. Mountains give way to low hills in the southern cone of the peninsula. Coastal plains extend along the Strait of Malacca on the west and the South China Sea on the east and are wider on the west. About two-thirds of the peninsula is less than 300 m above sea level.

The combination of an equatorial location, proximity to the sea, and low relief results in a climate that varies relatively little during the year or within the peninsula. Most of the peninsula receives more than 2,400 hours of bright sunshine per year. The mean annual temperature ranges from 26.1° to 27.2°C and is highest in the lowlands just inland from the west coast. The mean annual rainfall ranges from less than 1,800 to more than 3,600 mm. The wettest regions are the foothills near the east and northwest coasts. The peninsula has a weak monsoonal climate.

The peninsula's soils are heavily weathered (thus, they are often very deep), highly leached, and typically quite acidic (pH 4.2 to 4.8). They contain little organic matter and low levels of plant nutrients. Six of the 10 U.S. Department of Agriculture (USDA) soil orders occur: Entisols, Histosols, Spodosols, Oxisols, Ultisols, and Inceptisols (Tija, 1988). The last three types have good to excellent physical properties for agriculture.

Lee and Panton (1971 [cited in Ariffin and Chan, 1978]), drawing on the work of Wong (1971) (according to Soong et al., 1980), proposed a soil suitability classification for Peninsular Malaysia that continues to be used for land use planning. The system divides the peninsula's soils into five suitability classes (classes I–V), which are differentiated by the number of limitations in using the soils for agriculture. Ariffin and Chan (1978) suggest that potential agricultural land should best be confined to soils with, at most, one serious limitation for agriculture (classes I–III, which total 5.9 million ha). Barlow (1978), Lee (1978), and Aiken et al. (1982) suggest that 6.3 million to 6.5 million ha is suitable for agriculture. The Economic Planning Unit (Malaysia) (1980) of the Prime Minister's Department favors an estimate of 6.3 million ha.

Population

Peninsular Malaysia's population was estimated to be 14.7 million in 1990 (Department of Statistics [Malaysia], various issues). Approximately half of the population is Malay, one-third is Chinese, and one-tenth is Indian. The remainder includes the aboriginal people who preceded the Malays, the Orang Asli, whose population totaled only about 60,000 in 1980 (Rambo, 1988).

TABLE 1   Total and Urban Population Growth in Peninsular
Malaysia, 1835–1990

| | Total Population | | Urban Population | |
Year	Total (1,000s)	Growth Rate (percent)[a]	Total (1,000s)	Growth Rate (percent)[a]
1835–1836[b]	281	—	NA[c]	—
1891	944[d]	2.2	NA	—
1901	1,531[d]	4.8	NA	—
1911	2,339	4.2	250	—
1921	2,907	2.2	407	4.9
1931	3,788	2.6	571	3.4
1947	4,908	1.6	930	3.0
1957	6,279	2.5	1,667	5.8
1970	9,182	2.9	2,635	3.5
1980	11,437	2.2	4,251	4.8
1990	14,667	2.5	6,870[e]	4.8

[a]This is the growth rate during the interval since the preceding point estimate of population.
[b]Excludes Pinang, Melaka, and the Orang Asli.
[c]NA, Not available.
[d]Based on rates of growth for the Federated Malay States, given in Lim (1977: Appendix 1.2).
[e]Estimated by the authors by using the growth rate for the previous decade.

SOURCES: Aiken, S. R., C. H. Leigh, T. R. Leinbach, and M. R. Moss. 1982. Table 8.10 in Development and Environment in Peninsula Malaysia. Singapore: McGraw-Hill International; Department of Statistics (Malaysia). Various issues. Monthly Statistical Bulletin: Peninsular Malaysia. Kuala Lumpur: Department of Statistics; Ooi Jin Bee. 1976. Peninsular Malaysia. London: Longman.

Table 1 presents the growth trends for total and urban populations from 1883 to 1990. The rate of population growth in Peninsular
Malaysia was 2.2 percent/year during 1989–1990. The World Bank
(1990) projects the rate to remain at this level during 1988–2000 and
to fall to 1.2 percent/year during 2000–2025. The urban population
has been growing more rapidly than the rural population. Approximately 47 percent of the population lived in urban areas in 1990, up
from 27 percent at the time of independence, 1957. Better economic
opportunities (for example, manufacturing jobs) help to explain the
trend toward urbanization. Only 8 percent of households in urban
areas were classified as poor in 1984, whereas 25 percent of households in rural areas were classified as poor (Ministry of Agriculture
[Malaysia], various issues). In 1987, the mean annual gross house-

hold income was 72 percent higher in urban areas than it was in rural areas (Ministry of Agriculture [Malaysia], 1990).

Peninsular Malaysia's population density is low compared to that of most developing countries. In 1988, total land per capita was 0.95 ha, agricultural land in use was 0.28 ha per capita, and forest area was 0.45 ha per capita (Ibu Pejabat Perhutanan, Semenanjung Malaysia, 1990; Ministry of Agriculture [Malaysia], 1990).

Domestic Economy

The Malaysian government does not report all economic statistics separately for Peninsular Malaysia. For this reason, much of the information in this section pertains to Malaysia as a whole.

In 1988, Malaysia's gross national product (GNP) was 85.8 billion Malaysian dollars (M$; M$2.62 = US$1.00 in 1988). In per capita terms this was US$1,940, which makes Malaysia a middle-income developing country (World Bank, 1990). In 1988, exports equaled 64 percent (M$55.3 billion) of the GNP, while imports equaled 50 percent (M$43.3 billion) (Ministry of Agriculture [Malaysia], 1990). GNP per capita grew at an average rate of 4.0 percent/year during 1965–1988, which was tied for the highest rate among middle-income countries (World Bank, 1990). Continued strong economic performance is needed to enable Malaysia to service its debt. The country's long-term debt service as a percentage of GNP was 16.5 percent in 1988 (World Bank, 1990).

AGRICULTURE

Agriculture (including forestry and wood products) is a major sector of Peninsular Malaysia's economy and is important on a global basis as well. Malaysia is the world's largest producer and exporter of natural rubber (34 percent of global production and 40 percent of global exports in 1989), palm oil (59 percent of global production and 69 percent of global exports in 1989), and tropical logs and sawn wood (25 percent of global production and 78 percent of global exports in 1989) (Food and Agriculture Organization, 1991; Ministry of Primary Industries [Malaysia], 1990). With the exception of tropical logs, production and exports of these products are concentrated in Peninsular Malaysia. Agriculture, forestry, and fisheries accounted for 21 percent of Malaysia's GDP and employed 31 percent of Peninsular Malaysia's work force in 1988 (Ministry of Agriculture [Malaysia], 1990; World Bank, 1989).

Peninsular Malaysia's agriculture is based overwhelmingly on exotic crops: rubber, oil palm, rice, and cacao and, to a lesser extent, coffee, pineapple, tobacco, sugarcane, and maize (Hill, 1982). This is because the peninsula was among the last regions in Asia to be settled by agriculturalists. Production, exports, imports, and consumption of major agricultural products in Malaysia in 1989 are summarized in Table 2. Malaysia is unique among countries in southeast Asia in that rice is not its most significant crop in terms of either area cultivated or tonnage of output (Barlow and Condie, 1986). Cereal production—almost entirely rice—was only 0.10 metric tons per capita in Malaysia during 1986–1988 (World Resources Institute, 1990).

In 1988, Malaysia exported M$22.1 billion of food and agricultural products (including forestry and wood products) and imported M$7.8 billion of such products. This contributed to a net agricultural trade surplus of M$14.3 billion (Ministry of Agriculture [Malaysia], 1990), which was more than the country's total trade surplus in 1988. The export value of rubber and oil palm products alone totaled M$10.4 billion, more than the value of total imports of food and agricultural products. The export value of forestry and wood products totaled M$7.5 billion in 1988. Because of its diversified economy, food makes up a smaller share of Peninsular Malaysia's imports (8 percent in

A farm that grows mixed crops is situated on land cleared by slash-and-burn techniques in Malaysia. Credit: James P. Blair © 1983 National Geographic Society.

TABLE 2   Production, Consumption, and Trade of Major
Agricultural Products in Malaysia, 1989

Product	Metric Tons (1,000s)			
	Production	Exports	Imports	Consumption
Palm oil	6,055	4,948	41	1,148
Rubber	1,419	1,487	122	54
Rice	1,094^a	2^a	97^a	1,289^a
Palm kernel oil	965	634	0	331
Meat (including poultry)	438	6	53	485
Cacao	255	169	0	86
Pineapple	180	18	0	162
Copra and copra cake	91	42	6	55
Coconut oil	41	54	15	3
Edible vegetables, roots, tubers^a	NA^b	134	280	NA
Sugar and sugar products^a	NA	209	647	NA
Animal feed^a	NA	450	736	NA
Cereals and cereal preparations other than rice^a	NA	34	2,012	NA

^{a}1987 data.
^{b}NA, Not available.

SOURCES: Department of Statistics (Malaysia). Various issues. Ministry of Agriculture (Malaysia). 1988. Import and Export Trade in Food and Agricultural Products: Malaysia 1987. Kuala Lumpur: Ministry of Agriculture; Ministry of Agriculture (Malaysia). 1990. Agricultural, Livestock, and Fisheries Statistics for Management: Malaysia 1980–1988. Kuala Lumpur: Ministry of Agriculture; Ministry of Primary Industries (Malaysia). 1990. Profile: Malaysia's Primary Industries (Malaysia). 1990. Profile: Malaysia's Primary Commodities. Kuala Lumpur: Ministry of Primary Industries.

1988) than in the case of the average middle-income country (11 percent) (Department of Statistics [Malaysia], various issues; World Bank, 1990).

MANUFACTURING

Peninsular Malaysia's economy is increasingly dominated by the output of manufacturing sectors. Although Malaysia's agricultural GDP grew 3.7 percent/year during 1980–1988, its manufacturing GDP grew 7.3 percent/year (World Bank, 1990). In 1988, manufacturing and services accounted for 64 percent of Malaysia's GDP (World Bank, 1989). Nearly all of Malaysia's output of manufactured goods is produced in Peninsular Malaysia.

Land Use

Table 3 summarizes land use in Peninsular Malaysia in 1966, 1974–1975, and 1981 and provides information about land use for agriculture in 1988. The Ministry of Land and Cooperative Development (Malaysia) plans to carry out an updated land use survey under the Sixth Malaysia Development Plan (which covers the period 1991–1995).

The area in agricultural use increased from 21 percent of the peninsula's land area in 1966 to 31 percent in 1988. Most of the increase had occurred by 1981, and most was due to the expansion of tree crop plantations. The four major tree crops—rubber, oil palm, coconut, and cacao—covered 16 percent of the peninsula's land area in 1966 and 26 percent in 1988. The agricultural area in 1988—over 4 million ha—was about two-thirds of the area considered suitable for agriculture in the peninsula.

Most of the agricultural conversion by the early 1970s had taken place in the southern and western lowlands, where rubber was concentrated. Since then, extensive conversion to oil palm has occurred in the eastern lowlands, and oil palm has replaced much of the rubber in the western lowlands.

Forests

The lowland forests of Malaysia, Brunei, the Philippines, and western Indonesia are dominated by tree species in the family Dipterocarpaceae. According to Whitmore (1988:21): "There are no other forests anywhere in the world which have so many genera and species of a single tree family growing together in the same place." This ecologic characteristic, coupled with wood properties that allow the many species to be aggregated into a relatively small number of commercial groups with broadly similar properties, helps to explain why the timber harvested from these forests has dominated world trade in tropical timber since the end of World War II (Laarman, 1988).

FOREST FORMATIONS

Whitmore (1988) classified Peninsular Malaysia's forests into 10 forest formations. The small area of northwestern Peninsular Malaysia, which has a seasonally dry climate—climatically atypical for Peninsular Malaysia—is where (1) semievergreen rain forests, more common in Thailand and Burma, are found.

Forests on permanently wet soils include (2) mangroves on the coasts and (3) freshwater swamp forests and (4) peat swamp forests

TABLE 3   Land Use in Peninsular Malaysia, 1966–1988[a]

Land Use	Hectares (1,000s)			
	1966	1974–1975	1981	1988
Urban and associated areas	134	199	251	—
Agriculture	2,736	3,565	4,101	4,160
Perennial crops	2,092	2,782	3,340	3,485
Rubber	1,775	2,048	1,969	1,569
Oil palm	100	487	1,063	1,527
Coconut	176	203	179	210
Cacao	<1	4	34	142
Others	41	41	95	37
Paddy	412	424	432	474
Horticulture	200	274	243	179
Miscellaneous crops	32	81	70	21
Improved permanent pasture	<1	4	16	—
Forest	9,036	8,254	8,460[b]	—
Dryland forest	7,852	7,182	7,437[b]	—
Swamp/wetland forest[c]	1,176	1,070	1,023	—
Shifting cultivation[d]	8	2	—	—
Other	1,310	1,019	432[e]	—
Scrub forest	594	419	—	—
Grassland/scrub grassland	404	170	—	—
Recently cleared land	116	326	24	—
Unused land	62	6	—	—
Unclassified land	134	98	190[f]	—
Total	13,215	13,037	13,244	—

[a]Data for 1966, 1974–1975, and 1981 are based on land-use surveys. Data for 1988 are based on annual records on land alienation and development.

[b]Includes scrub forest, grassland/scrub grassland, and shifting cultivation.

[c]Includes mangroves.

[d]The estimates for 1966 and 1974–1975 exclude areas classified as scrub forest and grassland/scrub grassland. One might speculate that some of these areas were affected by shifting cultivation. If so, this could help explain why the forest inventories reported larger areas as being affected by shifting cultivation (see Table 4).

[e]Excludes scrub forest and grassland/scrub grassland.

[f]Includes unused land.

SOURCES: For 1966, Wong, I. F. T. 1971. The Present Land Use of West Malaysia (1966). Kuala Lumpur: Ministry of Agriculture and Lands; for 1974–1975, Economic Planning Unit (Malaysia). 1980. Land Resources Report of Peninsular Malaysia, 1974/ 1975. Kuala Lumpur: Prime Minister's Department; for 1981, Ministry of Agriculture (Malaysia). 1990. Agricultural, Livestock, and Fisheries Statistics for Management: Malaysia 1980–1988. Kuala Lumpur: Ministry of Agriculture; For 1988, Ministry of Agriculture (Malaysia). 1991. Perangkaan siri masa sektor pertanian. January 19, 1991. Memorandum.

in inland areas, depending on soil characteristics. (5) Woody beach vegetation is found in coastal areas.

The other five formations are found in inland areas that are not permanently wet. Most restricted in area are the (6) heath and (7) limestone forests. The remaining three formations account for the majority of the peninsula's forest area. Their distribution is largely determined by elevation. (8) Lowland evergreen rainforests once covered most of the peninsula up to an elevation of 750 m. Two floristic zones can be distinguished within this formation: lowland dipterocarp forests, which are found at elevations up to 300 m, and hill dipterocarp forests, which are found above these elevations. The demarcation is based on the distribution of seraya (*Shorea curtisii*), which dominates ridges in the hill dipterocarp zone. (9) Lower montane rain forests are found between elevations of 750 and 1,500 m. They have a smoother, lower canopy than do lowland rain forests. They, too, can be divided into two floristic zones: the upper dipterocarp forests, which are found at elevations up to 1,200 m, and the oak-laurel forests, which are found above 1,200 m but below 1,500 m. The final formation, (10) upper montane rain forests, is found above 1,500 m.

BIODIVERSITY

In terms of biodiversity, Peninsular Malaysia's rain forests are among the richest ecosystems in the world. Ng (1988) estimated that 2,650 tree species occur naturally in Peninsular Malaysia. The lowland dipterocarp forests are the richest of the peninsula's forest formations. Butterflies and moths provide one exception to this pattern: most of the peninsula's 1,014 species are found at elevations of 600–1,000 m, and only 23 species are endemic (Barlow, 1988). Many endemic plant species are found in limestone forests.

Wells (1988) reported that 282 of the 370 bird species that make heavy or exclusive use of forests or the forest fringe are associated with lowland dipterocarp forests. He cited studies, carried out at the Pasoh Forest Reserve in the state of Negeri Sembilan and at the Kerau Game Reserve in the state of Pahang, that recorded 196 and 202 bird species, respectively, in areas of 2 km^2 each. Yong (1988) reported that 33 families, 104 genera, and 203 species of mammals are native to Peninsular Malaysia. (In contrast, Denmark, which is also a peninsula and only slightly smaller in area, is home to only 13 families, 32 genera, and 45 species [Earl of Cranbrook, 1988].) Of the 203 mammal species, 194 have been sited in the forest, mainly the lowland dipterocarp forest (Earl of Cranbrook, 1988). According to Steven (1968 [cited in Earl of Cranbrook, 1988]), 78 percent of the mammal species other than bats are obligatory forest dwellers.

FOREST AREAS AND DEFORESTATION

The land use surveys (Table 3) are one source of information on forest areas in Peninsular Malaysia. The information they provide, however, is substantially less detailed and probably less accurate than the information generated by forest inventories carried out specifically to estimate forest areas and timber stocks. The most recent forest inventory in Peninsular Malaysia, Forest Inventory II, was carried out during 1981–1982 (Ibu Pejabat Perhutanan, Semenanjung Malaysia, 1987). It updated the peninsula's first forest inventory, Forest Inventory I, which was carried out during 1970–1972 (Food and Agriculture Organization, 1973) and superseded interim estimates for 1980 made by the Food and Agriculture Organization (FAO) of the United Nations (1981) from projections based on Forest Inventory I and the 1974–1975 land use survey. Forest Inventory III is planned to begin in the early 1990s.

Table 4 presents the original estimates of forest areas from the two inventories, revised estimates from Brown et al. (1991b) based on a GIS analysis of the inventory maps, and the estimate for 1980 from FAO (1981). (The definitions of forest types used here are the ones used in the forest inventories.) Although they were calculated by a cruder method, the original estimates from Forest Inventories I and II are quite close to the revised estimates of Brown et al. (1991b). The original estimates slightly understated virgin forest areas and slightly overstated logged-over areas. The FAO (1981) estimate of total virgin forest area in 1980 is very close to the original and revised estimates for 1981–1982, but the FAO estimate of total logged-over area is 27 percent higher than the revised estimate from Forest Inventory II. Given the similarities between the original and revised estimates, but the more precise method used to generate the latter, the revised estimates are regarded as the best estimates.

According to the inventories, forests declined from 62 percent of Peninsular Malaysia's land area during 1970–1972 to 52 percent during 1981–1982. Areas of forest decreased for all types except dryland and inland swamp forests that have been logged over since 1966. The loss in virgin forest area, 1.179 million ha, was equivalent to more than four-fifths of the aggregate decrease in forest area, 1.411 million ha. Brown et al. (1991b) found, however, that most (59 percent) of the decrease in virgin forest area represented forests that were logged over but not converted to nonforest uses. Most conver-

TABLE 4   Forest Areas in Peninsular Malaysia, 1970–1982

| Forest Type[a] | Forest Inventories (1,000 ha) | | | | |
| | Original | | Revised | | FAO |
	1970–1972	1981–1982	1970–1972	1981–1982	1980
Virgin	4,540	3,455	4,743	3,564	3,555
Below 1,000 m[b]	3,787	2,915	4,001	3,003	2,804
Superior	827	693	845	683	—
Good	1,150	848	1,160	916	—
Moderate	1,398	1,119	1,572	1,143	—
Poor	412	255	424	261	—
Above 1,000 m[b]	289	278	284	281	289
Inland swamp	464	262	458	280	462
Logged-over	3,332	3,046	3,175	3,042	3,875
Dryland	2,981	2,748	2,846	2,727	3,562
Before 1966	1,714	488	1,615	481	—
After 1966	1,267	2,260	1,231	2,246	—
Inland swamp	351	298	329	315	313
Before 1966	184	39	152	39	—
After 1966	167	259	177	276	—
Shifting cultivation[c]	261	220	317	216	—
Mangrove			155	121	149
Total[d] 8,131	6,721	8,233	6,822	7,430	7,430[e]

[a]Types are the categories used in the 1970–1972 and 1981–1982 forest inventories. The estimates of area of virgin forests in 1980 by the Food and Agriculture Organization (FAO) of the United Nations include forests termed "unproductive" by FAO (1981).

[b]The elevation boundary is 1,300 m for the 1980 FAO (1981) estimates. Superior, good, moderate, and poor refer to timber stocking.

[c]Includes only areas disturbed by shifting cultivation in 1966 or before.

[d]Excludes mangrove forests. Sum of subtotals might not equal stated totals because of rounding.

[e]Excludes shifting cultivation.

SOURCES: For the original forest inventories, Food and Agriculture Organization of the United Nations. 1973. A National Forest Inventory of West Malaysia, 1970–72. FO:DP/MAL/72/009, Technical Report No. 5. Rome, Italy: Food and Agriculture Organization of the United Nations and United Nations Development Program; Ibu Pejabat Perhutanan, Sememanjung Malaysia (Forest Department Headquarters, Peninsular Malaysia). 1987. Inventori Hutan Nasional II, Semenanjung Malaysia: 1981–1982. Kuala Lumpur: Ibu Pejabat Perhutanan, Sememanjung Malaysia; Food and Agriculture Organization of the United Nations. 1981. Malaysia. A. Peninsular Malaysia. Pp. 277–293 in Forest Resources of Tropical Asia. UN 32/6.1301-78-04, Technical Report No. 3. Rome, Italy: Food and Agriculture Organization of the United Nations.

sion to nonforest uses occurred in forests that were already logged over at the time of Forest Inventory I. The maps in Figure 2 show areas of primary (virgin) and disturbed (logged over) forests in 1972 (Figure 2A) and 1982 (Figure 2B). The maps are based on an analysis by Brown et al. (1991b) and show that most logging in primary forests from 1972 to 1982 occurred in the northern and eastern parts of the peninsula and that most deforestation occurred in the southeast.

According to the land use surveys (Table 3), the changes in forest area were from 68 percent of land area in 1966 to 63 percent during 1974–1975 to 64 percent in 1981. Although the estimate for 1974–1975 is comparable to the estimate for 1970–1972 from Forest Inventory I, the estimate for 1981 is substantially larger than the estimate for 1981–1982 from Forest Inventory II. One reason for the discrepancy is that the estimate of forest area in the 1981 land use survey included grassland/scrub grassland and scrub forest. These areas totaled 589,000 ha in 1974–1975, but this is much less than the discrepancy between the 1981 land use and 1981–1982 inventory estimates, 1.638 million ha.

Is it possible that the areas of grassland/scrub grassland and scrub forest increased nearly threefold from the mid-1970s to the early 1980s? Traveling around the peninsula, one observes few large areas of grassland, but one does encounter large areas of scrub forest and unproductive or idle land in several states. These areas result from abandonment of agricultural land, failure to develop land after it has been logged in preparation for agricultural conversion, and degradation of forests to scrub forests following intensive logging. Logging became increasingly intensive in the 1970s and 1980s as timber markets developed for an increasing percentage of the tree species found in Peninsular Malaysia's forests. According to Forest Inventory II, only 7.5 percent of the total timber volume (for trees with a minimum diameter at breast height of 30 cm) in superior, good, and moderate virgin forests was in trees classified as noncommercial species (Ibu Pejabat Perhutanan, Semenanjung Malaysia, 1987). Moreover, the minimum commercial log diameter is as low as 27 cm in Peninsular Malaysia today. It is generally believed that commercial logging alone does not cause deforestation (see, for example, Lanly [1982]); however, when selective logging approaches clear-felling as a result of extraction of a high proportion of small-diameter trees, clearly commercial logging is a decisive factor.

Another explanation might be that the 1981 land use survey underestimated areas in agricultural use. Although statistics compiled by the Ministry of Agriculture (Malaysia) (1991) and presented in Table 3 indicate that the area in agricultural use in 1988 was little

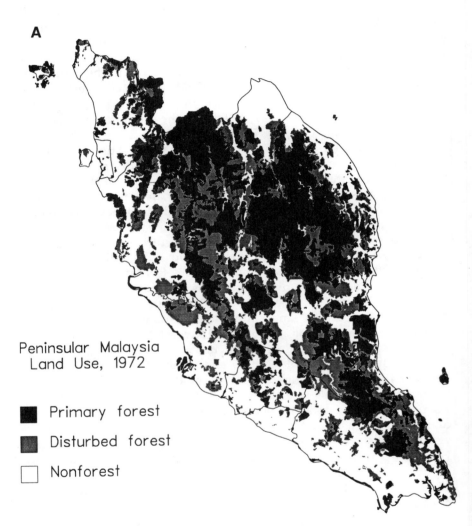

FIGURE 2   Areas of primary (virgin) and disturbed (logged over) forests in
Peninsular Malaysia in (A) 1972 and (B) 1982.   Source: Based on data from
Brown, S., L. Iverson, and A. E. Lugo. 1991b. Land use and biomass changes

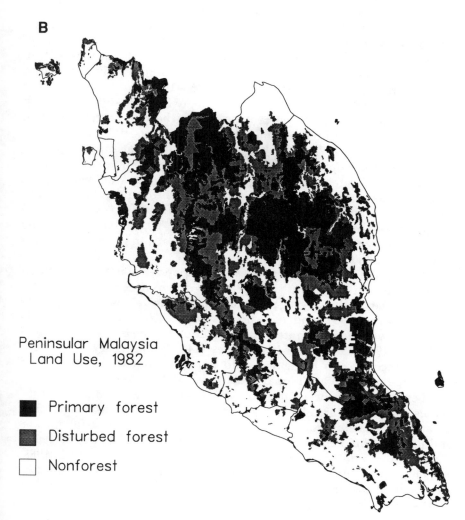

B

Peninsular Malaysia
Land Use, 1982

■ Primary forest
▨ Disturbed forest
□ Nonforest

of forests in Peninsular Malaysia, 1972–1982: Use of GIS analysis. Department of Forestry, University of Illinois, Urbana. Photocopy.

larger than the area indicated by the 1981 land use survey (4.1 million ha), Abu Bakar (1991) estimated the area in 1990 to be 4.8 million ha.

The 1966 land use survey and Forest Inventory I reported mid-year estimates for 1966 and 1972, while Forest Inventory II reported end-of-the-year estimates for 1981. Treating the estimates of total forest area from the three sources as point estimates for mid-1966, -1972, and -1982, the average annual rate of deforestation increased slightly from 134,000 ha/year during 1966–1972 to 141,000 ha/year during 1972–1982. The FAO's (1981) estimate for 1976–1980 was much lower, 90,000 ha/year. In percentage terms, the rate of deforestation rose slightly, from 1.55 percent/year during 1966–1972 to 1.88 percent/year during 1972–1982. The latter rate is about three times the average for the tropics estimated by Lanly (1982).

Table 3 indicates that the major cause of deforestation was expansion of lands in agricultural uses other than shifting cultivation. The expansion of land in agricultural use from 1966 to 1974–1975, 829,000 ha, just about matched the decrease in forest, 782,000 ha. Most of the increase in agricultural area was due to expansion of area in perennial crops, 690,000 ha.

As implied above, the statistical correspondence between agricultural expansion and deforestation broke down after the early 1970s. The increase in the aggregate area in agricultural use between the 1974–1975 and 1981 land use surveys was 89,000 ha/year, which is less than two-thirds the rate of deforestation on the basis of the 1970–1972 and 1981–1982 forest inventories (141,000 ha/year). For every hectare recorded as being put into agricultural use, slightly more than one-half of an additional hectare was deforested.

Between inventories, the Forestry Department of Malaysia estimates total forest area by using annual records on areas logged and cleared for development. The most recent estimate is for 1988, 6.288 million ha (Ibu Pejabat Perhutanan, Semenanjung Malaysia, 1990). This implies a deforestation rate of 89,000 ha/year during 1982–1988, which is substantially lower than the rate during 1972–1982 but much higher than the annual average increase in agricultural area during 1981–1988, 8,000 ha/year (Table 3).

PERMANENT FOREST AREAS

The high rates of deforestation in Peninsular Malaysia have raised concern about the area of land that will be permanently maintained under forest cover. If all the land that is suitable for agriculture is indeed ultimately developed for agriculture, then, at most, 6.7 million to 6.9 million ha of forest will remain. Some of this area will be converted to nonagricultural uses. Still, this constitutes 51 to 52 percent of the peninsula's land area, which is much larger than the 29 percent in Thailand or the probably overstated 37 percent in the Philippines (World Resources Institute, 1990).

The official government policy as of the mid-1980s was to maintain at least 4.75 million ha as permanent forest estate (PFE) (Thang, 1986). Sixty percent of the PFE, or 2.85 million ha, would be productive forests, which would be managed for commercial timber production on a sustainable basis. The remainder would be protective and amenity forests, which would not be logged. Protective and amenity forests would protect watersheds, protect wildlife habitat, and provide recreational opportunities. Outside the PFE, an additional 0.59 million ha would be in national and state parks and wildlife reserves.

As of December 31, 1988, some 4.9 million ha were either classified or in the process of being classified as PFE (Ibu Pejabat Perhutanan, Semenanjung Malaysia, 1990). Although this figure makes it appear that the 4.75 million ha target has already been exceeded, little information is available on the actual forest cover on these lands. Illegal land clearing is known to have occurred within the PFE; moreover, land within the PFE is often legally declassified by state governments for development.

In 1988, the one national park in Peninsular Malaysia, Taman Negara, covered 0.43 million ha, while wildlife and bird sanctuaries covered 0.31 million ha (Kiew, 1991). About two-thirds (0.19 million ha) of the sanctuaries were within the PFE, so protected areas outside the PFE covered 0.55 million ha. This is slightly less than the target of 0.59 million ha. However, some 0.65 million ha has been proposed to be added to the park and sanctuary systems. The proposed area includes a second national park, at Endau-Rompin.

COMMERCIAL LOGGING

Commercial logging is a major source of degradation of virgin forests. Moreover, heavy, repeated logging of forests that results in conversion of the residual stand to scrub forest might explain why deforestation has evidently exceeded agricultural expansion since the early 1970s. Although logging as a source of forest degradation is an important issue, the more relevant issue here is whether logging and agricultural expansion are connected. Background information is provided in this section; most is taken from Vincent and Binkley (1991).

Assuming a timber growth rate of 1.0 to 1.5 m^3/ha/year, the annual sustained yield from Peninsular Malaysia's productive PFE is in the range of 2.85 million to 4.28 million m^3. In contrast, the harvest in 1990 was 10.6 million m^3 (Ibu Pejabat Perhutanan, Semenanjung Malaysia, 1990). Why is the harvest so much larger than the sustained yield? Conversion of forests outside the PFE to tree crop plantations is one reason. Another is harvesting of virgin forests within the PFE, since virgin forests generally carry higher stocks of commercial timber than do second-growth forests. Even if only an area consistent with sustained yield were harvested each year, harvests would exceed sustained yields until all virgin forests within the PFE had been logged over.

For these reasons, a rate of harvest that exceeds sustained yield does not necessarily imply that forest sector development is on an unsustainable trajectory. As timber becomes more scarce, rising stumpage values (log prices minus logging costs) should cause investments in forest management to increase and demand for timber to decrease. If these supply-and-demand adjustments occur, then the rate of harvest should decline and eventually stabilize at the sustained yield level.

These adjustments have been hindered in Peninsular Malaysia by the combination of low timber fees and insecure concession tenure. The fees that states levy on timber concessionaires—which include a combination of royalties assessed on extracted logs and premiums assessed on concession areas—drastically understate stumpage values (Gillis, 1988; Sulaiman, 1977; Teo, 1966; Vincent, 1990). Vincent (1990) estimated that forest revenue systems in Peninsular Malaysia captured only about one-fifth of the stumpage value of forests harvested during 1966–1985. Timber fees in most states of Peninsular Malaysia remained virtually unchanged from the early 1970s until the mid-1980s, despite evidence of rising stumpage values. Hence, these fees failed to signal increasing timber scarcity to State Forestry Offices and the federal Forestry Department, and they failed to generate the revenue needed for public forest management efforts. They also made available to land developers huge profits when forests were clear-felled in preparation for conversion to agriculture and other uses.

TREE CROPS VERSUS NATURAL FORESTS

Expansion of tree crop plantations has been the major cause of deforestation in Peninsular Malaysia (Table 5 and Figure 3). The expansion has occurred in three distinct phases: a rapid phase (49,000 ha/year) during 1904–1932, led by rubber; a slower phase (24,000 ha/year) during 1932–1966, also led by rubber; and the most rapid phase of all (57,000 ha/year) during 1966–1988, led by oil palm. Expansion by private estates dominated the first phase, while expansion by independent small landholdings was more important in the second. By 1961, the area of rubber in small landholdings exceeded the area in estates. Small landholdings in government-sponsored

TABLE 5   Average Annual Changes in Rubber
and Oil Palm Areas in Peninsular Malaysia,
1900–1988

Decade	Hectares per Year (1,000)	
	Rubber	Palm Oil
1900–1910	22	0
1910–1920	66	0
1920–1930	36	2
1930–1940	14	1
1940–1950	5	1
1950–1960	11	2
1960–1970	18	23
1970–1980	–3	63
1980–1988	–21	67

SOURCES: Barlow, C. 1978. Table 2.2 and Appendix Table
3.2 in The Natural Rubber Industry: Its Development, Technology, and Economy in Malaysia. Kuala Lumpur: Oxford
University Press; Department of Statistics (Malaysia). Various issues. Monthly Statistical Bulletin: Peninsular Malaysia. Kuala Lumpur: Department of Statistics; Khera, H. S.
1976. The Oil Palm Industry of Malaysia. Kuala Lumpur:
Penerbit Universiti Malaya; Ministry of Agriculture (Malaysia). Various issues. Statistical Handbook, Agriculture: Malaysia. (Before 1979: Statistical Digest, Ministry of Agriculture and Lands, Peninsular Malaysia.) Kuala Lumpur: Ministry
of Agriculture.

land development schemes, particularly those under the Federal Land
Development Authority (FELDA), became a significant source of expansion during the latest phase. Expansion by private estates was
also important during the latest phase. Vincent and Hadi (1991) provide a brief review of these historical developments; Barlow (1978)
and Bauer (1948) provide more detailed accounts for rubber, while
Barlow (1986) and Khera (1976) do the same for oil palm.

From a policy standpoint, the bottom-line issue is whether Peninsular Malaysia has made itself better off, in the long run, by converting its natural forests to rubber and oil palm plantations. The following sections address four questions pertinent to this issue.

Are Tree Crop Plantations a Sustainable Land Use?

There is ample evidence that rubber and oil palm plantations can
produce stable, in fact, increasing, yields on a long-term basis in Pen-

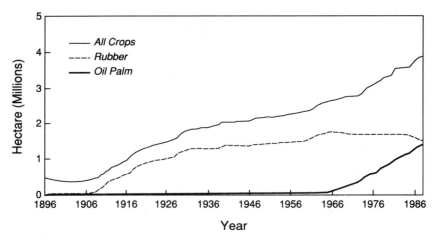

FIGURE 3   Trends in land use for Peninsular Malaysia. Sources: Barlow, C. 1978. Table 2.2 and Appendix Table 3.2 in The Natural Rubber Industry: Its Development, Technology, and Economy in Malaysia. Kuala Lumpur: Oxford University Press; Department of Statistics (Malaysia). Various issues. Monthly Statistical Bulletin: Peninsular Malaysia. Kuala Lumpur: Department of Statistics; Khera, H. S. 1976. The Oil Palm Industry of Malaysia. Kuala Lumpur: Penerbit Universiti Malaya; Ministry of Agriculture (Malaysia). Various issues. Statistical Handbook, Agriculture: Malaysia. (Before 1979: Statistical Digest, Ministry of Agriculture and Lands, Peninsular Malaysia.) Kuala Lumpur: Ministry of Agriculture.

insular Malaysia.   Rubber has been grown on some sites for nearly 100 years, and oil palm for more than 70 years. Yields of both crops continue to increase as a result of the research efforts of the Rubber Research Institute of Malaysia (RRIM) and the Palm Oil Research Institute of Malaysia (PORIM).

Average yields of rubber rose from 492 kg/ha/year (estates and small landholdings combined) during 1929–1930 to 1,103 kg/ha/year for independent small landholdings and 1,428 kg/ha/year for estates in 1982 (Barlow, 1978; Barlow and Jayasuriya, 1987; see also Ministry of Primary Industries [Malaysia], 1990). Barlow (1978) reported rubber yields approaching 2,400 kg/ha/year for available varieties under good management in the 1970s, and he forecast potential yields of 3,500 kg/ha/year. Other investigators have predicted higher yields. Future increases in yields are probable because of the long time between the initiation of research to develop an improved variety and the commercial availability of improved planting stock.

Combined yields of palm oil and palm kernel oil on estates rose from 1,850 kg/ha/year in 1960 to 4,155 kg/ha/year in 1982 (Barlow and Jayasuriya, 1987). Average yields for oil palm would be even higher if many of the prime coastal plain sites were not already under rubber (Hill, 1982). Average yields will rise as these sites are converted to oil palm.

Development of higher yielding varieties is the major reason for these increases, but improved management—planting and harvesting techniques, fertilization, and pest control—and, for rubber, use of chemicals that stimulate higher flows of latex have also been important (Barlow, 1978; Ministry of Primary Industries [Malaysia], 1990; Ng, 1983).

As with any agricultural crop, plantations need inputs to retain their productivity. Fertilizer is a key input for rubber and is even more important for oil palm (J. K. Templeton, World Bank, personal communication, 1990). Both RRIM and PORIM have carried out numerous studies of the responses of these crops to fertilization. Results of these studies can be found in Ng and Law (1971) [cited in Ooi, 1976]. Foster et al. (1985a,b), and Ahmad Tarmizi et al. (1986).

Ng (1983) expressed concern that the mechanical clearing and burning used during replanting of oil palm and the associated erosion and runoff might degrade the long-term productivity of the Ultisol and Oxisol soils. Can the productivity of Peninsular Malaysia's soils for growing rubber and oil palm be maintained in perpetuity through the application of fertilizers and other management inputs? To date, no obvious basis for answering in the negative has become apparent.

Are Tree Crop Plantations Economically Feasible?

Natural forests in Peninsular Malaysia appear to have been replaced by agricultural systems that have a long-term usefulness to humans. Conversion of forests to tree crop plantations could still be undesirable, however, if the economic costs of conversion exceed the economic benefits.

The growth of rubber small landholdings before independence, when government policies discriminated in favor of estates, and the growth of oil palm estates since the mid-1960s offer market-based evidence of the financial feasibility of these two crops. More formal evidence is provided by benefit-cost analyses of estates, independent small landholdings, and land development schemes. Benefit-cost analyses typically distinguish between financial and economic returns. Financial (sometimes termed "private") returns are calculated with costs and benefits measured by market prices. Economic (sometimes termed "social") returns are calculated with costs and benefits measured by shadow prices. Shadow prices perform two functions: (1) they adjust market prices to remove distortions caused by policies or market imperfections, and (2) they quantify (value) the economic importance of goods and services that lack market prices altogether. In theory, shadow prices should reflect environmental impacts. Various aspects of the application of benefit-cost analysis techniques to land development schemes in Peninsular Malaysia have been discussed by Dixon (1977).

Estimates of financial and economic rates of return for investments in rubber and oil palm plantations are presented in Tables 6 and 7. The studies cited used shadow prices, primarily to adjust the costs of labor, capital, and other inputs. None of the studies included shadow prices for environmental impacts. The moderate to high financial rates of return for rubber and oil palm estates and indepen-

TABLE 6   Internal Rates of Return for Rubber Plantations

Plantation Type and Reference	Percent Return	
	Economic	Financial
Estates		
Goering (1968 [cited in Lee, 1978])	—	9.4, 10.9
Ariffin (1977 [cited in Ariffin and Chan, 1978])	—	10.4
Bevan and Goering (1968 [cited in Khera, 1976])	15.4, 25.5	10.9
Goering et al. (1969 [cited in Khera, 1976])	—	10.9, 13.8
Ng (1971 [cited in Barlow, 1978])	—	12.4
Barlow (1978)	31.5	14.3
Pushparajah et al. (1974 [cited in Ariffin and Chan, 1978; Barlow, 1978])	—	23.5, 25.8
Independent smallholdings		
Barlow (1978)	27.1	15.9
Land development schemes		
FELDA		
Syed Hussain (1972)	9, 10	—
Thillainathan (1980)	13.5	4.4–19.9
Lim (1976)	23.3	8.9
State		
Syed Hussain (1972)	14–18	—
Lim (1976)	22.2	7.1
Fringe		
Lim (1976)	22.7	6.0

NOTE: Entries with more than one value are ones for which rates of return were given under various assumptions.

TABLE 7 Internal Rates of Return for Oil Palm Plantations

Plantation Type	Percent Return	
	Economic	Financial
Estates		
Little and Tipping (1972)	13.2–19.4	8.2
Goering et al. (1969 [cited in Khera, 1976])	20.8, 31.0	9.1–16.9
Ariffin (1977, cited in Ariffin and Chan, 1978])	—	12.0
Khera (1976)	18.9–41.2	14.0–22.6
Goering (1968 [cited in Lee, 1978])	—	14.1–16.9
Bevan and Goering (1968 [cited in Khera, 1976])	23.0, 34.2	16.9
Pushparajah et al. (1974 [cited in Ariffin and Chan, 1978; Barlow, 1978])	—	21.6, 26.7
Ng (1971 [cited in Barlow, 1978])	—	29.4
Barlow (1986)	20–25	—
Land development schemes (FELDA)		
Barlow (1986)	15–18	—
Syed Hussain (1972)	22	—
Thillainathan (1980)	28.1	16.2–35.7

NOTE: Entries with more than one value are ones for which rates of return were given under various assumptions.

dent rubber small landholdings explain why the private sector has historically been interested in investing in these crops. The tendency of both financial and economic rates of return to be higher for oil palm than for rubber indicates why, from the 1960s onward, many estates converted from rubber to oil palm and land development schemes increasingly emphasized oil palm. The positive and moderate to high economic rates (at least 10 percent in all but one instance) for land development schemes indicate that, on paper at least, the schemes earned an acceptable rate of return on public investment funds. However, the rates of return tended to be lower than those for other types of ownership, particularly in the case of rubber.

How Have Policies Affected Expansion of Tree Crop Plantations?

The inherent economic feasibility of tree crop plantations in Peninsular Malaysia indicates that fundamental economic forces, not misguided policies, were primarily responsible for conversion of forests to agriculture. Three policies not specific to rubber and oil palm had a crucial, positive impact on the responses of smallholders and estates to these economic forces. First, the government invested in infrastructure, which enabled farmers to transport their products to markets and gain access to goods they did not produce themselves. Hence, farmers were not restricted to subsistence agriculture. Second, the government made it possible to obtain secure land title, generally under permanent or long-term leases (Barlow, 1978). This gave farmers confidence that they would reap the returns of the labor and capital they invested in tree crop plantations. Third, the government organized one of the most productive agricultural research systems in the tropics. As noted earlier, research at RRIM and PORIM is largely responsible for the increasing yields of rubber and oil palm, which has maintained the economic viability of these crops.

Policies within the rubber and oil palm sectors in Peninsular Malaysia were designed, for the most part, to make estates and smallholders pay their own way. In some instances, policies might even have forced plantations to bear more than a fair share of development costs. Evidence on these points in the case of rubber has been provided by Barlow (1978, 1984), Barlow and Drabble (1983), Barlow and Jayasurija (1986), and Power (1971 [cited in Barlow, 1978]).

The government levied taxes on rubber exports and tacked on additional cesses to raise the funds for rubber research and replanting grants (Barlow, 1984). Hence, the research and replanting that rejuvenated the industry after World War II "did not represent a transfer of resources from other sectors but, in effect, financing provided by the industry itself" (Lee, 1978:222). Even land development schemes had aspects of self-financing, as settlers were expected to pay back, with interest, the greater portion of government investments made on their behalf (Barlow, 1978). Singh (1968) estimated that 64 to 67 percent of government expenditures in three FELDA rubber schemes would be recovered from settlers by loan repayments and taxes.

Independent smallholders received few benefits from the government. During the 1920s–1940s, first the Stevenson Scheme and then the International Rubber Regulation Agreement hindered their expansion. According to Barlow (1978), "Without the[se] schemes the area of small landholdings in the Malay Peninsula would certainly have expanded far more" (p. 72). All small landholdings up to the mid-1950s were established without subsidies (Barlow and Jayasurija, 1986). Although they became eligible for replanting grants at that time, they appear to have borne a disproportionate share of the financing burden relative to the grants they received (Barlow, 1978).

In the 1960s, state and federal governments largely excluded small landholdings from developing new land because the government favored land development schemes (Barlow, 1984).

There were also restrictions placed on estates. After World War II, and particularly after independence in 1957, the government took steps to prohibit foreign-owned estates from acquiring new land, in an effort to create opportunities for local ownership, particularly by Malays (Barlow, 1984). This effort became more aggressive in the late 1970s when the National Equity Corporation began buying out foreign shares of estates (Barlow, 1984).

Thus, there is little doubt that policies hindered expansion of plantations by the private sector, especially smallholders. This contrasts with the active government promotion of land development schemes beginning in the late 1950s. These schemes might have earned double-digit economic rates of return, but were they necessary, and did they earn the maximum rates of return? Lee (1978) claimed that there was no indication that the private sector had an inability or unwillingness to undertake large land development schemes. Although Barlow (1986) and Barlow and Condie (1986) acknowledged that rural credit markets might have been unable or unwilling to provide the long-term credit needed by smallholders to establish sizable plantations, Barlow (1978) doubted that centralized development schemes were the only way to overcome this problem. He also disputed the argument that schemes were justified because of increasing returns to scale, particularly in the case of oil palm (Barlow, 1986). He argued that land development programs based on assisting independent smallholders would have been less costly for the government (the costs per hectare for rubber could have been reduced to two-thirds those of FELDA), would have enhanced efficiency and flexibility by placing more decisions in the hands of smallholders, and would have increased household income and economic independence.

Schemes did not always achieve their projected high rates of return. Fringe rubber schemes suffered from widespread abandonment and helped prompt the creation of the Federal Land Consolidation and Reclamation Authority (FELCRA) (Barlow, 1978). The poor performance of schemes has been linked to political motivations for land development (Guyot, 1971; Syed Hussain, 1972). Guyot (1971) concluded that a major reason why schemes were more successful in the state of Johor than in the state of Terengganu was because there was a greater tendency to develop land primarily to recruit and reward political supporters in Terengganu. Consequently, the sites were chosen poorly, and the state government provided little technical support to the settlers, who generally had little experience growing tree crops. Of 54 state and 43 fringe schemes for which land was alienated in Terengganu, only 22 and 1, respectively, were actually developed to the planting stage.

In addition to vote-seeking, schemes were sometimes motivated by rent-seeking. Land alienation for proposed schemes has been used as an excuse to grant timber concessions and thereby capture windfall stumpage values. Lee (1978) claimed that "There had been obvious cases of abuse in that the recipients of alienated land were more interested in removing the timber on the land rather than in its subsequent development" (p. 406), and he backed up this claim with data showing that only 58 percent of the land designated for agriculture during 1961–1970 was actually developed. No other historical data have ever been compiled on the amount of land that was alienated and logged, but never developed, in Peninsular Malaysia.

Although land development schemes do not appear to have been the most economically efficient means of promoting smallholder rubber and oil palm, this does not necessarily imply that less forest would have been converted in the absence of land development schemes. For example, estates and independent smallholders might have picked up the slack if policies had been less discriminatory toward them.

What Are the Environmental Impacts of Conversion of Natural Forests to Tree Crop Plantations?

The most obvious omission from benefit-cost analyses of rubber and oil palm plantations is the environmental impact. Environmental impacts of converting natural forests to tree crop plantations include increased soil erosion, increased variability of stream flows, and loss of biodiversity. Efforts to quantify these environmental impacts in economic terms and, thereby, to incorporate them directly into benefit-cost analyses remain rudimentary not only in Malaysia but also in most other tropical countries. Although resource economists have developed an array of nonmarket valuation techniques during the past 3 decades, these methods have received little application in developing countries. There are exceptions, however (Dixon and Hufschmidt, 1986; Dixon and Sherman, 1990; Hufschmidt et al., 1983; Vincent et al., 1991).

If the net environmental impacts of conversion of natural forests to tree crop plantations are negative, then market forces might lead to excessive expansion of plantations by the private sector (estates and independent smallholders). Furthermore, if decisions about land development schemes are based on project appraisals that ignore these impacts, the decisions would be biased toward acceptance of the schemes. Hence, in the presence of negative environmental impacts and incomplete benefit-cost analysis, market and policy failures are created; these failures, in theory, might lead to the excessive conversion of natural forests to tree crop plantations.

The sections below review the physical information on the environmental impacts of conversion of natural forests to tree crop plantations. Although the information is not presented in economic terms, it does provide insights into the extent to which plantations provide environmental services comparable to those provided by natural forests. The review focuses on three services for which the most information is available: soil conservation, protection of water systems, and preservation of biodiversity. Most of the information pertains to rubber plantations, for which environmental impacts have been the most studied (Aiken et al., 1982). Brown et al. (1991a,b) provide information on a fourth service, the sequestration of carbon in woody biomass in natural forests (but not tree crop plantations).

SOIL CONSERVATION

For rubber and oil palm, the risk of soil erosion is greatest during plantation establishment and replanting. First, the natural forest (or the old plantation) is logged if commercial timber is present in sufficient quantities. Then, the remaining vegetation is allowed to dry; when it is sufficiently dry, it is pushed into piles and burned. Finally, heavy machinery is used to terrace the site (if it is a new plantation) and prepare it for the planting of ground covers and rubber trees or oil palms. Typically, several months elapse from the time the site is logged until ground cover is established, and several years pass before the tree canopy closes. The amount of erosion that occurs depends on the erosivity of the rainfall, the erodibility of the soil, and the speed at which ground cover is established and the tree crop canopy closes.

Mean annual erosivity exceeding 15,000 J/m² places the entire east coast, the portion of the rubber belt on the west coast from Kuala Lumpur to Pinang, and most of the land development schemes in the states of Terengganu, Pahang, and Johor at high risk for soil erosion (Morgan, 1974 [cited in Soong et al., 1980]; Morgan, 1979). Policies and practices in Peninsular Malaysia recognize that erodibility increases with increasing slope. The Conservation Enactment requires that lands alienated for agriculture have a slope of less than 18.3°. Since World War II, terracing has become a standard practice for plantations established on slopes (Aiken et al., 1982).

Erosion was a greater problem earlier in the twentieth century because of the rubber estates' policy of "clean weeding"—removing all surface growth at the time of planting and keeping the soil surface clear even after trees became established. The rationale was that clean weeding would make nutrients more available and would inhibit diseases. Instead, it created serious erosion problems. Fermer (1939 [cited in Aiken et al., 1982]) estimated that rubber estates that were clean weeded lost an average of 8 cm of topsoil during 1902–1939. As a consequence, the productivity of vast areas was seriously reduced, and some plantations were abandoned (Barlow, 1978). Clean weeding began to be replaced after the mid-1920s by the planting of various types of leguminous ground covers soon after clearing (Barlow, 1978). Ground covers can reduce soil loss by 35 to 87 percent compared with the amount of soil lost from bare soils (Ling, 1976 [cited in Aiken et al., 1982:Table 7.2]). Moreover, the nitrogen provided by legumes saves on fertilizer expenses (Ti et al., 1971 [cited in Soong et al., 1980]). Although planted ground covers die off as the shade from rubber trees increases, they are replaced by natural ground covers that also control erosion (Rubber Research Institute of Malaysia, 1973 [cited in Soong et al., 1980]).

Even with a well-established ground cover and even after rubber trees have become established, soil erosion is greater than that in a natural forest. This is despite evidence that canopy interception of rainfall by mature rubber plantations is similar to that by natural forests (Aiken et al., 1982:Table 7.3). Morgan (1979) found that suspended sediment transport (for a strip 1 cm wide and 10 m long) in a rubber plantation was nearly double that in a natural forest. Aiken et al. (1982:Table 7.5) found that suspended sediment transport ranged from 0.058 to 2.63 cm^3/cm/year for a rain forest to 6.66 to 41.89 cm^3/cm/year for a rubber plantation. Whether these rates are sufficiently high to undermine the long-term sustainability of rubber plantations is not clear.

Soil loss is probably less in oil palm plantations, even though the canopy remains more open for a longer period of time, because oil palm plantations tend to be established on slopes that are less steep (Aiken et al., 1982; Soong et al., 1980).

PROTECTION OF WATER SYSTEMS

Conversion to tree crop plantations has three principal environmental impacts on water systems: increased sedimentation, increased flooding, and increased pollution. Sedimentation increases because of greater soil erosion. Clean weeding caused particularly acute sedimentation problems. Aiken et al. (1982) indicated that, "During the early 1930s, 2,835 ha (7,000 acres) of paddy land along the Malacca River had to be abandoned because of inundation by silt eroded from rubber estates upstream" (p. 122). They also indicated that the harbor at the mouth of the Malacca River was so badly affected by siltation that it had to be dredged on a regular basis. The off-site impacts of clean weeding led to the Silt Control Enactment of 1917, which "empowered the State Resident to take action against any person who allowed sediment eroded from his land to damage or interfere with the cultivation of neighbouring land" (Aiken et al., 1982).

Today, because of improved management practices, rates of soil erosion from agricultural land are generally much lower than those common in the early decades of the twentieth century, but because of the increased areas affected, river sediment loads may be no lower (Aiken et al., 1982). The sediment load in the Pahang River, which is the largest river in Peninsular Malaysia, more than tripled from the start of the twentieth century to 1975 because of increased logging and land conversion (Australian Engineering Consultants, 1974, cited in Aiken et al. [1982:Table 7.15]).

Rainfall runoff increases because of lower rates of canopy interception (at least in immature plantations), more compacted soil, and reduced humus. Increases in total annual runoff are relatively modest, about 10 percent (Tan, 1967; Hunting Technical Services et al., 1971 [both cited in Aiken et al., 1982]). Most of the increase comes during periods of peak rainfall, which increases the frequency and magnitude of floods (Aiken et al., 1982) and might diminish the recharging of aquifers. Daniel and Kulasingham (1964 [cited in Soong et al., 1980]) found that peak runoff per unit area was about twice as large in a catchment largely converted to rubber and oil palm as in one that was undisturbed natural forest. In a similar comparison of catchments, Hunting Technical Services et al. (1971 [cited in Aiken et al., 1982]) reported increases in peak runoff ranging from 34 to 140 percent during six periods of high rainfall in 1970.

Plantations contribute to water pollution as a result of fertilizer, pesticide, and herbicide runoffs and processing wastes. Maene et al. (1979 [cited in Ng, 1983]) estimated that runoff and leaching from oil palm plantations resulted in the loss of 17 percent of the nitrogen, 10 percent of the phosphorus, and 9 percent of the potassium fertilizers applied. Pesticide and herbicide use has been regulated more strictly since the Pesticides Act of 1974.

The major source of pollution related to tree crops is effluent from processing mills. Processing of rubber and palm oil requires large amounts of water, and the effluent contains organic and inorganic compounds that lead to high chemical and biological oxygen demand (BOD) (Aiken et al., 1982). According to Gill (1978 [cited in Hill, 1982:205]), in the 1970s "oil palm factories contribute[d] 80 percent of all pollutants to rivers in Peninsular Malaysia." The biological oxygen demand produced by palm oil processing mills in 1978 was estimated to be equivalent to the amount produced by domestic sewage from 15.9 million people—a population greater than that of Peninsular Malaysia today (Abdul Aziz bin Ahmad, 1974 [cited in Aiken et al., 1982:Table 7.11]). Discharge of effluents was reduced dramatically following the passage of the 1974 Environmental Quality Act, which incorporated an innovative combination of a regulatory standard and a market-oriented discharge fee (Panayotou, 1992). By 1984, the total biological oxygen-demand load released from palm oil mills was less than 1/40th the load in 1978 (Ong et al., 1987 [cited in Panayotou, 1992]).

PRESERVATION OF BIODIVERSITY

The number of forest-dwelling species that can survive in tree crop plantations is small. Wells (1988) estimated that "fewer than 20 [species of] birds of inland forest have effectively established themselves beyond the limits of original habitat" (p. 193). He claimed that no monocultural agricultural system has yet been shown to support a breeding population of forest-dwelling birds. Yorke (1984) estimated that about 50 percent fewer bird species were recorded in rubber plantations than in neighboring primary forests and that most of the species were more typical of disturbed habitat than primary forests. The Earl of Cranbrook (1988) pointed out that small indigenous mammals that adapt to early successional stages of forest regeneration thrive as pests in tree crop plantations, but he concluded that most forest-dwelling mammal species cannot exist outside mature natural forests. Steven (1968 [cited in Earl of Cranbrook, 1988]) estimated that only 10 percent of the mammal species other than bats in Peninsular Malaysia can subsist in cultivated areas. Fifty-two percent of the mammal species in Peninsular Malaysia are native to forests below 300 m, which is where most plantations are found (Aiken and Leigh, 1985; Aiken et al., 1982).

Potential decreases in biodiversity because of agricultural expansion are not limited to terrestrial ecosystems. Pollution and sedimentation have reduced fish populations in streams and coastal areas (Aiken and Moss, 1976 [cited in Aiken et al., 1982]). Siltation and sedimentation might have contributed to the disappearance of the dugong in coastal waters, the decline of the river terrapin in the

Perak River, and the decline of coral reefs (Aiken and Leigh, 1985; see also Langham, 1976; Lulofs, 1974 [both cited in Aiken et al., 1982]).

Rubber plantations appear to have greater value as wildlife habitat than do oil palm plantations (Duckett, 1976). Rubber plantations tend to contain more pockets of remnant natural forest, generally wet areas where rubber trees grow more poorly than oil palm trees do. Crowns of rubber trees provide better nesting conditions for birds and small mammals and are disturbed less by the collection of latex than are oil palm crowns by the collection of fruit bunches. On the other hand, oil palm fruits are more attractive to wildlife.

Plantations have better value as buffer zones around remaining natural forests than do annual agricultural fields, since they shade the edge of the forest. To a certain extent, plantations can also serve as corridors between patches of natural forest for certain species, but their effectiveness as corridors decreases sharply as the distance between the forest patches increases (J. Wind, National Park Development Project, Bogor, Indonesia, personal communication, 1990). Moreover, the species-richness of many of Peninsular Malaysia's remnant patches of lowland forest has diminished as these patches have become increasingly isolated and reduced in size in a landscape dominated by rubber and oil palm.

FUTURE PROSPECTS

How much further is agricultural expansion likely to proceed in Peninsular Malaysia? As noted earlier, land in agricultural use covered 4.2 million ha in 1988. Because 6.3 million to 6.5 million ha of soils is suitable for agriculture, agricultural expansion could theoretically result in a maximum of 2.1 million to 2.3 million ha of deforestation in the future. On the basis of soil suitability, both rubber and oil palm could expand well beyond their current areas. In 1988, 1.6 million ha was in rubber; 3.6 million to 5.7 million ha is suitable for the crop (Ariffin and Chan, 1978; Barlow, 1978). Some 1.5 million ha was in oil palm, and 3.3 million to 5.0 million ha is suitable (Ariffin and Chan, 1978; Barlow, 1978; Lee, 1978; Ng, 1968 [cited in Ooi, 1976]).

Recent Developments in the Tree Crop Sector

Recent developments suggest that neither crop is likely to expand to cover all the area for which it is technically suitable. The rate of expansion for the four major tree crops decreased from 83,000 ha/year during 1975–1981 to 34,000 ha/year during 1981–1988 (Table 3). A number of factors are responsible for dampening the rate. Rising scarcity of rural labor is perhaps the most important. Estates, particularly rubber estates (which are more labor intensive), have suffered increases in labor costs as rural people have migrated to urban areas (Barlow, 1984; Barlow and Condie, 1986; Barlow and Jayasurija, 1986; Ministry of Primary Industries [Malaysia], 1990). Immigrant workers from other Asian countries, Indonesia and the Philippines in particular, have provided an important source of replacement labor (Barlow and Jayasuriya, 1987; Tsuruoka, 1991). One source estimated that 300,000 Indonesians worked in the palm oil industry in Malaysia (mainly eastern Malaysia) in 1991 (Tsuruoka, 1991). FELDA and FELCRA schemes have also faced labor shortages (Barlow, 1986), as the migration of the population out of rural areas has reduced the number of potential new smallholders and reduced the work force in existing smallholder households. By the 1980s, rises in the opportunity cost of rural labor had cut the economic rates of return to rubber schemes to a borderline level (Barlow and Jayasuriya, 1987).

Two additional factors are government revenue and commodity prices. Expansion of land development schemes in the 1970s benefited from a windfall of government revenue created by oil production (Malaysia is a net petroleum exporter). This source of funds was reduced sharply in the 1980s when oil prices fell. Government expenditure is also constrained by Malaysia's debt burden, although this is lightening because of continued strong economic performance and financial measures by the government.

Although rubber and palm oil prices boomed after 1972, more recently they have dropped and appear to have resumed their long-term decline in real (inflation-adjusted), if not nominal, terms. Natural rubber faces competition from synthetic rubber, whose price is heavily dependent on the price of petroleum. Hence, low petroleum prices negatively affect the economics of rubber schemes in two ways. Malaysian palm oil faces competition not only from palm oil produced in Indonesia (where labor costs are much lower) but also from a host of other fats and oils.

To some degree these three negative factors are offset by research that improves the economic returns to tree crop cultivation (Ministry of Primary Industries [Malaysia], 1990). Both RRIM and PORIM are conducting research on mechanization and other means of reducing labor needs, including less frequent tapping systems for rubber. Efforts are under way to reduce the period of immaturity for both crops and to develop intercropping systems that provide additional economic returns. Wood from rubber trees has become an internationally valuable furniture wood, so much so that the Malaysian government recently imposed levies to restrict the export of logs and lumber from rubber trees. The development of commercial uses of oil palm trunks is more difficult because of their monocotyledonous wood anatomy, but pilot projects are under way. More promising is the development of new industrial products from palm oil. By-products of palm oil processing are increasingly used as an inexpensive fertilizer, which also helps to reduce pollution problems. An oleochemical industry is developing; detergents, lubricants, pharmaceuticals, and polyurethane are among the products that can be made from palm oil (Tsuruoka, 1991).

In spite of this, even the Malaysian government doubts that these research advances can fully offset the negative impacts of labor scarcity, limited public funds, and commodity price declines. The Ministry of Primary Industries (1990) projects that the area of rubber plantations will continue to decline marginally in both the estate and smallholder sectors. The Ministry expects growth in the area of oil palm plantations to slow as estates and smallholders emphasize upgrading existing plantations by replanting with improved varieties. The Ministry of Rural Development recently announced that the government will not open additional land for new agricultural schemes (New Straits Times [Kuala Lumpur], ca. July 15, 1991). In line with this new policy, the Ministry of Rural Development has proposed that FELDA, FELCRA, and the Rubber Industry Smallholder Development Authority (RISDA) be merged and reoriented toward land rehabilitation, market assistance, and enhancing the productivity of existing land development schemes.

Although significant additional expansion of rubber and oil palm plantations is not anticipated, it is conceivable that a new tree crop could follow oil palm and lead a new burst of agricultural expansion. Cacao is the crop that has expanded most rapidly recently, partly because its price trend has been more favorable. Soils suitable for cacao, however, overlap those where rubber, oil palm, and coconut plantations are already established. In 1988, cacao covered only 142,000 ha (Table 3), and it is the optimal crop on only 708,000 ha (Ariffin and Chan, 1978). Because of the peninsula's rural labor shortage, it seems unlikely that there is a tree crop that could generate sufficient economic returns to justify the establishment of plantations in newly cleared areas of forests.

Deforestation Projections

Deforestation for the period 1990–2030 was forecast by using a regression equation that compared the area under agricultural use

TABLE 8 Deforestation Scenarios

| Scenario | Hectares (1,000) | | | Ending Decadal | |
	Beginning Forest Cover	Annual Loss	Decadal Loss	Forest Cover	Percent Loss
Base case[a]					
1990–2000	6,110	33	334	5,776	5.5
2000–2010	5,776	34	343	5,433	5.9
2010–2030	5,433	37	373	4,687	6.9
Worst case[b]					
1990–2000	6,110	62	622	5,488	10.2
2000–2010	5,488	64	637	4,851	11.6
2010–2030	4,851	69	694	3,463	14.3
Best case[c]					
1990–2000	6,110	30	302	5,808	4.9
2000–2010	5,808	16	158	5,651	2.7
2010–2030	5,651	–3	–60	5,711	–0.5

[a]Rural population growth rate is assumed to equal 0.83 percent/year, and deforestation is assumed to equal agricultural expansion.

[b]Rural population growth rate is assumed to equal 0.83 percent/year, and deforestation is assumed to be 1.86 times agricultural expansion.

[c]Rural population growth rate is assumed to equal 0.53 percent/year during 1990–2000 and –0.45 percent/year during 2000–2030, and deforestation is assumed to equal agricultural expansion.

from 1904 to 1988 to logged area and rural population growth rate (Vincent and Hadi, 1991). Three scenarios were considered: scenario 1, the base case, in which the rural population grows at 0.83 percent/year and the area deforested equals the area of agricultural expansion; scenario 2, the worst case, in which the rural population grows at 0.83 percent/year and the area deforested equals 1.86 times the area of agricultural expansion; and scenario 3, the best case, in which the rural population grows at 0.53 percent/year from 1990 to 2000 and –0.45 percent/year from 2000 to 2030 and the area deforested equals the area of agricultural expansion.

The 0.83 percent/year rural population growth rate is the rate during the 1980s. The 0.53 and –0.45 percent/year rates are based on the World Bank's (1990) projections of the overall population growth rate. The factor 1.86 is based on the ratio of the area deforested to the area of agricultural expansion during 1972–1982.

The projections are presented in Table 8. The estimate of forest area in 1990, 6.11 million ha, is based on the Forestry Department's estimate for 1988, 6.288 million ha (Ibu Pejabat Perhutanan, Semenanjung Malaysia, 1990), reduced by the annual rate of deforestation (89,000 ha) during 1982–1988 calculated from the 1988 estimate of area and the estimate of Brown et al. (1991b) for area in 1981–1982.

In the base-case scenario, annual deforestation during 1990–2030 is less than half that during 1982–1988. The level keeps rising, however, because of the steadily growing rural population. In 2030, the amount of remaining forest is comparable to the target area of the PFE (4.75 million ha). Because of continuing population growth, deforestation continues beyond 2030.

In the worst-case scenario forests remain in 2030, but the area is less than three-fourths of the target area of the PFE. As in the base-case scenario, the level of deforestation keeps rising beyond 2030.

The best-case scenario is similar to the base-case scenario until 2000. After 2000, the rate of deforestation slows and then goes to zero in 2016. Aggregate deforestation is negative during 2010–2030, indicating that forest area increases because of net abandonment of agricultural land. In 2030, Peninsular Malaysia would have only 6.5 percent less forest than it did in 1990.

The best-case scenario is the most likely. Stabilization of Peninsular Malaysia's forest area is under way because of the region's sustainable tree crop industries, which make land developed for agriculture permanently productive, and because of the growth in its economy's nonagricultural sectors, which leads to urbanization and declines in rural population growth. This conclusion is in contrast to that of another recent study of Peninsular Malaysia by Brookfield et al. (1990), which warns that "It seems not improbable that worse is to come before improvement" (p. 507).

SUMMARY

Deforestation in Peninsular Malaysia during the twentieth century demonstrates that shifting cultivation is not a necessary ingredient for extensive conversion of forests in the humid tropics and that sustainable agriculture is possible even on nutrient-poor tropical soils. It also demonstrates that the creation and adoption of sustainable agricultural systems will not, on their own, forestall the expansion of agriculture into undisturbed forests. In fact, the sustainability of rubber and oil palm plantations is a fundamental reason why their area has expanded: their ability to produce ongoing yields increased the area where they earned minimum acceptable economic returns. Deforestation might have been even greater, however, if farmers in Peninsular Malaysia had not had the option of tree crop farming and had resorted to shifting cultivation instead. In recent years, rapid industrialization has created off-farm employment opportunities that have led to labor shortages in rural areas and thus decreased agricultural expansion. The phase of land development marked by deforestation appears to be coming to a rapid close in Peninsular Malaysia.

Expansion of plantations has not resulted from government policies that subsidized the expansion. Rather, it has been driven by the moderate to high financial returns (for estates and small landholdings) and economic returns (for land development schemes) earned by the plantations. The fundamental economic feasibility of plantations has been buttressed by government policies to develop infrastructure, promote secure land tenure, and support agricultural research. Although many policies probably discriminated against expansion by estates and small landholdings during most of the century, policies related to land development schemes have created economic inefficiencies.

Although rubber and oil palm plantations appear to provide sustainable uses of converted forestland, environmental costs have been incurred during the conversion process. The failure of markets (for estates and small landholdings) and project appraisals (for land development schemes) to account for environmental impacts suggests that the area of plantations might have expanded too far. The economic data needed to evaluate these impacts and to determine whether overexpansion affected a significant area do not exist. Nevertheless, sufficient information is available to cast doubt on the contention of some authors that conversion of forests to tree crops in Peninsular Malaysia has been an environmental disaster (Aiken et al., 1982; Aiken and Leigh, 1985; Brookfield et al., 1990). Soil erosion and water-related problems have lessened over time because of better conservation practices (ground cover management, terracing) and increasingly stringent water pollution policies. Although populations of many species are shrinking as the few remaining areas of lowland rain forests are converted to other uses, there is little evidence of large-scale extinctions. Moreover, environmental impacts surely would have been greater if farmers in Peninsular Malaysia had lacked the option of sustainable tree crop plantations and had practiced shifting cultivation instead.

Research Needs

Several research needs emerge from the study of Peninsular Malaysia. First, the discrepancy between estimates of agricultural expansion from land use surveys and estimates of deforestation from forest inventories needs to be explained. Perhaps the next forest inventory will help in this regard, but what is truly needed is an updated, comprehensive, detailed land use inventory. Second, areas that were alienated for agriculture and then logged but never developed need to be studied to understand better the political economy of agricultural expansion, particularly in the case of land development schemes. Third, benefit-cost analyses that incorporate values for environmental impacts need to be carried out for private and public plantation investments. Such analyses would provide better estimates of the net benefits of past agricultural expansion and would help to ensure that future expansion creates net benefits.

Replicating Peninsular Malaysia's Success

The possibility of replicating Peninsular Malaysia's twofold success—enhancing rural standards of living from the use of perennial crops and slowing deforestation by the combination of sustainable agriculture technologies and reductions in rural population growth—needs to be studied by careful comparison of Peninsular Malaysia's ecologic, social, and economic conditions with those of other regions in the humid tropics. The factors involved in Peninsular Malaysia's success included an active research program that raised yields and reduced the costs of growing tree crops (and thereby offset declines in product prices), public investments in infrastructure that enabled growers to get latex and palm oil to markets efficiently and to purchase food and other supplies they did not produce themselves, and land tenure policies that enabled estates and smallholders to obtain secure, long-term leases or outright ownership. Although some might argue that other tropical countries lack the financial resources to replicate the first two factors, the research effort was financed by taxes paid by the tree crops sector itself. Land titling in Peninsular Malaysia was facilitated by the peninsula's low population density, but forested areas in many other humid tropical countries are also lightly populated.

Other countries might also face stiffer competition in entering rubber and palm oil markets than did Peninsular Malaysia because Peninsular Malaysia entered the markets early on in their development. This timing issue is a less important factor in Peninsular Malaysia's success, however, than was the effort it put into research, infrastructure, and land titling. Moreover, market opportunities for other countries might be created as Peninsular Malaysia's competitive position in rubber and palm oil continues to be eroded by rising labor costs.

REFERENCES

Abu Bakar, b. M. 1991. Kesan pembangunan pertanian terhadap sumber tanah perhutanan di Semenanjung Malaysia. Paper presented at Seminar Kebangsaan Berkenaan dasar Perhutanan Negara, Economic Planning Unit, Kuala Lumpur, January 21–22, 1991.

Ahmad Tarmizi, M., H. L. Foster, Z. Z. Zakaria, and C. S. Chow. 1986. Statistical and Economic Analysis of Oil Palm Fertiliser Trials in Peninsular Malaysia. PORIM Occasional Paper No. 22. Kuala Lumpur: Palm Oil Research Institute of Malaysia.

Aiken, S. R., and C. H. Leigh. 1985. On the declining fauna of Peninsular Malaysia in the post-colonial period. Ambio 14(1):15–22.

Aiken, S. R., C. H. Leigh, T. R. Leinbach, and M. R. Moss. 1982. Development and Environment in Peninsular Malaysia. Singapore: McGraw-Hill International.

Ariffin, b. M. N., and H. Y. Chan. 1978. Strategical Changes and Optimum Land Use Alternatives for Perennial Agriculture in Peninsular Malaysia. Kuala Lumpur: Research Institute of Malaysia.

Barlow, C. 1978. The Natural Rubber Industry: Its Development, Technology, and Economy in Malaysia. Kuala Lumpur: Oxford University Press.

Barlow, C. 1984. Institutional and policy implications of economic change: Malaysian rubber, 1950–1983. Department of Economics, Research School of Pacific Studies, Australian National University, Canberra. Photocopy.

Barlow, C. 1986. Oil Palm as a Smallholder Crop. PORIM Occasional Paper No. 21. Kuala Lumpur: Palm Oil Research Institute of Malaysia.

Barlow, C., and C. Condie. 1986. Changing economic relationships in Southeast Asian agriculture and their implications for small farmers. Department of Economics, Research School of Pacific Studies, Australian National University, Canberra. Photocopy.

Barlow, C., and J. Drabble. 1983. Government and the emerging rubber industries in the Netherlands East Indies and Malaya, 1900–40. Revised version of paper presented at the Conference on Indonesian Economic History in the Dutch Colonial Period, Canberra, December 16–18, 1983. Photocopy.

Barlow, C., and S. K. Jayasurija. 1986. Stages of development in smallholder tree crop agriculture. Dev. Change 17:635–658.

Barlow, C., and S. K. Jayasuriya. 1987. Structural change and its impact on traditional agricultural sectors of rapidly developing countries: The case of natural rubber. Agric. Econ. 1(2):159–174.

Barlow, H. S. 1988. Forest lepidoptera. Pp. 212–224 in Key Environments: Malaysia, Earl of Cranbrook, ed. Oxford: Pergamon.

Bauer, P. T. 1948. The Rubber Industry: A Study in Competition and Monopoly. Cambridge, Mass.: Harvard University Press.

Brookefield, H., F. J. Lian, L. Kwai-Sim, and L. Potter. 1990. Borneo and the Malay Penninsula. Pp. 495–512 in The Earth as Transformed by Human Action, B. L. Turner II, W. C. Clark, R. W. Kates, J. F. Richards, J. T.

Mathews, and W. B. Meyer, eds. Cambridge, U.K.: Cambridge University Press, with Clark University.

Brown, S., A. J. R. Gillespie, and A. E. Lugo. 1991a. Biomass of tropical forests of south and southeast Asia. Can. J. Forest Res. 21(1):111–117.

Brown, S., L. Iverson, and A. E. Lugo. 1991b. Land use and biomass changes of forests in Peninsular Malaysia, 1972–1982: Use of GIS analysis. Department of Forestry, University of Illinois, Urbana. Photocopy.

Department of Statistics (Malaysia). Various issues. Monthly Statistical Bulletin: Peninsular Malaysia. Kuala Lumpur: Department of Statistics.

Dixon, J. A. 1977. Some Economic Aspects of Rural to Rural Migration and Land Settlement in East Asia. Ph.D. dissertation. Harvard University, Cambridge, Massachusetts.

Dixon, J. A., and M. M. Hufschmidt, eds. 1986. Economic Valuation Techniques for the Environment. Baltimore: Johns Hopkins University Press.

Dixon, J. A., and P. B. Sherman. 1990. Economics of Protected Areas. Washington, D.C.: Island.

Duckett, J. E. 1976. Plantations as a habitat for wild life in Peninsular Malaysia with particular reference to the oil palm (*Elaeis guineensis*). Malayan Nature J. 29(3):176–182.

Earl of Cranbrook. 1988. Mammals: Distribution and ecology. Pp. 146–166 in Key Environments: Malaysia, Earl of Cranbrook, ed. Oxford: Pergamon.

Economic Planning Unit (Malaysia). 1980. Land Resources Report of Peninsular Malaysia, 1974/1975. Kuala Lumpur: Prime Minister's Department.

Food and Agriculture Organization (FAO). 1973. A National Forest Inventory of West Malaysia, 1970–72. FO:DP/MAL/72/009, Technical Report No. 5. Rome, Italy: Food and Agriculture Organization of the United Nations and United Nations Development Program.

FAO. 1981. Malaysia: A. Peninsular Malaysia. Pp. 277–293 in Forest Resources of Tropical Asia. UN 32/6.1301-78-04, Technical Report No. 3. Rome, Italy: Food and Agriculture Organization of the United Nations.

FAO. 1991. 1989 Yearbook of Forest Products: 1978–1989. Rome, Italy: Food and Agriculture Organization of the United Nations.

Foster, H. L., K. C. Chang, M. T. H. Dolmat, A. T. Mohammed, and Z. Z. Zakaria. 1985a. Oil Palm Yield Responses to N and K Fertilizers in Different Environments in Peninsular Malaysia. PORIM Occasional Paper No. 16. Kuala Lumpur: Palm Oil Research Institute of Malaysia.

Foster, H. L., M. T. H. Dolmat, and Z. Z. Zakaria. 1985b. Oil Palm Yields in the Absence of N and K Fertilizers in Different Environments in Peninsular Malaysia. PORIM Occasional Paper No. 15. Kuala Lumpur: Palm Oil Research Institute of Malaysia.

Gillis, M. 1988. Malaysia: Public policies and the tropical forest. Pp. 115–164 in Public Policies and the Misuse of Forest Resources, R. Repetto and M. Gillis, eds. Cambridge, U.K.: Cambridge University Press.

Guyot, D. 1971. The politics of land: Comparative development in two states of Malaysia. Pacific Affairs 44(3):368–389.

Hill, R. D. 1982. Agriculture in the Malaysian Region. Budapest: Akademiai Kiado.

Hufschmidt, M. M., D. E. James, A. D. Meister, B. T. Bower, and J. A. Dixon. 1983. Environment, Natural Systems, and Development. Baltimore: Johns Hopkins University Press.

Ibu Pejabat Perhutanan, Semenanjung Malaysia (Forestry Department Headquarters, Peninsular Malaysia). 1987. Inventori Hutan Nasional II, Semenanjung Malaysia: 1981–1982. Kuala Lumpur: Ibu Pejabat Perhutanan, Semenanjung Malaysia.

Ibu Pejabat Perhutanan, Semenanjung Malaysia (Forestry Department Headquarters, Peninsular Malaysia). 1990. Perangkaan Perhutanan, Semenanjung Malaysia: 1986–1990. Kuala Lumpur: Ibu Pejabat Perhutanan, Semenanjung Malaysia.

Khera, H. S. 1976. The Oil Palm Industry of Malaysia. Kuala Lumpur: Penerbit Universiti Malaya.

Kiew, B. H. 1991. The national parks and wildlife sanctuaries of Malaysia. The Star (Kuala Lumpur), September 14, 1991, p. 8.

Laarman, J. G. 1988. Export of tropical hardwoods in the twentieth century. Pp. 147–163 in World Deforestation in the Twentieth Century, J. F. Richards and R. P. Tucker, eds. Durham, N.C.: Duke University Press.

Lanly, J. P. 1982. Tropical Forest Resources. FAO Forestry Paper No. 30. Rome, Italy: Food and Agriculture Organization of the United Nations.

Lee, H. L. 1978. Public Policies and Economic Diversification in West Malaysia, 1957–1970. Kuala Lumpur: Penerbit Universiti Malaya.

Lim, S. C. 1976. Land Development Schemes in Peninsular Malaysia: A Study of Benefits and Costs. Kuala Lumpur: Rubber Research Institute of Malaysia.

Lim, T. G. 1977. Peasants and Their Agricultural Economy in Colonial Malaya: 1874–1941. Kuala Lumpur: Oxford University Press.

Little, I. M. D., and D. G. Tipping. 1972. A Social Cost Benefit Analysis of the Kulai Oil Palm Estate, West Malaysia. Development Center Studies, Series on Cost-Benefit Analysis, Case Study No. 3. Paris: Organization for Economic Cooperation and Development.

Ministry of Agriculture (Malaysia). Various issues. Statistical Handbook, Agriculture: Malaysia. (Before 1979: Statistical Digest, Ministry of Agriculture and Lands, Peninsular Malaysia.) Kuala Lumpur: Ministry of Agriculture.

Ministry of Agriculture (Malaysia). 1988. Import and Export Trade in Food and Agricultural Products: Malaysia 1987. Kuala Lumpur: Ministry of Agriculture.

Ministry of Agriculture (Malaysia). 1990. Agricultural, Livestock, and Fisheries Statistics for Management: Malaysia 1980–1988. Kuala Lumpur: Ministry of Agriculture.

Ministry of Agriculture (Malaysia). 1991. Perangkaan siri masa sektor pertanian. January 19. Memorandum.

Ministry of Primary Industries (Malaysia). 1990. Profile: Malaysia's Primary Commodities. Kuala Lumpur: Ministry of Primary Industries.

Morgan, R. P. C. 1979. Soil Erosion. London: Longman.

Myers, N. 1978. Conversion of Tropical Moist Forests. Washington, D.C.: National Academy of Sciences.

New Straits Times (Kuala Lumpur). ca. July 15, 1991. Ghafar: Felda, Felcra and Risda under one administration.

Ng, F. S. P. 1988. Forest tree biology. Pp. 102–125 in Key Environments: Malaysia, Earl of Cranbrook, ed. Oxford: Pergamon.

Ng, S. K. 1983. Advances in Oil Palm Nutrition, Agronomy and Productivity in Malaysia. PORIM Occasional Paper No. 12. Kuala Lumpur: Palm Oil Research Institute of Malaysia.

Ooi, J. B. 1976. Peninsular Malaysia. London: Longman.

Panayotou, T. 1992. By stick or by carrot: Economic instruments for environmental management in developing countries. Unpublished manuscript. Harvard Institute for International Development, Cambridge, Massachusetts.

Rambo, A. T. 1988. People of the forest. Pp. 273–288 in Key Environments: Malaysia, Earl of Cranbrook, ed. Oxford: Pergamon.

Singh, S. 1968. An evaluation of three land development schemes in Malaysia. Malayan Econ. Rev. 13(1):89–100.

Soong, N. K., G. Haridas, Y. C. Seng, and T. P. Hua. 1980. Soil Erosion and Conservation in Peninsular Malaysia. Kuala Lumpur: Rubber Research Institute of Malaysia.

Sulaiman, b. H. N. 1977. A method of forest revenue assessment based on inventory data. Malaysian Forester 40(3):144–159.

Syed Hussain, W. 1972. Land development strategies in Malaysia: An empirical study. Kajian Ekonomi Malaysia 9(2):1–28, 10(2):1–50.

Teo, P. C. 1966. Revision of royalty rates. Malaysian Forester 29(4):254–258.

Thang, H. C. 1986. Can the existing wood resources meet future domestic requirements? Paper prepared for Institut Rimbawan Malaysia Symposium, Kuala Lumpur, September 13, 1986. Photocopy.

Thillainathan, R. 1980. Discriminatory allocation of public expenditure benefits for reducing inter-racial inequality in Malaysia—An evaluation. Dev. Econ. 18(3).

Tija, H. D. 1988. The physical setting. Pp. 1–19 in Key Environments: Malaysia, Earl of Cranbrook, ed. Oxford: Pergamon.

Tsuruoka, D. 1991. Plantation pressures. Far Eastern Econ. Rev. (August 22):41–44.

Vincent, J. R. 1988. Malaysia: Key player in international trade. J. Forestry 86(12):32–35.

Vincent, J. R. 1990. Rent capture and the feasibility of tropical forest management. Land Econ. 66(2):212–222.

Vincent, J. R., and C. S. Binkley. 1991. Forest-Based Industrialization: A Dynamic Perspective. Development Discussion Paper No. 389. Cambridge, Mass.: Harvard Institute for International Development.

Vincent, J. R., and Y. Hadi. 1991. Deforestation and Agricultural Expansion in Peninsular Malaysia. Development Discussion Paper No. 396. Cambridge, Mass.: Harvard Institute for International Development.

Vincent, J. R., E. W. Crawford, and J. P. Hoehn. 1991. Valuing Environmental Benefits in Developing Economies. Special Report 29, Michigan Agricultural Experiment Station. East Lansing: Michigan State University.

Wells, D. 1988. Birds. Pp. 167–195 in Key Environments: Malaysia, Earl of Cranbrook, ed. Oxford: Pergamon.

Whitmore, T. C. 1988. Forest types and forest zonation. Pp. 20–36 in Key Environments: Malaysia, Earl of Cranbrook, ed. Oxford: Pergamon.

Wong, I. F. T. 1971. The Present Land Use of West Malaysia (1966). Kuala Lumpur: Ministry of Agriculture and Lands.

World Bank. 1989. Malaysia: Matching Risks and Rewards in a Mixed Economy. A World Bank Country Study. Washington, D.C.: World Bank.

World Bank. 1990. World Development Report 1990: Poverty. Washington, D.C.: World Bank.

World Resources Institute. 1990. World Resources: 1990–91. New York: Oxford University Press.

Yong, H. S. 1988. Mammals: Genetic diversity and evolution. Pp. 138–145 in Key Environments: Malaysia, Earl of Cranbrook, ed. Oxford: Pergamon.

Yorke, C. D. 1984. Avian community structure in two modified Malaysian habitats. Biol. Conserv. 29:245–362.

Mexico

Arturo Gómez-Pompa, Andrea Kaus, Juan Jiménez-Osornio, David Bainbridge, and Veronique M. Rorive

In tropical Mexico and throughout the nation, deforestation is not only an ecologic concern but also an indicator of much wider social, political, and economic factors. It is the result of ecologic conditions combined with land use patterns as well as human decisions and the consequent actions on the tropical environment. These decisions are influenced by internal and external social and environmental factors, from local land tenure to national politics and from local soil conditions to widespread natural disasters. This profile briefly reviews the social and economic contexts in which deforestation occurs and discusses land use patterns, forest resources and rates of deforestation, and sustainable resource management.

Arturo Gómez-Pompa is a professor of botany and plant sciences at the University of California, Riverside, California, and is director of the University of California Institute of Mexico and United States; Andrea Kaus is codirector of Groundworks International, Inc., Riverside, California; Juan Jiménez-Osornio is a professor of ecology and coordinator of the Tropical Natural Resources Management and Conservation Program at the Autonomous University of Yucatán, México; David Bainbridge is restoration ecologist in the Biology Department at San Diego State University, San Diego, California; and Veronique M. Rorive is research assistant to Arturo Gómez-Pompa at the University of California, Riverside, and codirector of Groundworks International, Inc., Riverside, California.

THE SOCIAL AND ECONOMIC CONTEXT

Past Population and Land Use in the Mexican Tropics

Demographic change in Mexico from the time of contact with Europeans to the present has been a subject of study and debate by many scientists and scholars. Cook and Borah (1980) estimated that the native Indian population of Central Mexico in 1518 was 25.1 million people. Yet, by 1620 only 750,000 people remained. Diseases and war had reduced the population to a fraction of its former size.

The depopulation of Mexico after conquest by the Europeans was followed by the introduction of large-scale agricultural activities in the tropical forests. Cattle ranching, in particular, has become a major factor in the economy and ecology of present-day Mexico. The replacement of traditional tropical land use practices with techniques and agricultural models imported from temperate zones and Western European experience has led to cultural degradation along with the loss of biologic and genetic diversity.

The food production systems found in pre-Hispanic times were more efficient than the systems found there today. In pre-Hispanic times, intensification of agricultural production was well developed. According to Gliessman et al. (1983), Gómez-Pompa (1987a), Siemens (1983), and Turner (1974), the principal subsistence systems known to have existed were shifting agriculture (probably very intensive with short rotations and carefully managed fallows), tree orchards (including cacao with leguminous trees), different types of extensive and diverse forest gardens, terraces, and intensive hydraulic agriculture in lowlands and swamps.

The most notable examples of intensive hydraulic systems in the archaeological record are the raised fields of the Maya lowlands. These are thought to have provided a highly sophisticated agricultural system based on intensive human labor combined with the efficient use of water and renewable biological resources (Denevan, 1970; Gliessman et al., 1983; Gómez-Pompa and Jiménez-Osornio, 1989; Siemens and Puleston, 1972). The ancient Maya also hunted and gathered in the noncultivated areas and may have managed the mature vegetation to improve the level of production from forest resources.

Despite discrepancies and gaps in the available data, it is increasingly evident that present-day rural lands once contained urban centers and human populations larger than those supported today by modern land use practices. Furthermore, areas now considered to be "virgin" forest or "pristine" ecosystems were previously inhabited and, in many cases, still support indigenous populations and their traditional forms of agriculture (Gómez-Pompa and Kaus, 1992).

At present, Mexico has millions of farmers who belong to more than 50 ethnic groups, each with their own language, traditions and land use practices. Loss of the cultural diversity once found in the tropical forests means a loss of the opportunity to understand and learn from the experiences of others who live and work in tropical regions (Bennett, 1975). The value of traditional land use practices for agricultural development and conservation efforts under current socioeconomic conditions is often underestimated because of two principal myths: (1) the myth that the *campesino* (peasant) or Indian is ignorant of "modern" problems (Redford, 1990; Wilken, 1987); and (2) the myth that shifting cultivation is the sole cause of deforestation (Repetto, 1990).

Tropical deforestation occurs as a result of Western, indigenous, and mestizo land use practices. However, much can be learned from the failures as well as the successes. Traditional land use practices, that is, the techniques developed over generations in a given region, provide examples of time-tested experiments of human ingenuity in linking the natural and social environments. The added benefit is that these practices are not rigidly fixed and can adjust to and even alter environmental trends based on farmers' predictions and evaluations of future change.

Present Socioeconomic Trends in Mexico

In Mexico there are several nonecologically based trends that both contribute to tropical deforestation and indicate the need to create incentives that will alter the present predominance of unsustainable land use policies and practices. This situation is not only critical for reasons of environmental degradation but also for the well-being of Mexico's citizens.

At present there is a low density of inhabitants in the tropical regions of Mexico in comparison with estimations of the densities during the pre-Hispanic era. According to the latest census by the Instituto Nacional de Estadística Geografía e Informática (INEGI), the population of Mexico was 81,140,922 in 1990 (National Institute of Statistics, Geography and Information, 1990a). The World Bank (1990), however, estimated that Mexico had a population of 87,262,000 in 1990. The estimates of the World Bank were based on 1980 census figures; and the newest INEGI census produced figures that cannot be explained, for example, a decrease in the population of the Federal

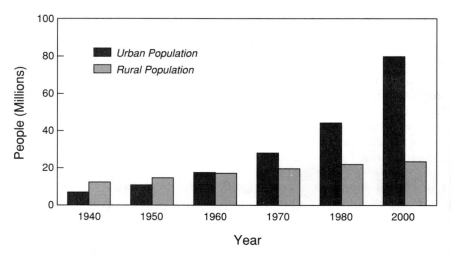

FIGURE 1   The urban and rural populations of Mexico, from 1940 to 2000, as estimated.

District from 8,831,079 to 8,236,960 inhabitants, which is highly unlikely.

According to INEGI (1990a), the population of Mexico increased sixfold during the twentieth century, from 13,607,272 to 81,140,922 inhabitants, and continued increases are projected in the future (Figure 1).   These population increases will likely add to the already increasing population density in tropical regions of Mexico.

According to Cabrera (1988), the debate on population growth dates back to the early 1960s. In 1963, the Bank of Mexico produced the first long-term projections of population growth and the potential impact on various economic areas, particularly the agricultural sector. In the early 1970s the Mexican government reacted by proposing the General Law on Population, which was approved in 1973. The law stated the need to regulate population growth to obtain a just and equitable distribution of the benefits of economic and social development. This was the beginning of the family planning programs of the Mexican government, whose goals in 1977 were to diminish population growth to 1 percent annually by the end of the century. The programs were well received. By 1988, annual population growth had been reduced to 2 percent. The goal of 1 percent annual population growth by the year 2000 appears to be feasible.

More than one-third of the present population of Mexico, however, is less than 15 years old, and the labor force (those 15 to 64 years of age) continues to grow at a rate of 3.5 percent per year (Ministry of Finance and Public Credit, 1991) requiring at least 800,000 new jobs each year. Since neither opportunities nor jobs are being provided by the agricultural sector in rural areas, many workers migrate to the major urban centers. The industrial sector has been unable to employ this growing work force. In 1988 unemployment reached a level of 24.5 percent (6.5 million people were unemployed and 20.1 million people were employed) (Calva, 1988).

Forty-five percent of the agricultural population of southeastern Mexico can be classified as infrasubsistence farmers, that is, those who do not produce enough food to sustain their own households. An inadequate food supply in Mexico is not a matter of inadequate food production. It is related to unequal income distributions and flawed food distribution policies. Mexico has initiated many efforts to address the constant problems of unequal food distribution and poor living conditions in rural areas. Yet, they have not solved the underlying discrepancies in income and wealth distribution.

One of the key components for a sustainable land use strategy in a peasant economy is food self-sufficiency, allowing, at the very least, for a family to sustain itself on the same plot of land over time (Calva, 1988; Comisión Económica para la América Latina, 1982, Cordera and Tello, 1981; Toledo et al., 1985). In the early 1980s, the Mexican government initiated SAM (Sistema Alimentario Mexicano [Mexican Nutrition System]), a program for food self-sufficiency. The main objective of SAM was to make Mexico self-sufficient in basic grain production within 2 years. This was possible, given that funds were available for credit, fertilizers were provided, no constraints were placed on the use of livestock pastures for growing crops, and the producers were able to make a good profit. The program was so successful in terms of production that the country was not prepared for the surplus. Thousands of metric tons of maize spoiled because of a lack of storage capacity in Veracruz or were used as fodder for cattle. In 1982, however, a combination of late rains and the devaluation of the Mexican peso reduced the grain yield and the ability of the government to invest heavily in the program. The program was terminated with the change in Mexican presidents in the same year (Riding, 1989).

Results of the SAM program show that distribution, storage, and access to land suitable for crop production are more important for low-income families than is increased production for improving the lives of people in Mexico. The experience of SAM also shows the potential capacity of agricultural lands and Mexican farmers to produce food surpluses if farmers are given sufficient means and incentives. The failure of the SAM program shows the dependence of adequate land use strategies on the social, economic, and political factors that exist external to the region of production.

Food production for external markets is different from production of basic commodities for use by farm households, and they must be examined from different perspectives. In much of Mexico, local peasant farmers do not concentrate on producing basic items like maize and beans, but produce specialty items like fruits and vegetables for a market that demands a wide range of products. The present infrastructure in Mexico cannot deal with the development of small-scale production of various specialty items because of transportation, storage, processing, marketing, and credit limitations, although small-scale production is an integral part of the peasant economy and a starting point for building equity into agricultural systems.

There was some concern in the past that Mexico's need to be self-sufficient in food production would take away from its ability to export agricultural products. However, these two forms of land use and priorities represent two types of production that commonly use different types of land. They need not be mutually exclusive. In Mexico's agricultural boom of the early 1960s, 1,549,577 ha (13.7 percent of the cultivated land at that time) was used to grow crops for export. By 1979, this amount had dropped to 1,224,697 ha, at the same time that Mexico lost its self-sufficiency in food production. In fact, over the past 2 decades, Mexico has increasingly relied on food imports rather than internal production. From 1966 to 1987, average maize imports increased 17-fold (from an average of 157,103 metric tons between 1966 and 1970 to 2,821,860 metric tons between 1983 and 1987). Wheat imports, on the other hand, increased nearly 300-fold (from an average of 1,157 metric tons between 1966 and 1970 to 345,501 metric tons between 1983 and 1987) (Calva, 1988).

A new trend in Mexico is to advocate food self-reliance. The objective is to produce 75 to 80 percent of the basic grains (maize, rice, and wheat) within Mexico (Calva, 1988). Mexico has the agricultural capacity for increased internal production without losing export potential (a considerable amount of land now used for livestock grazing could also be used to grow crops for export) (Table 1). However, little new agricultural land is available for extensive production. A 1987 evaluation by the Secretary of Agriculture and Hydraulic Resources of Mexico shows that Mexico has an agricultural reserve of 9.5 million ha and a total of 32.7 million ha with agricultural potential (Calva, 1988). Half of the 9.5 million ha is forested; the other half is used for cattle grazing. More than half (5.2 million ha) of this total is in the humid tropics and would require drainage and irrigation for agricultural use.

TABLE 1   Area Planted for Consumption and Export Crops,
1960–1979

Year	Area (in hectares) Human Consumption	Export	Forage
1960	9,163,406	1,549,577	320,266
1965	12,033,043	1,458,733	567,265
1971	12,270,642	1,174,372	1,325,813
1979	9,919,403	1,224,697	2,508,991

SOURCE: Calva, J. L., ed. 1988. Crisis Agrícola y Alimentaria en México: 1982–1988. México, D.F.: Fontamara 54.

The potential for improved production still exists for land that is already in agricultural use. Food self-reliance can be obtained by increasing the level of production per ha without using any more land. Maize production alone could be increased from 1.6 to 3.2 metric tons/ha by using already available technologies. These higher yields do not necessarily require increases in purchased or nonrenewable inputs, as the high production from some traditional farming systems shows (Wilken, 1987). Often, better knowledge is the only thing required to obtain better yields. A. Turrent and associates from the National Institute of Forestry, Agriculture and Animal Husbandry Research (INIFAP) have shown increased productivity from local farmers' fields through the use of simple technologies and techniques such as alley cropping, terracing, intercropping, and in situ postharvest seed conservation.

Past efforts for improved production in Mexico have not considered the various production components of Mexican small farms. Labor-intensive practices such as terrace construction, intercropping, soil improvement by nonchemical means, pest management, or simple irrigation techniques that rely on hand-carried water are often overlooked (for a full discussion of these methods, see Wilken, 1987). The female sector of the work force is typically forgotten or undervalued, even though the household economy often depends on their contribution to child care, gardening, small livestock production, firewood collection, food processing and preparation, and carrying water. Also overlooked is the value of the work done by children and elderly members of the household, whose contributions through experience or basic labor can be important for the family. However, the lack of recognition of traditional farming techniques, the contributions of various household members, or even self-sufficiency is not the only

TABLE 2   Contribution of Natural Renewable Resources to the Gross National Product (GNP), 1987

Sector	Value[a]	Percentage of GNP
Agriculture	242,419	5.06
Animal husbandry	132,945	2.77
Silviculture	20,616	0.43
Hunting and fishing	15,460	0.32
Wood industry	37,953	0.79
Paper products and printing industry	61,303	1.28

[a]Millions of 1980 pesos.

SOURCE: Department of Agriculture and Hydraulic Resources. 1987. Inventario Cartográfico de Recursos Agropecuarios y Forestales y Clasificación Agrológica Estatal Sobre Frontera Agrícola y Capacidad de Uso del Suelo. México, D.F.: Secretary of Agriculture and Hydraulic Resources.

gap in present and past efforts to alleviate problems of low levels of agricultural production and poverty in the Mexican tropics. None of the programs will improve without the participation of farmers in the decisions that affect their work and living conditions or without their direct control of production (Chambers et al., 1989).

The agricultural sector remains an important contributor to the Mexican economy, but it is underdeveloped (Table 2). Forestry has played a very minor role in the economy, but it could contribute more if it were developed to its full potential and properly managed for its long-term production capability. In 1989, forestry's contribution was only 1.9 percent of the gross national product (GNP). Wood production has been maintained at a level of 9 million m^3/year, which is only 23 percent of the potential level of production by a recent estimate (Comisión Nacional Forestal, 1988). At the same time, Mexico has imported an average of US$228 million of wood products per year over the past 10 years (Comisión Nacional Forestal, 1988).

Land Use

The present socioeconomic trends in the agricultural sector of Mexico coupled with increasing environmental degradation indicate the urgent need for alternatives in resource management. These alternatives should provide for the basic needs of peasant households without depleting the natural resources on which both the households and the national economy rely. The resource management options available in the Mexican humid tropics are similar to those available in other tropical regions of the world and are dependent on the land area that is to be managed, the available capital and infrastructure, and knowledge of the available technologies and potential markets.

In tropical Mexico, as in other tropical countries, two types of agricultural producers can be found on either end of a gradient (Table 3): (1) a large group of infrasubsistence farmers who practice traditional agriculture on small parcels of land, mainly for their own subsistence, and (2) a much smaller group of farmers who run large businesses that produce goods for regional, national, and international markets. CEPAL (1982) refers to these producers as peasant agriculture and commercial agriculture, respectively. Farmers who practice agricultural methods between these two extremes are called transitional farmers.

Peasant agriculture is practiced by 88 percent of the farmers on 57 percent of the country's agricultural lands. It relies primarily on household labor. Within the peasant agricultural sector, infrasubsistence farmers make up 45 percent of the agricultural producers in tropical Mexico. On average, their parcels are less than 4 ha. In contrast, commercial producers represent only 2 percent of the agricultural sector in the southeastern states of Mexico and hold 21 percent of the agricultural lands in that region, with average parcel sizes of more

TABLE 3   Types of Agricultural Producers (in Percent)

Producer Type	Number of Producers	Agricultural Area	Work Days in the Field (per year)
Infrasubsistence	55.7	10.8	29.6
Subsistence	16.2	11.1	13.4
Stationary	6.5	7.4	6.1
Excedentaries	8.2	27.5	9.2
Transitional	11.6	22.4	28.4
Small business	1.1	7.2	5.7
Medium-sized business	0.4	5.0	2.6
Large business	0.3	8.6	5.0

NOTE: "Number of producers" is not stated in whole numbers because many of these producers must be classified in more than one of these categories.

SOURCE: Comisión Económica de la América Latina. 1982. Economía Campesina y Agricultura Empresarial. México, D.F.: Siglo XXI Editores.

TABLE 4 Types of Agricultural Producers, by State (in Percent)

State	Infra-subsistence	Subsis-tence	Stationary	Surplus	Transi-tional	Small Business	Medium-Sized Business	Large Business
Campeche	59.0	22.9	4.6	6.0	6.2	0.7	0.3	0.3
Quintana Roo	90.4	2.4	0.4	0.3	6.1	0.3	0.1	—
Tabasco	25.3	21.6	14.8	27.9	8.4	1.2	0.5	0.3
Veracruz	39.2	16.1	9.5	13.0	19.9	1.5	0.5	0.3
Yucatán	82.4	7.2	1.8	1.5	5.8	0.5	0.3	0.5

SOURCE: Comisión Económica de la América Latina. 1982. Economía Campesina y Agricultura Empresarial. México, D.F.: Siglo XXI Editores.

than 12 ha (Table 4). They also have rights to 42 percent of the irrigated lands, whereas the peasant agricultural sector has rights to only 10.4 percent of the irrigated lands (Comisión Económica de la América Latina, 1982; Volke Haller and Sepúlveda González, 1987). Although the irrigated districts have attracted agriculturalists, there has also been a general trend of migration out of the region. One contributing factor is that the mechanization of agriculture associated with large-scale irrigated agriculture has replaced hand labor (Cabrera, 1979).

For the development of sustainable agricultural systems that integrate the concepts of agroecology with available information on alternative cropping systems, an agricultural model based on small-scale farmers who farm small parcels of land would have excellent potential. Small-scale producers already play an important role in export crop production in the Mexican humid tropics. For example, most coffee producers are not large-scale landholders, although coffee is a lucrative export crop (Nolasco, 1985). Sixty percent of the coffee plantations in Mexico are between 1 and 5 ha, and coffee plantations of this size account for 31 percent of the total area devoted to coffee plantations and 30 percent of total coffee production (Mexican Institute of Coffee, 1974).

Scherr (1985) noted that in the 1970s the average size of cacao farms in Tabasco was less than 3 ha. The parcel size is dependent on the availability of family labor and has likely averaged from 4 to 6 ha for centuries (Scherr, 1985). A frequent strategy of cacao and coffee growers is to have an interim phase of subsistence crop production while waiting for the cacao harvest. A sociodemographic survey of Tabasco showed that only 30 percent of the farmers planted cacao alone; the remainder planted maize, bananas, coconut or sugarcane, or included cattle production. Farmers with less than 2 ha of land were more likely to produce cacao alone or to grow only maize as a secondary crop (Scherr, 1985).

Improved production and self-sufficiency among small-scale landholders hold the potential for reducing destructive agricultural practices in tropical areas of Mexico. The agricultural practices of small- and large-scale landholders and long-term residents as well as recent immigrants contribute to the real and potential destruction of tropical forests. However, the greatest population concentration is found among small-scale landholders and recent colonists (immigrants who have claimed land they settled on). People in these two groups are often blamed for causing deforestation and for practicing unsustainable agricultural techniques. They also represent the people with the least means and support for improving their agricultural practices. Yet, they could be an underestimated ally in the use of sustainable agricultural systems and conservation practices in the humid tropics.

Small-scale farmers have much to gain from programs that enhance their self-sufficiency in food production and the security of their land tenure. In turn, long-term residents have much to contribute to current research on sustainable agriculture based on their intimate knowledge and experience with the land and on both their successes and failures with different techniques or crops. However, the means for sustainable agriculture are not attainable for the majority of these farmers. Credit, infrastructural support (for example, equipment, machinery, transportation), and adequate technology and information are usually not available; and those government credit, development, or agricultural programs that do exist often advocate unsustainable land use practices. Most small-scale farmers are more concerned about short-term production practices with the means available to them than about investing capital or labor in unpredictable and uncertain high-yield, technology-intensive practices. Sustainable agricultural systems need to be designed so that the small-scale farmers of Mexico can be included in the efforts to halt tropical deforestation. However, sustainability is not confined only to ecologic continuity; sustainable agricultural systems must also be economically viable and culturally acceptable if they are to be supported by the majority of the small-scale farmers. New initiatives must also take into consideration income and land distribution inequities along with insecure land tenures. Failure to take these factors into account led to the high social cost of the green revolution's technological package. Despite dramatic increases in food production, the green revolution provided greater benefits for the large-scale producers and landholders and provided few benefits for the small-scale farmers (Dahlberg, 1990; Perelman, 1976).

An emphasis on production, a belief in the neutrality of technology, and a poor accounting of the environmental and social costs have encouraged the replacement of ecologically complex farming systems with extensive monocultural systems. Plant breeding efforts that focus on grain have neglected a wide range of products that small-scale farmers need, such as thatch and fodder. Increases in crop yields generally require irrigation and high levels of fertilizer inputs (Stewart, 1988). The high-yielding crop varieties that respond well to high inputs of fertilizer and water are often less pest and drought resistant than traditional varieties, and their cultivation, combined with the overuse of chemical pesticides, leads to the emergence of new pests as a result of the elimination of natural predators (Perelman, 1976; Van den Bosch, 1980).

The development of these extensive monocultural systems has also had profound effects on small-scale landholders and farm laborers, many of whom have been displaced by land consolidation and mechanization. For small-scale farmers, the new seeds and technological inputs are expensive. Farmers often apply for rural credit from banks or aid from government programs, increasing the risk of the agricultural venture for the household while transferring control to the bank or the government.

Control over land use by small-scale farmers is further complicated by the nature of land tenure in Mexico. At present, the principal forms of land tenure are federally owned land, private properties, *ejidos*, and *comunidades*. *Comunidades* are the least common, referring to villages whose usufruct rights (the legal right to use and enjoy the fruits or products belonging to somebody else) have been restored for land used before the Mexican Revolution (1910–1920). *Ejidos* are the most common form of land tenure and refer to lands where the usufruct rights have been given to a collective of Mexican citizens as part of the land reform established after the Mexican Revolution (Sanderson, 1984; Yates, 1981). The land itself, however, remains the property of the Mexican government. Private properties with land areas that exceed the amount established by the Mexican Constitution are also at risk of expropriation by the government, usually for redistribution to landless peasants as *ejidos*. *Ejidos* may be worked individually or collectively, but the responsibility for the *ejido*, in terms of management and administration, is collective. The stability of the entire *ejido* system has been thrown into doubt with the remarkable and unanticipated government regulatory changes of 1991 that allow the sale of *ejido* land and use of *ejido* land as collateral for loans. The full implications of these changes will not be apparent for some time but the goal has been to increase efficiency in agricultural production.

The *ejidatarios* (the beneficiaries of *ejidal* grants) must maintain the productive use of their land in order to retain their right to use it; however, they often do not have the capital or infrastructure to do so. No credit or income is gained from conservation practices, despite the fact that many *ejidos* are in marginal, nonarable environments where conservation practices are necessary for the sustainability of the ecosystems and agricultural production. Instead, the incentives, opportunities, and loans offered by government programs, private landowners, or entrepreneurs advocate unsustainable practices for their short-term gain at the *ejidatarios*'s and land's expense.

A new type of agricultural revolution is needed to benefit the small-scale farmers of Mexico. Without changing the overall objective to produce food for all, the emphasis needs to be on equity and distribution, self-sufficiency, and sustainable land use practices rather than on higher levels of food production. In addition, these efforts need to take into consideration small-scale farmers' needs and aspirations and integrate their knowledge of the agricultural capacity of the local area with conventional scientific research and technological applications. Many traditional practices on rainfed parcels could enhance present research efforts to increase the agricultural capacity of nonirrigated land without degrading the environment. The emphasis must be on sustainable use and land tenure security for the land already under cultivation and the inhabitants already in residence. The pressure to clear the remaining tropical forests will not diminish as long as the surrounding land continues to lose its ability to provide for its poorest inhabitants and as long as those inhabitants are at risk of displacement by extensive land use systems such as cattle ranching. For these reasons, the agricultural capacities of cleared and degraded lands need to be increased or restored, as do the value of small-scale farmers' production and their role in caring for the land for the next generation.

THE FOREST RESOURCES AND DEFORESTATION

The tropical forests of Mexico occur in the coastal lowlands along the Pacific coast between the states of Sinaloa and Chiapas and along the Caribbean Sea from Quintana Roo to the coastal states on the Gulf of Mexico (Tables 5 and 6). Ecologists have described the forests in the tropics of Mexico and have classified them as several different types (Table 7). The vegetation types in the lowlands range from low thorny tropical forests (less than 10 m high) to the tall evergreen rain forests (more than 30 m high). In the highlands, the vegetation ranges from the tropical cloud forests to the low evergreen tropical forests, also known as *elfin forests*.

The majority of tropical forests that remain in Mexico can be found on *ejidal* lands or federal property (Tables 8 and 9). The states of Campeche, Quintana Roo, Yucatán, and Tabasco were chosen for this analysis because they are not mountainous and contain only tropical forests. The extent of forests on private or government property can be deduced from the data in Table 10. The distinction between private and government property is important because strategies for conservation and sustainable development may be very different for these two main types of land ownership—private and *ejidal*.

Strategies for developing sustainable land use practices for the tropical forest area of Mexico should be focused on the *ejidal* lands.

TABLE 5   Area Covered by High and Medium-Sized
Tropical Forest Trees, by State (in Thousands of
Hectares)

| State | Area (ha) | |
	High[a]	Medium[b]
Campeche	100	2,700
Chiapas	800	1,100
Quintana Roo	350	1,200
Oaxaca	300	500
Veracruz	80	300
Tabasco	10	40
Total	1,640	5,840

[a]More than 30 m in height.
[b]10–30 m in height.

SOURCE: Comisión Nacional Forestal. 1988. Hacia un Programa de
Acción Forestal Tropical en México. Propuesta para la Conservación
y el Desarrollo de las Selvas del Sureste. México, D.F.: Secretary of
the Agrarian Reform, Secretary of Agriculture and Hydraulic Resources, and Secretary of Urban Development and Ecology.

They include more of the forestland and represent the greater challenge for the sustainable development of forestlands in Mexico. Nonetheless, the private lands should not be ignored; improved management practices that include the conservation of ecosystems, flora, and
fauna may increase the profitability of these lands while providing
conservation benefits.

   In this discussion two tropical forest types are relevant: the tall
evergreen forests (evergreen forests taller than 30 m) and the tall or
medium-height semideciduous forests (forests with some deciduous
species taller than 15 m) (Pennington and Sarukhán, 1989). These are
the most abundant forests and are the most threatened by agricultural
activities. All other forest types cover less land area, although they
may be more important from a conservation perspective (Rzedowski,
1978). However, conventional means of protecting areas (for example,
parks, reserves, refuges) are more applicable for preservation of these
areas than is the development of better systems of conservation and
sustainable use.

   The state of Chiapas is considered to be one of the greatest centers of biodiversity in northern tropical America because of the quantity (50 percent) of tall tropical forests that remained in 1988 (Toledo,
1988). In the southeastern states of Mexico (21 percent of the country), for example, there are some 7.7 million ha of tropical forests,

TABLE 6   Forest Area by State and Vegetation Type
(in Thousands of Hectares)

| | Area (ha) | |
| | High | Medium-Sized |
State	Tropical Forests[a]	Tropical Forests[b]
Campeche	126	2,836
Chiapas	899	1,226
Colima	0	98
Guerrero	0	244
Hidalgo	0	11
Jalisco	0	161
Michoacan	0	320
Nayarit	0	320
Oaxaca	53	921
Puebla	0	124
Quintana Roo	462	1,206
San Luis Potosi	0	5
Sinaloa	0	980
Tabasco	61	179
Tamaulipas	0	6
Veracruz	513	357
Yucatán	0	298
Total	2,114	9,292

NOTE: Data presented here are based on data from studies done
between 1965 and 1987.

[a]More than 30 m in height.
[b]10–30 m in height.

SOURCE: Department of Agriculture and Hydraulic Resources. 1989.
México Forestal en Cifras. 1987. México, D.F.: Secretary of Agriculture and Hydraulic Resources.

from which 1.214 million m^3 of forest products are produced each
year and from which 7 million m^3 of firewood is obtained for consumption each year (Comisión Nacional Forestal [National Forestry
Commission], 1988). Most of the forests in that region are not well
preserved, however (Table 11). The most important remnants of high
tropical evergreen forests are found in the Lacandon forest of Chiapas,
including the region of Marquéz de las Comillas on the border with
Guatemala, where a battle to save the remaining forests is being fought.
At present, the winners are cattle ranching and secondary vegetation
(Table 11). On the other hand, Campeche contains 46 percent of the
medium-size forests and Tabasco has been totally deforested in the
last few decades.

A Definition of Deforestation

The various definitions of deforestation have a lot to do with the different estimates and perceptions of the process (Grainger, 1984; Lugo and Brown, 1981; Melillo et al., 1985). One view refers to the conversion of mature (older) forest ecosystems to less diverse ecosystems, which may mean the loss of pristine forests or virtually undisturbed forests. These mature forest ecosystems contain the greatest biodiversity in the tropics.

A second definition of deforestation includes the conversion of any forest ecosystem to nonforest ecosystem. This includes the conversion of secondary forests, agroforestry lands, and forest plantations to nonforest ecosystems, such as grasslands or other treeless agricultural systems. The concern is more for the known and potential roles that forest ecosystems play—in soil conservation, provision of forest products, and the earth's carbon dioxide balance—than for the roles they play in conserving biodiversity. This second type of deforestation is usually less important in the humid tropics, since it can be reversed in many cases. Forested land cleared for shifting agriculture can again become forest in a few years.

Evaluation of deforestation is difficult, however, because most studies are done by using aerial photographs or satellite images, and the distinction between the two types of deforestation given above is difficult to make by using aerial photographs or satellite images. The only clear distinction that can be made is that between forested and

TABLE 7   Ecosystems of Mexico for 1500 and 1985

Ecosystem	Total Area of Mexico (%)		
	1500	1985	Percent Change
High and medium-sized tropical forests	15	3	−80
Low tropical and thorn forest	14	20	+43
Pine-oak forest	20.4	15	−26
Mesophyll forest	0.5	0.1	−80
Pasture/grasslands	10	15	+50
Desert	40	47	+18

NOTE: High tropical forests, more than 30 m in height; medium-sized tropical forests, 10–30 m in height.

SOURCE: Fundación Universo Veintiuno. 1990. Desarrollo y Medio Ambiente en México. Diagnóstico, 1990. México, D.F.: Friedrich Ebert Stiftung.

TABLE 8 *Ejidal* Land in Mexico

State	Area of State (km²)	Number of Ejidos	Area (in hectares)				
			Ejidal Land	Agriculture	Forests	Grasslands and Secondary Forest	Cities, Roads, etc.
Campeche	50,812	344	3,115,750	339,722	1,651,522	1,092,536	31,970
Quintana Roo	50,212	270	2,743,286	339,352	1,698,890	703,825	1,219
Tabasco	25,267	694	1,011,991	232,189	94,684	568,080	117,038
Yucatán	38,402	718	2,162,147	561,450	270,168	1,071,637	258,892

SOURCE: National Institute of Statistics, Geography, and Information (Instituto Nacional de Estadística Geografía e Informática). 1990a. XI Censo General de Población y Vivienda. Aguascalientes, México: Instituto Nacional de Estadística Geografía e Informática.

TABLE 9   Distribution of Tropical Forests by State, 1987 (in Thousands of Hectares)

State	Area (ha)	
	High Tropical Forests[a]	Medium-Sized Tropical Forests[b]
Campeche	100	2,700
Chiapas	800	1,100
Quintana Roo	350	1,200
Oaxaca	300	500
Veracruz	80	300
Tabasco	10	40
Yucatán	—	200
Sinaloa	—	700
Nayarit	—	300
Michoacan	—	250
Guerrero	—	200
Jalisco	—	100
Puebla-Queretaro	—	100
Hidalgo	—	10
San Luis Potosi	—	5
Tamaulipas	—	5
Total	1,640	7,710

[a]More than 30 m in height.
[b]10–30 m in height.

SOURCE: Comisión Nacional Forestal. 1988. Hacia un Programa de Acción Forestal Tropical en México. Propuesta para la Conservación y el Desarrollo de las Selvas del Sureste. México, D.F.: Secretary of the Agrarian Reform, Secretary of Agriculture and Hydraulic Resources, and Secretary of Urban Development and Ecology.

unforested lands, that is, the degree of forest cover. The process is also complicated by the rapid succession rate that is possible in the humid tropics. Within 10 to 15 years, it is possible to develop a forest that is dominated by secondary-growth trees on cleared land (Gómez-Pompa and Vázquez-Yanes, 1981). The process is continuous in these areas, and the changes through time can be dramatic (Estrada and Estrada, 1983). For these reasons, deforestation and reforestation figures should be considered as approximations. Ground surveys are essential for more accurate assessments of the nature and type of deforestation and the changes in species composition that are occurring. In this profile, deforestation rates are mostly derived from the literature and include both types of deforestation described above.

TABLE 10   Ownership of Tropical Forests in Selected Tropical
States (in Hectares)

State	Total Area of Tropical Forests (1970)[a]	Ejido Forests (1988)[b]	Private Ownership and Estate Communities
Campeche	2,642,000	1,651,522	990,478
Quintana Roo	3,358,000	1,698,890	1,659,110
Tabasco	358,000	94,684	263,316
Yucatán	454,000	270,168	1,245,832

NOTE: Data are calculated from vegetation map and land use map produced by Annex 2.1 in Secretaria de Programación y Presupuesto. 1981. Carta de vegetación y uso actual del suelo esc. 1:100,000. In Atlas Nacional del Medio Físico. México, D.F: Secretaría de Programación y Presupuesto.

[a]SOURCE: Toledo, V. M., J. Carabias, C. Toledo, and C. González-Pacheco. 1989. La Producción Rural en México: Alternativas Ecológicas. Número 6. México, D.F.: Siglo XXI Editores.

[b]SOURCE: National Institute of Statistics, Geography, and Information. 1990b. México, D.F.: El Sector Alimentario en México. Instituto Nacional de Estadistico. Geograria e Informática and Comisión Nacional de Alimentación.

The information available from these sources is sufficient to evaluate the degree of conversion and to estimate the rates of deforestation.

Current Estimates of Deforestation

Although Mexico is always included in the list of countries with the most rapid rates of deforestation, precise data to support this claim do not exist. The best-known source to date has been a report of the Food and Agriculture Organization (FAO) of the United Nations and United Nations Environment Program (UNEP) (1981), which places Mexico third in Latin America with a deforestation rate of approximately 500,000 ha/year from 1981 to 1985.

Toledo's estimates (1988), which are probably the best available, challenged the FAO and UNEP estimates, arguing that the growth rate of cattle grazing areas and the expansion of the agriculture frontier is much greater than the FAO and UNEP figures suggest. Using the information from the 1980 census (Toledo, 1988) and inventories of land use and cattle grazing, Toledo projects a deforestation rate of about 1.1 million ha/year. If the areas destroyed by forest fires and forestlands cleared for new agricultural activities are added, defores-

TABLE 11 Changes in Land Use in the Lacandon Rainforest of
Chiapas

Land Use	Area in ha (percent of total area)		
	1973	1981	Percent Change
Well-conserved vegetation	76,526 (35)	—	−100
Secondary vegetation	59,963 (26)	141,500 (65)	148
Agriculture	67,388 (31)	32,700 (15)	−52
Cattle ranching	—	43,500 (20)	100

SOURCE: Department of Agriculture and Hydraulic Resources. 1984. Comisión del
Plan Nacional Hidráulico. Desarrollo Rural Integrado de la Selva Lacandona. México,
D.F.: Secretary of Agriculture and Hydraulic Resources.

tation could reach 1.5 million ha/year, which is 3 percent of the total
forestland in Mexico (Toledo, 1988).

The current total forest area in Mexico is unknown. In the 1970s,
Mexico had 80 million ha of basically unperturbed forest (Toledo,
1988). If Toledo's estimates are correct, the total forest area of approximately 80 million ha in the 1970s was reduced to 65 million ha
by 1990 and will drop to 35 million ha by the end of the century if the
trend is not slowed, stopped, or reversed.

Of its total land area, Mexico has 30,870,555 ha of tropical forests
(INEGI, 1990b). They include forests that range from low deciduous
tropical forests to tall evergreen tropical forests. There are, however,
different estimates of the forested area and the deforestation rate in
Mexico, as follows:

• Rzedowski (1978) estimated that 90 percent of the forests in
the lowland humid tropics of Mexico were eliminated by the 1970s.

• According to Toledo et al. (1985), these forests probably occupied 15 million ha—approximately 8 percent of the total land area of
Mexico—in the past.

• The best-known figures are those published in 1990 by the World
Resources Institute (WRI). The data is based on the and other reports
(Food and Agriculture Organization and United Nations Environment
Program, 1980, 1981, 1988; Lanly, 1982, 1989). According to these
reports, in 1980 the forest resources of Mexico covered 48,350,000 ha,
including 46,250,000 ha of closed-canopy forests and 2,100,000 ha of
open-canopy forests. The annual deforestation rate was 615,000 ha,
or 1.3 percent of the total forest. The average annual area reforested
was only 28,000 ha/year in the 1980s (World Resources Institute, 1990).

TABLE 12 Land Use in Humid and Subhumid Tropics in Mexico, 1981 (in Hectares)

Ecological Zone	Total	Agriculture	Forests	Cattle Grazing	Uses	Combined Others
Humid tropics	20,144,137	20,143,445	11,696,263	3,789,220	724,972	1,213,000
Subhumid tropics	32,576,050	32,575,823	14,578,467	2,867,426	6,594,553	788,000
Total	52,720,187	52,719,268	26,274,730	6,656,646	7,319,525	2,001,000

NOTE: Data are abstracted from the VI General Agricultural, Cattle and *Ejidal* Census, 1981.

SOURCE: Toledo, V. M., J. Carabias, C. Toledo, and C. González-Pacheco. 1989. La Producción Rural en México: Alternativas Ecologiás. No. 6. México, D.F.: Siglo XXI Editores. Taken from Annex 6.1, p. 293.

• According to the Tropical Forest Action Plan for Mexico (Comisión Nacional Forestal, 1988), there was 37 million ha of forested areas in Mexico between 1986 and 1987, which was nearly 11 million ha less than in 1980. Of these, 9.3 million ha is tropical forest. Of this area, 6 million ha is considered productive, with potential for exploitation. The other 3.3 million ha has an ecologic rather than an economic value and includes parks, reserves, and other protected areas.

No reliable figures for the forests in the Mexican humid tropics can be found, but combining the values of humid and subhumid forests from Table 12, more than 26 million ha of tropical forests existed in 1981, nearly half of the forested land in Mexico at that time. This estimate may be misleading because the deforestation rate has been faster in the tropics than in the other climatic regions of Mexico. By using Toledo's (1988) indirect method, it can be estimated that the number of cattle in the tropical states of Campeche, Chiapas, Quintana Roo, and Tabasco increased markedly since the 1970s to the 1980s. The deforestation cycle of lumber or mineral extraction followed by colonization, land acquisition, and conversion to pasture for cattle is well known in Mexico (Gómez-Pompa, 1987b). For this reason, an increase in the number of cattle entering the tropics implies that the deforestation rate in tropical areas is greater than the deforestation rate in all of Mexico.

The area deforested in the states of Chiapas, Tabasco, Campeche, and Quintana Roo between 1984 and 1989 was approximately 1 million ha of a total forested area of approximately 20 million ha, an average annual loss of 167,000 ha of forest. This is in contrast to Toledo's (1988) estimate of 1.5 million ha per year for the entire forested area of Mexico (200 million ha). The states considered here make up one-tenth of the country's total area (20 million ha), and the deforestation rate in the tropics (5 percent) is slightly higher, yet it is consistent with those for the country as a whole (4.5 percent). Although these calculations need to be checked against aerial photography or satellite images, they correspond well with more qualitative estimates and document the amount of deforestation in the Mexican humid tropics.

If these estimates are correct, the small remnants of tropical forests that existed in Tabasco, Veracruz, and Oaxaca in the late 1970s have vanished. This conclusion can only be confirmed if reliable forest inventories are undertaken. An indirect way to document environmental change would be to ask people, who live and routinely travel in these areas, about changes in the forest cover over time.

The loss of species in the humid tropics is also debatable, since no reliable inventories or national biological surveys exist. Toledo

(1988) suggests that deforestation along with selective extraction of rare plant species for the international market has led to the extinction of at least 17 species and that 477 species are currently endangered. This represents 17 percent of the flora of Mexico.

Causes of Deforestation

Deforestation is the consequence of many processes and actions. The predominant factors, which are described in detail below, include cattle ranching, colonization projects, forest fires, disputes over tree ownership and land tenure, national security, local farming and intensive commercial agriculture, timber exploitation, and road building and other engineering works.

CATTLE RANCHING

Cattle ranching has been the most important cause of deforestation (Table 13) (Denevan, 1982; Myers, 1981; Shane, 1980). In Mexico, the following are some of the avenues and incentives by which forests are converted to pastures for cattle grazing (Denevan, 1982):

- Direct clearing,
- Contracted shifting cultivation,
- Contracted deforestation,
- Land consolidation,
- Small ranches,
- Large ranches,
- Land tenure (in Mexico, by law, there is a maximum number of hectares of agricultural or cattle land that can be allotted to any one person),
- Economic viability of the land,
- Poor understanding of environmental processes and actions,
- Inadequate enforcement of regulations for environmental protection,
- National markets,
- U.S. and international markets,
- National financial incentives, and
- International financial incentives.

Cattle raising activities have been a key factor in deforestation for several reasons:

- There is an open market (national and international) for beef products, which creates increased incentives for conversion of forests to pasture land.

TABLE 13   Grazing Areas and Cattle in Chiapas

Year	Number of Head	Grazing Area (ha)	Variation in Grazing Area from Previous Year
1982[a]	3,391,839	4,409,391	—
1983[a]	3,422,141	4,448,783	39,392
1984[a]	3,056,998	3,974,097	—
1985	3,072,954	3,994,840	20,743
1986	3,104,083	4,035,308	40,468
1987	3,150,644	4,095,837	60,529
1988[b]	2,942,103	—	—

NOTE: Total area of Chiapas is 7,441,500 ha.

[a]SOURCE: Instituto Nacional de Estadística Geografía e Informática. 1990b. El Sector Alimentario en México. México, D.F.: Instituto Nacional de Estadística Geografía e Informática and Comisión Nacional de Alimentación.

[b]SOURCE: Instituto Nacional de Estadística Geografía e Informática. 1990c. Anuario Estadístico del Estado de Chiapas. México, D.F.: Instituto Nacional de Estadística Geografía e Informática.

• Cattle ranching is a relatively simple operation that can be managed by only a few people per hectare and administered from a distant location.

• Lines of credit for cattle ranching are available and offered as incentives.

• Cattle ranching enterprises are given preferential treatment in government regulations and protected by the government.

• A long cultural tradition—with roots in Spain and Portugal—identifies cattle ranchers as persons of status and respect, regardless of their actual production and profit.

COLONIZATION PROJECTS

The perception of tropical forested areas as agricultural frontiers has strongly influenced development policies in Mexico (Department of Agriculture and Hydraulic Resources, 1987; Parsons, 1976; Partridge, 1984). At one time, many deforested lands were federal lands that the government used to alleviate the need for land by landless peasants (Gómez-Pompa, 1987b).

The new areas of colonization are "prepared" for the peasants by use of government funds (frequently backed by loans from international banks) that give concessions for valuable wood to selected contractors. These contractors construct or use the roads paid by the government, "mine" the wood, and sell it on the national market. In the past, the wood was also sold on the international market.

The areas granted to the *campesinos* are the ones where the valuable wood has already been extracted. Land that is not distributed to potential *ejidatarios* is signed over to a forest-clearing contractor through the National Commission of Deforestation of the federal government. By this process, many new lands also fall into private hands or are left unassigned. Squatters take temporary possession of unassigned lands for subsistence agricultural activities. The land is later converted to private land, primarily for cattle ranching. At times, the beneficiaries are the squatters, but more frequently they are influential people—state governors, military officials, local political strongmen, and "city cattlemen" (those who run cattle ranching operations from urban areas).

FOREST FIRES

Natural and anthropogenic forest fires also contribute to deforestation in the Mexican humid tropics. A fire in Chiapas during 1982 burned 600,000 ha, and another in Quintana Roo in 1988 burned 1,200 ha (López Portillo et al., 1990).

Fire is often the cheapest, most efficient tool available to small-scale farmers for clearing an area for agriculture. Farmers can be divided into two main categories: those who have legal rights to their property and those who do not (Gómez-Pompa, 1987b). The first group usually uses fire as part of their shifting cultivation activities. They have detailed knowledge of when and how to use the fire, how to burn the slash and the fallen trees, and the necessary techniques to guide and control the fire. It is rare for a forest fire set by shifting cultivators to extend outside the area of the forest that has been cut.

Agriculturalists without legal rights to the land realize that the area does not belong to them and that there is a high probability that they will lose it. Therefore, their burning is done with little care or foresight. These farmers are usually newcomers to the area and have limited knowledge of management that is appropriate for the area. Their primary goal is simply to produce enough food for their families to survive. The clearing of trees provides a "cleaned" area for the cattle ranchers when the colonizers abandon or are evicted from the land. Forest burning by shifting cultivators has received most of the publicity and blame for deforestation. The International Rice Research Institute (1992) claims that shifting cultivation accounts for an estimated 30 percent of deforestation in Latin America; but government policies and the interests of cattle ranchers are behind this process.

PROBLEMS OF TREE OWNERSHIP AND LAND TENURE

One of the most neglected issues with regard to deforestation is tree and forest ownership and land tenure. Tree ownership is a long tradition in many non-Western cultures, but it is not well recognized or accounted for in development programs. According to Fortmann and Bruce (1988:5),

> Most forestry and agroforestry initiatives are based either on the premise that rural people will plant trees or that they will preserve and protect trees planted by someone else including the government. However, people do not preserve, protect or plant trees nor allow others to, if doing so is costly to them personally. Tree species planted by government offices are unlikely to have a high survival rate on private or community land.

Home gardens, consisting of tropical forest trees, are often the only forested areas left. These trees are planted, maintained, managed and protected by the people in whose household gardens the trees grow. The key component of the home gardens is that they belong to the household, and household members select and manage the trees they want.

Who, ultimately, has the tenure rights to the forests? Local inhabitants of the forest have always believed that the forest "belongs" to them because they have the same rights to use it as their parents and grandparents had before them. However, they are now learning that the land and its resources belong to the nation and that the government is empowered to give concessions for timber extraction or other uses to outsiders. In Mexico, this forest tenure conflict has been resolved by applying a fee per hectare or per volume of wood that is paid by concessionaires to individuals or communities as forest rights. This is usually only a token offering when compared with the value of the tropical woods on the national and international markets.

NATIONAL SECURITY

Mexico has cleared extensive areas of forest on its border with Guatemala to facilitate colonization of those areas. These colonies form a human shield to protect and buffer the country from political refugees fleeing the Guatemalan army.

AGRICULTURE

Some large-scale agriculture projects and the consequent clearing of large tracts of land have been of importance in the Mexican tropics. The only extensive agricultural system involved in discussions of deforestation is shifting agriculture. However, it should not be assumed that shifting agriculture is a cause in deforestation; rather, it should be considered a silvicultural technique when it is practiced under the appropriate conditions (for details see Gómez-Pompa et al., 1991; Ramakrishnan, 1984). Shifting cultivators who have ample knowledge of local conditions and species, skilled labor, and a commitment to long-term maintenance of their families and communities may also play a key role in the implementation of sustainable resource management practices.

INTENSIVE COMMERCIAL AGRICULTURE

Intensive commercial agriculture plays a minor role in deforestation when one considers the total land area covered. It typically involves commercial farming—usually perennial bush and tree crops—on permanent fields (Denevan, 1982). The major crops grown on these fields include coffee, cacao, rubber, sugarcane, pineapple, cotton, coconut, and mango. During the late 1800s, considerable areas were cleared for henequen (a fiber used to make binder twine). The amount of land used to grow avocado, melon, pineapple, watermelon, coconut, lemon, mango, orange, and banana was 372 ha in 1970 and 503 ha in 1980. Total production of these crops was 3.98 and 6.32 million metric tons in 1970 and 1980, respectively.

TIMBER EXPLOITATION

The valuable tropical woods of Mexico have already been largely depleted. For example, only in the remote and inaccessible areas—which are rarely found—is it possible to find mahogany. The contribution of timber exploitation to deforestation is not so much from the select logging of valuable trees as from the roads timber exploitation creates and the secondary damage that results from harvesting the desired species. Therefore, the starting point of deforestation is timber extraction, which is followed by the clearing of the remaining trees for agricultural fields by incoming landless peasants. These fields eventually become grasslands or secondary forests.

ROAD BUILDING AND OTHER ENGINEERING WORKS

Another cause of deforestation is the opening of new roads for oil exploration, lumber extraction, communication, or domination. Roads allow improved access to forested lands for colonizers.

Protected Areas

The protected areas of Mexico did not include tropical forest areas until the late 1970s. At present, the total area of protected closed forests has been estimated to be 360,000 ha (World Resources Institute, 1990). In 1989, Mexico had six biosphere reserves—Sian Ka'an, Montes Azules, El Cielo, Sierra de Manatlán, Mapimí, and Michilía—encompassing 1,288,454 ha, and 47 protected areas (excluding the Marine and Coastal protected areas) covering 5,582,625 ha (World Resources Institute, 1990). A recent survey (Ecosfera, 1990) showed a total of 308 protected areas in the Maya region. Seven percent of the total land area is under some form of protection as parks, reserves, or refuges. However, designation as a protected area does not necessarily ensure that it will be protected. The areas that are actually protected, in terms of the prevention of deforestation in core or buffer zones, is considerably below 7 percent.

SUSTAINABLE RESOURCE MANAGEMENT

Sustainable resource management activities range from gathering forest products at one extreme to a conventional agricultural system that is energy and petrochemical intensive at the other. Many of the changes and improvements that have or will be developed and tested will be of value to farmers across this full spectrum.

A Definition of Sustainability

There is no universally accepted definition of sustainable resource management. Some definitions are philosophically based, others address economic issues, whereas others specify management practices. Resource management can be said to be *economically sustainable* when supply matches demand and reasonable profits are made; *ecologically sustainable* when practices are environmentally sound and enhance rather than degrade the natural resource base; and *culturally sustainable* when farmers, families, communities, and the fabric of rural life remain viable. (For a more detailed discussion of the many different concepts and practices, see Bainbridge and Mitchell, 1988.)

Although economics and ecologic sustainability are often the only components discussed in sustainable resource management, a definition that also includes cultural sustainability is better because the maintenance of a viable culture, although perhaps the most challenging element, is in many ways the most important one. Farmer's knowledge, effort, and investment of energy and time are critical to sustainable resource management, and in return for their efforts they should be able to anticipate a better future. The family, the community, transportation links, and suppliers are also essential for sustainable resource management.

The ecologic basis for sustainability is also critical. If the ecologic foundation deteriorates, there is little chance of maintaining a long-term production capability. Although some restoration efforts have been successful, the rehabilitation costs can be many times higher than the immediate economic return. Therefore, it is much easier to avoid ecologic decline than to reverse it.

Environmentally sound production practices will help to bring real production costs down and improve profitability. This can provide farm families and communities with the incomes they need to survive and provide the stability needed to improve rural services—education, health care, transportation, utilities, and water.

Sustainable resource management can be achieved with existing equipment and facilities, conventional crops, and traditional markets. It requires more accurate knowledge and precise management of on-farm and off-farm resources to minimize production costs, maximize production efficiency, improve quality and grade of products, and reduce adverse environmental impacts. Improved planning and marketing will more closely match production to demand and will enable farmers to retain a larger share of retail cost rather than lose much of the value of their products to middlemen—transporters, distributors, storage, and retailers.

Small-scale subsistence farmers are concerned with sustaining their households, usually under severe economic constraints. Whereas large-scale commercial farmers are concerned with maximizing profits, small-scale farmers are often more concerned with minimizing risk. For each type of farmer, the importance and consequences of sustainability will be different. For subsistence farmers, a sustainable agriculture system must include self-sufficiency in the production of food and a variety of other products they and their household need (for example, firewood); sustainability for commercial farmers implies continued profitability through the extensive production of foods or commodities for sale to large markets. Each type of farmer usually allocates a wide range of resources—time, labor, and capital—very efficiently in pursuit of these goals. Revising incentives and benefit structures to reward sustainable practices can lead to the rapid adoption of new technologies for both subsistence and commercial farmers.

The international food and commodity markets are highly competitive, and government interventions and regulations distort prices and production in most countries. Prices of commodities are often subsidized at levels above competitive world prices, and commodities purchased by the government under these programs may be dumped in the world market and sold at prices far below the actual cost of production. Improving the accounting practices to include environmental costs, for example, erosion and land degradation, would do much to improve resource production and move production to areas where it is most efficient economically and ecologically.

Agricultural sustainability must be addressed not only from the personal perspective of the farmer's needs and resources but also from the national perspective of the country's needs and resources. Many small-scale solutions will eventually combine to contribute to global agricultural sustainability.

Sustainable Resource Management Practices in the Mexican Humid Tropics

An evaluation of sustainability can be made for virtually any resource management practice in the humid tropics of Mexico, from extensive cattle ranching on cleared forestlands to cattle production in feedlots, or from the manual labor of shifting agriculture to equipment-intensive timber production (Table 14). Sustainability is not inherent in scale, labor input, or management intensity, but rather reflected in the combined effects of many aspects of a particular agricultural system. The application of biodegradable pesticides by peasants without suitable protection (respirators and protective clothing) and management of contaminated containers and waste material cannot be considered sustainable because of the high risk to human health. Yet this same material could be used in a sustained manner if the materials were carefully controlled and the users and community were properly protected. In terms of shifting agriculture, short fallow periods are likely to be unsustainable as soil fertility gradually declines; but shifting agriculture with a sufficient fallow period (often 10 to 15 years) can be maintained indefinitely as the leguminous trees and shrubs restore soil fertility. Raised field beds in swampy areas could be sustainable, but only with a corresponding master hydraulic plan to regulate water quality and water levels.

The sustainability of any agricultural system can be enhanced by

TABLE 14 Comparison of Four Primary Production Systems

Production System	Annual Production	Value (millions of 1986 pesos/yr)	Number of Jobs Created	Characteristics
Shifting agriculture	180,000 metric tons of maize	1,800	50,000	High number of nonsalaried jobs; provides subsistence to high numbers of people; if done well, has high conservation and ecologic values
Cattle ranching	7,800 metric tons of cattle	4,600	300	Low number of jobs generated; benefits a reduced number of people; low conservation or ecologic values
Potential forest production	200,000 m^3 unprocessed trunks	6,000	3,500	Potential for new production jobs; increased monetary rewards for owners; trunks high conservation and ecological values if done properly
Sustainable agriculture	—	—	—	Potential for generating productive jobs, capitalization of farmers, if done properly; has high ecological value

SOURCE: Modified from Comisión Nacional Forestal. 1988. Hacia un Programa de Acción Forestal Tropical en México. Propuesta para la Conservación y el Desarrollo de las Selvas del Sureste. México, D.F.: Secretary of the Agrarian Reform, Secretary of Agriculture and Hydraulic Resources, and Secretary of Urban Development and Ecology.

using appropriate techniques. In some cases this may require the use of organic fertilizers; in others, chemical fertilizers. It may also include biologic controls instead of chemical pesticides. In some systems intercropping and rotations might be appropriate, whereas in other systems several combinations of mixed cropping in time and space may be appropriate. It is often easier to balance energy and nutrient demands and flows in mixed cropping systems that include animals and poultry than it is to balance those with only plants.

Some agricultural systems are easier to make sustainable than others, but those systems may not meet the basic needs of the household or the nation. For example, the extractive uses often mentioned as an option for forest reserves may provide limited resources for a few people but not for a larger population. Each agricultural system has its idiosyncracies and should be treated differently. It is more challenging to develop sustainable systems for an area with 500 people per km² than it is for an area with only 5 people per km². An appropriate set of management strategies and practices (for example, crop selection, markets, inclusion of large or small livestock, and labor requirements) must be developed for each agricultural system, although there may be some overlap between related production systems in similar environmental settings.

This leads to the larger problem of developing and managing the biosphere in a sustainable manner. The pollution of air, land, and water; the depletion of biological diversity; and increased deforestation indicate that modern society has not mastered resource management (McNeeley et al., 1990). The loss of biodiversity will not be solved by recommendations for sustainable agricultural approaches or major reforestation programs because the loss of biodiversity and other problems in the biosphere are affected by the cumulative effect of individual actions and responses to the economic and political incentives for clearing and using forested land. The reports and programs are essential, but they must be linked and related more directly to market incentives and factors that influence individual decision making at the most basic level of the smallest farm and family plot.

Lessons from Traditional Resource Management

There are already many traditional resource management approaches that can help in the search for sustainable agricultural production in Mexico (Altieri, 1987; Wilken, 1987). The relationship between traditional cropping practices and the control of pests—both insects and weeds—has been discussed in numerous articles (Altieri and Merrick, 1987; Gliessman et al., 1981). Management of organic matter (mulches, compost, and manures) helps to conserve nutrients, as do traditional methods of soil and water conservation (Wilken, 1987).

Many of these practices can be improved with scientific knowledge and technology and should be considered in the development of viable alternatives. It is essential, however, to begin with a detailed understanding of the motivations, practices, and needs of the local people. Only then can appropriate technologies begin to be developed. This is in contrast to the typical approach, by which the technology is developed first, without considering the cultural aspects.

It is also important to acknowledge environmental constraints. Traditionally, enormous expenditures have been made to fit the environment to the crop. The growing recognition of a wide range of useful crops (local, traditional, and global), however, makes it increasingly easy to select a crop that fits the environment. In addition, more research is needed to explore the wide range of potential products that can be extracted from the tropical forests of Mexico.

One of the most striking features that has emerged from research in the humid tropics of Mexico is the importance of human intervention and management in the development of the forests in that region, which were previously considered untouched, pristine, and certainly unmanaged (Gómez-Pompa and Kaus, 1990). These traditional agroforestry systems are valuable resources that have been developed and refined over the centuries. They are invaluable knowledge banks for understanding and improving tropical forestry management and should be studied before they disappear. Some traditional systems have been studied (Alcorn, 1984, 1990; Flores Guido and Ucan Ek, 1983; Gómez-Pompa, 1987a; Nations and Komer, 1983), but much remains to be learned.

THE LOWLAND MAYA

An alternative approach to tropical forest management, described in this profile, has been shaped by on-going work with Maya groups in Mexico (Gómez-Pompa, 1987a; Gómez-Pompa and Bainbridge, 1991), whose ecologically sophisticated forest management practices have provided many important lessons based on long-term experience with the surrounding ecologic and sociocultural systems. The ecologic complexity of the Maya forests is clear, both in the numbers of species and in their temporal and spatial arrangement (Gómez-Pompa, 1987b; Rico-Gray et al., 1988). Many Maya farmers have detailed knowledge of plants and soils and the regeneration process, which they use in their management of trees and forests.

Evidence from archeological and historical research suggests that in ancient times, agroforestry (combining trees that provide food, fodder, medicine, and building materials with annual and perennial crops, animals, and poultry) may have provided much of the basic needs of people in the densely populated regions of the Yucatán Peninsula. Forest management by the Maya included a variety of methods and techniques, many of which are still practiced. They do not, however, practice the integrated systems believed to have existed in pre-Hispanic times (Gómez-Pompa, 1987a). Past and present Maya agroforestry consists of the protection, cultivation, selection, and introduction of trees in the milpas, fallows, plantations, natural forests, forest gardens (a combination of trees, annual crops, and animals within a limited area around the house), as well as protected forest networks along trails, cenotes (sink holes in limestone with a pool at the bottom, found especially in Yucatán), and towns (Gómez-Pompa, 1987a; Gómez-Pompa and Kaus, 1987, 1990; Lundell, 1938).

One of the most striking features of present day Maya towns is the abundance of useful trees in the forest gardens: approximately 60 to 80 species in a family plot and some 100 to 200 species in a village (Herrera Castro, 1990). The trees of the forest gardens provide building materials, firewood, food and beverages, medicine, and fodder. Many of the more common trees are the same species found in the surrounding natural forests, although new species—such as papaya, guava, banana, lemon, orange, and other citrus fruit trees—have also been incorporated. Both indigenous and exotic species of herbs, shrubs, vines, and epiphytes grow in the patches of sunshine on the ground or in the shade of the trees. Useful wild species that appear in managed areas are often not weeded out and become established in these gardens. The importance of forest gardens in Yucatán can be calculated as follows. Approximately 25 percent of the Yucatán population has a forest garden. The average plot size is 400 m². Thus, the combined forest area of these gardens may be more than 25,000 ha, adding almost 10 percent to the forested area of Yucatán.

The Maya also plant or protect trees along the edges of or scattered throughout their agricultural fields. Many of these trees are nitrogen-fixing species (for example, *Acacia* spp., *Leucaena* spp., *Mimosa* spp.), and the abundance of these species may reflect centuries of human selection and protection (Flores Guido, 1987). These nitrogen-fixing trees provide most of the nitrogen required to maintain soil fertility under intensive high-yield cultivation practices.

The use of leguminous trees as shade trees for cacao was a pre-Hispanic practice that is now used on coffee plantations (Cardós, 1959; Jiménez and Gómez-Pompa, 1981). Shaded coffee plants produce fewer coffee beans on an annual basis, yet the shade adds many years to the useful life of the coffee plants.

Other agroforestry techniques are also incorporated into the management of milpas, including the selection and protection of useful individual plants on the site selected for cultivation. The protected species are determined by the interest, knowledge, and needs of the farmer, a factor that helps to explain the high level of biodiversity found on fallow lands and older (20 to 50 years) secondary forests. Even the manner in which trees are cut affects the survival of the forest. If regrowth is allowed to begin from a high trunk (coppicing), the survival rate is improved and is a key factor in the succession process. Although only about 10 percent of the trees may be coppice starts, they may account for more than 50 percent of the biomass during the recovery phase (Illsley, 1984; Rico-Gray et al., 1988).

The conservation of strips of forest along trails and surrounding milpas is also important. This strip probably plays an important role in the regeneration of fallow lands (Remmers and de Koeyer, 1988), provides valuable shade on the trails, and interlinks fragments of the forest so that wildlife has access to all parts of a forest. Studies by Thomas Lovejoy in the Amazon have shown that links between patches of forest increase the effective size of the forest and help to maintain species diversity. They may also play a critical role in maintaining deer, birds, and other game, which are valuable food sources for Maya hunters.

Although some researchers (Abrams and Rue, 1988; Morley, 1946) contend that the collapse of the Maya was caused by misuse of the environment, recent research (Barrera-Vázquez, 1980; Bowers, 1989) supports Thompson's (1954) earlier suggestion that the collapse of the Maya resulted from increased hostilities and warfare. Trees would be vulnerable to intense warfare. Present-day practices that are similar to those used by the Maya during the pre-Hispanic era indicate that sustained use of the tropical forest would have been possible for a long period of time.

The regeneration of the ecosystems of the Maya area after successive abandonments, the last one occurring after the Spanish conquest, was possible only because seed banks existed in the managed and protected natural ecosystems of the area (Gómez-Pompa et al., 1972), and land use did not cause irreparable harm to the soils.

THE LACANDON MAYA

The forest management of the Lacandon Maya incorporates many of the same practices incorporated by the Yucatec Maya. In the midst of the forest can be found complex agroforestry systems that may include 75 crop species, including fruit trees, in multicanopied single hectare plots (Nations and Komer, 1983). The plot is repeatedly harvested until it is engulfed by the forest, and then a new milpa plot is started.

THE HUASTEC MAYA

The Huastec Maya of northeast Mexico manage the humid forest in a manner that combines commercial and subsistence production (Alcorn, 1984). As many as 300 species may be found in a plot that provides food, construction materials, fuelwood, fodder, medicine, and chemicals. The forest plots are an important adjunct to the agricultural enterprise and buffer the farmer against market fluctuations and failures of single crops.

Lessons from Development and Conservation Programs

Several different programs for small-scale producers on mostly nonirrigated lands were implemented in the past.

PLAN PUEBLA

The most important project of this type was Plan Puebla, which was initiated in 1967. The plan recommended the following components as part of a sustainable agricultural system: increased use of chemical fertilizers, timely application of fertilizers, and carefully determined densities of different races of maize. Plan Puebla provided credit and advice and was successful in improving maize productivity, which went from 1,330 kg/ha in 1967 to 3,000 kg/ha in 1981 (Volke Haller and Sepulveda-González, 1987).

An evaluation of this plan after 15 years, however, showed that the complete system was adopted by only 0.8 percent of the producers, and in turn, 0.6 percent decided not to follow any of the suggested techniques (González-Pacheco, 1983). Fifty-seven percent of the producers adopted only 30 to 70 percent of the techniques recommended by Plan Puebla.

It is important to examine the reasons producers had for not following the techniques recommended by Plan Puebla because they represent many of the points that need to be addressed in future recommendations for sustainable agricultural techniques. The principal ones mentioned by Volke Haller and Sepulveda-González (1987) are as follows:

- Lack of knowledge of the new technology;
- Greater economic risk from using the recommended technology;
- Aversion to the credit needed to obtain the recommended technology and the paperwork needed to apply for credit;
- Deficiencies in the insurance included with the loan (insurance usually does not pay in case of natural disasters);
- Delays in fertilizer deliveries;
- Competing opportunities for income outside the field of agriculture;
- Small field sizes (the smaller the field, the lower the adoption of the technology); and

- Other causes, such as the age, education, and family size of the producer and the complexity of the technology.

Gladwin (1976) stressed that the critical factors that limited the adoption of one recommendation of the program were not necessarily the critical factors that limited the adoption of other recommendations. For instance, fertilizer use was limited by credit ineligibility, whereas different planting techniques to increase plant populations were not adopted because of the lack of knowledge of the specific recommendations.

TROPICAL CHINAMPAS

Another case worth reviewing is the transfer of chinampa (agricultural production in raised fields surrounded by water) technology to the tropical lowlands of Tabasco and Veracruz during the 1970s. Although experience was gained from this project, the transfer was successful only in the pilot demonstration plots. The structure (raised fields) of the technology was transferred to the swamps of Tabasco (the Camellones Chontales project), but the agricultural component was not (Gómez-Pompa and Jiménez, 1989). This was mainly because the need to intensify agricultural activities was not identified by the farmers, the time required to maintain the system was much more than the time normally devoted to agricultural activities by local farmers, the lack of markets for the proposed products provided little incentive for its adoption, and no credit was available to the farmers.

One of the most important reasons that the majority of small-scale farmers gave for not adopting new technologies or new crops was frequently ignored: the uncertainty of the market. These farmers were aware of the experiences of other small-scale farmers who embarked on projects that left them in debt or with products they could not sell. Technologies that may improve the productivity of the fields without the risk of putting the farmer into debt would likely have more followers than would technologies that are capital intensive.

SECONDARY FORESTS OF VERACRUZ

Because most of the tropical rain forest in Mexico has disappeared, it is important to use and manage secondary forests so they may provide a wide range of agricultural products—from vegetables to timber. In situ experimental research on secondary forest has been undertaken in Uxpanapa, Veracruz, where the secondary vegetation has been used as a substrate for newly introduced, valuable species (del Amo, 1991). A variety of agroforestry systems have been evaluated, including a diverse milpa and an enriched 11-year-old secondary forest (*acahual*). The project demonstrated that use of combinations of various types of crops and arrangements—in patches—of different systems like diversified milpa, orchard, and agroforestry are possible alternatives for the Veracruz region.

Tropical Forest Action Plan (PAFT)

Mexico has joined an international effort headed by FAO to develop a worldwide Tropical Forestry Action Plan. Several versions of an action plan for Mexico (Plan de Acción Forestal del Trópico [PAFT]) have been produced by the undersecretary of forestry of the Secretary of Agriculture and Hydraulic Resources of Mexico (Comisión Nacional Forestal, 1988). PAFT follows the same unsuccessful lines that Mexico has been using for some time: calling for the management of forest resources without specifying what type of management or what will ensure the plan's continuation after the initial funding for development is gone.

The first draft of PAFT is discussed here because of the amount of effort and the resources that may be allocated to it. Several points of the first draft of the proposed action plan can be criticized:

- PAFT recommends the establishment of forest plantations without specifying the species, areas, or techniques that should be used and, most important, without the participation of the private sector or local communities.

- The conservation of genetic resources could be a significant contribution from PAFT, but the plan does not specify how this will be done or who will be responsible for protecting genetic resources.

- There are no specifications for collaboration with the research institutions or nongovernmental organizations that made PAFT a reality.

- The strengthening of education and research is a necessary and fundamental action, but the action plan provides no guidelines on how this will differ from the education and research elements of the present programs in agriculture, forestry, agroforestry, or resource management, which are inadequate.

- No opportunities for independent research organizations have been created, even though several such organizations and institutions have ongoing tropical forest research projects.

• PAFT proposes to undertake the inventories needed for planning, but there is no mention of the relationship of PAFT with other development plans in the agriculture, animal husbandry, or oil exploration sectors. The need for coordination and development of a land use system with enforced zoning is not discussed; yet, without this, the inventories are of little use except for monitoring deforestation.

• The development of roads as a result of the recommendations of the action plan will only contribute to more deforestation—a common consequence of development programs.

• The project includes the temperate pine-oak forests of the Sierra de Juárez (Oaxaca).

• In the past, the treatment of "sick" forest stands, known as "forest health" activities (*"sanidad" forestal*) has received large financial expenditures, although there is little scientific basis for the forest health program. Inclusion of forest health in PAFT seems dubious.

• The identification of rare and endangered plants and animals is of great importance, but PAFT does not indicate that this be will accomplished or what will be done with the information if identification is accomplished.

• The restoration of lands deforested by shifting agriculture seems the most appealing project, but PAFT provides no information on how this will be accomplished or what role the shifting cultivators would have in the plan. The same applies to the management of secondary forests proposed by PAFT.

• The establishment of pilot projects for the integral management of natural resources also has great potential and has been tried several times in the recent past. There is no information as to why these pilot projects should succeed while others have failed.

The Tropical Forest Action Program (PROAFT), a new tropical forest action plan, which will attempt to rectify these problems, is currently under way in Mexico.

Sustainable Food and Commodity Production

An ecologic approach to food and commodity production is important to the tropical environment in Mexico because it is essential to develop food production systems that depend less on inputs, particularly import inputs (for example, reliance on outside production). Many of the traditional cultural practices commonly used by local farmers may contain important ecologic attributes that contribute to sustainable agricultural yields. The problems that small-scale producers face, however, have made many traditional practices inappropriate for the sustained production to meet current market demands. Nevertheless, Gliessman et al. (1983) and Gómez-Pompa (1978) have provided good examples of how the strengths of traditional agricultural systems can be retained in a system that is modified to meet contemporary needs.

Sustainable food production can be tailored to fit each unique situation. A sustainable low input system with intensive hand labor may share a few characteristics with a high-input highly mechanized system in the same area with both having a limited impact on the environment. The hard labor-intensive system may rely on biological fertilizers (for example, organic matter and fallow) while the highly mechanized input system may rely on carefully placed chemical fertilizers with more limited use of organic fertilizer—yet, if each is done well, they may be equally sustainable, technically.

Agroforestry for Mexico

Despite recent advances in tropical forest ecology and forest management, deforestation continues virtually unabated. Reforestation efforts are insignificant and the area of humid tropical forest under management that will maintain productivity and profitability is growing slowly, if at all. Improving forest management is perhaps the best and only hope for saving and restoring the tropical forests of the Mexican humid tropics, maintaining the productivity of these often fragile lands, and improving the quality of life for the residents of those areas.

The loss of tropical forests in the Mexican humid tropics is more than an ecologic tragedy. Tropical forests play an important role in regional and global scales in ecologic and economic terms. Ecologically, tropical forests are a primary factor in the carbon dioxide balance in the atmosphere. Economically, they contain many species of economic importance (timber, fruits, nuts, gums, medicines, understory plants, birds, and animals). Thousands of yet undiscovered or unstudied species have potential economic value, including species with future value for genetic engineering.

For the humid tropics of Mexico and Central America, agroforestry is receiving attention as a method of resource management that efficiently uses resources and that is environmentally positive (Adelhelm and Kotshci, 1985; Alcorn, 1984, 1990; Gómez-Pompa, 1987a; Lagemann, 1982; Nations and Komer, 1983; Vergara and Briones, 1987), but it will take time to develop skilled agroforesters. There are few people adequately trained in this field to teach new practitioners and even with adequate support, it may take 10 years to provide a sufficient number of practitioners and instructors to meet the demand. The rapid increase in interest and promotion of agroforestry has not yet been accompanied by well-funded interdisciplinary research to better understand how traditional agroforestry systems work, how to improve the methods of teaching agroforestry, and how to improve demonstration and development projects.

Agroforestry research is, by necessity, slow and complex (Cannell and Jackson, 1985). This makes the study of traditional agroforestry systems extremely valuable. The lessons that have been learned from successful and failed agroforestry systems are equally important.

The advantages and the potential of the complex, traditional types of forest management are clear for the humid tropics (Bene et al., 1977), but forest management has not been improved. Although numerous factors account for the disparity between the promise and reality of agroforestry, ignorance is the greatest problem (Bainbridge, 1987a). The complex forest management practices that must be used do not fall under either conventional forestry or agricultural systems. As a result, they were ignored until recently (see, for example, Winterbottom and Hazelwood [1987] and Shepherd and Stewart [1988]).

Most of the research in traditional forestry management in the humid tropics of Mexico has been done by anthropologists and ethnobotanists. The International Center for Research on Agroforestry (ICRAF) was established in 1977, but a comprehensive work program for the center was not developed until 1982 (Lundgren and Raintree, 1982), and the location of ICRAF (Nairobi, Kenya) has led to an emphasis on Africa. Work in other areas of the world, such as the humid tropics of Mexico, has been very limited.

Although the Centro Agronómico de Investigación y Enseñanza (CATIE) (Costa Rica) has been active and effective with limited resources, it has not been able to effectively contribute to the improvement of resource management in Mexico and other Latin American countries with tropical forests. It is most unfortunate that a comprehensive plan for preserving traditional knowledge and for developing education programs, demonstration plots, research programs, and data bases for the many ecosystems and cultures of the humid tropics of Mexico has not been developed.

In addition to the traditional methods of forest management in the Mexican humid tropics, there are many potentially valuable methods and crops from comparable humid tropical zones. One of the most promising of these combines strips of trees with agricultural crops (alley cropping). The most common trees for these alley cropping systems are fast-growing, multipurpose, nitrogen-fixing trees that, through root symbioses, make atmospheric nitrogen available to the tree and, subsequently, to other plants and crops) (Torres, 1983; Wilson et al., 1986; Yamoah and Burleigh, 1990). These systems provide fuelwood, building materials, and fodder while they increase and maintain the productivity of the agricultural crops and provide other ecologic and environmental benefits including slope stabilization, erosion control, and habitat for wildlife (Ehui et al., 1990).

Sustainable Livestock Production

Because cattle ranching has been the most important cause of deforestation (Denevan, 1982; Myers, 1981; Shane, 1980), cattle production must be improved on lands where the forest has been removed. Sufficient work has been done to suggest some of the possibilities for sustainable livestock production in the humid tropics (Murgueitio, 1988, 1990; Preston and Leng, 1987). The rapid growth of fodder trees, including nitrogen-fixing species, makes it possible to improve cattle production with trees (Preston and Murgueitio, 1987). Unpublished work from researchers at the Postgraduate College of Chapingo in Veracruz indicate that aquatic plants and other nonconventional plants can be used as fodder for cattle. The Australians have adapted ruminant microflora to better utilize *Leucaena* spp. (Reid and Wilson, 1985).

Sustainable Management of Biodiversity

A sustainable approach for the conservation of biodiversity in tropical forests is to protect forests from human actions that threaten diversity. One alternative would be to use protected areas where the ecosystems are managed and used rather than just preserved.

PROTECTED AREAS AND BUFFER ZONES

Protected areas are not islands but, rather, areas within larger ecologic and social systems. Management of these areas requires continual adjustment to external social, political, and economic pressures; otherwise, they run the risk of being engulfed by unsustainable practices. This type of management could include the selective and careful extraction of valuable woods, prescribed burning of land, hunting, ecotourism, even forest restoration, provided that it is done in a manner that will enhance the ecosystem's sustainability. If local people are to protect these areas, they should be provided with jobs and benefits, whether directly through employment in management work (protection and restoration) or more indirectly (through tourism or reserve support) and improved access to resources essential to survival (for example, fuelwood and food) (McNeely, 1988). Integrating the needs of these people into reserve management plans is not only challenging but also essential. Local people cannot, and should not, be expected to bear the cost of conservation.

PRESERVING BIODIVERSITY IN MANAGED AREAS

Other approaches to protecting plant biodiversity might include identifying new markets for rare landraces or traditional crop varieties, subsidies to farmers who cultivate important landraces, or support for more traditional methods of species protection in botanical gardens and gene banks (Altieri and Merrick, 1987). These alternatives might provide jobs and resources for a limited number of local people.

STRATEGIES FOR IMPROVING RESOURCE MANAGEMENT

Improvement of resource management systems to protect and restore the humid tropical forests will require a variety of strategies and programs involving policy, research, education, demonstration, and implementation. These strategies and policies offer the best hope for conserving the existing forests, improving management of the existing forests, promoting reforestation, and improving living conditions for the local people. If they are ignored, the forest area will decline, extraction of forest products from biologically and culturally rich areas will continue, invaluable species and traditional knowledge will be lost, and poverty levels among the local people will increase. As Janzen (1988:243) stated, "Restriction of conservation to the few remaining relatively intact habitat patches automatically excludes more than 90 percent of tropical humanity from its direct benefits; restoration is most needed where the people live."

It is a mistake to continue to underestimate the skills and knowledge of the local people. In many cases they have managed the forests in a sustainable manner for hundreds of years. If Mexico fails to adopt an ecologic and cultural approach to sustainable resource management, funding and energy will be expended to protect forest areas with little hope for success. Present conservation management approaches continue to ignore the fact that the forests were, are, and will continue to be inhabited. A wiser approach is needed to protect the needs of both the environment and the people and should involve the local inhabitants in the protection and management of the environment.

Policy

The following are some suggestions related to policy issues for improving sustainable resource management in the Mexican humid tropics.

• *A Policy Prospectus Is Needed* A policy statement that explicitly states the importance of sustainable management in resource planning for the humid tropics of Mexico is needed for the areas of agriculture, forestry, and associated land-use practices.

• *Incentives for Sustainable Land Use Are Needed* Sustainable land management will develop only if it is profitable in economic and social terms and only if people receive a benefit from doing what is appropriate for long-term use (Carpenter, 1989; Murray, 1989). Research should include an evaluation of tax policies and possible incentives to promote long-term planning for sustainable agricultural practices, which often provide large profits over the long-term but low immediate returns. As Repetto (1990) observed, institutional factors often drive the system toward ecologic and economic disaster.

The results can be striking when local people are involved in the planning process and receive immediate benefits. For example, Haitians voluntarily planted more than 5 million trees on their own land in the first 2.5 years of a project that incorporated local people into the planning process (Murray, 1989). Success was attributed to more than just profitability. Project planners consciously tried to introduce trees that could be integrated into the farmers' existing cropping systems, which is important for ensuring the acceptance of any innovation (Evans, 1988). The rapid expansion of intercropping in China, from 20,000 ha intercropped with *Paulownia* trees in 1973 to more than 1.5 million ha in 1988, was made possible by an equally well-designed program (Zhaohua, 1988).

This effort should also include the development of incentives for the sustainable management of tropical forests. This is the best way to ensure the survival of large areas of forest. Methods and techniques are available; long-term commitments by government and private industry are needed.

• *Incentives to Conserve Biodiversity Are Needed* Initiatives for conserving biodiversity and for small-scale farmers to use sustainable resource management practices should be developed and promoted. These incentives should include actions that improve the quality of life for people in the local communities.

Three-way alliances for conservation and sustainable agriculture could be established. In these alliances, the central party is the community or local people integrated with a second party consisting of a research organization (private or governmental). The third party would be a funding agency (governmental or nongovernmental) that would facilitate and support the activities. Through such an alliance, long-term agreements to protect small or large areas in small-scale farming communities could be established.

• *New Policies for Conserving Biodiversity Are Needed* A working network of reserves based on the biologic importance of different areas needs to be established. Although most reserves have been placed where a large piece of less disturbed forest exists or an important archeological site or place of beauty can be found, the importance of the biodiversity of various regions has rarely been taken into account. Areas of special biodiversity must be identified and protected.

A new network of protected areas and protected agricultural systems needs to be developed to conserve important landraces of cultivated plants, especially plant material related to the major food crops: such as maize, cassava, and beans. The sustainability of future production may depend on this.

Ex situ genetic banks of valuable, rare landraces and other important crop relatives should be established.

Management plans for all existing reserves should be prepared. These should be designed to conserve the biodiversity, to favor its enrichment, to follow and guide natural changes, and to allow for experimentation.

In the conservation of biodiversity, incentives should be developed for the participation of the private sector and those who own large areas of land. One option is to use tax breaks. This may encourage the creation of small to large reserves in Mexico as well as provide financing. In addition, those who own large areas of land need to take responsibility for the potential effects of their own agriculture and ranching activities on the land they own and on the ecosystems that surround their land.

More attention and research needs to be focused on buffer zones (areas surrounding or adjacent to important protected areas). Well-managed buffer zones could provide models for the integration of conservation and sustainable land use practices to other regions of Mexico.

• *Institutional Barriers Need To Be Broken* Studies of the humid tropical forests of Mexico should include a detailed review of institutional needs and limitations, so that projects can proceed with minimal interference and maximum support from government regulatory and administrative programs (local to national scale). This review should include the needs and limitations of government departments, owners of large areas of land, and land managers. It should also take into account the market system for tropical forest products, from the producer to eventual retail outlet; the commercial sector, including alternatives to rain forest products; schools; religious groups; and the economic community (banks and lenders, etc.).

• *Local Land and Tree Tenure Considerations Need To Be Reviewed* One of most important and sensitive issues in resource management is the insecurity of land tenure. Even if resource management systems protect the soil, conserve nutrients, and provide food and income, farmers have little or no motivation to invest in agricultural activities with long-term benefits unless there is some certainty of reaping the benefits (Fortmann, 1985). In some areas, tenants risk eviction if they improve the land they farm; if the land becomes too productive, the landlords may claim it and farm it themselves (for further information on tenure, see Fortmann and Bruce, 1988; Fortmann and Riddell, 1985; Labelle, 1987; Raintree, 1987a). The separate problems of security for the *ejido, ejidatarios'* households, and *ejidal* lands need to be examined to develop policies that are not contradictory and that are specific to the needs of people in different regions.

Research

Research is needed in many areas. The following are strategies for improving future research.

• *Traditional Knowledge Should Be Documented by Working with Local People and Communities* Because it may prove to be difficult to match the ecologic and cultural adjustments achieved by traditional farmers after centuries of trial and error, the development of detailed data on traditional agroforestry systems is of paramount importance, especially since detailed knowledge of the local environment is vanishing along with the forests (Gómez-Pompa, 1987b; Gómez-Pompa and Bainbridge, 1991; Raintree, 1982; Raymond, 1990). This research should involve multidisciplinary teams and must include the people from the local communities involved. Multidisciplinary, mixed-gender teams of local students, faculty, and international collaborators are preferred for the development of detailed information on the full ecologic and cultural complexities of these systems. The decision-making processes of farmers should be an important part of this research. Gladwin (1976, 1979, 1983) has laid the groundwork for an appraisal of why farmers and foresters plant and harvest specific crops and why they do or do not accept recommended changes.

The need to include indigenous peoples in research and development programs has been emphasized in numerous case studies and reports on the process of development (Richards, 1985). The lack of this kind of input places in question the sustainability of any introduced change, despite the best intentions of those involved in development (Chambers, 1987). Although it is often assumed that people will accept an innovation because "it is good for them," to succeed, a program must meet the real and perceived needs of the people involved and fit the social and cultural setting (Leeger, 1989). Research done in collaboration with local people provides the groundwork for successful development and demonstration projects.

The successes and the failures of traditional agricultural systems must be evaluated. The objective is to understand the ability of a given agroecosystem to meet environmental and sociocultural needs in a given region. The integration of experienced folk knowledge with conventional scientific knowledge of agricultural, silvicultural, and cattle production systems can serve as a powerful base for designing improved agroecosystems and assessing the potential for technology transfer.

• *Research Incentives that Include Basic and Applied Management Considerations, Farmer-to-Farmer Exchanges, and Farmer-Managed Research Should Be Developed*   The case study approach is one of the best ways to teach agroforestry and to encourage agroforestry research (Bainbridge, 1990a,b; Huxley, 1987). In academic settings, the system for meritorious recognition should be restructured to ensure that research solutions for real-world problems are given at least as much consideration as peer-reviewed journal articles. The role of farmers in this work must be expanded because farmers are often excellent teachers and extension workers (Gómez-Pompa and Jiménez-Osornio, 1989; Springborg, 1986) and are often better able to discuss issues and give demonstrations than are extension agents and researchers.

• *Support for Long-Term Research Should Be Increased*   The short-term nature of most research programs discourages and impedes agroforestry research. Typical funding cycles of 1, 3, or (more rarely) 5 years are incompatible with agroforestry research projects that may take 10, 20, 50, or more years. The importance of long-term funding has been recognized in only a few programs, most notably the Long-Term Ecological Research Program of the National Science Foundation (Callahan, 1984).

• *Support for Long-Term Monitoring Should Be Provided*   It is difficult to plan a research program without accurate information of current and past land use and environmental trends. The monitoring of the social, economic, and ecologic variables that are thought to contribute to the deforestation process is needed. The human component in environmental monitoring is often forgotten, even though social and ecologic factors are obviously mutually driven and intertwined. This implies that not only inventories of flora, fauna, soils, and air quality be collected over time but also corresponding temporal data on human population and distribution, land use, and market patterns for economically valuable natural resources be collected as well. It requires accounting for environmental subsidies (for example, soil erosion and declining soil fertility). The information could be integrated and computerized in a geographic information system data base that could be used as a basis for future planning and recommendations.

• *A Regional Data Base of Mulitpurpose Tree Species Should Be Developed* Creation of a regional data base of tree species, particularly multipurpose trees, deserves special priority. Multipurpose trees are of particular value in sustainable resource management for both subsistence and market production activities (Bainbridge, 1987b; Von Carlowitz, 1984). For example, the bread nut tree (*Brosimum alicastrum*) is a widespread species with multiple uses (food, fodder, and fuel) in the humid tropics of Mexico. It is thought to have been a vital food resource of the ancient Maya (Puleston, 1982) and may again return to prominence as a vital part of sustainable agricultural systems for the humid tropics (Pardo-Tejeda and Sánchez-Muñoz, 1980).

• *A Regional Data Base of Nitrogen-Fixing Tree Species Should Be Developed* A data base of nitrogen-fixing trees, which are effectively used in traditional agroforestry systems and of special value in maintaining fertility and restoring degraded lands needs to be developed (Flores Guido, 1987; Ngambeki, 1985; Virginia, 1986).

• *A Regional Network of Resource Research Groups and Institutions Should Be Established* A regional network of research groups and institutions modeled after the regional cooperative research and food production program known as Precodepa (Regional Cooperative Potato Program [see Niederhauser and Villarreal, 1986]) should be established. Precodepa's emphasis has been on building national research capabilities to provide a regional base of specialization and to transfer technology along with distributing shared information. This has enabled each participating country to take control of the program in their country and take pride in the achievements. Precopeda has been effective in maximizing the benefits gained from the limited funding for potato research. Funds are competitively allocated regionally, allowing specialization in various aspects of potato production and utilization in different countries along with excellent distribution of shared information.

Forest management and information also need to cross over sociopolitical boundaries. Regional networking increases the effectiveness of research and allows more information and progress to be obtained from the limited funding available (Piñeiro et al., 1987). Inclusion of nongovernmental organizations is of special importance because their contributions to solutions to the problems of deforestation and sustainable resource management have proportionally been far greater than the funding they receive.

Education

The following are strategies for improving education in sustainable resource management.

• *Local Farmers Should Be Included as Teachers in Educational Efforts* The knowledge and wisdom of local farmers need to be included in educational curricula and resource management studies (Gómez-Pompa and Kaus, 1992). This knowledge has been ignored by experts, and this has been a persistent problem in both agriculture and forestry research and extension (Bainbridge, 1987a).

• *Educational Programs that Encompass the Full Range of Resource Management Issues and Address Integrated Resource Management Should Be Developed* Schools of agriculture, veterinary medicine, human medicine, anthropology, biology, engineering, and economics should be involved in and include resource management issues.

Agroforestry systems, which are not part of conventional forestry or agricultural systems, are often considered primitive and have been ignored. Instead, intensive high-input systems have been emphasized despite their repeated failures. In many cases, these high-input systems perform poorly while local people continue to survive with long-established (but unstudied) agroforestry systems with native trees.

The educational systems of the United States and Mexico have emphasized a narrow vision of forestry that prepares students for intensive industrial production of monocultures (for example, pine and eucalyptus trees) but that ignores agroforestry applications (Bainbridge, 1987a). In the index of the major North American forestry journal, the *Journal of Forestry*, for example, there was not a single listing for agroforestry in 1990. Most agroforesters have remedied the failures of the United States and Mexican educational systems by working with farmers who use traditional agricultural systems—a necessary step but one that could be much more valuable with appropriate training in the classroom.

The educational systems in both the United States and Mexico must be revised to introduce the complexity and interaction of ecologic and cultural systems (Bainbridge, 1985, 1990b; Bawden et al., 1984; Chowdry, 1984). This has become much easier with the development of educational materials at ICRAF (for example, see Zulberti, 1987) and CATIE (Major et al., 1985), but there is still a shortage of material in Spanish. There is little or no information in the Maya language or in a pictorial format suitable for use by the people, many of whom are illiterate, who are expected to do the hands-on work or adopt the proposed forestry or agricultural programs. In addition, most of the education represents urban perceptions of the environment and neglects rural knowledge, experience, needs, and aspirations (Gómez-Pompa and Kaus, 1990).

• *Information About Different Approaches for Sustainable Resource Management in the Tropics, from Shifting Agriculture to Grain-Fed Cattle Ranging, Needs to Be Disseminated* The public needs to understand that virtually all food and natural resource production practices can be sustainable if the correct approach is used. Pollution is not necessarily a synonym for modern agriculture, and traditional agriculture is not a synonym for low productivity. The conventional myths of agriculture, forestry, and conservation need to be dispelled before public pressure will lead to policies and practices that are appropriate to the realities of working and caring for the land.

Demonstration Projects

The following are suggestions for demonstration projects of sustainable resource management in the Mexican humid tropics.

• *Demonstration Projects Need to Be Developed in Local Communities* Demonstration projects should be one of the first priorities for future funding. There is no shortage of potential sites, but there is a lack of trained personnel. Demonstration projects can provide much needed training in project management. Janzen's effort to reforest the Guanacaste National Park in Costa Rica (Murphy, 1987) is a worthy model. Many projects with scopes and visions similar to those of Janzen's project are needed in the humid tropics of Mexico.

By necessity, the development and testing of agroforestry systems for the humid tropics of Mexico must begin before all the desired information on tree species and traditional agroforestry practices is available. Fortunately, as has been learned in ecologic studies, it is possible to make advances without a complete understanding of each component of the agroecosystem. Agroforestry development and implementation are rarely simple, and new tools may have to be developed to properly consider the complex ecologic and social factors involved (Raintree, 1982, 1987b).

• *Natural Forest Management Projects Should Be Developed*   There are no large areas of managed tropical forests in Mexico. Although the management of natural forests is an important alternative that has been neglected in most tropical areas (Gómez-Pompa and Burley, 1991), there are methods for doing it (Schmidt, 1991). These methods could be demonstrated on small private and government-owned forests.

• *Plantation Designs Should Be Improved and Tested*   The establishment of tree plantations by private groups is also suggested to meet the demand for wood products. Research in this area is also of great importance. Trees are as challenging to grow as other agricultural crops and merit long-term research efforts to improve tree production and marketing and protection of trees from pests and diseases.

• *Restoration Reserves Should Be Established*   Experimental reserves for the restoration of biodiversity are needed (Bainbridge, 1990b) as are research and information on the restoration of degraded or impoverished tropical ecosystems. The degraded ecosystems are predominant, yet they hold great potential for the future. Restoration reserves should include sound agricultural, silvicultural, and animal husbandry activities that are compatible with sustained use of the area's resources. This research is challenging and the magnitude of the task should not be underemphasized. Although it is not possible to state that the full complex community of the humid tropical rain forest can be restored, many important species and functions of the forest can be reestablished in areas that are now degraded and very unproductive.

Implementation of Sustainable Resource Management

High priority should be given to the *ejido* (peasant) sector of the Mexican population. These rural populations may better understand new sustainable approaches of resource management because of their firsthand experience with similar, traditional agricultural practices. This should include education, extension activities, financing, marketing, and in particular, in-situ research. It is strongly suggested that any activity in this sector involve local people because without their participation the program is bound to fail. Use of the alliance approach outlined above for the conservation of biodiversity would also be a good strategy. This calls for a new green revolution of small-scale agricultural landholders. The rewards could be extraordinary.

The following are suggestions for implementing sustainable resource management.

• *Development-Oriented Projects with Local People Should Be Developed* In addition to demonstration projects done on a local level, larger development-oriented projects for the protection and restoration of humid tropics must also be established. Experience has demonstrated that expert recommendations for development are often of little value to local people because the recommendations commonly reflect the goals of the experts, not the local people whose needs are complex and who are adverse to risk (Edwards, 1989). It is essential to determine what people are doing and using, what they need and want, and why (Gómez-Pompa and Bainbridge, 1991; Gómez-Pompa and Jiménez-Osornio, 1989; Jecquire, 1976; Raintree, 1987b; Retiere, 1988). Improved tropical forestry management cannot be imposed from above or abroad. It must be developed by working with local communities and people.

• *Participation of Women in Education, Research, Extension, and Development Should Be Increased* The role of women is important and should not be ignored or neglected (Charlton, 1984; Fortmann and Rocheleau, 1985; Rocheleau, 1988). In some countries, more than half of the agricultural labor force is composed of women and from 40 to 80 percent of agricultural products are produced by women (Boulding, 1977; Howell, 1978). There are few data on the contributions of women to resource management in Mexico. It is known that women are very active in food production (commonly in the homegarden), in raising small livestock and poultry, and in gathering fuelwood. Their importance is greater than these data imply, however.

• *Rural Appraisal or Evaluation Forms Should Be Developed and Survey Materials Should Be Made Available to Researchers, Educators, and Communities so that They Can Understand Existing Practices and Land Use Allocations and Develop More Sustainable Management Packages* The AFRENA (Agroforestry Research Network for Africa) survey (Scherr, 1987) is a useful starting point for development projects, but it should be augmented with more detailed ethnobotanical, ecologic, and cultural surveys.

Farmers are "inventive, but development agencies rarely harness this inventiveness because they misunderstand the nature of both the agriculture and the politics of communities where food production is a major interest" (Richards, 1985:192). Intimate knowledge of a community and its culture is a prerequisite to any work that is intended to aid that community (de Wilde, 1967). To be successful the project must meet local needs, fit the local environment, and provide sufficient benefits so that action will be taken (Murray, 1989). If these requirements are met, the techniques will spread.

  • *A Program to Help Local Communities Plan and Implement Appropriate Development Programs Should Be Developed* Working with local people, information on planning and implementing appropriate development programs could then be used to develop a set of management goals and objectives. These would include economic (cash crops), subsistence (food, fodder, medicine), and environmental objectives. Planning should include long-term (10-, 20-, 50-, 100-year) objectives and project future demands based on population growth (Gómez-Pompa and Bainbridge, 1991). To reduce the risk from such activities, emphasis should be given to native species and, preferably, local ecotypes in mixed stands rather than monocultures. Ecologic succession can be used to reduce the cost and uncertainty of establishing a program in harsh and difficult environments (Khoshoo, 1987).

  • *Innovative Investment Programs Should Be Developed* Access to credit or capital is often the factor that limits improvements in resource utilization. Loans or small grants (less than US$200) may be catalysts for change, as the innovative small-loan program of the Grameen Bank in Bangladesh has shown (Yunus, 1990). Targeting investments to remove infrastructural constraints (for example, transport and storage problems) may be more important than making investments at the farm level. One way to stimulate diversified activities is to connect *campesinos* with markets (Brannon and Baklanoff, 1987) or to help develop local markets.

Plan for Success

If sustainable agriculture options are successfully implemented (as they can be), secondary problems may arise. For example, if farmers are successful, they may build up their equity and begin efforts to increase the size of their farms. If this demand for new lands is not met by converting existing grazing lands, it will put additional pressure on the few remaining forested lands and on farmland operated by less productive farmers. It will be important to develop and implement policies carefully to restrict the use of forested lands for agricultural expansion and to establish policies that will encourage the use of grazing lands for agriculture. Cattle production can easily be accommodated by using more efficient cattle feeding methods (Caesar, 1990; Preston, 1990).

Another possible consequence of a successful transition to sustainable resource management may be more efficient systems that require less hand labor, therefore providing more opportunities for the family to find other jobs in more urban areas. This trend may need to be addressed by government policies that will improve opportunities for displaced farmers or their children to obtain education or jobs or both in towns and cities. Many of these jobs may be provided by new processing and manufacturing facilities that use new forest products.

SUMMARY

These are the priorities for reversing current trends of deforestation and the use of unsustainable agricultural practices in the Mexican humid tropics. They are many and complex, and there is no single answer to the deforestation problem.

The best solutions are with small-scale farmers who have the experience, know the terrain, and have the most to gain. The responsibility of the nonrural sector—researchers, educators, industry, funding agencies, governments, and policy makers—lies in developing the infrastructure necessary to solve the problems. This will include better information, education, research, technological assistance, and credit incentives that help farmers build equity. The rural sector cannot respond to opportunities in the market without the means to adjust their production levels in terms of equipment, labor, market access, and knowledge. Investment in small-scale farmers at this very basic level, coupled with preparation for the consequences, can bring deforestation into check and can make agricultural practices in the Mexican humid tropics sustainable.

ACKNOWLEDGMENTS

The authors thank Silvia del Amo, Marlene de la Cruz, José Gonzalez, Steve Mitchell, Edward O. Plummer, and William W. Wood, Jr., for comments and suggestions on the first draft of this report. Reports from research undertaken under the Maya Sustainability Project (which is sponsored by the MacArthur Foundation) were used to supplement the available literature.

REFERENCES

Abrams, E. M., and D. J. Rue. 1988. The causes and consequences of deforestation among Prehistoric Maya. Hum. Ecol. 16(4):377–395.

Adelhelm, R., and J. Kotshci. 1985. Development and introduction of self-sustaining agricultural practices in tropical smallholder farms. Entwicklung und Landlicher Raum 19(4):17–20.

Alcorn, J. B. 1984. Development policy, forests and peasant farms: Reflections on Huastec managed forests' contribution to commercial production and resource conservation. Econ. Bot. 38(4):389–406.

Alcorn, J. B. 1990. Indigenous agroforestry systems in the Latin American tropics. Pp. 195–210 in Agroecology and Small Farm Development, M. Altieri and S. B. Hecht, eds. Boca Raton, Fla.: CRC Press.

Altieri, M. 1987. Agroecology: The Scientific Basis of Alternative Agriculture. Boulder, Colo.: Westview.

Altieri, M., and L. C. Merrick. 1987. In situ conservation of crop genetic resources through maintenance of traditional farming systems. Econ. Bot. 41(1):86–96.

Bainbridge, D. A. 1985. Ecologic education—Time for a new approach. Bull. Ecol. Soc. Amer. 66(4):461–462.

Bainbridge, D. A. 1987a. Agroforestry and the need for institutional reform. Cookstove News 7(3):9–20.

Bainbridge, D. A. 1987b. Multi-Purpose Tree Crops. Bibliography No. 2. Riverside, Calif.: Dry Lands Research Institute, University of California.

Bainbridge, D. A. 1990a. The Systems Approach to Complex Environmental Problem Solving. Elgin, Ariz.: Ecocultura.

Bainbridge, D. A. 1990b. The restoration of agricultural lands and dry lands. Pp. 4–13 in Environmental Restoration, J. Berger, ed. Washington, D.C.: Island.

Bainbridge, D. A., and S. M. Mitchell. 1988. Sustainable Agriculture for California: A Guide to Information. Davis, Calif.: University of California Sustainable Agriculture Research and Education Program.

Barrera-Vázquez, A. 1980. Esbozo de antecedentes etnicos en Yucatán. Pp. 21–38 in Seminario de Producción Agrícola en Yucatán, E. Hernández X, ed. Chapingo, México: Colegio de Postgraduados.

Bawden, R., R. D. Macadam, R. G. Packham, and I. Valentine. 1984. Systems things and practice in the education of agriculturalists. Agric. Syst. 13:205–225.

Bene, J. G., H. W. Beall, and A. Cote. 1977. Trees, Food, and People—Land Management in the Tropics. Ottawa, Canada: International Development Research Center.

Bennett, C. F. 1975. The advantages of cultural diversity. Unasylva 27(110):11–15.

Boulding, E. 1977. Women in the Twentieth Century World. New York: Wiley, for Sage Publications.

Bowers, B. 1989. Classic Maya fight to their finish. Sci. News 136(22):365.

Brannon, J., and E. N. Baklanoff. 1987. Agrarian Reform and Public Enterprise in Mexico. Tuscaloosa, Ala.: University of Alabama Press.

Cabrera, G. 1979. Especialización económica y movimientos migratorios en México. Pp. 215–216 in Crecimiento de la Población y Cambio Agrario, V. L. Urquidi, and J. B. Morelos, eds. México, D.F.: El Colegio de México.

Cabrera, G. 1988. La política de población en el contexto de las perspectivas de largo plazo del desarrollo nacional. Pp. 41–53 in México: El Desafío del Largo Plazo, G. Bueno, ed. México, D.F.: Limusa.

Caesar, K. 1990. Developments in crop research for the third world. Ambio 19(8):353–357.

Callahan, J. T. 1984. Long term ecological research. BioScience 34:363–367.

Calva, J. L., ed. 1988. Crisis Agrícola y Alimentaria en México: 1982–1988. México, D.F.: Fontamara 54.

Cannell, M. G. R., and J. E. Jackson, eds. 1985. The Attributes of Trees as Crop Plants. Huntington, U.K.: Institute of Terrestrial Ecology.

Cardós, A. 1959. El comercio entre los mayas antiguos. Acta Antropol. 2:50.

Carpenter, B. 1989. Faces in the forest. US News World Rep. 108(22):63–69.

Chambers, R. 1987. Sustainable Rural Livelihoods: A Strategy for People, Environment and Development. Commissioned Study No. 7. London: Institute of Development Studies.

Chambers, R., A. Pacey, and L. A. Thrupp. 1989. Farmer First: Farmer Innovation and Agricultural Research. London: Intermediate Technology Publications.

Charlton, S. E. M. 1984. Women in Third World Devleopment. Boulder, Colo.: Westview.

Chowdry, K. 1984. Agroforestry, the rural poor and institutional structures. Pp. 11–19 in Social, Economic and Institutional Aspects of Agroforestry, J. K. Jackson, ed. Tokyo: United Nations University.

Comisión Económica de la América Latina. 1982. Economía campesina y agricultura empresarial. México, D.F.: Siglo XXI Editores.

Comisión Nacional Forestal. 1988. Hacia un Programa de Acción Forestal Tropical en México. Propuesta para la Conservación y el Desarrollo de las Selvas del Sureste. México, D.F.: Secretary of the Agrarian Reform, Secretary of Agriculture and Hydraulic Resources, and Secretary of Urban Development and Ecology.

Cook, S. F., and W. Borah. 1980. Ensayos sobre Historia de la Población: México y California. México, D.F.: Editorial Siglo XXI.

Cordera, R., and C. Tello. 1981. México, La Disputa por la Nación. México, D.F.: Siglo XXI Editores.

Dahlberg, K. A. 1990. The industrial model and its impacts on small famers: The green revolution as a case. Pp. 83–90 in Agroecology and Small Farm Development, M. Altieri and S. B. Hecht, eds. Boca Raton, Fla.: CRC Press.

del Amo, R. S. 1991. Management of secondary vegetation for artificial creation of useful rain forest in Uxpanapa, Veracruz, Mexico. Pp. 343–350 in Rain Forest Regeneration and Management, A. Gómez-Pompa, T. C. Whitmore, and M. Hadley, eds. Park Ridge, N.J.: Parthenon; and Paris: United Nations Educational, Scientific, and Cultural Organization.

Denevan, W. M. 1970. Aboriginal drained-field cultivation in the Americas. Science 169:647–654.

Denevan, W. M. 1982. Causes of deforestation and forest and woodland degradation in Tropical Latin America. Pp. 168–171 in Assessment of Technologies to Sustain Tropical Forest and Woodland Resources. Report to the Office of Technology Assessment, U.S. Congress. Washington, D.C.: Government Printing Office.

Department of Agriculture and Hydraulic Resources. 1984. Comisión del Plan Nacional Hidráulico. Desarrollo Rural Integrado de la Selva Lacandona. México, D.F.: Secretary of Agriculture and Hydraulic Resources.

Department of Agriculture and Hydraulic Resources. 1987. Inventario Cartográfico de Recursos Agropecuarios y Forestales y Clasificación Agrológica Estatal Sobre Frontera Agrícola y Capacidad de Uso del Suelo. México, D.F.: Department of Agriculture and Hydraulic Resources.

Department of Agriculture and Hydraulic Resources. 1989. México Forestal en Cifras. 1987. Dirección General de Política Sectorial. México, D.F.: Department of Agriculture and Hydraulic Resources.

de Wilde, J. C. 1967. Experiences with Agricultural Development in Tropical Africa, Vol. 1: The Synthesis. Baltimore: John Hopkins University Press.

Ecosfera. 1990. Análisis preliminar de las áreas silvestres de la zona Maya. Chiapas, México: San Cristobal de las Casas.

Edwards, M. 1989. The irrelevance of development studies. Third World Quart. 11(1):116–135.

Ehui, S., B. Kang, and D. Spencer. 1990. Economic analysis of soil erosion effects in alley cropping. No-till and bush fallow systems in South Western Nigeria. Agric. Syst. 34(4):349–368.

Estrada, A., and R. Coates-Estrada. 1983. Rain forest in Mexico: Research and conservation at Los Tuxtlas. Oryx 17:201–202.

Evans, P. T. 1988. Designing agroforestry innovations to increase their adoptability: A case study from Paraguay. J. Rural Studies 4(1):45–55.

Food and Agriculture Organization (FAO) and United Nations Environment Program (UNEP). 1980. An interim report on the state of forest resources in the developing countries. Food and Agriculture Organization of the United Nations, Rome, Italy.

FAO and UNEP. 1981. Tropical Forest Resources Assessment Project, Vol. 1. Rome, Italy: Food and Agriculture Organization of the United Nations.

FAO and UNEP. 1988. An interim report on the state of forest resources in the developing countries. Forest Resources Division, Food and Agriculture Organization of the United Nations, Rome, Italy.

Flores Guido, J. S. 1987. Yucatán, tierra de las leguminosas. Rev. Universidad Autónoma de Yucatán 163(Oct/Nov):33–37.

Flores Guido, J. S., and E. Ucan Ek. 1983. Nombres usados por los Mayas para designar la vegetación. Cuadernos de Divulgación INIREB (Insituto Nacional de Investigaciones Sobre Recursos Bióticos) 10:1–33.

Fortmann, L. 1985. The tree tenure factor in agroforestry with particular reference to Africa. Agrofores. Sys. 2:229–251.

Fortmann, L., and J. W. Bruce. 1988. Whose Trees? Proprietary Dimensions of Forestry. Boulder, Colo.: Westview.

Fortmann, L., and J. Riddell. 1985. Trees and Tenure: An annotated bibliography for agroforesters and others. Madison, Wis.: Land Tenure Center, University of Wisconsin; and Nairobi, Kenya: International Center for Research in Agroforestry.

Fortmann, L., and D. Rocheleau. 1985. Women and agroforestry: Four myths and three case studies. Agrofores. Sys. 2(4):254–272.

Fundación Universo Veintiuno. 1990. Desarrollo y Medio Ambiente en México. Diagnóstico, 1990. México, D.F.: Friedrich Ebert Stiftung.

Gladwin, C. H. 1976. A view of the Plan Puebla: An application of hierarchical decision models. Amer. J. Agric. Econ. 59(5):881–887.

Gladwin, C. H. 1979. Cognitive strategies and adoption decisions: A case study of non-adoption of an agronomic recommendation. Econ. Dev. Cult. Change 28(1):155–173.

Gladwin, C. H. 1983. Contributions of decision tree methodology to a farming systems program. Hum. Org. 42(2):146–157.

Gliessman, S. R., R. García, and M. Amador. 1981. The ecological basis for the application of traditional agricultural technology in the management of tropical agroecosystems. Agro-Ecosystems 7:173–185.

Gliessman, S. R., B. L. Turner II, F. J. Rosado-May, and M. F. Amador. 1983. Ancient raised field agriculture in the Maya lowlands of southern Mexico. Pp. 97–111 in Drained Field Agriculture in Central and South America. BAR International Series 189. Oxford, England: BAR International.

Gómez-Pompa, A. 1987a. On Maya silviculture. Mexican Studies 3(1):1–17.

Gómez-Pompa, A. 1987b. Tropical deforestation and Maya silviculture: An ecological paradox. Tulane Studies Zool. Bot. 26(1):1–17.

Gómez-Pompa, A., and D. A. Bainbridge. 1991. Tropical forestry as if people mattered. In A Half Century of Tropical Forest Research, A. E. Lugo and C. Lowe, eds. New York: Springer-Verlag.

Gómez-Pompa, A., and F. W. Burley. 1991. The management of natural tropical forests. In Rain Forest Regeneration and Management, A. Gómez-Pompa, T. C. Whitmore, and M. Hadley, eds. Park Ridge, N.J.: Parthenon; and Paris: United Nations Educational, Scientific, and Cultural Organization.

Gómez-Pompa, A., and J. J. Jiménez-Osornio. 1989. Some reflections on intensive traditional agriculture. Food and Farm: Current Debates. Monogr. Econ. Anthropol. 7:221–253.

Gómez-Pompa, A., and A. Kaus. 1987. The conservation of resources by traditional cultures in the tropics. Paper presented at the World Wilderness Congress, Estes Park, Colo.

Gómez-Pompa, A., and A. Kaus. 1990. Traditional management of tropical forests in Mexico. Pp. 45–67 in Alternatives to Deforestation: Steps Toward Sustainable Use of the Amazon Rain Forest, A. B. Anderson, ed. New York: Columbia University Press.

Gómez-Pompa, A., and A. Kaus. 1992. Taming the wilderness myth: A View of Environmental Education from the Field. BioScience 42(2):271–279.

Gómez-Pompa, A., and C. Vázquez-Yanes. 1981. Successional studies of a rain forest in Mexico. Pp. 246–266 in Forest Succession: Concepts and Application, D. C. West, H. H. Shugart, and D. B. Botkin, eds. New York: Springer-Verlag.

Gómez-Pompa, A., C. Vázquez-Yanes, and S. Guevara. 1972. The tropical rain forest: A non-renewable resource. Science 177:762–765.

Gómez-Pompa, A., T. C. Whitmore, and M. Hadley, eds. 1991. Rain Forest Regeneration and Management. Park Ridge, N.J.: Parthenon; and Paris: United Nations Educational, Scientific, and Cultural Organization.

González-Pacheco, C. 1983. Capital Extranjero en la Selva de Chiapas, 1863–1982. Ira, ed. México, D.F.: Instituto de Investigación Economicas, UNAM.

Grainger, A. 1984. Quantifying changes in forest cover in the humid tropics: Overcoming current limitations. J. World Forest Resource Manag. 1:3–23.

Herrera Castro, N. 1990. Estudios ecológicos en los huertos familiares Mayas. Report to Maya Sustainability Project, Riverside, California.

Howell, B. 1978. Women in Development. Bread for the World Background Paper 29. November. Washington, D.C.: Bread for the World.

Huxley, P. A. 1987. A combined systems/case study approach for agroforestry teaching. Pp. 122–127 in Professional Education in Agroforestry, E. Zulberti, ed. Nairobi, Kenya: International Center for Research in Agroforestry.

Illsley, C. 1984. Vegetación y producción de la milpa bajo roza, tumba y quema en el ejido de Yaxcabá, Yucatán, México. Tesis profesional. Escuela de Biología, Universidad de Michoacana de San Nicolás de Hidalgo, Hidalgo, México.

Instituto Méxicano de Café. 1974. Tecnología Cafetalera Méxicana. México, D.F.: Instituto Méxicano de Café.

Instituto Nacional de Estadística Geografía e Informática (INEGI, National Institute of Statistics, Geography, and Information). 1986. Datos Básicos sobre la Población de México. México, D.F.: Instituto Nacional de Estadística Geografía e Informática.

INEGI. 1990a. XI Censo General de Población y Vivienda. Aguascalientes, México: Instituto Nacional de Estadística Geografía e Informática.

INEGI. 1990b. El Sector Alimentario en México. México, D.F.: Instituto Nacional de Estadística Geografía e Informática and Comisión Nacional de Alimentación.

INEGI. 1990c. Anuario Estadístico del Estado de Chiapas. México, D.F.: Instituto Nacional de Estadística Geografía e Informática.

International Rice Research Institute. 1992. IARCs and national systems seek alternatives to slash and burn farming. IRRI Hotline 2(3):1.

Janzen, D. H. 1988. Tropical ecological and biocultural restoration. Science 239:243–244.

Jecquire, N. 1976. Appropriate Technology: Problems and Promises. Part I. The Major Policy Issues. Paris: Development Center of the Organization for Economic Cooperation and Development.

Jiménez, E., and A. Gómez-Pompa. 1981. Estudios Ecológicos en el Agroecosistema Cafetalero. México, D.F.: Instituto Nacional de Investigociones Recursos Bióticos.

Khoshoo, T. N. 1987. Ecodevelopment of Alkaline Land. Lucknow, India: National Botanical Research Institute.

Labelle, R. 1987. Agroforestry: General Concepts, Early Work and Current Initiatives—A Review of the Literature. Nairobi, Kenya: International Center for Research in Agroforestry.

Lagemann, J. 1982. Problems of agricultural production in humid tropical lowlands. Entwicklung und Landlicher Raum 16(3):15–17.

Lanly, J. P. 1982. Tropical Resources. Forestry Paper 30. Rome, Italy: Food and Agriculture Program, Food and Agriculture Organization of the United Nations.

Lanly, J. P. 1989. The status of tropical forests. In A Half Century of Tropical Forest Research, A. E. Lugo and C. Lowe, eds. New York: Springer-Verlag.

Leeger, B. W. 1989. Agroforestry: Its effect on food security, risk taking, and third world community development. Master's thesis. William Carey International University, Pasadena, California.

López-Portillo, J., M. R. Kayes, A. González, E. Cabrera, and O. Sánchez. 1990. Los incendios de Quintana Roo: ¿Catástrofe ecológica o evento periódico? Ciencia y Desarrollo 15(91):43–57.

Lugo, A., and S. Brown. 1981. Tropical ecosystems and the human factor. Unasylva 33(133):45–52.

Lundell, C. L. 1938. The 1938 botanical expedition to Yucatán and Quintana Roo, Mexico. Pp. 143–147 in Carnegie Institute of Washington Yearbook. Washington, D.C.: Carnegie Institute of Washington.

Lundgren, B., and J. B. Raintree. 1982. Agroforestry. Paper presented at the Conference of Directors of National Agroforestry Research Systems in Asia, Jakarta, Indonesia.

Major, M., G. Budowski, and R. Borel. 1985. Manual of teaching methods for use in agroforestry short courses. Turrialba, Costa Rica: Centro Agronómico Tropical de Investigación y Enseñanza.

McNeely, J. A. 1988. Economics and Biological Diversity: Developing and using economic incentives to conserve biological resources. Gland, Switzerland: International Union for the Conservation of Nature and Natural Resources.

McNeely, J. A., K. R. Miller, W. V. Reid, R. A. Mittermeier, and T. B. Werner. 1990. Conserving the World's Biological Diversity. Gland, Switzerland: International Union for the Conservation of Nature and Natural Resources, World Resources Institute, and World Wildlife Fund; and Washington, D.C.: World Bank.

Melillo, J. M., C. A. Palm, R. A. Houghton, et al. 1985. A comparison of two recent estimates of disturbance in tropical forests. Environ. Conserv. 12(1):37–40.

Méxican Institute of Coffee. 1974. Méxicano Cafetalera Méxicano. México, D.F.: Méxican Institute of Coffee.

Ministry of Finance and Public Credit (Secretaria de Hacienda y Credito Publico). 1991. Mexico: A New Economic Profile. México, D.F.: Ministry of Finance and Public Credit.

Morley, S. G. 1946. The Ancient Maya. Palo Alto, Calif.: Stanford University Press.

Murgueitio, E. 1988. Los Arboles Forrajeros en la Alimentacion Animal. Cali, Colombia: Centro Internacional de Agricultura Tropical.

Murgueitio, E. 1990. Intensive sustainable livestock production: An alternative for deforestation. Ambio 19(8):397–400.

Murphy, J. 1987. Growing a forest from scratch. Time 128(26):65.

Murray, G. F. 1989. The domestication of wood in Haiti: A case study in applied evolution. Pp. 148–156 in Applying Anthropology: An Introductory Reader, A. Podolefsky and P. J. Brown, eds. Mountain View, Calif.: Mayfield.

Myers, N. 1981. Deforestation in the tropics: Who gains, who loses?. Pp. 1–21 in Where Have All the Flowers Gone? Deforestation in the Third World, V. H. Sutlive, N. Altshuler, and M. Zamora, eds. Studies in Third World Societies, Pub. No. 13. Williamsburg, Va.: Department of Anthropology, College of William and Mary.

Nations, J. D., and D. I. Komer. 1983. Central America's tropical forests: Positive steps for survival. Ambio 12(5):233–239.

Ngambeki, D. S. 1985. Economic evaluation of alley cropping Leucaena with maize-maize and cowpea. Agric. Syst. 17 (4):243–258.

Niederhauser, J. S., and V. Villarreal. 1986. Precodepa, A successful model for a new concept in regional cooperation for international agricultural development. Amer. Potato J. 63:237–240.

Nolasco, M. 1985. Café y Sociedad en México. México, D.F.: Centro de Ecodesarrollo.

Pardo-Tejeda, E., and C. Sánchez-Muñoz. 1980. *Brosimum alicastrum* (breadnut tree): A Neglected Tropical Forest Resource. Xalapa, Veracruz, México: Instituto Nacional de Investigaciones Recursos Bióticos.

Parsons, J. J. 1976. Forest to pasture: Development or destruction? Rev. Biol. Tropical 24(Suppl.1):121–138.

Partridge, W. L. 1984. The humid tropics cattle ranching complex: Cases from Panama reviewed. Hum. Org. 43(1):76–80.

Pennington, T. D., and J. Sarukhán. 1969. Manual para la Identificatión de Campo de los Principles Árboles Tropicales de México. México, D.F.: Instituto Nacional de Investigaciones Forestales, Secretaria de Agricultura y Recursos Hidraulicos.

Perelman, M. 1976. The green revolution: American agriculture in the third world. Pp. 111–126 in Radical Agriculture, R. Merrill, ed. New York: Harper & Row.

Piñeiro, M. E., T. V. R. Pillary, F. Torres, D. L. Winkelmann, and E. Gastal. 1987. Networking as a means of increasing efficiency of agricultural research. Pp. 89–147 in Impact of Research on National Agricultural Development, B. C. Webster, C. Valerde, and A. J. Fletcher, eds. The Hague, Netherlands: International Service for National Agriculture.

Preston, T. R. 1990. Future strategies for livestock production in tropical third world countries. Ambio 19(8):390–393.

Preston, T. R., and R. A. Leng. 1987. Matching Ruminant Production Systems with Available Resources in the Tropics and Subtropics. Armidale, New South Wales, Australia: Penambul Books.

Preston, T. R., and E. Murgueitio. 1987. Tree and shrub legumes as protein sources for livestock. Pp. 94–104 in Forage Legumes and Other Local Protein Sources as Substitutes for Imported Protein Meals, D. Walmsley, ed. Wageningen, Netherlands: CTI.

Puleston, D. E. 1982. The role of Ramon in Maya subsistence. Pp. 353–366 in Maya Subsistence, K. V. Flannery, ed. New York: Academic Press.

Raintree, J. B. 1982. Readings for a socially relevant agroforestry. Paper presented at the International Workshop on Tenure Issues in Agroforestry, Nairobi, Kenya.

Raintree, J. B., ed. 1987a. Land, Trees, and Tenure: Proceedings of an International Workshop on Tenure Issues in Agroforestry. Madison, Wis.: Land Tenure Center, University of Wisconsin; and Nairobi, Kenya: International Center for Research in Agroforestry.

Raintree, J. B., ed. 1987b. D & D User's Manual: An Introduction to Agroforestry Diagnosis and Design. Nairobi, Kenya: International Center for Research in Agroforestry.

Ramakrishnan, P. S. 1984. The science behind rotational bush fallow agricultural system (jhum). Proc. Indian Acad. Sci. 93(3):379–400.

Raymond, C. 1990. Researchers see loss of cultural diversity in destruction of world's rain forests. Chron. Higher Educat. 37(15):5, 8–9.

Redford, K. H. 1990. The ecologically noble savage. Orion 9(3):25–29.

Reid, R., and G. Wilson. 1985. Agroforestry in Australia and New Zealand. Box Hill, Victoria, Australia: Goddar and Dobson.

Remmers, G., and H. de Koeyer. 1988. El "tolche" en pixoy. Master's thesis. University of Wageningen, Wageningen, Netherlands.

Repetto, R. 1990. Deforestation in the tropics. Sci. Amer. 262(4):36–42.

Retiere, A. 1988. Nadie desarrola a nadie: No one is developed by anyone: Buscando reinventar el papel técnico en la comunidad. San Cristobal de las Casas, Chiapas, México: Instituto de Asesoria Antropológica para la Región Maya.

Richards, P. 1985. Indigenous Agricultural Revolution: Ecology and Food Production in West Africa. Boulder, Colo.: Westview.

Rico-Gray, V., J. G. García-Franco, A. Puch, and P. Sima. 1988. Composition and structure of a tropical dry forest in Yucatán, Mexico. Int. J. Ecol. Environ. Sci. 14:21–29.

Riding, A. 1989. Distant Neighbors: A Portrait of the Mexicans. New York: Vintage Books.

Rocheleau, D. E. 1988. Yours, mine and ours. Paper presented at the Second Kenya National Seminar in Agroforestry. International Center for Research on Agroforestry, Nairobi, Kenya.

Rzedowski, J. 1978. Vegetación de México. México, D.F.: Editorial Edición Limusa.

Scherr, S. J. 1985. The Oil Syndrome and Agricultural Development: Lessons from Tabasco, Mexico. New York: Praeger.

Scherr, S. J. 1987. AFRENA worksheets for land use system description. Pp. 69–105 in D & D User's Manual: An Introduction to Agroforestry Diagnosis and Design, J. B. Raintree, ed. Nairobi, Kenya: International Center for Research in Agroforestry.

Schmidt, R. C. 1991. Tropical rain forest management: A status report. Pp. 181–207 in Rain Forest Regeneration and Management, A. Gómez-Pompa, T. C. Whitmore, and M. Hadley, eds. Park Ridge, N.J.: Parthenon; and Paris: United Nations Educational, Scientific, and Cultural Organization.

Secretaria de Programación y Presupuesto. 1981. Carta de vegetación y uso actual del suelo esc. 1:100,000. In Atlas Nacional del Medio Físico. México, D.F.: Secretaría de Programación y Presupuesto.

Shane, D. R. 1980. Hoofprints on the forests: An inquiry into the beef cattle industry in the tropical forest areas of Latin America. Manuscript prepared for Office of Environmental Affairs, U.S. Department of State, Washington, D.C.

Shepherd, G., and J. Stewart. 1988. Poor people's forestry. Appro. Technol. 15(1):1–4.

Siemens, A. H. 1983. Wetfield agriculture in prehispanic Mesoamerica. Geograp. Rev. 73(2):166–181.

Siemens, A. H., and D. E. Puleston. 1972. Ridged fields and associated features in southern Campeche: New perspectives on lowland Maya. Amer. Antiq. 37:228–239.

Springborg, R. 1986. Impediments to the transfer of Australian dryland farming technology to the Middle East. Agric. Ecosyst. Environ. 17(3/4):229–251.

Stewart, J. I. 1988. Response Farming in Rainfed Agriculture. Davis, Calif.: Wharf Foundation.

Thompson, J. E. 1954. The Rise and Fall of Maya Civilization. Norman: University of Oklahoma.

Toledo, V. M. 1988. La diversidad biológica de México. Ciencia y Desarrollo 81:17–30.

Toledo, V. M., J. Carabias, C. Mapes, and C. Toledo. 1985. Ecología y Autosuficiencia Alimentaria. México, D.F.: Siglo XXI Editores.

Toledo, V. M., J. Carabias, C. Toledo, and C. González-Pacheco. 1989. La Producción Rural en México: Alternativas Ecológicas. Número 6. México, D.F.: Siglo XXI Editores.

Torres, F. 1983. Potential contribution of Leucaena hedgerows intercropped with maize to the production of organic nitrogen and fuelwood in the lowland humid tropics. Leucaena Res. Rep. 4:50–53.

Turner, B. L., II. 1974. Prehistoric intensive agriculture in the Maya lowlands. Science 185:118–124.

Van den Bosch, R. 1980. The Pesticide Conspiracy. New York: Doubleday.

Vergara, N. T., and N. D. Briones. 1987. Agroforestry in the Humid Tropics: Its Protective and Ameliorative Roles to Enhance Productivity and Sustainability. Honolulu, Hawaii: East-West Center and Southeast Asian Regional Center for Graduate Study and Research in Agriculture.

Virginia, R. A. 1986. Soil development under tree legume canopies. Forest Ecol. Manag. 16:69–79.

Volke Haller, V., and I. Sepúlveda González. 1987. Agricultura de Subsistencia y Desarrollo Rural. México, D.F.: Editorial Trillas.

Von Carlowitz, P. G. 1984. Multipurpose tree yield data—The current state of knowledge. Agrofores. Syst. 4:291–294.

Wilken, G. C. 1987. Good Farmers. Berkeley: University of California Press.

Wilson, G. F., B. T. Kang, and K. Mulongoy. 1986. Alley cropping: Trees as sources of green manure and mulch in the tropics. Biol. Agric. Horticulture 3(2/3):251–267.

Winterbottom, R., and P. T. Hazelwood. 1987. Agroforestry and sustainable development: Making the connection. Ambio 16(2/3):100–110.

World Bank. 1990. World Development Report 1990: Poverty. Washington, D.C.: World Bank.

World Resources Institute. 1990. World Resources 1990–91. A Guide to the Global Environment. New York: Oxford University Press.

Yamoah, C. F., and J. R. Burleigh. 1990. Alley Cropping Sesbania sesban (L) Merrill with food crops in the highland region of Rwanda. Agrofores. Syst. 10(2):169–181.

Yates, P. L. 1981. México's Agricultural Dilemna. Tucson: University of Arizona Press.

Yunus, M. 1990. Credit as a human right: A Bangladesh bank helps poor women. New York Times. April 2, 1990. 139:A13, A17.

Zhaohua, Z. 1988. A new farming system—Crop/paulownia intercropping. In Multipurpose Tree Species for Small Farm Development. IDRC/Winrock.

Zulberti, E., ed. 1987. Professional Education in Agroforestry. Nairobi, Kenya: International Center for Research in Agroforestry.

The Philippines

*Dennis P. Garrity, David M. Kummer, and
Ernesto S. Guiang*

This profile focuses on the most pressing issues of sustainable natural resource management in the sloping upland areas of the Philippines. It begins with an analysis of the historical and current dimensions of land use in the upland ecosystem, reviews and critiques proposed actions, and recommends solutions within an overarching strategy that builds on the linkages that exist between farming and forestry systems.

The upland ecosystem must be addressed as a distinct entity. The uplands are rolling to steep areas where both agriculture and forestry are practiced on slopes ranging upward from 18 percent. The sloping uplands occupy about 55 percent of the land surface of the country (Cruz et al., 1986) and have an estimated population of 17.8 million. The upland population is projected to be 24 million to 26 million in the year 2000, with a density of 160 to 175 persons per km². Upland inhabitants are primarily poor farming families with insecure land tenure. Subsistence food production rather than forestry is their over-

Dennis P. Garrity is an agronomist/crop ecologist with the International Rice Research Institute, Los Baños, Philippines; David M. Kummer is a visiting assistant professor with the Graduate School of Geography, and a research associate with the George Perkins Marsh Institute, Clark University, Worcester, Massachusetts; Ernesto S. Guiang is a community forest management specialist with the Department of Environment and Natural Resources, Quezon City, Philippines.

riding priority. The paramount objective for public intervention in upland management is that of obtaining the greatest good for the greatest number of people in ways that are consistent with the long-term sustainability of the productive capacity of the ecosystem.

Forest denudation is at an advanced stage in the Philippines. Total forest cover shrank from 10.5 million ha in 1968 to 6.1 million ha in 1991. The remaining old-growth forest covered less than 1 million ha in 1991 and possibly as little as 700,000 ha. At current rates of logging, nearly all vestiges of the country's primary dipterocarp forest biota may be depleted in the next 10 to 15 years. The will of the people and government to effectively address the Philippine deforestation problem is growing, but it is still weak.

There have been several recent reviews concerning natural resource management in the Philippines. These reviews examined government policy, the political climate, and the institutional framework and made numerous specific recommendations for a major reorientation. In addition, the Master Plan for Forestry Development (Department of Environment and Natural Resources, 1990) has recently been issued by the Philippine government. It lays out a framework for forestland management over the next 25 years. It sets a detailed, optimistic agenda that adopts a strategy of reduced public management in favor of increased private management of forest resources through people-oriented forestry.

Although this profile focuses on the dynamics of upland agricultural technology in relation to deforestation, many factors other than agricultural technology have a stronger direct influence on the rate and extent of forest depletion or conversion. These factors include inappropriate forest policy, poor policy implementation, and the insecurity of land tenure among upland farm populations. Commercial logging (legal and illegal) directly caused the majority of old-growth forest depletion during the past half century, and it continues to do so today. The accessibility to remote forestlands brought about by the opening of logging roads stimulated the settlement of small-scale farmers and resulted in the subsequent conversion of depleted forests to farms.

The initial sections of this profile examine the present state of the natural resource base of the uplands and past trends in resource degradation. The profile then reviews the importance of land and forest resources to the political economy of the Philippines and the failure of development in the Philippines in the post-World War II period. This is followed by an analysis of potential solutions to the problems identified. The solutions to the upland resource management and subsistence crises fall into a general strategy with three essential components: land tenure, resource management technology, and infrastructure delivery. The final section outlines a proposed action strategy in terms of these three components.

THE STATE OF THE PHILIPPINE
UPLAND ECOSYSTEM

This section analyzes the important factors that have determined the development of land use systems in the Philippines uplands. The major forces and constraints that directly affect upland agriculture and forestry are emphasized.

Physical Environment

The Philippines is an archipelago with a total land area of 30 million ha. Although it encompasses more than 7,000 islands, the majority of these are insignificant in terms of size and population. The 15 largest islands make up 94 percent of the total land area. Luzon and Mindanao occupy about 35 and 32 percent of total land area, respectively. The Philippines is a physically fragmented state, and separateness is a major feature of its geography and culture. The island nature of the country gives it a very long coastline relative to its size. No inland area is far from the ocean.

The country has a complex geology and physiography. Although Luzon and Mindanao have major lowland areas, most of the islands have relatively narrow coastal plains. The Philippines as a whole is characterized by high relief. Steep upland areas with greater than 18 percent slope make up about 55 percent of the total area (Cruz et al., 1986). The climate is humid tropical. However, because of the mountainous terrain, the occurrence of typhoons in the northern half of the country, and the effects of two separate monsoon seasons, there is striking micro- and macrovariation in the seasonal distribution and amount of precipitation. Within-season droughts and the limited length of the growing season are common constraints, but the total quantity of precipitation is abundant: 90 percent of the country receives at least 1,780 mm per year (Wernstedt and Spencer, 1967).

The high relief, the relatively high levels of precipitation, and the frequent extreme concentration of rainfall in short periods because of typhoons contribute to serious soil erosion problems. Given the complex geology and geologic history, the soils of the Philippines are varied but are generally not as weathered as most humid tropical soils because of their relatively younger age. The inherent soil properties are limiting in many sloping upland areas (particularly where extensive

erosion and land degradation have occurred), but the Philippines
has a comparatively favorable soil base for a country in the humid
tropics.

Land Use

In the Philippines today, about half the land is classified as alienable and disposable. This land may be privately owned. The other
half, which mostly has slopes of greater than 18 percent, is classified
as public forestland. Only 6 million ha has significant tree cover and
less than 1 million ha of old-growth or primary forest remains (Table
1). In comparison, there was 10 million ha of old-growth forest in the
1950s. The extent of this forest conversion has reduced to critically

TABLE 1   Forest Cover in the Philippines as Determined by
Various Inventories (in Thousands of Hectares)

Forest Cover	Swedish Space Corporation (1988)	German Inventory[a]	LANDSAT 1980[b]	Official 1981[c]
Pine	81	239	227	193
Mossy or unproductive	246	1,681	1,320	1,759
Dipterocarp	6,629	4,403	6,304	6,588
Closed	2,435	1,042	2,940	2,794
Open	4,194	3,361	3,363	3,794
Mangrove	149	—	175	112
Other	—	—	121	—
Total	7,105	6,323	8,146	8,652

[a]The Philippine–German Forest Resources Inventory Project (Forest Management
Bureau, 1988) covers only lands it has classified as forestlands, which would exclude
as much as 1.4 million ha of "forest" on alienable and disposable lands. Forest cover
in mangroves was not reported.

[b]Open canopy was synonymous with "residual stands" or "young growth." Mangrove includes both mature and residual stands, as does pine. "Brushland" was not
counted as "forest."

[c]Official data were based on continuous updating of earlier estimates of inventory
data, including older aerial photos. "Brushland" was excluded from "forest."

SOURCES: Swedish Space Corporation. 1988. Mapping of the Natural Conditions of
the Philippines. Solna: Swedish Space Corporation; German inventory: Philippine-
German Forest Resources Inventory Project. 1988. In Results of the Forest Resources
Inventory Project, C.V. Gulmatico, ed. Unpublished paper. Forest Management Bu-
reau, Dilimän, Quezon City, Philippines; LANDSAT: Unpublished computer print-
out. Forest Management Bureau, Dilimän, Quezon City, Philippines; Official: World
Bank. 1989a. Annex 3, Table 1, in Philippines: Environment and Natural Resource
Management Study. Washington, D.C.: World Bank.

TABLE 2 Land Use in the Philippines
(in Thousands of Hectares)

Land Cover	Area
Forest	7,226
Pine	81
Mossy or unproductive	246
Dipterocarp	6,629
Closed	2,435
Open	4,194
Mangrove	149
Other	121
Extensive cultivation	11,958
Open in forest	31
Grassland	1,813
Mixed[a]	10,114
Intensive cultivation	9,729
Plantation	5,336
Coconut	1,133
Other	90
Coconut and cropland	3,748
Other and cropland	365
Cropland	4,393
Fish ponds	205
Fish ponds created from mangroves	195
Other fishponds	10
Other land or lakes	542
Unclassified area	546
Total	30,206

[a]Mixed grass, brush, plantation, and other crops.

SOURCE: Swedish Space Corporation. 1988. Mapping of the Natural Conditions of the Philippines. Solna: Swedish Space Corporation.

low levels the habitat of the many species of flora and fauna endemic to the Philippines.

Recently, the Swedish Space Corporation (1988) completed a study—the first and only one to cover all types of land uses—of the natural vegetation in the Philippines (Table 2). On the basis of that survey, the World Bank (1989a) calculated that cultivated land covers 11.3 million ha, or 38 percent of the total land area. Cultivated area in the uplands is about 3.9 million ha.

The *1980 Census of Agriculture* (National Census and Statistics Office, 1985) estimated the area of cultivated land to be 9.7 million ha in 1980. If these data and World Bank estimates are correct, then the area of cultivated land increased by more than 1.6 million ha between 1980 and 1987, an annual increment of 229,000 ha/year. The average annual rate of deforestation between 1980 and 1987 was 157,000 ha/year. Although direct conversion from forestlands to croplands cannot be inferred, it appears that large areas of grasslands are now being converted to agricultural uses, increasing the pressure on the limited land resources.

Population Growth

Rapid population growth in the past half century is widely acknowledged as a major force in the accelerated deterioration in the country's natural resources (Porter and Ganapin, 1988). The 1990 population of the Philippines was estimated to be 66.1 million and was increasing at an annual rate of 2.6 percent (Population Reference Bureau, 1990).

Table 3 presents Philippine population data since 1948. Although the rate of growth of the Philippine population declined slowly from the 1948–1960 period to the 1975–1980 period, the population growth rate remains the highest of any country in Southeast Asia. The current population density is second only to that of Singapore (Population Reference Bureau, 1990). The rural population, as a percentage of the total population, has been declining, but at a slow rate (from 73 percent in 1948 to 63 percent in 1980). Urban growth is predominantly in the city of Manila (Pernia, 1988).

The Philippines has a serious population growth problem, but acceptance of this fact has been fairly recent. As late as 1969, Duckham

TABLE 3   Philippines Population Data, 1948–1980

Year	Population (1,000s)	Average Annual Rate of Increase over Previous Date (percent)	Population Density (Number of persons/ km²)	Urban Population (1,000s)	Rural Population (1,000s)
1948	19,254	—	64.1	5,184	14,050
1960	27,085	3.06	90.3	8,072	19,015
1970	36,681	3.01	122.3	11,678	25,007
1975	42,070	2.79	140.2	14,047	28,024
1980	48,097	2.71	160.3	17,944	30,155

SOURCE: National Census and Statistics Office. 1980. Population, Land Area, and Density: 1970, 1975, and 1980. Manila: National Census and Statistics Office.

and Masefield stated that the Philippines had a low population density and "no real pressure of population on resources" (p. 417). This assessment seems almost naive today, suggesting how fast the settlement frontier closed in recent years and the inertia in public recognition of the current situation.

The availability of areas with low population densities and available agricultural lands has induced interregional migration in the Philippines since World War II (Abad, 1981; Abejo, 1985; Concepcion, 1983; Institute of Population Studies, 1981; Zosa-Feranil, 1987). Since 1948 the major migration patterns have been toward the frontier, primarily to Mindanao, and toward urban areas, particularly the metropolitan Manila area. Although migration to urban areas has been particularly pronounced since 1960, movement to frontier or upland areas continues (Cruz et al., 1986). Between 1975 and 1980, the destination of almost one-fourth of all interregional migrants was the uplands (Cruz and Zosa-Feranil, 1988). The major out-migration areas have been the Visayas and the Bicol and Ilocos regions of Luzon. Although substantial differences persist among some areas, the population has become more evenly distributed since 1948 (Herrin, 1985).

The upland population was estimated by Cruz and Zosa-Feranil (1988) to have reached about 17.8 million by 1988. This included an estimated population of 8.50 million people who reside on public forestlands. This population includes 5.95 million members of indigenous cultural communities and 2.55 million migrants from lowland groups (Department of Environment and Natural Resources, 1990). One-third of the upland forest inhabitants are displaced lowland farmers who do not have long-standing land use traditions such as those commonly observed among indigenous communities, which have a better grasp of the fragile nature of the ecology of their lands (Sajise, 1979). The displaced population is also growing faster. The University of the Philippines Population Institute projects that the upland population will grow at a rate of 2.72 to 2.92 percent during the next 25 years, increasing by the year 2015 to a density of 371 persons per km^2, which is a high population for sloping marginal lands.

Current and projected trends in the economy, social attitudes, and government commitment to effective delivery of family planning services may succeed in reducing national population growth rates. Even so, there is little likelihood that the upland population will participate significantly in this transition. The upland rural population has the least access to family planning programs and is least likely to accept the notion that limiting family size is in its best interest. Wherever open access to public lands prevails, children are viewed as additional labor to clear and cultivate more land.

Agriculture and the Uplands

Agriculture continues to play a major role in the Philippine economy. The Agricultural Policy and Strategy Team (1986) states:

> [N]o significant structural transformation has taken place over the past 25 years. Despite the strong industrial orientation of past economic policies, agriculture, fisheries, and forestry continue to employ half of the labor force, contribute about a quarter of the gross domestic production, and earn two-fifths of export revenues. Over 60 percent of our population lives in the rural areas. Our country remains today as it has been in the past, a predominately rural society composed of small farmers, agricultural laborers, fishermen, pedicab drivers, and others.

Agriculture's share of the total economy declined slowly in the postwar period, from 36 percent of net value added in 1955 to 29 percent in 1980 (David, 1983). Agriculture's share of the Philippine gross domestic product in 1987 (28.5 percent) was almost the same as it was in 1970 (World Bank, 1989b).

Between 1972 and 1980, the ratio between the price of rice and the non-food price index declined from 1.0 to 0.59 (Hill and Jayasuriya, 1984). The growth that did occur in the agricultural sector came not as the result of but despite government policies (David, 1982; Rocamora, 1979).

Landlessness and near landlessness in rural areas has been reported to be more than 75 percent (Rosenberg and Rosenberg, 1980), and landlessness among the agricultural farm population is almost 50 percent (Agricultural Policy and Strategy Team, 1986; Porter and Ganapin, 1988). Land reform has largely been ineffective in transferring land to the tenant cultivators because of bureaucratic delays and widespread erosion of the spirit of the agrarian reform laws (Carroll, 1983; International Labour Office, 1974; Kerkvliet, 1974; Tiongzon et al., 1986; Wurfel, 1983).

Has the limited effectiveness of land reform resulted in further concentration of control over agricultural lands? In Mindanao, commercial agricultural plantations are expanding. This expansion forces poorer farmers onto marginal lands, particularly in association with the banana and pineapple industries (Agricultural Policy and Strategy Team, 1986; Costello, 1984; Tiongzon et al., 1986; van Oosterhout, 1983). Krinks (1974) showed that there was an increasing concentration of poor farmers in a frontier region in southern Mindanao. Commercial use of agricultural land and the increased concentration of poor farmers on agricultural lands in lowland areas in Leyte has decreased the amount of land available for poor farmers, forcing poor farmers to initiate farming in upland areas (Belsky and Siebert, 1985). The expansion of land for raising sugarcane in the western Visayas from 1960 to 1975 was also primarily at the expense of small-scale upland rice and maize production (Luning, 1981). As effective control of agricultural land becomes more concentrated in the hands of wealthier farmers and corporations, small farms are becoming smaller (Luning, 1981), a process that has been accelerated by the subdivision of property through inheritance. The end result has been increasing landlessness for the rural poor (Cruz and Zosa-Feranil, 1988).

Arable land that can be sustainably farmed on an annual basis with minimal investment in land conservation covers 8.4 million ha, or 28 percent of the country (Bureau of Soils, 1977). Most of the increase in farm area since 1960 has been on nonarable land, as defined by the Bureau of Soils (1977).

Kikuchi and Hayami (1978) argued that the Philippines shifted from extensive to intensive cultivation between 1950 and 1969. As the land/labor ratio declined, the rate of increase in the amount of cultivated land slowed and the Philippine government was forced to invest in irrigation. Hooley and Ruttan (1969) proclaimed the closing of the land frontier in the 1960s.

There was widespread agreement that by the late 1960s or early 1970s, the Philippines had reached the limits of its land frontier and that future growth of agricultural output would have to come from increases in productivity rather than from increases in the area of production. Agricultural output and productivity did increase, but the area under cultivation also increased considerably. From 1970 to 1980, the number of farms increased by 1.06 million (45.3 percent) and farm area (Table 4) increased by 1.23 million ha (14.5 percent). As a result, the average farm size decreased 21 percent, from 3.61 to 2.84 ha. The continued decrease in forest area in the 1980s also implies that the area of farmland continues to increase. Thus, the notion of a land frontier based on arable, safely cultivated land is not appropriate for conditions in the Philippines (Cruz and Zosa-Feranil, 1988; Gwyer, 1977; National Economic Development Authority, 1981). In 1982, 2.5 million ha of cropland was on upland areas (Agricultural Policy and Strategy Team, 1986).

Upland Migration

Cruz et al. (1986) estimated that 14.4 million people lived in the uplands in 1980, and 77 percent of those people lived on lands officially classified as public forestlands. From 1948 to 1980, the upland population grew at a rate of 2.5 to 2.8 percent per year. This is less

TABLE 4   Deforestation and Its Relationship to Increases in Population and Farmland in the Philippines, 1948–1980

Period	Increase in Farmland (km²)	Increase in Population (millions)	Loss of Forest Cover (km²)	Area Deforested per Person Increase in Population (km²)	Ratio of Area Deforested Per Increase in Farm Area
1948–1960	20,459	7,813	25,073	0.32	1.2
1960–1970	7,212	9,596	22,465	0.23	3.1
1970–1980	12,315	11,416	21,032	0.18	1.7

NOTE: The area of forest cover in 1948 was assumed to be 150,000 km². Forest cover in 1960 was determined by the straight-line method by using National Economic Council (1959) data for 1957 and Forest Management Bureau (1988) data for 1969. Forest cover for 1970 was determined by the straight-line method by using Forest Management Bureau (1988) data for 1969 and 1980.

SOURCES: Forest Management Bureau. 1988. Natural Forest Resources of the Philippines. Manila: Philippine–German Forest Resources Inventory Project; National Economic Council. 1959. The Raw Materials Resources Survey: Series No. 1, General Tables. Manila: Bureau of Printing.

than the national rate because of the higher mortality and the lower birth rates in the upland areas than in the lowland areas (M. C. Cruz, College of Development Economics and Management, University of the Philippines, Quezon City, personal communication, 1990).

Migration accounted for the bulk of the population growth in the upland areas (Cruz et al., 1986). Of the 18.6 million people who lived in the uplands in 1988, 6 million had lived there before 1945, 2 million had migrated there between 1945 and 1948, and 10 million had migrated there since 1948 (Lynch and Talbott, 1988). In addition, high rates of migration to the uplands continued in the 1980s (World Bank, 1989a). The highest rates of population growth in the uplands were in municipalities with logging concessions (Cruz and Zosa-Feranil, 1988).

Most observers agree that migration occurs because of the lack of opportunities in the lowlands. Poor people are forced to the uplands because they have no other suitable choices. Cruz and Zosa-Feranil (1988) estimated that 70 percent of all upland migrants were landless lowlanders. These poor farmers may be referred to as shifting or slash-and-burn cultivators (Westoby, 1981).

Intensification of Rice Production in the Lowlands

Lowland rice fields in the Philippines are about half irrigated and half rainfed. Initially, the green revolution (the breakthroughs in rice varietal technology in the late 1960s) increased labor use intensity in rice production (Otsuka et al., 1990). More rice crops were produced each year (two instead of one), and more intensive management was applied. But rainfed rice farming did not experience the extent of technical change that occurred in irrigated rice farming or the same gain in productivity. Therefore, the economic disparity between the irrigated and rainfed rice fields increased (Otsuka et al., 1990).

The increased labor demand for irrigated rice accelerated the migration of labor from rainfed to irrigated areas. The intensity of labor use in irrigated rice production plateaued, however, and in many areas it declined as labor-displacing technologies gained widespread use. The technologies included broadcast seeding rather than transplanting of seedlings and herbicide application rather than weeding by hand. This reduced the labor absorption potential and the returns to labor, particularly landless labor. The income-earning prospects of the landless labor pool has declined, as exemplified by the evolution of labor arrangements that are progressively less favorable.

There is some potential for further intensification of rice cropping in irrigated areas and diversification to alternative higher income crops, including grain legumes, and tree crops. It is unlikely, however, that these changes will proceed fast or far enough to substantially increase the amount of labor that can be absorbed in lowland rice farming activities in the future, suggesting a continued rapid increase in the number of underemployed or unemployed families in lowland rural areas.

Upland Farming Systems

One of the most serious gaps in understanding land use in the uplands, particularly agriculture-forest interactions, relates to shifting (slash-and-burn) cultivation. Agriculture in the uplands consists of traditional shifting cultivation (long fallow periods), nontraditional or migrant shifting cultivation (short fallow periods), permanent or intensive agriculture, backyard gardens, pastoral systems, or any combination of these. There is no reliable information on the extent of these forms of agriculture or the proportion of shifting cultivation in grasslands or secondary or primary forests. There are also no data at the national or provincial level on how often farmers shift their plots, although case studies do exist (Barker, 1984; Conklin, 1957). Vandermeer

(1963) in a study of Cebu province, which is now entirely deforested, points out that what had originally been a shifting system of maize cultivation has now been transformed into permanent, sedentary farming. The main impetus for the change was increasing population density. Table 4 notes the relationships among deforestation, increases in population, and increases in the amount of farmland.

Analysis of an upland area in Mindanao from 1949 to 1988 revealed a dynamic land use transition from fallow rotation to permanent open-field and perennial crop systems (Garrity and Agustin, In press). The evolution of permanent, mixed agricultural systems in a pioneer community in the mountains of Laguna province dominated by shifting cultivation was documented by Fujisaka (1986) and Fujisaka and Wollenburg (1991). The planting of trees and perennial crops was observed by Cornista et al. (1986) as a typical stage in the evolution toward more permanent cultivation in communities throughout the Philippines.

Agricultural expansion has resulted in a net reduction in the country's grassland area. Data from an historical study of land use changes for an upland community in Mindanao from the immediate postwar period to the present illustrates this trend (Garrity and Agustin, In press). The area of cultivated land increased at a much faster rate than the loss of forest cover from 1949 to 1987. The steady decline in the grassland area provided the major source for the expansion of the area devoted to crops (Figure 1).

DEFORESTATION IN POSTWAR PHILIPPINES

There are few reliable historical data on forest cover in the Philippines. Many of the records that did exist have been lost. The Spanish forest records were consumed in a Manila fire in 1897 (Tamesis, 1948), the records of the Bureau of Forestry in Manila and the College of Forestry in Los Baños were destroyed during fighting in 1945 (Sulit, 1947), and the comprehensive Mindanao forest survey of 1954–1961 (Agaloos, 1976; Serevo et al., 1962) has disappeared. The authoritative source of current forest cover data is the Philippine–German Forest Resources Inventory Project (Forest Management Bureau, 1988).

Forest Types

Philippine forests are usually divided into six types: dipterocarp, molave, beach, pine, mangrove, and mossy. Dipterocarps account for more than 90 percent of all commercial forest products in terms of economic value (Agaloos, 1984). Some 89 percent of the total log

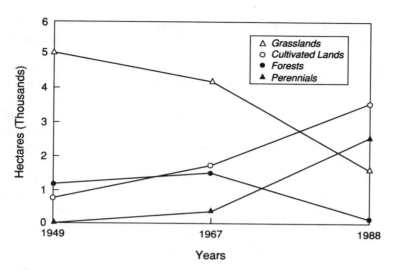

FIGURE 1   Comparative changes in major land use areas between 1949 and 1987. Claveria, Misamis Oriental Province (Mindanao), Philippines. Source: Garrity, D. P., and P. Agustin. In press. Historical Land Use Evolution in a Tropical Acid Upland Agroecosystem. Agric. Ecosyst. Environ.

production in the Philippines comes from the species *Shorea almon* (almon), *Dipterocarpus grandiflorus* (apitong), *Parashorea plicata* (tikan), *S. plicata* (mayapis), *S. negrosensis* (red lauan), *S. polysperma* (tanguile), and *Pentacme contorta* (white lauan). The largest timber volume comes from red lauan.

The molave forest, a dry, monsoon forest found only in the western Philippines, makes up only 3 percent of the total forest area of the Philippines (Agaloos, 1984) and is usually included in the dipterocarp category (Umali, 1981). Beach forests formerly grew in coastal areas as a transition between mangrove and other inland forests, but they have been virtually eradicated in the Philippines (Agaloos, 1984) and Southeast Asia (Whitmore, 1984). Two types of pine are native to the Philippines—Benguet pine (*Pinus kesiya*), found in northern Luzon, and Mindoro pine (*P. merkusii*), found in parts of Mindoro and the Zambales Mountains in western Luzon. Pine forests occupy less than 1 percent of the total land area (Forest Management Bureau, 1988).

Mangrove forests are restricted to coastal fringes and tidal flats and occupy about 139,000 ha (Forest Management Bureau, 1988), less than 0.5 percent of the total land area. They have been subjected to intense logging pressure because woods that grow in mangrove forests are valuable for fuel (charcoal) and thatch. As a result many mangrove forests have been converted to fish ponds (Gillis, 1988; Johnson and Alcorn, 1989).

Mossy forests are stunted forests with no commercial value (Agaloos, 1984; Weidelt and Banaag, 1982). They are referred to in the literature as mountain or cloud forests and as unproductive forest by the Forest Management Bureau. They are found at higher elevations (usually above 1,800 m) throughout the Philippines and cover about 4 percent (1.14 million ha) of the total land area (Forest Management Bureau, 1988).

Some 92 percent of the decrease of forest types since 1969 has been accounted for by the loss of old-growth dipterocarp forests (Forest Management Bureau, 1988). Destruction of mangroves has been rapid and dramatic as well, but the area involved is insignificant compared with the area of dipterocarps lost. The major cause of the decline of primary forests has been logging (World Bank, 1989a).

Forest Cover Before 1950

Deforestation in the Philippines has not occurred only in the twentieth century. Wernstedt and Spencer (1967) reported that forest cover declined from about 90 percent of the total land area at the time of the first contact with the Spanish in 1521 to about 70 percent by 1900. The major causes were likely to have been the steady increase in population and the spread of commercial crops (primarily abaca [a fiber from the leafstalk of banana—*Musa textilis*—native to the Philippines], tobacco, and sugarcane) as the Philippines slowly became integrated into the world economic system (Lopez-Gonzaga, 1987; Roth, 1983; Westoby, 1989).

Reliable statistics on forest cover before 1950 do not exist; thus, a discussion of forest cover and its decline must be based on estimates made by contemporary observers. Comparisons between the various estimates are problematic. Therefore, the estimates presented in Table 5 are meant to be broadly indicative. The area of the Philippines covered by forests declined from 70 percent in 1900 to just below 60 percent in 1939. Logging increased rapidly after 1945 and was back to pre-World War II production levels by 1949 (Poblacion, 1959; Tamesis, 1948). In addition, farming in the forests increased after the war because of continuing food shortages (Sulit, 1963; Tamesis, 1948). The overall extent of deforestation was estimated by Myers (1984) to be 55 percent in 1950. A figure closer to 50 percent for 1950 is probably more appropriate based on subsequent estimates.

TABLE 5   Estimates of Forest Cover in the Philippines, 1876–1950

Date	Percent Forest Cover	Source
1876	68	U.S. Bureau of the Census (1905)
1890	65	Bureau of Forestry (1902)
1900	70	Wernstedt and Spencer (1967)
1903	70	U.S. Bureau of the Census (1905)
1908–1910	50[a]	Whitford (1911)
1910	66	Zon (1910)
1911	64	Talbot and Talbot (1964)
1918	68	Census Office of the Philippine Islands (1920)
1919	67	Wernstedt and Spencer (1967)
1923	50	Zon and Sparhawk (1923)
1929	57[a]	Borja (1929)
1934	58	Revilla (1988)
1937	57	Tamesis (1937)
1937	58	Pelzer (1941)
1939	60	Food and Agriculture Organization of the United Nations (1946)
1943	60	Dacanay (1943)
1944	60	Allied Geographic Section (1944)
1945	66	Hainsworth and Moyer (1945)
1948	59	Food and Agriculture Organization of the United Nations (1948)
1948	59	Tamesis (1948)
1950	55	Myers (1984)

[a]Data are for commercial forests only.

Forest Cover Changes, 1950–1987

Since 1950 there has been a continuous decline in forest cover in the Philippines. In absolute terms, deforestation in the 1950–1969 and 1969–1987 periods were about the same (Table 6). On a percent basis, deforestation was more rapid from 1969 to 1987 than it was from 1950 to 1969, with the highest rates occurring from 1976 to 1980 (Table 7). The very high rates of deforestation observed for the 1976-1980 period were associated with the peak period of martial law, when large-scale corruption in timber extraction was prevalent (Alano, 1984; Aquino, 1987).

Although data are not strongly reliable, the rate of deforestation apparently slowed in the 1980s because the remaining forests became much less accessible. If the rate of deforestation estimated to have occurred from 1980 to 1987 continued to 1991, the Philippines had about 6.03 million ha of forest cover in 1991, about 20 percent of the country's total land area.

The Master Plan for Forestry Development (Department of Environment and Natural Resources, 1990) estimated total forest cover to be 6.69 million ha. The area of old-growth dipterocarp forests was projected to be only 949,000 ha. However, if the old-growth dipterocarp forest has continued to decline at the 1969-1987 rate of deforestation, then only 409,600 ha of this forest type would have remained in 1991. If this rate of decline continues, old-growth dipterocarp forests will disappear entirely by 1995—long before effective management systems to preserve them can be put into place. Thus, one of the major issues confronting Philippine forestry is how to manage secondary dipterocarp forests on a sustainable basis, for which there is little proven experience.

The calculated rates of annual deforestation differ widely, depending on the data sets chosen for analysis (Table 8). The 1980 forest data are from the Forest Development Center (1985) and the Philippine–German Forest Resources Inventory Project (Forest Management Bureau, 1988), which were projected back from deforestation data for 1987. The 1987 data are from the Swedish Space Corporation and the Philippine–German Forest Resources Inventory Project. There are large discrepancies in deforestation rates among the four possible combinations of the two surveys each for 1980 and 1987. Between the smallest and largest rates of deforestation, the difference is more than 200 percent. A reasonable estimate is that deforestation

TABLE 6   Forest Cover in the Philippines, 1950–1987

Date	Percentage of Land Area	Source
1950	49.1	Projection from 1969[a]
1957	44.3	National Economic Council (1959)[a,b]
1969	34.9	Forest Management Bureau (1988)[a]
1976	30.0	Bonita and Revilla (1977)[a,c]
1980	25.9	Forest Development Center (1985)[a]
1987	23.7	Swedish Space Corporation (1988)[a,d]
1987	22.2	Forest Management Bureau (1988)[a,e]

[a]Includes forestland and nonforestlands.
[b]Does not include brushlands or marshes or swamps.
[c]Since the original figures included approximately 10 percent brushland (Revilla, 1988), the total was reduced by 10 percent.
[d]Does not include land area that was not classified.
[e]Data from 1988 were projected back to 1987.

TABLE 7 Deforestation Rates in the Philippines, 1950–1987

Period	Average Annual Change		Source
	km^2	Percent	
1950–1957	2,210	1.6	Projection and National Economic Council (NEC) (1959)
1957–1969	2,262	1.9	National Economic Council (1959) and Forest Management Bureau (1988)
1969–1976	2,081	2.1	Forest Management Bureau (1988) and Bonita and Revilla (1977)
1976–1980	3,048	3.6	Bonita and Revilla (1977) and Forest Development Center (1985)
1980–1987	1,570	2.2	Forest Development Center (1985) and Forest Management Bureau (1988)
1950–1969	2,243	1.8	Projection and Forest Management Bureau (1988)
1969–1987	2,103	2.5	Forest Management Bureau (1988)
1950–1987	2,175	2.0	Projection and Forest Management Bureau (1988)

NOTE: Deforestation rates were calculated from the data presented in Table 6.

in the 1980s was about 155,000 ha/year. The World Resources Institute (1990) estimated that deforestation is about 143,000 ha/year. This issue is discussed more thoroughly in the section on future scenarios.

The Deforestation Process in the Philippines

Figure 2 is a simplified model of the major forces that have led to deforestation in the Philippines. Although some deforestation has been caused by other factors, for example, the use of trees to make charcoal and the conversion of mangrove forests to fish ponds, the two most important activities leading to deforestation were logging (legal and illegal) and the expansion of agriculture. Both of these factors must be considered together, along with rural poverty and the open-access nature of forests (Gillis, 1988). The deforestation process in the Philippines since World War II can be characterized by two major activities: the conversion of primary to secondary forests by logging activities and the removal of secondary forest cover by the expansion of agriculture. In most cases, roads provide access to the forest for both types of activities.

Logging does not necessarily result in deforestation; rather, selective logging, properly practiced, converts a primary forest into a degraded secondary forest (Figure 2). Clear-cutting is known to have been practiced in certain areas, but this has been relatively rare in Southeast Asia (Gillis, 1988), and data on the relative extent of clear-cutting versus selective logging in the Philippines do not exist. Selective logging results in some deforestation, given the extensive road networks and collection and loading areas needed for capital-intensive logging and the extensive damage to forests reported to occur as a result of some logging operations (Blanche, 1975; Burgess, 1971, 1973; Egerton, 1953; Gillis, 1988; Philippine Council for Agriculture and Resources Research and Development, 1982; World Bank, 1989a).

The relationship between logging and the conditions of primary and secondary forests is a dynamic one. As logging converts primary forests to secondary forests, loggers move on to new primary forests. Implicit in this scheme is the notion that secondary forests do not return to a state suitable for a second harvest, although several concessionaires in the Philippines are known to have returned for a second cut. Concessionaires have not, in general, engaged in protection of secondary forests, enrichment planting, or reforestation (Food and Agriculture Organization and United Nations Environment Program, 1982). Overall, it appears that there has been minimal protection of forests in the Philippines.

Expansion of agriculture takes place primarily in secondary forests. Logged forests are more likely than primary forests to be penetrated by roads, and roads greatly facilitated the expansion of agriculture (Asian Development Bank, 1976; Edgerton, 1983; Food and

TABLE 8   Annual Rates of Deforestation in the Philippines Between 1980 and 1987 Based on Different Forest Inventories

1980 Data	1987 Data	Annual Deforestation Rate	
		km^2	Percent
FDC	FMB	1,571	2.2
FDC	SSC	951	1.3
FMB	FMB	2,103	2.8
FMB	SSC	1,483	2.0

NOTE: Deforestation rates were calculated from the data in Table 6. The annual decline in forest area (km^2) was determined as the difference in forest area between 1980 and 1987 using the respective estimated data sources for each year referenced in columns 1 and 2. FDC, Forestry Development Center; FMB, Forest Management Bureau; SSC, Swedish Space Corporation.

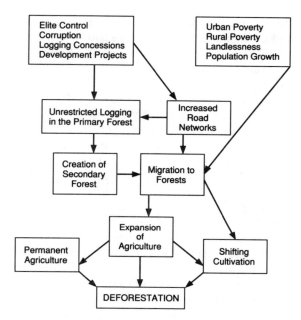

FIGURE 2   Model of deforestation in the Philippines. Source: Kummer, D. 1992. Deforestation in the Postwar Philippines. Chicago, Ill.: University of Chicago Press.

Agriculture Organization and United Nations Environment Program, 1981: Hackenberg and Hackenberg, 1971; Segura-de los Angeles, 1985; Vandermeer and Agaloos, 1962; van Oosterhaut, 1983). Also, it is much easier for poor farmers to clear secondary forests than it is for them to clear primary forests (Byron and Waugh, 1988). In an economic sense, logging lowers the costs of clearing the land by settlers (Southgate and Pearce, 1988). The majority of logged-over forestlands have been converted to grasslands or are used for agriculture (Hicks and McNicoll, 1971).

Natural forest regeneration is prevented by a range of prevailing factors: fire in uncultivated logged-over areas and ranch areas, grass succession and loss of tree seed in shifting cultivated areas, and permanent conversion to agricultural fields in intensively farmed areas. The relationships among the expansion of agriculture, the creation of secondary forests, and deforestation are also dynamic. Preceding logging and the expansion of agriculture is the construction of roads (Hackenberg and Hackenberg, 1971). These roads are primarily the result of development considerations by provincial or national government or are built by loggers who have concessions. The roads vary from little more than dirt tracks to paved highways. They facilitate the spread of agriculture by opening up new areas; this occurred in parts of Mindanao in the 1950s and early 1960s (Vandermeer and Agaloos, 1962; Wernstedt and Simkins, 1965). In addition, logging provides jobs and, thus, directly leads to population increases. The relationship between new roads and deforestation has been clearly made by Thung (1972) for Thailand and by Fearnside (1986) for Brazil.

The expansion of agricultural activities onto forested lands is driven by two forces: increases in population and widespread poverty. In addition, the expansion of agriculture in some areas is promoted by wealthier people who open up forestlands for perennial crop production or cattle grazing or simply to establish a land claim. This is often accomplished through support for poor farmers who are subsidized to clear the land. The overriding goal of the low-income households in upland regions is to produce or earn enough to eat. Food income provides basic security (U.S. Agency for International Development, 1980). Poor people are forced to engage in subsistence agriculture because it is often the only option available (Gwyer, 1978). Segura-de los Angeles (1985), in a case study of an upland agroforestry project in Luzon, noted that 88 percent of all those surveyed consumed all of the rice they produced and did not have a marketable surplus. Although upland farmers in Davao grew some commercial crops, their primary crops were rice and maize (Hackenberg and Hackenberg, 1971).

Timber Concessions

The granting of timber concessions occurred for two reasons: the legitimate desire of the Philippine government to foster development and the granting of political favors to either Philippine elites or multinational corporations (primarily U.S. corporations in the 1950s and 1960s). Postwar Philippine governments do not appear to have been concerned with development in the forest sector; rather, it appears that forests are viewed as an asset whose benefits should flow mainly to politicians and well-connected individuals (Ofreno, 1980; Palmier, 1989). As Hackenberg and Hackenberg (1971) pointed out in their study of Davao City, Mindanao, "The basis of wealth is lumber, and the profits are instantaneous for those with political connections to secure a concession" (p. 8). In fact, it is difficult to distinguish between politicians and loggers, since loggers contribute heavily to political campaigns and many politicians control logging concessions (The Economist, 1989). It is now generally accepted that commercial forest resources were vastly underpriced throughout the postwar period and that the high rents flowed to a small group of people (Boado, 1988; Cruz and Segura-de los Angeles, 1984; Power and Tumaneng, 1983; Repetto, 1988).

Factors Associated with Deforestation

Deforestation in 67 provinces was analyzed statistically from 1970 to 1980 (Kummer, 1990). The study used data on the annual allowable cut, which was greater than legally reported logging and may more accurately reflect the actual volume of timber harvested, considering the additional timber that is extracted illegally. Deforestation from 1970 to 1980 was positively related to the annual allowable cut in 1970 and to the absolute change in the area devoted to agricultural activities (Kummer, 1990). The distance from Manila was not significantly related to the deforestation rate, but in those areas of the Philippines where logging was banned during the reign of Ferdinand E. Marcos (1965–1986), the logged area determined from the rates of deforestation were actually higher than the rates where logging was allowed (Schade, 1988).

Postwar discussions of deforestation in the Philippines have tended to blame either loggers or migrant farmers in frontier areas engaged in nontraditional shifting cultivation for the decline in forest cover. These two agents cannot be considered separately; rather, they are linked. The Philippines has recently completed the Master Plan for Forestry Development (Department of Environment and Natural Resources, 1990). The plan articulates a people-oriented forestry program that is sensitive to the current understanding of the complex underlying determinants of deforestation. The policy prescriptions and implementation devices presented in the plan are analyzed later in this chapter.

APPROACHES TO LAND USE SUSTAINABILITY IN THE UPLANDS

This section evaluates current and potential directions for formulating concrete solutions to deforestation and sustainable land use. It examines the determinants of sustainable agricultural systems and forest systems within each of the three major land use subecosystems in Philippine uplands. The approach emphasizes the interrelatedness of social and technical issues and the importance of an integrated social-technical approach to forest and agricultural development.

A large and rapidly expanding portion of the upland landscape is being converted to areas that are permanently farmed. These farms

are found in the more relatively accessible sloping areas that are closest to the lowlands and nearest to roads. They are predominantly
cultivated with subsistence food crops, particularly maize and upland rice, but they are partly used for perennial crop plantations,
especially coconut plantations. At increasing elevations and more
remote locations that are difficult to access, the land predominantly
contains grasslands and brushlands. The remaining forested areas
are generally the secondary forest remnants of previous logging activities or localized unlogged areas, which are found at the highest
elevations and on the steepest slopes.

These three broad land use types (permanently farmed sloping
lands, grasslands, and forested lands) tend to form distinct entities
that flow into each other. The permanently cultivated lands expand
into the grasslands as shifting cultivation on the grassland margins
intensifies, and the grasslands advance at the expense of the forested
lands as settlement and the relentless use of fire open and transform
the forests. The human and natural ecology of each of these three
entities is distinct, and technology and policy instruments must be
adapted to the realities of each one.

Permanently Farmed Sloping Lands

The major issue in permanently farmed sloping lands is how to
sustain and increase farm productivity to improve the welfare of the
farm population and thereby reduce the rate of migration into the
remaining forested lands. Increase in and sustainability of farm productivity may be achievable through policy reform and technological
changes in agricultural activities, but the development of more successful farming systems in sloping settled lands will not eliminate
the migratory pressure on forested lands. Technical change could
make forested lands more valuable for agriculture, thus encouraging
further migration. It is also evident, however, that if the current
upland populations cannot become more successful in sustaining their
incomes and increasing their employment opportunities, more farmers and their families will be forced to migrate from unproductive
farms that can no longer support them, resulting in more rapid and
destructive misuse of forestlands.

This suggests that sustainable upland agricultural production systems are necessary to alleviate many problems of human welfare in
the uplands and lowlands and ensure more effective forest conservation, but such changes are not sufficient to solve the problem of the
conversion of forests to agricultural uses. The essential elements of a
strategy for upland development are the same as those that would

apply in lowland areas. They include the need for a positive incentive framework and the availability of appropriate technical solutions. Agricultural technology can provide a crucial, supporting role in solving the forest conversion problem. Progressive policies in forestry, agriculture, land tenure, and general economic development will impinge greatly on the effectiveness and appropriateness of potential technologies.

There are many factors that limit the stability, productivity, and sustainability of upland farms, including climatic variations, biologic stresses, and social and economic uncertainties. A fundamental factor is the nature and rapidity of soil degradation.

The sloping upland soils in the Philippines fall into three contrasting types: acidic, infertile soils; young, relatively fertile volcanic soils; and calcareous soils. The strongly acidic, infertile soils, which are low in available phosphorus, are predominant. The young, more fertile volcanic soils cover large areas in the southern Tagalog and Bicol regions, on Negros Island, and in some areas of Mindanao. These have been the most successfully developed upland agricultural areas. Calcareous upland soils are found on the central Visayan islands of Cebu and Bohol. Restrictions on the available phosphorus also tend to be pronounced in calcareous soils.

In addition to the three basic classes of soils, the immense and localized variations in rainfall patterns because of the diverse topography of the Philippines, and the frequency and severity of damage from catastrophic typhoons affect the sustainable management of upland agricultural systems. Farming systems must be adapted to take into account these various conditions.

Philippine upland farmers face a diversity of land types and high levels of risk, yet they have limited access to credit and marketing resources. Under these conditions, agricultural technologists must be able to offer practical, low-cost farming practices that are viable under a wide array of conditions or that are more specifically tailored to a few conditions but that produce results quickly.

CONTOUR HEDGEROW SYSTEMS

Research on upland agroforestry in the Philippines is limited. Agriculturalists and foresters have few technical tools to cope with the enormous variety of circumstances that require attention. Gibbs et al. (1990) pointed out that the highly inadequate knowledge of agroforestry techniques was probably the weakest aspect in the successful evolution of the government's Integrated Social Forestry Program.

Leucaena Hedgerows   Leucaena (*Leucaena leucocephala*) is common in rural areas with less acidic soils. It was indigenously grown in fencerows as a fodder source for cattle. The National Research Council (1977) indicated that the tree showed promise as a hedgerow intercrop that could supply large quantities of nitrogen and organic matter to a companion food crop. Those observations stimulated applied research on hedgerow intercropping in several locations around the Philippines. Guevara (1976) reported that hedgerow intercropping produced crop yield increases of 23 percent. Vergara (1982) cited experiments in which yields increased by about 100 percent, with no advantage of inorganic nitrogen application beyond the nitrogen supplied by green leaf manure. Alferez (1980) observed a 56 percent yield increase when upland rice was grown in alleys between hedgerows of *Leucaena*.

Hedgerows of *Leucaena* provided a barrier to soil movement on sloping lands. Data from studies on a steeply sloping site in Mindanao indicated a dramatic reduction in both runoff and soil loss (O'Sullivan, 1985). In that study, O'Sullivan (1985) also observed a consistent yield advantage over a 4-year period with maize fertilized by the *Leucaena* prunings obtained from adjacent hedgerows.

By the early 1980s, hedgerow intercropping was advocated by the Department of Agriculture as a technology that was better able to sustain permanent cereal cropping with minimal or no fertilizer inputs and as a soil erosion control measure for sloping lands. The extension of this system among Filipino farmers was encouraged by the work of the Mindanao Baptist Rural Life Center (MBRLC), a nongovernmental organization (NGO) that began working with *Leucaena* in the mid-1970s (Watson and Laquihon, 1987). MBRLC developed a 10-step program for farmer implementation of *Leucaena* hedgerows that was designated sloping agricultural land technology (SALT). SALT recommended that every third alleyway between the double hedgerows of *L. leucocephala* be planted with perennial woody crops, such as coffee trees, with the majority of the alleys maintained by continuous cropping with annual food crops. This concept offered the possibility of more diversified sources of farm income and improved soil erosion control.

By the mid-1980s, SALT was adopted by the Philippine Department of Agriculture as the basis for its extension effort in the sloping uplands. The Department of Environment and Natural Resources also used it as the technical basis for its social forestry pilot projects. A training effort for extension personnel was launched, and demonstration plots of SALT were installed on farmers' fields throughout the country. Several publications have been developed to spread practical information about the SALT system (Celestino, 1984, 1985; Philippine Council for Agriculture and Resources Research and Development, 1986).

Some adoption of *Leucaena* hedgerows occurred in high-intensity extension projects, but there was little evidence of widespread farmer interest in the SALT system. The lack of secure land tenure was implicated as a constraint to the implementation of this or any long-term land improvement system among tenant farmers or occupants of public lands. Among farmers with secure land tenure, however, the large initial investment of labor, the difficulty in obtaining planting materials, and the technical training and information required for sustained implementation were serious constraints to initiating SALT systems. In addition, the labor needed to manage the hedges, particularly to prune them 3 to 10 times each year, depending on the management system, was found to absorb a large proportion of the household's available labor. This labor investment tended to compete with other income-generating tasks and may have limited the area that could feasibly be farmed in this manner (S. Fujisaka, Social Sciences Division, International Rice Research Institute, Los Baños, Philippines, personal communication, 1989).

Hedgerows of Other Species The extension effort on *Leucaena* hedgerows suffered a major setback in 1985 when the exotic psyllid leafhopper (*Heteropsylla cubana*) invaded the Philippines, attacking hedgerows and killing or stunting trees throughout the country. This forced a search for replacement hedgerow tree species. *Gliricidia sepium* has been the most common replacement, but it must be propagated from cuttings in most areas, increasing the labor investment to establish hedgerows. Other species that have shown promise in hedgerow trials include *Flemingia congesta*, *Acacia vellosa*, *Leucaena diversifolia*, and *Cassia spectabilis* (Mercado et al., 1989; H. R. Watson, Mindanao Baptist Rural Life Center, Bansalan, Philippines, personal communication, 1989). *Alnus japonica* is used in the acid soil highlands in northern Luzon (Barker, 1990).

Pava et al. (1990) compared the changes in crop yields associated with planting a double row of leguminous hedgerows by a group of 10 farmers who adopted the system and a control group of farmers who did not. Over the 2-year interval of monitoring, maize yields increased by both methods, but the greatest increase was among the control group of nonadopters. Fertilizer use among both groups was very similar. When queried about the perceived value of the hedgerows, the farmers who adopted leguminous hedgerows emphasized that their investment in hedgerows was long-term insurance that their children could continue to farm the land.

Contour Bunding with Hedgerows   World Neighbors, another NGO, made a substantial contribution during the past decade (Granert, 1990; Granert and Sabueto, 1987).   The World Neighbors approach was oriented toward the development of a high degree of direct participation by farmers in devising and implementing local solutions to the perceived dominant constraints to crop cultivation on steeply sloping lands.   A system of contour bunding was developed.   The bunds provided a base for the establishment of double-contour hedgerows of leguminous trees or forage grasses and a barrier to surface runoff, which is carried off the field in contour ditches.

The contour hedgerow concept was applied to the strongly acidic upland soils by the International Rice Research Institute (IRRI) and the Philippine Department of Agriculture (Fujisaka and Garrity, 1988). Although these soils are generally deep, soil loss is a problem because it exposes a very acidic subsoil with toxic levels of aluminum. After 3 years of hedgerow intercropping, there was a striking natural development of terraces (Figure 3).   Modest yield benefits were ob-

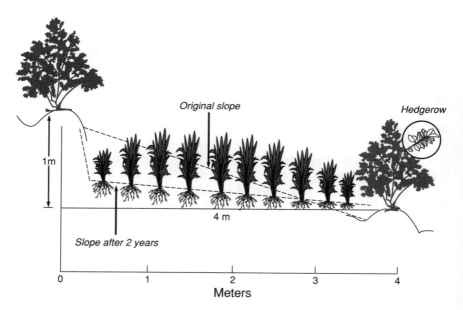

FIGURE 3   Terrace formation and crop growth in a contour hedgerow system of upland rice and leguminous trees on strongly acidic Oxisol soils. Source: Basri, I., A. Mercado, and D. P. Garrity. 1990. Upland rice cultivation using leguminous tree hedgerows on strongly acid soils. Paper presented at the Annual Meeting of the American Society of Agronomy, San Antonio, Texas, October 21–26, 1990.

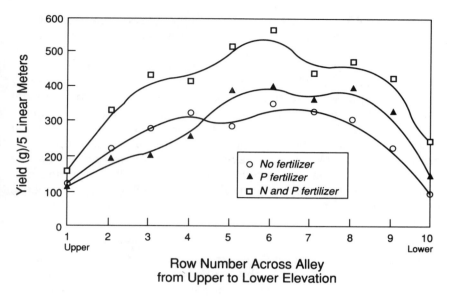

FIGURE 4   Yield (on a row-by-row basis) of upland rice grown in alleys between hedges of a leguminous tree, *Cassia spectabilis*, that supplied green, leaf manure for the rice crop. P, phosphorus; N, nitrogen. Source: Basri, I., A. Mercado, and D. P. Garrity. 1990. Upland rice cultivation using leguminous tree hedgerows on strongly acid soils. Paper presented at the Annual Meeting of the American Society of Agronomy, San Antonio, Texas, October 21–26, 1990.

served when upland rice was grown between hedgerows of *Cassia spectabilis*, a common non-nodulating leguminous tree (Basri et al., 1990). Yields of maize and rice were consistently increased when they were intercropped with hedgerows of *Gliricidia sepium* (Mercado et al., 1992). However, crop yields were seriously reduced in the rows adjoining the hedges, with or without the application of external nitrogen and phosphorous fertilizers (Figure 4). The primary roots of both tree species spread laterally into the alleyways at shallow depths (20 to 35 cm) immediately beneath the plow layer. Feeder roots were situated to explore and compete for nutrients and water in the crop root zone.

Sustainability in Alley Cropping Systems   The sustainability of crop yields in alley cropping systems is a major concern on all soil types. The work reviewed by Szott et al. (1991) raises particular questions about the viability of hedgerow intercropping on strongly acidic soils. The high level of exchangeable aluminum in the subsoil inhibits the deep tree-rooting patterns that are typically observed on higher-base-status soils. Phosphorus and other mineral elements are often more limiting than nitrogen in these soils. The acidity of the subsoil appears to promote intense competition among roots for mineral nutrients in the surface soil of the alleys and prevents nutrient pumping from the deeper soil layers. The organic matter inputs from hedgerow prunings of *Gliricidia* and *Cassia spectabilis* do not supply adequate quantities of phosphorus to meet the nutrient requirements of cereal crops (Basri et al., 1990). Furthermore, the prunings are composed of phosphorus that the tree may have captured predominantly from the crop root zone. The results obtained with other alley cropping systems on acidic Ultisols in Peru (Fernandes, 1990) and in Sumatra, Indonesia (Evensen, 1989), support the results obtained in Mindanao by IRRI.

Grass Strips   Grass strips have also received major attention as contour vegetative barriers for erosion control in different parts of the world (Lal, 1990). Considerable work has been done in the Philippines with napier grass (*Pennisetum purpureum*), guinea grass (*Panicum maximum*), and other grasses (Fujisaka and Garrity, 1988; Granert and Sabueto, 1987). The predominant attention has been given to the more vigorous forage grasses, since they tend to provide high levels of biomass for ruminant fodder. Therefore, they are presumed to serve as a beneficial way to use the area of the field occupied by hedgerows, which is lost to food crop production. Experimental data (Table 9) and field observations of plantings in various locations indicate that use of forage grasses for intercropping has the potential to markedly reduce erosion and rapidly develop natural terraces on slopes. Therefore, the establishment of forage grasses has been extended as an alternative to the use of leguminous tree species on contour bunds.

Two major problems have surfaced from the use of grass strips. Farmers have difficulty keeping the tall, rapidly growing tropical forage species trimmed to prevent them from shading adjoining field crops. The biomass productivity of grass hedgerows exceeds the fodder requirements of most small-scale farm enterprises, and it is a burden for farmers to cut the unnecessary foliage frequently. High levels of biomass production also tend to exacerbate competition for nutrients and water with the adjoining food crops and reduce cereal crop yields (D. P. Garrity and A. Mercado, International Rice Research Institute, unpublished data).

Intercropping with Noncompetitive Species   The constraints observed from intercropping with both trees and forage grasses have stimu-

TABLE 9   Soil Loss Affected by Contour Hedgerow
Grasses Vegetation

Hedgerow Species	Soil Loss (cm)
Gliricidia sepium and *Paspalum conjugatum*	0.38
Pennisetum purpureum	0.62
Gliricidia sepium and *Penisetum purpureum*	1.38
Gliricidia sepium alone	1.50
Open field (conventional practice)	4.20

NOTE: Monitoring was done in a large replicated trial on-farm in
Claveria, Misamis Oriental (Mindanao), Philippines, from August
1986 to April 1990.

lated an alternative concept of using hedgerows that contain noncompetitive or relatively inert species (Garrity, 1989). An inert species is one that has a short stature and a low growth rate, which
minimizes hedgerow-crop competition but provides an effective ground
cover for filtering out soil particles. This concept places primary
emphasis on the rapid and effective development of terraces to improve field hydrology and maximize soil and nutrient retention. *Vetiver
zizanioides* may exemplify an inert hedgerow species (Smyle et al.,
1990). *Vetiver* is found throughout the Philippines. It tends to form a
dense barrier and does not self-propagate to become a weed in cultivated fields. However, it must be propagated by vegetative tillers,
which is a laborious process.

Natural Vegetative Filter Strips An alternative approach that has
received little attention is the installation of natural vegetative filter
strips. These are narrow contour strips that are left unplowed and on
which vegetation is allowed to grow naturally. They may be established at the time that a piece of fallow land is brought into cultivation or during the interval between crops in a continuous cropping
system. The dominant species in natural vegetative filter strips are
native weedy grasses: *Imperata cylindrica, Paspalum conjugatum,
Chrysopogon aciculatus*, or others, depending on the location and the
management regime to which the strips are subjected. These natural
grasses can be suppressed by allowing cattle to graze them, cutting
them down, or mulching them with crop residues. Natural vegetative filter strips are capable of reducing soil loss at least as effectively
as commonly recommended introduced species (Table 9, *Paspalum
conjugatum* treatment). They are generally less competitive with food

crops than other hedgerow species, and they are adapted to local ecosystems and resilient in terms of longevity and reestablishment.

There have been some isolated observations of the indigenous development of natural vegetative barriers by upland farmers in the Philippines (Baliña et al., 1991; Fujisaka, 1990; Ly, 1990). However, research has not been targeted to exploit this option in Philippine uplands. (In the United States there has been extensive research on the use of natural vegetative filter strips for sediment and chemical pollution control [Williams and Lavey, 1986].)

Farm-level adoption of natural vegetative filter strips has been observed to be comparatively simple. Contour lines are laid out at the desired spacing. The field is plowed on the contour, allowing the designated strips to be left as fallow vegetation. In fields where the technique has been implemented, the soil in runoff water is deposited at the filter strip. This deposition, combined with the movement of soil down the slope during tillage operations, results in the rapid development of terraces of 30 to 70 cm deep within 2 years. The leveling effect of terrace formation evidently improves water retention in the field, and the loss of either applied or native soil nutrients is reduced. These effects need to be investigated under a range of field conditions.

The natural vegetative filter strip approach can be considered the initial stage in a long-term process of contour hedgerow development on farms. As terraces form, farmers may diversify the terrace risers for use in other enterprises by planting trees or perennial crops as they fit their management objectives. The natural vegetative filter strip concept may be a practical basis for the rapid, wide-scale dissemination of hedgerow technology. Therefore, a substantial effort in both strategic and farmer-participatory research on natural vegetative filter strips is warranted.

Cash Crop Production in Hedgerows may also be suitable for the production of perennial cash crops. Some perennial crops that have been used in these systems include coffee, papaya, citrus, and mulberry. The suitability of the perennial species is limited by the degree of shading of the associated food crops. The cash income that can be made is a major advantage of using perennial crops. Erosion control may not be provided by the perennial crop, but it may be provided by grass that occupies the area between the widely spaced plants.

Cattle Production Backyard production of cattle has become an important enterprise in some densely settled upland areas, particularly Batangas province. A trend toward more intensive small-scale beef and goat production is now under way in many parts of the country.

This trend is stimulated by historically high meat prices. Leguminous tree species, particularly *Leucaena leucocephala* and *Gliricidia sepium*, are widely used as high-protein forages, especially in the dry season. Backyard ruminant production will stimulate more intensive husbandry of manure. An important model of the development of leguminous trees in hedgerows is the use of prunings as a source of animal feed, either for on-farm use or off-farm sales (Kang et al., 1990). Harvesting of fodder potentially increases the value of the hedgerow prunings, but it also depletes soil nutrient reserves more rapidly because the nutrients contained in the prunings are removed from the field before they can provide their nutrients to the crop. Unless this manure is spread back on the land or replaced, and nutrient supplements provided in the form of fertilizer, the rate of soil depletion may be accelerated. Currently, the use of green leaf manure is insignificant in upland cropping systems.

The experience of the past 15 years with alley cropping and the use of contour hedgerows suggests that appropriate solutions must be tailored to the diverse soil and environmental conditions, farm sizes and labor availabilities, markets, and farmer objectives. The tendency for a package approach to be applied by extension systems must be replaced with a model that recognizes a wide range of possible hedgerow species and management systems (Garrity, 1989). There has been little attempt to clarify the appropriate hedgerow technologies for the range of specific local physical and institutional settings.

REDUCED-TILLAGE SYSTEMS

Clean cultivation is the universal soil management practice of Filipino upland farmers whether they use animal power or hand tillage on steep slopes. Crop residues are plowed under, burned, or removed and used as fodder. Retention of surface residues through conservation tillage systems is unexploited, although the value of such practices in reducing soil erosion is profound on tropical sloping uplands (Lal, 1990). Many studies have shown significant benefits from maintaining a surface mulch. Thapa (1991) found that soil loss was reduced by 90 percent by the presence of a vegetative barrier, but the maintenance of crop residues on the soil surface reduced soil loss by more than 98 percent. It has been shown (R. Raros, Visayas State College of Agriculture, Baybay, Leyte, Philippines, personal communication, 1989) that upland rice can be dependably established in thick residues without tillage in a hedgerow system, and the yields of a system with three continuous crops per year can be sustained.

At present, no practical approach has been developed to satisfactorily cope with weeds in reduced-tillage systems. Broad-spectrum herbicides such as glyphosate are beginning to be used on a limited basis by small-scale farmers, but the intense weed pressures on upland farms and the tendency for weed species to shift rapidly to resistance to herbicides has severely constrained the development of herbicide-based solutions.

The possibility of successfully using a reduced-tillage system has been reinforced by recent observations on a farmer-evolved system of maize production in Mindanao (D. P. Garrity, International Rice Research Institute, unpublished data). The system involves a crop sequence of three crops of maize monoculture per year but only one primary tillage operation annually. Interrow cultivation and late weeding during the maize grain-filling period enable the second and third crops to be planted on the day of harvest without tillage and with low weed pressures. This unconventional approach provides interesting prospects for practical techniques for reducing the tillage needed for food crop farming with limited resources.

NUTRIENT SUPPLY

External fertilizer use on food crops by upland farmers is seldom important. This is due to their severe capital constraints, transport difficulties, and low returns from fertilizer use. Therefore, a long-term decline in yields is typically observed (Fujisaka and Garrity, 1988). It is widely believed that the sustainability of food crop production could be enhanced by improved retention of crop residues and by the adoption of more diverse crop rotations that include nitrogen-fixing legumes (McIntosh et al., 1981). The limited work done to date has shown that there are mixed benefits from these practices. The practical constraints to the implementation of improved nutrient cycling practices are often considerable.

Leguminous grains play an insignificant role in upland cropping systems. Mung beans (*Phaseolus aureus*) and soybeans (*Glycine max*) are adapted to neutral and slightly acidic soils, whereas cowpeas (*Vigna sinensis*, also known as black-eyed peas) are more suited to highly acidic soils (Torres et al., 1988). When leguminous grains are inserted into cereal crop-based rotations immediately before upland rice or maize is planted, the legume improves the nutrient balance of the next cereal crop (Magbanua et al., 1988; Torres et al., 1989). Intercropping of cereals and legumes may increase their combined productivities, but it does not increase the net availability of nitrogen to the cereal crop (Aggarwal et al., 1992).

Farmers who cultivate grain legumes do so as an income or food source, but they do not usually observe better cereal crop performance as a result of the legume's inclusion as a second crop in cereal-based rotations (International Rice Research Institute, 1991). This appears to be due to the low biomass production by tropical leguminous grains that mature early and to nitrogen losses during the long fallow period between the time that the legume is harvested and the establishment of the following wet season crop.

Forage legumes have greater longevity in the field than do leguminous grains, and they produce large amounts of nitrogen-rich biomass. On high base-status soils, viny legumes such as lablab (*Lablab purpureus*) or siratro (*Macroptilium atropurpureum*) can be intercropped with upland rice or maize. They produce 100 to 200 kg of nitrogen/ ha in plowed down green manure during the dry season for the succeeding wet season cereal crop (Aggarwal and Garrity, 1989; Torres and Garrity, 1990). They also provide high-quality forage during the dry season. Lablab also provides a nutritious and marketable food legume for humans (Torres and Garrity, 1990).

On strongly acidic soils, most of the forage legumes have slow establishment rates, are not resilient to pruning, and do not accumulate substantial amounts of biomass during the dry season. This may be attributed to poor rooting and nodulation in the presence of high levels of exchangeable aluminum and low amounts of available phosphorus in the soil. Their inclusion within annual crop sequences therefore often appears to be impractical without the application of lime or phosphorus or both.

PHOSPHORUS AS A CRITICAL CONSTRAINT

The acidic upland soils of the Philippines are predominantly fine-textured, with organic carbon contents of 2 to 3 percent and with a moderate level of total nitrogen. Phosphorus deficiency is frequently the most limiting nutritional problem (International Rice Research Institute, 1987) and often must be overcome before any response to nitrogen is observed (Basri et al., 1990; Garrote et al., 1986). Phosphorus pumping from the deeper soil layers is limited by subsoils with toxic levels of aluminum and low phosphorus reserves. Since constant nutrient removal or offtake is occurring, crop yield sustainability and significant biologic nitrogen fixation will depend on the importation of mineral nutrients, particularly phosphorus and lime. Greater appreciation of the importance of importing these nutrients in upland agroecosystems with acidic soils is needed.

Deposits of phosphate rock in the Philippines are an efficient source of both phosphorus and calcium (Atienza, 1989; Briones and

Vicente, 1985). The exploitation of phosphate rocks for farm use has been neglected and could be expedited. This would require greater government and commercial recognition of the fundamental importance of these minerals to permanent upland agricultural system.

PERENNIAL CROPS

Coconuts are the dominant plantation crop in the Philippines, which has the world's largest area devoted to this crop, covering nearly one-sixth of the land surface (4.88 million ha [Swedish Space Corporation, 1988]). In addition, there are about 100,000 ha of plantations of rubber and other estate trees.

Coconut trees occupy much of the steepest nonarable land at lower elevations. Although the canopy of a coconut plantation is relatively open, the land on which coconut is grown provides satisfactory soil protection against erosion when an appropriate grassy or leguminous ground cover is established. Much of the land on which coconut is grown is owned by wealthier families but is managed in smallholdings by tenants or caretakers. The livelihoods of millions of the poorest families and the economic future of many parts of the uplands are heavily dependent on the health of the coconut industry. A long-term decline in the world market demand for coconut oil is projected because of the increasing worldwide preference for vegetable oils, which have a lower saturated fat content.

Land tenure is the dominant barrier to more productive management of the lands on which coconut is grown. Landlords generally prohibit understory cropping to avoid future claims to permanent occupancy. However, numerous crop species thrive under coconuts (Paner, 1975). Multistory cropping systems—with a two- or three-tiered canopy that may include fruits, vegetables, and food crops—improve farm income and are observed in some areas. It is unclear whether the planned extension of agrarian reform to the areas planted in coconuts, which was indicated in the 1987 Comprehensive Agrarian Reform Program legislation, will have any effect in overcoming this land tenure barrier. The titling of lands on which coconut is grown to tenant farmers would result in a dramatic increase in land use intensity for coconut. This would significantly alleviate the high degree of income uncertainty for tenant farmers who grow coconuts.

FARM FORESTRY

The concept of farmers producing fast-growing trees as crops was popularized in the mid-1970s by the Paper Industries Corporation of the Philippines, which set up woodlots on farms to grow trees for pulpwood production (World Bank, 1989a). The practice has gained momentum in recent years, as the depletion of old-growth hardwood forests sent domestic timber prices steeply upward. Substantial numbers of small-scale farmers in northern Mindanao now plant in short rotations and then sell gmelina (*Gmelina arborea*) and falcata (*Albizia falcataria*) as timber. *G. arborea* is harvested and coppiced in up to three 10-year cycles. Fast-growing hardwoods such as gmelina are also integrated into contour hedgerow systems. The Master Plan for Forestry Development (Department of Environment and Natural Resources, 1990) places emphasis on contract forestry with private individuals and communities and is supported by a loan from the Asian Development Bank. Development of these systems would be greatly accelerated if credit for contract tree growing is extended to small-scale farmers and hardwood production in hedgerows is encouraged.

DIVERSIFICATION

The most plausible model of sustainable smallholder farming in the uplands is one of diversification into mixed farming systems. Given the exceptionally high production and marketing risks in the uplands and the generally low marginal returns, a number of alternative enterprises must be undertaken on upland farms to provide stability (Chambers, 1986) and to take maximum advantage of the complementarities that occur among income-generating activities (for example, leguminous trees for fodder, green leaf manure, and fuelwood; cattle for labor, cash income, and manure).

Upland farm families must place primary or exclusive emphasis on subsistence food crop production. The land use systems that result from the pursuit of these needs, however, are the least ecologically sustainable alternatives. The issue from policy, research, and extension perspectives is how to enable the farm enterprise to move profitably along a trajectory that will continually increase the area devoted to perennial plants and decrease the area devoted to annual plants (Figure 5). The gradual expansion of home gardens, ruminant livestock production, and plantation and timber tree crops will contribute to this end. Greater private and public sector support for the development of these enterprises will be essential. However, this must be linked with the improvement of methods for greater sustained food crop production per unit area to release land and labor for other cash-generating activities.

The Philippine Department of Agriculture has only recently begun to give significant attention to the task of understanding upland

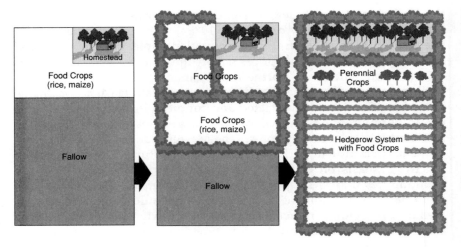

FIGURE 5 Model of the evolutionary development of a small-scale upland farm on sloping land.

agricultural technologies. Upland agricultural systems are in stark contrast to the less heterogeneous lowland systems that have histori- cally received overwhelming attention. Therefore, a major reorienta- tion of both the research and extension approach is under way. This reorientation involves the decentralization of operations to the local level. The Department of Agriculture has adopted a farming systems research and development model for technology generation in the uplands, with strong emphasis on farmer-participatory research (Dar and Bayaca, 1990). To be effective, this transformation must be pur- sued more vigorously and will require major increases in staff capa- bility and mobility.

The Grasslands and Brushlands

The most common form of vegetation in the Philippine uplands is grass, predominantly *Imperata cylindrica* (cogon) or *Themeda triandra* (samsamong, silibon, or bagocboc) or, at higher elevations, *Miscanthus japonicus* (runo). The rhizomes of these perennials are highly resis- tant to fire, but the shoots are flammable during dry periods. They readily invade abandoned swiddens, land cleared of forests, and for- est openings. A small portion of the grassland area may be a result of natural disturbances, but the overwhelming majority owe their existence to repeated disturbance by fire, which is usually started by humans to obtain game or fodder or to clear land (Bartlett, 1956).

At the turn of the twentieth century, 40 percent of Luzon and extensive areas of other Philippine islands were covered with grass. The land classification of 1919 estimated that grassland covered 19 percent of the country, a figure that stayed roughly constant through 1957 (Roth, 1983). An analysis (Swedish Space Corporation, 1988) of Philippine land use estimated the area of pure grassland to be 1.8 million ha, with an additional 10.1 million ha in extensive cultivation mixed with grasslands and brushlands (that is, about 33 percent of the country's land surface). This suggests that more than 20 percent of the surface area of the country is covered by grasslands (see Table 2). The grasslands appear to have served as an intermediate zone—a portion continually being transformed into permanent croplands or plantations—for a long period of time, whereas new area is created as the forest withdraws. In some intensive grass-fallow rotation systems, fire climax savannah is used indefinitely as the fallow species (for example, see Barker [1984]).

The cogon grasslands are commonly used as pasture, but they have a carrying capacity that is probably lower than 0.25 animal units (0.3 cattle) per ha (World Bank, 1989a). Cogon grass is suitable as a forage only during early growth, so the range is regularly burned toward the end of the dry season, which contributes to wildfires that penetrate and further destroy forestlands. Range management by private ranchers is generally poor, and improved management practices have not resulted in competitive economic returns. Overgrazing during the regrowth period reduces ground cover and makes grassland the most significant source of soil erosion in the Philippines. Thus, the net social returns from cattle ranching are low, and justification of this form of land use is questionable.

There has been a precipitous decline in ranching during the past 15 years. A major factor has been the communist insurgency, which targeted its operations against ranches. Associated with this has been a 50 percent decline in the size of the national cattle herd during this 15-year period.

What should be done about the grasslands? They continue to function as a migratory sink for the settlement of landless and jobless families, and in this sense, they are still a frontier. The social value of these lands, however, is greatly constrained by government land use policy and a regressive pattern of formal and informal land tenure. Although the land is publicly administered as forestland by the Department of Environment and Natural Resources (DENR), wealthy families (pseudo-landlords) have laid claim to large areas, relegating settler families to tenancy.

Small-scale farming in grasslands is predominantly practiced with animal labor. Settlers initially practice a migratory system of farming, shifting their farm area as necessary to sustain crop yields. The greater population densities necessitate rotating the fallow areas of fields within permanent farm boundaries. As the farm size decreases, permanent cropping evolves, in many cases with extremely low comparative yields (Vandermeer, 1963).

SECURITY OF LAND TENURE

Since 1894, the Philippine state has proclaimed about two-thirds of the country's area as public forestland. In 1975, all land with a slope of 18 percent or greater was proclaimed by legislation to be part of the public domain. Subsequent legislation further eroded the rights of occupant families to the land on which they lived. Although the legislation was ostensibly intended to strengthen the state's ability to conserve the forests, its unanticipated effect was to greatly weaken occupants' interest in any long-term forms of sustainable land management.

Later, the realization grew that the upland populations were going to be permanent and were increasing rapidly. This led to a succession of weak programs that involved occupancy permits and communal tree farming contracts. The Integrated Social Forestry Program (ISFP) arose in the early 1980s as an extension of the earlier approaches. It was based on a Certificate of Stewardship Contract (CSC), which grants leasehold occupancy rights for up to 7 ha of land to a family for a 25-year period and is renewable for another 25 years (Department of Environment and Natural Resources, 1990). CSC holders are obligated to use conservation farming practices, plant at least five trees per hectare, and assist in protecting adjacent forest areas. The ISFP promotes agroforestry practices, particularly contour hedgerow farming.

Although the CSC is aimed at strengthening the land tenure security of upland farm families, it is a weak instrument for doing so. Many poor farmers and their families face substantial problems in asserting a CSC claim against the claim of more powerful but absentee pseudo-landlords. The CSC lease is nontransferable and, thus, cannot be used as collateral for loans for investing in farm improvements. The CSC lease may be canceled at the discretion of the Forest Management Bureau, and it is heritable only within the 25-year lease period.

The speed of implementation of ISFP has been disappointing. Only 2.5 percent of the upland area has so far been included in stewardship leases. The Master Plan for Forestry Development (Department of Environment and Natural Resources, 1990) targeted CSCs to be issued to 626,700 families during the 10-year period from 1988 to 1997. This would cover an estimated 1.88 million ha of public land. Assuming an average of six persons per family, this would involve a population of 3.76 million. These targets appear to be overly optimistic unless major new funding and staffing becomes available.

Secure land tenure in the uplands would decrease the number of large land claims by elite individuals who use poor families as tenants. Many poor families are part of a well-organized effort of occupation of forestlands carried out by wealthier individuals who hope to lay claim to the land by paying taxes on it. Under such arrangements, the agricultural inputs of the cultivator may be subsidized by the pseudo-landlord and personal credit may be advanced to the cultivator, or the cultivator may be contracted to plant perennial crops for an agreed price per plant and permitted to grow food crops on the young plantation until the trees become established. Then, the cultivator must move on to a new area to renew the cycle or may be hired to care for the plantation.

CSC leaseholds provide a mechanism that serves as a counterweight to the grip of local elites. Effective independence for the cultivator will depend, however, on the infrastructure and support services that will make it possible to earn a viable living from the land without the patronage of landlords. The sense of security that the CSC provides to powerless migrant farmers was explored by Pava et al. (1990). The granting of CSCs will encourage more migration into the uplands. This will happen even if recent migrants are excluded from the program. It will be especially pronounced in areas where the bulk of the fertile lowlands are controlled by a few landed elites.

FALLOW IMPROVEMENT SYSTEMS

There are a variety of farming systems in the grasslands, ranging from shifting cultivation to permanent cultivation systems. The technology appropriate for a shifting cultivation system differs from that for a permanent field cultivation system because of the major differences in labor and land use intensity required for each system. As Raintree and Warner (1986) pointed out, shifting cultivators maximize their returns to labor rather than to land and resist inappropriate labor-intensive technologies. Hedgerow farming is a solution that is suitable to the more intensive stages of permanent cultivation. A more relevant concern in shifting systems is management of fallow fields.

Barker (1984) analyzed the role of fallow fields in shifting cultivation. A crop that improves fallow fields must yield higher nutrient
levels and accumulate more organic matter than the natural fallow it
is to replace. Little work has been done on practical methods of
rapidly regenerating soil fertility in fallow fields of the Philippines.
Fallow fields are usually burned or subjected to intensive grazing.
Farmers acknowledge that these practices are often ineffective in regenerating fertility, and this has been corroborated by sampling the
nutrient status of fields (Fujisaka, 1989).

Leguminous cover crops have been proposed as candidates for
managed fallow fields, but empirical evidence of their practical utility is sparse. The ubiquitous presence of dry season grassland fires
and the difficulty in preventing fires on the grasslands will limit this
practice. Protection from communal grazing is also a constraint in
many areas. Problems of seed supply and seed collection limit the
adoption of leguminous cover crops, but a system for marketing cover
crops is rapidly developing (P. C. Dugan, Department of Environment and Natural Resources, personal communication, 1990). A much
greater research effort is needed at national and local levels, particularly regarding species that can be used as food for humans (for example, *Psophocarpus palustris* [siratro]).

Systems for enhancing fallow fields with leguminous trees have
been demonstrated. MacDicken (1990) described an indigenous planned
fallow that has evolved on steep slopes in Cebu since before 1900.
Dense stands of naturally reseeded *Leucaena leucocephala* are used in
the fallow portion of the cycle. When the *Leucaena* trees are cut, the
stems are placed on the contour and staked to create contour bunds.
A fallow period of 3 to 7 years is followed by several years of cereal
cropping. The concept of naturally reseeded fallow fields deserves
serious attention as an alternative fallow for both grassland and forest agroecosystems, where natural woody plant regeneration after
cropping is suppressed. Tree species that are suited to strongly acidic
soils and are prolific in seed production also need to be identified.
Flemingia congesta is a candidate species for medium-elevation sloping acid soils, and *Alnus japonica* is a candidate species for the highlands.

A tree fallow system for shifting cultivation on the island of Mindoro,
which used cuttings of *Leucaena* that was intercropped with the food
crops, allowed development of a tree cover on fallow land after the
cropping cycle (MacDicken, 1990). The value of such systems remains unconfirmed. There are also uncertainties in applying these
systems—or variations of them—to the diverse range of fallow environments on grasslands or forestlands. Exclusion of fire will also be

a dominant concern in successful implementation of such systems. A major sustained research effort on managed fallows is critical.

REFORESTATION EFFORTS

The grassland areas have been a major target of Philippine government reforestation efforts for the past 30 years (Department of Environment and Natural Resources, 1990). Official forestry statistics indicate that about 1 million ha of tree plantations was planted between 1960 and 1989. This effort was managed by the Forest Management Bureau.

In most ongoing reforestation contracts, fast-growing and leguminous hardwoods are planted as nurse trees to form a protective canopy, with a few premium species planted as the climax crop. Foremost among the nurse trees are *Acacia mangium*, *Acacia auriculiformis*, *Leucaena diversifolia* (psyllid-resistant strains of *L. leucocephala*), and *Gliricidia sepium*. The major premium quality species include *Swietenia* species and *Pterocarpus grandiflorus*. Other species that can grow in areas dominated by *Imperata cylindrica* are *Gmelina arborea*, *Eucalyptus camaldulensis*, and leguminous pioneer species. Sometimes, contractors mechanically till the areas to be planted and seed leguminous cover crops during the first year to improve the soil microenvironment. In most projects, nursery-grown plantings are used.

The success record, however, has been disappointing. In a recent nationwide inventory of the status of plantations (Forest Management Bureau, 1988), the actual extent of surviving trees was found to be only 26 percent. In the central and western areas of the country, which have prolonged dry seasons, the situation was more dismal. For example, Reyes and Mendoza (1983) found that after an intensive reforestation effort in the watershed containing the Pantabangan Reservoir, the survival of replanted trees was only 10 to 15 percent because of poor weed control, pests and diseases, and fire.

Control of fires on newly established plantations is difficult and costly. Public reforestation projects are given neither adequate incentives nor appropriate management capabilities to provide protection from fires. In fact, many plantations were deliberately torched by local people who saw that there was nothing to be gained from the presence of a government plantation in their area.

CONTRACT REFORESTATION

The overwhelming failure of reforestation efforts managed by the Forest Management Bureau has recently prompted a major redirection in approach. The approach is called contract reforestation, by which DENR plans to establish artificial forests via contracts with families, communities, local governments, the private sector, and NGOs on about 630,000 ha by the year 2015 (Department of Environment and Natural Resources, 1990). Contracting consists of a two-phase strategy. First, DENR contracts for the establishment, maintenance, and protection of artificial forests for a 3- to 4-year period. If the contractors perform well in meeting the provisions of the reforestation contract, they can apply for a Forestland Management Agreement (the second phase). This entitles them to harvest, process, and sell or otherwise use the products grown on their reforested areas. The private forestland manager, however, must pay the government a share of the income from sales of production output. This share is equivalent to the amount of money needed to reforest 1 ha of denuded area when 1 ha of 3- to 4-year-old trees is cut. Harvesting and other thinning activities are done in accordance with a DENR-approved management plan.

The majority of the lands targeted for the contract reforestation program are relatively degraded or remote. Because of low profitability and high interest rates, private firms are hesitant to invest their own corporate funds to establish industrial tree plantations (Domingo, 1983; Guiang, 1981). The funds that the government has designated for this program are largely from international donors, particularly the Asian Development Bank.

DENR hopes to generate reforestation funds from production shares under the Forestland Management Agreement. In this way, DENR could spread the financial and environmental benefits of reforestation activities. It is presumed that managers have strong incentives to protect and manage their artificial forests, since they reap the major profit from the sale of the tree crops. They can also plant and intercrop cash crops, fruit trees, and other agricultural crops to augment their incomes and to provide additional incentive for protecting, replanting, or enriching the plantation forests. DENR has also provided an indirect subsidy for rehabilitating grasslands and brushlands that are not profitable under the industrial tree plantation scheme. Enthusiasm for contract forestry is tempered by apprehension about constrictive regulatory controls. If the regulatory attitude prevails during implementation of the program, as is typical of DENR programs, progress will be disappointing.

A major factor in the success of the contract forestry program is the assumption that independent managers will strive to protect their investment from fire. The excellent fire control technologies of indigenous peoples, for example, methods used on the 15,000-ha ancestral lands of the Kalahan Education Foundation, Nueva Viscaya, can be more widely disseminated (Barker, 1990).

THE ECOLOGY AND MANAGEMENT OF FIRE

When an area is cleared of tropical forest it changes from an ecosystem essentially immune to fire to one in which fires are extremely common. J. B. Kauffman's research (cited in Savonen [1990]) showed that rain forests are capable of catching fire only on an average of 1 day each 11 years, but partially logged areas burn after an average of only 6 rainless days. Grassland areas are flammable after only 1 rainless day.

Repeated burning kills potential tree propagules in fallow fields and favors grasses, in particular *Imperata cylindrica*, over perennials. When burning or other disturbance is halted, *I. cylindrica* is rapidly invaded and shaded out by taller, woody species. If the area is large enough, however, *I. cylindrica* grass may persist for decades, even after the fires have stopped, because the propagules of other plants have been eliminated.

All aspects of this discussion on technology for more productive uses of grasslands for agriculture and forestry emphasize the dominance of fire as a debilitating constraint. Determined ecologic and farm-level management research on fire control will be essential to achieve progress in the better use of grasslands. Identification of practical and cost-effective tactics will require a systems approach. A national research project on the ecology and management of fire could collate the knowledge on the subject that can be provided by indigenous peoples, design a comprehensive framework for investigation, and assist regional and local research teams in undertaking work in this area within the respective land use system research programs.

LOCAL ORGANIZATION FOR CONSERVATION AND SUSTAINABLE AGRICULTURE

During the past decade, social forestry research has provided much insight into the complex constraints in the evolution of effective community organizations to sustainably manage local upland resources (Borlagdan, 1990). Many of these organizations will be needed to serve the needs of upland farmers in thousands of villages throughout the Philippines. The initiation of farmers' organizations has so far been limited to specific project sites. Careful consideration must be given to the development of a structure that will link these organizations at the provincial, regional, and national levels. Such a structure might draw on some of the experiences of the conservation districts in the United States (Cook, 1989). These independent units of local government, of which there are more than 3,000, regulate resource use and assist farmers in implementing conservation practices. Conservation districts are created through a referendum involving all occupants of the land. They are governed by an elected board that enlists the skills and services of government agencies at all levels to advance conservation programs in the district.

Saving and Rebuilding the Remaining Natural Forests

The commercially exploitable old-growth dipterocarp forests in the Philippines are nearly exhausted. The Master Plan for Forestry Development (Department of Environment and Natural Resources, 1990) estimates their extent at slightly less than 1 million ha. We estimate that the actual extent may be closer to 700,000 ha—or lower. Nearly all of this area is to be protected under recently enacted DENR policies banning logging in old-growth forests. Therefore, DENR anticipates that further declines in forested areas will be slight (Department of Environment and Natural Resources, 1990). It appears to be optimistic to assume that commercial logging will stop immediately, that illegal logging can be controlled (since it has been resilient in the past), and that indigenous communities and migrants to the forest will not further convert significant areas of the forest to permanent agricultural uses.

The Philippine government has now acknowledged that it is incapable of managing forestlands on its own (Department of Environment and Natural Resources, 1990). DENR recognizes the logic of community control in managing forest resources. The issue now is whether DENR mechanisms set in place to implement this concept will be sufficient to address the needs.

THE ROLE AND RIGHTS OF INDIGENOUS COMMUNITIES

The people of the indigenous communities differ in their willingness to accept the concept of stewardship leases rather than full titling of the land to the community. Their reasons fall into three categories, depending on the community's circumstances:

• Ethnic communities that have been able to maintain secure control of their land: Forest-dwelling ethnic minorities of the Cordillera who have staunchly protected their land fear that acceptance of stewardship leases will mean that they must give up their claim to ownership.

- Communities that have traditionally possessed land but whose lands are under strong encroachment pressure from lowland settlers or plantation expansion: Groups such as the Ikalahans and Mangyans struggled successfully over a long period of time to obtain a lease and consider stewardship leases to be the best practical means for trying to maintain the integrity of their land.
- Communities that have been displaced from their traditional lands: These communities, such as the T'boli, have been forcibly dispossessed and inhabit new locations where they do not have a basis for traditional land claims. Others, such as the Bilaan, have been completely dispossessed of any land and live in squalid refugee camps. These groups are desperately seeking some form of land tenure security and are highly receptive to leasing arrangements.

The predominant concern of many communities regarding land tenure is encroachment by outside interests. The first Communal Forest Lease was obtained in 1974 by the Ikalahan in Nueva Viscaya (Cornista and Escueta, 1990). The major land threat was from lowland farmers and elites from the nearby municipality who claimed land on the Ikalahan's traditional reservation. By 1988, a total of nine communal leases ranging from 50 to more than 15,000 ha were issued to a variety of groups.

An organizing force was critical to the eventual development of these leases. This was usually provided by an NGO. Developing community leadership to manage the process was an essential and often difficult process. Many failures in community management can be anticipated; therefore, a heavy investment in management skills will be essential within DENR, NGOs, and the communities.

COMMUNITY-BASED FOREST MANAGEMENT

In 1989, DENR moved to implement the Community Forestry Program (CFP) (Department of Environment and Natural Resources, 1990). This allows organized cooperatives of forest occupants and upland farmers to extract, process, and sell forest products in exchange for the community's commitment to protect, manage, and enrich the residual forest. DENR provides 25-year wood utilization permits to organized communities under a Community Forestry Management Agreement, which is renewable for another 25 years. The change in policy was intended to democratize access to forest resources, generate employment in the uplands, and manage the remaining production forests in a sustainable manner.

Under DENR's Master Plan for Forestry Development (1990), a

total of 1.5 million ha is targeted for community-based forest management. The forests classified for CFP are generally fragmented, inadequately stocked, part of canceled concession areas, near rural communities, and unprofitable for large-scale commercial extraction and processing. In 1990, 26 percent of the forests classified for CFP were in good condition, 40 percent were in fair condition, and 34 percent were in poor condition. Only small-scale and labor-intensive types of forest extraction and processing will result in profitable operations in these forests.

The CVRP-1 Social Forestry Project (1984–1989) was the first test of the community-based forestry concept (Dugan, 1989). The project was located on a 17,000-ha site on Negros Oriental island that had 4,500 ha of forest and about 17,500 inhabitants. The area had been under a logging ban since 1979, but illegal deforestation continued at an annual rate of about 1,360 ha. Eighteen Forest Stewardship Associations composed of forest occupants and farmers were initiated. They assumed responsibility for managing and conserving designated portions of the forest under the guidance of the project staff. The rate of forest destruction declined abruptly—by 92 percent—as the cooperatives began policing their zones, and it remained at only 100 ha annually through 1989. Shifting (slash-and-burn) cultivation in the forest was drastically curtailed. Large-scale illegal logging was eliminated. Using labor-based technology, the cooperative members participated in limited wood extraction, which increased their incomes far beyond what they had earned previously. These projects were proven successes that supported the hypothesis that the deforestation process can be controlled only when the forest occupants have a direct stake in the enterprise.

Nevertheless, some serious deficiencies in community organization, training, and cooperative management were observed. These deficiencies led to confusion in the cooperatives, and instances of corruption and abuses of forest regulations were uncovered. The need for a major reorientation of the skills and attitudes of the foresters involved in a community-based management setting was also highlighted. Success of the approach will be possible only with a large core of committed and competent people. Currently, no organized pool of people has such expertise. The limitation of human resources in the communities and in DENR will make the rapid expansion of community-based forestry uncertain. To date, DENR's experience with implementation of CFP has been limited to the selection of NGOs to operate the program and site identification, but inadequate attention has been given to organizing and training members of the community (Guiang, 1991; Guiang and Gold, 1990). Therefore, emphasis on training programs that can teach the required managerial skills will be needed.

The technical, managerial, social, marketing, and financial management requirements of community-based forest management projects are enormous. Most NGOs, which have strong community-organizing capabilities, must strengthen their capabilities in taking resource inventories, preparing management plans, harvesting methods, marketing, processing, and managing finances.

Under a 1989 DENR directive, part of the money from the sale of products extracted from residual forests should be invested in systems that provide forest dwellers with alternative livelihoods. These systems must not be dependent on forest resources. A key need is for investment in village nurseries that will supply perennial and timber seedlings to individuals on a sustained basis.

SUSTAINED-YIELD FORESTRY

Little is known about the ecology of dipterocarp forests. It is not possible to say with confidence that any selective cutting system will ensure the sustained development and harvest of dipterocarp wood. Therefore, maintenance of the remaining fragments of lowland and upland old-growth dipterocarp forests is of the highest priority. Much more research into the ecology and physiology of dipterocarp forests is essential if the remaining fragments are to be expanded into viable forests. Previous efforts to establish dipterocarp forests have generally failed, but there have been a few cases of dipterocarp forest survival on plantations (Department of Environment and Natural Resources, 1990). The factors that govern such successes need to be investigated more thoroughly.

LABOR-BASED TIMBER EXTRACTION

Some foresters argue that sustained-yield timber extraction is highly feasible when native-style logging exclusively is used by local communities (Dugan, 1989). The experience gained from the CVRP-1 Social Forestry Project lends strong support to this contention. Timber extraction is naturally limited by the lower technical efficiency of carabao (water buffalo) logging, but the economic efficiency and profitability for both local harvesters and sawmills is attractive as compared with mechanized logging. Mechanized logging is skewed toward once-over extraction of the 150-plus-year-old virgin trees, with a return harvest expected after some 30 to 100 years, assuming that forest destruction in the logging operation did not permanently disrupt the ability of the valued timber species to regenerate.

Indigenous logging methods emphasize repeated extraction of small amounts of timber and other forest products. These labor-based systems may allow an incremental annual extraction, determined on the basis of the annual accumulation of wood that can be harvested. This would provide continuous income from a limited tract of land and would be less destructive to the environment than capital- and machine-intensive systems. Employment in forest industries may quadruple if indigenous systems are adopted (P. C. Dugan, Department of Environment and Natural Resources, personal communication, 1990).

FOREST ENRICHMENT

As communities manage forests to achieve sustainable yields, there will be a tendency to extract the higher quality species, which will eventually lead to species impoverishment—a major concern. Enrichment planting of valuable timber species is a method that has been proposed to avoid impoverishment of economically valuable species in selectively or severely logged forests. There are virtually no data, however, to verify the effectiveness of enrichment techniques or to address the numerous practical questions that arise in their implementation. A strong research effort involving species establishment and ecologic studies in the field is urgently needed. Strategic research will need to be complemented with in-depth surveys of the methods of indigenous farmers and evaluations by participating farmers from multiple locations in forests representing wide ecologic gradients.

FUTURE IMPERATIVES FOR SUSTAINABLE UPLAND FARMING AND FORESTRY

The phenomenal depletion of natural resources in the Philippines reflects major deficiencies in the country's development efforts since its independence in 1946. The outstanding characteristics of the lack of development are the failure to create jobs and raise the living standards of the majority of Filipinos as well as the large inequalities in the distribution of wealth and access to financial and social resources. Therefore, a critical consideration in an assessment of future scenarios of forestry and agriculture in the Philippine upland ecosystem must include accurate prediction of trends in the political economy.

There is a no lack of detailed studies of the state of the Philippine environment or suggestions as to what should be done. Such studies include Dames and Moore International et al. (1989), Fay (1989), Porter and Ganapin (1988), World Bank (1989a), and the Master Plan for Forestry Development (Department of Environment and Natural Resources, 1990). The major structural problem in the Philippines has been the inequality of income and wealth. Most observers agree that land reform in postwar Philippines has failed to reduce the power of the landed elites or to transfer substantial amounts of land to tillers. Implementation of the current agrarian reform program is clouded by similar doubts.

Another dominant structural problem is the failure of the industrial sector to provide new jobs at a rate fast enough to absorb the burgeoning labor pool. Upland agricultural and environmental problems cannot be solved as long as the mass of Filipinos are unemployed or underemployed and earn less than a subsistence wage. There must be a structural shift away from agriculture. The upland sustainability crisis is strongly interconnected with national political, economic, and ecologic stability. The strategy for attacking it must be bold, but it must be sensitive to the realities of these aspects.

Elements of a Strategy

There are three overarching elements to a comprehensive strategy for evolving sustainable land use systems in the Philippine uplands: tenure, technology, and delivery. Tenure encompasses human populations and their relationship to the land. Technology covers the technical solutions and the institutional capabilities to develop them. Delivery involves the mechanisms that government institutions and the private sector use to deliver the policy and various infrastructural supports to facilitate and guide the process of change.

TENURE: PEOPLE AND EMPOWERMENT

Reduce Population Growth Rates Any strategy to address the sustainable management of upland resources must include a reduction in the rate of population growth. This must be powered by a national consensus on the need for a vigorous population control program. National and international efforts could vigorously pursue that policy dialogue by supporting the call by a group of Filipino development specialists for a new national consensus on establishing the two-child family (Porter and Ganapin, 1988).

The poorest households in the rural uplands have the highest birth and mortality rates. Government must redirect health care programs to ensure that there are greater investments in village-level health and paramedical personnel, and family planning support and education should be an integral part of the effort. The cost and political risks from embarking on a vigorous population control program will necessitate strong and sustained international support. Demographic goals and an effective organization to meet those goals must be highlighted as a fundamental component of such support.

Reform Land Tenure to Reinforce Local Stewardship Future success in bringing sustainable land use to the uplands is fundamentally dependent on major changes in the ways that public lands are managed. The Philippine government has proved to be incapable of managing the country's land area. The area under direct central government control must be decreased rapidly. Although this is a declared intention of government policy, progress has been slow.

To harness the energies of upland populations in creating sustainable land use systems and to ensure the success of reforestation and forest remnant conservation efforts, the national government must establish a new political relationship with the upland population. It must recognize the boundaries of the lands held by the indigenous occupants and move to recognize their full ownership rights. The dominant issue is empowerment of the upland people so that they can have a secure stake in the land.

The Philippine Constitution restricts leaseholds on public lands to terms of 25 years, which are renewable for another 25 years. However, further definition of the terms of the lease is at DENR's administrative discretion. As of 1988, only 2.2 percent of publicly owned forestlands were placed under leasehold arrangements; thus, only a fraction of the upland farming population has been affected. The Master Plan for Forestry Development (Department of Environment and Natural Resources, 1990) projects a large increase in leaseholds, but DENR has not allocated budgetary support and does not have the implementation capacity to effectively carry out an aggressive program.

In addition, the form of land tenure security in the Certificate of Stewardship Contract (CSC) now being issued will not be adequate to foster viable farm operations with the degree of land stewardship needed. The CSC must be amended to enable it to be transferable and so that farmers can use it as collateral to obtain credit. The transferability of the CSC should, however, apply only to actual land occupants, to avoid an eventual concentration of landholdings.

These provisions should be interpreted as the initial stages that will eventually lead to unrestricted land titles. They give the occupants time to demonstrate their capacity to develop a sustainable land management system. Complete title to the land would then become an incentive to practice conservation farming methods and to be a good steward of the land. Granting of immediate and unconditional titles to the land is not practical because of the immense administrative work load it would entail.

A comprehensive government response must be initiated to deal with the existence of tenancy in the uplands resulting from the land claims of pseudo-landlords. Although they are illegal, these claims result in nominal tax revenues for local governments, which otherwise have very limited sources of income. It is essential that local governments realize that the changes in land tenure in the uplands will be to their benefit through taxes, income, and social stability. Therefore, the national government must make provisions for local governments to receive alternative sources of income. The 1991 Local Government Code began the process of enabling local governments to obtain local tax revenue. Two additional mechanisms that can be implemented are the allocation of authority for local governments to levy modest taxes on individual leaseholds and to undertake contractual forestry activities on the public lands in their jurisdiction.

Recognize the Ancestral Rights of Indigenous Occupants There is a strong legal basis granting ownership rights to indigenous peoples who have historically inhabited the land (Lynch and Talbott, 1988). Recognition of these rights has so far been ignored by DENR, but we believe it is a crucial element in the sustainable management of upland resources. In general, the optimum mechanism by which these rights can be recognized is a community title. The precise instrument by which secure tenure should be granted, however, may have to vary somewhat for different communities. Direct titles to the land should immediately be given to indigenous communities that have strong and cohesive leadership, particularly in the autonomous regions in Muslim Mindanao and the Central Cordillera area of northern Luzon, which have legislative power over ancestral domains and natural resources.

Initially it is not necessary that all those with ancestral property rights receive titles that recognize those rights. The most immediate need is for the delineation of the ancestral domains by survey teams, so that a common basis of understanding exists between the national government and the communities (Lynch and Talbot, 1988) and so that communities can exercise effective control over their domains.

An important activity in developing an instrument of land tenure should be the formulation of a management plan than contains flexible but comprehensive mechanisms for allocating land among the inhabitants and for applying sound land management practices. As public land management is progressively privatized, it will be necessary to give local governments the authority to apply zoning restrictions so that they can control private land usage. These functions will strengthen local governments and overcome the strong objections from some quarters that the titling of public lands will lead to abuse of the land.

TECHNOLOGY: DEVELOPMENT AND DISSEMINATION

Research Upland Agriculture New technologies will be critical to the development of sustainable agriculture in the uplands, but the technologies being extended have not been proved in the diverse environments and for the variety of circumstances farmers face in the uplands. Two issues must be addressed: What will it take to make small-scale farming permanently sustainable? What will it take to improve fallow-rotation systems where they are still practiced?

Permanent small-scale upland farming systems are evolving in the sloping upland areas and are gradually replacing shifting cultivation. Acceleration of the trend toward permanent agricultural systems will fundamentally require simple, effective soil erosion control on open fields by use of vegetation barriers and residue management; mineral nutrient importation to balance the uptake of nutrients by crops and to stimulate greater biological nitrogen fixation; and diversification toward mixed farming systems that include perennials and ruminant animals, in addition to subsistence food crops. The technologies needed to meet these needs are known. Some fulfill multiple requirements (for example, trees in contour hedgerows may provide erosion control, fodder, and crop nutrients). But knowledge of how to adapt them to the wide array of diverse ecologic niches encountered by upland farmers is still inadequate. Much can be done now to take specific action to implement these concepts. The work must rely on farmer-participatory experimentation to refine specific solutions for local conditions.

The major innovation for farming on sloping lands has been the sloping agricultural land technology (SALT) that uses hedgerows of leguminous trees. A serious constraint of SALT is its high labor requirement. On acidic soils, there are questions concerning negative crop-hedgerow interactions. A major extension problem is the lack of hedgerow planting materials of forages, multipurpose trees, and perennial crops.

Because of the limitations of trees and introduced forage grasses in hedgerows, serious efforts should be invested in the refinement and dissemination of simpler methods, including natural vegetative filter strips. The advantages of natural vegetative filter strips are their simplicity of installation, their low labor requirements, and their excellent erosion control and terrace formation capabilities. They also provide a good foundation for soil conservation efforts, so that farmers may subsequently diversify into more labor-intensive hedgerow enterprises, including those that grow perennials, leguminous trees, and improved forages.

The importation of mineral nutrients will be essential to the development of sustainable food crop production on permanent farms in the uplands. Because the majority of soils are strongly acidic, phosphorus is usually the most limiting nutrient, and lime application is often necessary to lower the soil's acidity and alleviate aluminum toxicity. Programs that help upland farmers reduce soil degradation should also consider how to provide supplies of phosphorus and lime at the most favorable prices and provide instruction as to their most efficient use. Nitrogen fertilizer is an important tool that can be used to familiarize lowland rice farmers with nutrient use and bolster national rice sufficiency.

In areas that use fallow rotation systems, there is hope for improved fallow management if fire can be controlled. The use of trees planted in fallow fields has been demonstrated successfully in systems without animal labor. Little research has been directed to the agronomics of trees in fallow fields. In systems that use animal labor, forage legumes have been tested as an alternative to natural *Imperata cylindrica* infestation, but their effects are poorly documented. Much more research will be needed to refine the agronomic practices used in managed fallows in different environments.

Other top research priorities for sloping lands involve the development of appropriate small-scale mixed farming systems, such as those that include animal, perennial, and tree production, to gradually reduce reliance on food crops. Systems research will be essential for making more rapid progress in diversifying small-scale upland farms. Many NGOs are active in promoting sustainable low-input agricultural systems in the Philippines (Garcia-Padilla, 1990) and will play an important role in adapting solutions to specific local conditions.

Integrate Livestock into Upland Farming Systems There must be greater emphasis on ruminant livestock in achieving sustainability in mixed farming systems. Most hedgerow systems supply the farm with increased quantities of legumes or grass forage. Hedgerow farming enables larger livestock populations and contributes to alleviating the deficit in ruminant meat production.

There is an opportunity for greater investment in NGO-operated programs to distribute ruminants (cattle, goats, and sheep) to small-scale upland farmers for cut-and-carry production systems. Animals would be distributed to farmers who have succeeded in installing hedgerows that contribute to conservation practices. The incentive would popularize the use of contour hedgerows and make it economically attractive to practice conservation. Farmers would receive parent animals and then retain female animal offspring, returning the parent animals so that the program rolls over and expands. International donors may also find such a program to be a sound investment, if it is well managed.

Reorient Forestry Research and Development Forestry in the Philippines will change dramatically in the next 20 years. The extraction of high-value timber from old-growth dipterocarp forests will disappear as the few remaining forests vanish or become protected. The reorientation of forestry to the development of sustainable management systems for secondary forests should begin in earnest. Interest in rehabilitating degraded forests will grow as the real value of timber rises. Tree plantations and farm forestry can then become viable income-producing activities.

Management systems in forestry must be drastically altered, but the technical knowledge base to support these changes is extremely weak. Research on both technical solutions and management systems must be accelerated to provide a sound basis for new directions. Major research efforts will be needed in the following areas: the ecology and management of dipterocarp forests for sustained production, community-oriented forest management, restoration systems for degraded secondary forests, the ecology and management of fire, the impact of policy changes on the supply of wood, and plantation and farm forestry issues. The research must be strongly oriented to the social as well as biologic sciences and requires a systems approach. The development of joint international collaboration will be important to the acceleration of forestry research.

Develop a Research Methodology It is at the interface between forestry and farming that the major future research and development challenges will be encountered (Figure 6). The forestry sector must engage in forestry for the benefit of the land and the people, and the agricultural sector must do the same, thereby creating sustainable upland farming and forestry. An understanding of the constraints and solutions is needed before upland farming populations and government can become effective partners in conserving, managing, and replanting forests while meeting basic subsistence food production needs. Teamwork is essential.

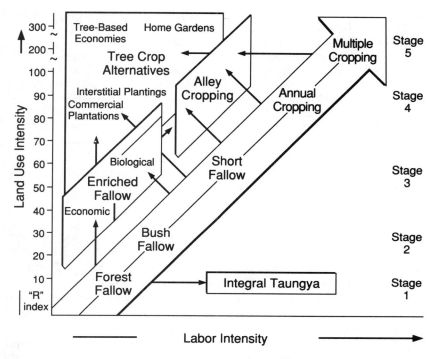

FIGURE 6   Evolution of a more integrated approach to sustainable land use in sloping uplands.

Farming systems research evolved as a framework for a more comprehensive, multidisciplinary attack on the complex constraints in agroecosystems (Harrington et al., 1989). Ecosystem-based research should be targeted to the broader continuum that includes forest management and agriculture. Such work needs a methodology that provides foresters and agriculturalists a common framework within which to interact.

Hart and Sands (1991) have proposed a sustainable land use-systems research strategy based on a farming systems approach that may provide a starting point. It applies a farming systems perspective to the land use system, targeting the land management unit within the context of its biophysical and socioeconomic environments and emphasizes the ecosystem as the starting point of problem analysis and research design (Figure 7).

The watershed is the natural unit on which to base a systems research effort because of the interconnected nature of all land uses

FIGURE 7 A research and development process that could be used by a multidisciplinary team as a guide in the development of an appropriate sustainable land use systems research framework. Source: Hart, R. D., and M. W. Sands. 1991. Sustainable land use systems research. In Sustainable Land Use Systems Research, R. D. Hart and M. W. Sands, eds. Kutztown, Pa.: Rodale Institute.

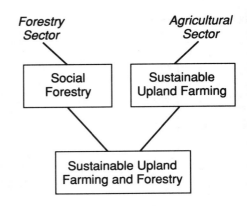

within a water catchment area, particularly the interplay between uplands and lowlands. The most technically and economically efficient approach would focus on site-specific conservation-oriented farming and forestry technologies. The watershed framework ensures that the social, economic, and political linkages between upstream and downstream lands are not neglected in the analyses (Magrath and Doolette, 1990).

Institutional mechanisms and project structures need to be evolved to make it feasible for the forestry and agricultural sectors to jointly participate in common research and extension work. Professionals in both sectors—long separated by administrative barriers and divergent academic traditions—need to recognize the improved research that can be the result of working together. International donors can assist in generating research opportunities; for example, the Ford Foundation has provided support to a team of foresters and agriculturalists at Central Mindanao University to develop methods of farmer participation in generating practical solutions for sustainable hillside cultivation (Pava et al., 1990).

Colleges of agriculture and forestry need to be encouraged to set up joint academic and research programs targeted to upland ecosystems. The recent initiation of the Committee on Agroforestry at the University of the Philippines, Los Baños, is a step in this direction (R. del Castillo, Agroforestry Program, University of the Philippines, Los Baños, personal communication, 1990). Mechanisms for research collaboration between professionals in DENR and the Department of Agriculture are urgently needed. These may be fostered by an expansion in scope and the participation of the Upland Working Group of DENR (Gibbs et al., 1990). Explicit linkages between the Ecosystems Research and Development Bureau of DENR and the Department of Agriculture's research programs, particularly key community-based forestry and contract reforestation projects, would generate a greater focus on the constraints to using various land use systems in deforested areas. The Philippine Council for Agricultural and Resources Research and Development, which has responsibility for approving and encouraging both agriculture and forestry research, will play a central role in expanding resource management-oriented research.

The Philippines needs a more definitive network of on-farm (field) laboratories in carefully selected watersheds where multidisciplinary research and development teams can focus their efforts. These field laboratories need sustained support with a budget structure that keeps team members working together. These sites would be linked to the less intensive applied research and extension programs carried out by NGOs and government departments. Research should be particularly sensitive to the use of techniques that enhance participatory approaches to rural development, drawing strongly on the technical knowledge of indigenous people in all phases of research (S. Fujisaka, Social Science Division, International Rice Research Institute, Los Baños, Philippines, personal communication, 1989).

Support International Research   The complex upland sustainability issues faced by the Philippines are common to most countries in Southeast Asia. Because the problems transcend national boundaries, stronger international mechanisms that provide efficient research and development support to the respective nations are needed. A number of institutions and networks are involved with upland resource management (Garrity and Sajise, 1991), including the Southeast Asian Universities Agroecosystems Network, the Asian Rice Farming Systems Network, the International Board for Soils Research and Management (IBSRAM) Sloping Lands Network, and the Multipurpose Tree Species (MPTS) Network.

The major challenge is to evolve new institutional arrangements that direct research toward the upland ecosystem as a totality. A focus on the Southeast Asian upland ecosystem does not fall within the mandate of any of the Consultative Group on International Agricultural Research (CGIAR). But there are major CGIAR initiatives in forestry (Center for International Forestry Research) and agroforestry (the Southeast Asian regional program of the International Centre for Agroforestry Research). Nevertheless, there remains concern that such efforts may address only components of the upland ecosystem, whereas the key to eventual success lies in coping with the interrelatedness of the problems across sectors and in developing the capacity to strengthen each country's research and development institutions to conceptualize, plan, and implement interventions that are appropriate to each ecosystem. This will require a novel upland ecosystem-based approach to international research. The evolving concept of ecoregional research (Consultative Group on International Agricultural Research, Technical Advisory Committee, 1991), under which a consortium of international centers is planning a joint long-term effort to develop alternatives to shifting (slash-and-burn) agricultural systems, represents a promising mechanism for providing this leadership.

DELIVERY: INSTITUTIONAL CHANGE, PROGRAMS, AND POLICY

Implement Institutional Changes   DENR has recognized that its future role will be primarily in development, replacing its historical role as a regulatory agency. It acknowledges that development and management of production forests and plantation forestry are the domain of the private sector and that it should support and guide this transition (Department of Environment and Natural Resources, 1990). Such a role will require a fundamental restructuring of DENR's administration, policy framework, and staff technical capabilities and attitudes. The recent enactment of the Local Government Code requires the transfer of many DENR functions to local government units, decentralizing resource management and giving much greater authority to local leaders.

The redirection of DENR must specifically include a systematic strengthening of forestry policy and planning capabilities, for which there is substantial support expected from international donors. Operations will need to be further decentralized, with much greater accountability and resources at the local level.

DENR has consistently claimed exclusive control over public forestlands, 55 percent of the land area of the country. However, the majority of that land is devoted to agricultural pursuits, not forestry. The development of sustainable upland agricultural systems is a task for which the Department of Agriculture has a much stronger capability. DENR should recognize the potential role of the Department of Agriculture in providing agricultural and agroforestry research and extension services. Within the past several years, the Department of Agriculture has reoriented its priorities to give much greater attention to upland agriculture. A much greater level of support for upland technology development and extension is required to widen this role.

Vigorously Implement the Master Plan for Forestry Development
The Master Plan for Forestry Development (Department of Environment and Natural Resources, 1990) marks a fundamental turning point in the philosophy and methodology of forest management in the Philippines. It provides a basis for a range of reforms and restructuring that is essential to future forest preservation and sustainable land use systems. The master plan contains unrealistically optimistic projections for trends in forest cover, but it provides a framework for the kind of comprehensive, directed effort that is necessary.

Enforce Timber Pricing Reform and Logging Ban   New fees for timber cutting based on recent legislation have been increased to 25 percent of the actual market price (for example, for logs with a price of P2,000 [US$80.00] per cubic meter, the fee is P500 [US$20.00]). It remains to be seen how effective the government will be in collecting the increased fees and using them to increase forest protection and management expenditures.

A major national debate on a total logging ban occurred in 1991. DENR directives in 1991 instituted a ban in old-growth dipterocarp forests. Logging in secondary-growth forests was restricted to lands with slopes of less than 50 percent and land less than 1,000 m above sea level. Enforcement of these policies will be impossible, however, unless greater investments in enforcement procedures are made and forest occupants are directly involved in forest preservation through limited use of the forests. An integrated protected area system for the conservation of the most important natural habitats is under development. NGOs are seen to be the key to the successful implementation of this effort. They will assume responsibility for the management of national parks, wildlife refuges, and other wild lands.

Give Priority to People-Oriented Forestry   Now that regulation of the forests by the national government has been acknowledged to be inadequate, forest protection through empowerment of people and their communities is officially accepted as the only workable model. Implementation of a successful community forestry program will be an immense organizational task that will require a strong commitment by the forest occupants and upland farmers. Capable NGOs will be a key to the program. If further conversion of forests to agricultural uses is blocked through effective community enforcement and shifting cultivation is to decline, there must be agricultural innovation to maintain viable farming systems on the lands surrounding the forests. The equitable capture of income from the limited harvest of forest products will be crucial to financing this transition.

The implementation of current policy will turn the primary responsibilities for forest protection, tree production, and land conservation over to upland communities, NGOs, and individuals. This grass roots approach will open a new era in the management of the uplands. However, it may not be any more effective in forest conservation than a top-down approach unless local management entities receive appropriate support to develop the complex skills needed to guide their efforts. Community-based organizations will require professional guidance to achieve even minimal management capabilities.

NGOs will be involved in implementing many of the new people-oriented forestry programs. They are working as partners with DENR in contracting reforestation and community forest management projects. Eventually, they might form local environment and natural resource centers that would assist the national government in training and on-farm research. Only a few NGOs are competent to handle community-based resource management on a large scale. A major priority of national and international support must be to strengthen NGOs.

The Timber License Agreements (TLAs), by which logging rights are allocated, need thorough reform. Long-term security is essential to engendering a sustainable management perspective among private forest managers. The national government, however, has the tendency to cancel leases on areas peremptorily, sometimes without due process. Many TLA holders continually fear the cancellation of their leases as political circumstances change, with the consequent loss of their fixed investments in processing plants, infrastructure, and forest development in their areas. Moreover, the total 50-year lease period (an initial 25 years that is renewable for another 25 years) does not provide sufficient time for responsible firms to practice sustained forest management. Dipterocarp forests require at least 30 to 40 years for each cutting cycle, and cutting cycles are often much longer. To overcome the destructive short-term perspective, longer lease periods will be necessary. However, these will be accompanied by much stricter enforcement of sustainable forestry practices, making the threat of cancellation solely contingent on quantifiable performance standards. TLAs will be given to only a few firms that demonstrate a people-oriented management focus.

Coordinate International Donor Imperatives   Foreign assistance has been critical in all facets of the change toward people-oriented forestry and forest policy reform that has emerged in the Philippines in the recent past. The Ford Foundation's sustained support for research on social forestry developed the knowledge and institutional base for government to test the concept. Innovative projects supported by the U.S. Agency for International Development (particularly the Rainfed Resources Development Project) and the World Bank enabled new models of upland management to be implemented on a trial basis. Because the administrative and policy environment has shifted in a favorable direction, international aid to ensure the success of new models will be even more crucial.

The overall effort needs a comprehensive blueprint for sustainable upland management. The Master Plan for Forestry Development (Department of Environment and Natural Resources, 1990) is an important step in this direction. A coordinated donor approach to upland development could assist in rationalizing the priorities and ensuring that the effort is comprehensive and consistent. The redirection of programs within DENR and the Department of Agriculture will place tremendous pressure on their limited staffs and resources. It is essential that staff supported by international projects be equally distributed among programs managed by the departments. However, project aid should be contingent on identifiable progress made by the national government in implementing policy and institutional change over a set period of time. NGOs are envisioned to assume a vastly greater role in upland development.

Deforestation Scenarios

The Master Plan for Forestry Development (Department of Environment and Natural Resources, 1990) is an appropriate starting point for anticipating future land use scenarios in the Philippine uplands. The plan recognizes the limitations of past forestry management and attempts to formulate a macrolevel plan to change the nature of the forestry sector. Specifically (Department of Environment and Natural Resources, 1990:60),

> the forestry sector of the country will be directed in the long run towards a condition whereby all of the forest resources will be under efficient and equitable management, conservation, and utilization, satisfying in appropriate ways and on a sustainable basis the needs of the people for forest-based commodities and services.

The master plan presents three scenarios to the year 2015 based on (1) a continuation of the status quo, (2) the implementation of a total logging ban, and (3) the implementation of the master plan. If implemented, the master plan would provide for extensive reforestation, continued logging of secondary forests on a commercial scale, and an aggressive integrated social forestry program. The estimated increase in total protection and production forests would be from 6.693 million ha in 1990 to 8.422 million ha in 2015.

Several major shortcomings of the plan have led to overly optimistic projections. The master plan states that total forest cover in 1990 was 6.694 million ha; however, the Philippine–German Forest Resources Inventory Project (Forest Management Bureau, 1988) concluded that forest cover in 1988 was only 6.461 million ha. The master plan may have started with a larger forest base than is justified.

The master plan assumed a deforestation rate of 88,000 ha/year in 1990. The Philippine–German Forest Resources Inventory Project (Forest Management Bureau, 1988) determined the deforestation rate to have been 210,300 ha/year between 1969 and 1988 and suggested a rate of about 130,000 ha/year in 1987–1988. Kummer (1990) calculated the rate to have been 157,000 ha/year from 1980 to 1987. It is likely that the current deforestation rate is significantly greater than the master plan's assumption.

The master plan indicates that reforestation increased from 40,000 ha in 1987 to 131,000 ha in 1989. Such a rapid increase appears optimistic, considering the actual maximum plantation survival rates of 50 to 70 percent. The sustainability of such rates is also uncertain. The master plan also assumes that secondary forests can be managed effectively to achieve sustained yields. Little evidence is available to support this, particularly the expectation that selectively logged forests can be returned to their full stocks in 20 to 40 years.

Overall, the master plan does not adequately address the numerous constraints that may limit its success. Given the past failure of Philippine forest management, the current political and economic uncertainties, and the sustained commitment of personnel and resources that is necessary, the master plan appears to be overly optimistic, even if one were to assume a best-case scenario.

Table 10 presents three scenarios of projected trends in the natural forest cover of the Philippines. These estimates were constructed to envelop the range of forested areas that may be expected. The baseline scenario assumes a current rate of forest loss of 125,000 ha/year that gradually decreases to 25,000 ha/year by 2015. It assumes that it will be about a decade before there is an effective capability to enforce policies that limit either old-growth or secondary forest loss and that a moderate rate of reforestation (75,000 ha/year) will begin to significantly reduce the pressure on the natural forest after 2000.

The worst-case scenario assumes that the political and economic fortunes of the Philippines will deteriorate during the 1990s. Reforestation rates would decline to 25,000 ha/year (Table 11). Natural forest cover loss would continue to exceed 100,000 ha/year into the first decade of the twenty-first century because of the lack of enforcement capability and political uncertainty. The natural forest cover would be reduced to 3.32 million ha by 2015.

TABLE 10   Scenarios of Natural Forest Cover in the Philippines, 1990–2015

Scenario and Year	Average Forest Loss Per Year (ha)	End of Period Forest (ha)	Annual Loss of Forest Cover (percent)[a]	Cumulative Loss of Forest Cover (percent)[a]
Baseline				
1990–1995	125,000	5,575,000	2.20	11
1995–2000	100,000	5,075,000	1.80	20
2000–2005	75,000	4,700,000	1.60	27
2005–2010	50,000	4,450,000	1.00	32
2010–2015	25,000	4,125,000	0.40	36
Worst case				
1990–1995	125,000	5,575,000	2.20	11
1995–2000	125,000	4,950,000	2.40	22
2000–2005	125,000	4,325,000	2.60	36
2005–2010	100,000	3,822,500	2.40	46
2010–2015	100,000	3,325,000	2.80	59
Best case				
1990–1995	100,000	5,700,000	1.60	8
1995–2000	75,000	5,325,000	1.40	15
2000–2005	50,000	5,075,000	1.00	20
2005–2010	25,000	4,950,000	0.40	22
2010–2015	10,000	4,900,000	0.20	23

NOTE: These scenarios are for all natural forests. They do not include plantations, and no attempt was made to provide detail on specific forest types. The total land area of the Philippines is approximately 30 million ha.

[a]Percent rates of change are calculated by dividing the absolute loss of forest cover by the average forest cover for the period in question; that is, the denominator is determined by adding forest cover at the beginning and end of the period and dividing by two.

TABLE 11   Alternative Reforestation Scenarios of Natural and Plantation Forests in the Philippines, 1990–2015 (Hectares)

Year	Baseline		Worst Case		Best Case	
	Ref	Def	Ref	Def	Ref	Def
1990	75,000	125,000	25,000	125,000	100,000	125,000
1995	75,000	100,000	25,000	125,000	100,000	75,000
2000	75,000	75,000	25,000	125,000	100,000	50,000
2005	75,000	50,000	25,000	100,000	100,000	25,000
2010	75,000	20,000	25,000	100,000	100,000	10,000
1990–2015	1,875,000	1,875,000	625,000	2,877,500	2,500,000	1,425,000

NOTE: Ref, net reforestation (area is established and viable); Def, net deforestation (net loss of natural forest cover).

TABLE 12   Estimates of Forest Cover in 2015 Based on Three Scenarios (in Hectares)

Forest Type	Baseline	Worst Case	Best Case	Master Plan
Natural	4,125,000	3,325,000	4,900,000	5,400,000
Plantation	2,275,000	1,025,000	2,900,000	3,000,000
Total	6,400,000	4,350,000	7,800,000	8,400,000

In the best case scenario, it is assumed that the master plan will be largely successful. Substantial annual reforestation (100,000 ha/year) will occur, and deforestation will drop to negligible levels by 2015. The natural forest cover at that time would be 4.90 million ha. This compares with the 5.40 million ha estimated to result from full implementation of the master plan (Table 12). The master plan assumes a confluence of numerous optimistic assumptions in limiting natural forest losses, for which the cumulative probability is low. However, the two scenarios provide similar estimates for the area of coverage achieved in forest plantations by the year 2015 (2.90 million versus 3.00 million ha), up from less than 0.50 million ha in 1991. The Philippines will be highly dependent on the successful expansion of plantation forestry to avoid the complete loss of natural forest cover.

SUMMARY

The next 30 years will be a crucial period for the Philippines. Recognition is dawning that many aspects of life will be changed. The land frontier that had always existed as a safety valve for poor and dispossessed people has disappeared during the present generation. The forest resources that had seemed virtually inexhaustible were expended in a prodigal manner. Yet, the population that relies on extractable resources from the uplands is growing as rapidly as ever. The ecologic balance has been lost, and national awareness of the dire implications of this loss is only beginning to emerge. It is difficult for a country to learn how to cope with circumstances in which all of the old assumptions are overturned. Such a serious crisis, however, also offers opportunities to take bolder steps than would be politically feasible in better times. It will be a period in which the willingness to experiment with new solutions will grow.

What is the desired vision of the state of the uplands in 2015 emerging from the current national debate? It is one of a much denser upland population than was previously anticipated. However, uplanders will be involved in managing forestlands and farmlands in novel ways. Families that occupy upland farms will have a form of secure land tenure by which they can gain credit to intensify and diversify their farming systems. Perennial and tree cropping systems will be common enterprises and will be integrated with livestock and food crop production. Cropping systems will use improved cultivars along with soil fertility-enhancing fertilizer and lime amendments and will be practiced on slopes that are naturally terraced with vegetative barriers. The structural transformation of the national economy will have occurred, and the population of the rural uplands will gradually have begun to decline.

In 2015, large areas of degraded grasslands will be managed as farm forests planted by individuals and communities under secure land tenure agreements. The natural production forests will be managed by local communities—with guidance from professional foresters—by using low-disturbance logging methods with animal labor. Indigenous communities will have secure control of their ancestral lands. The preservation forests and protected areas will be managed by communities and NGOs in collaboration with the national government. Much of the Philippines' remaining biodiversity will have been lost in this period, but protection will have stabilized some of the most representative habitats.

Such a picture of the future of the uplands may be overly optimistic. It embodies landmark changes in philosophy and policy that are now accepted by the national government and some that are already part of existing programs. The critical concern, however, is whether the political will and the management capacity can be developed to thoroughly implement the changes. During the years between now and then, judicious international assistance in research, training, policy, and financing will be critical.

REFERENCES

Abad, R. G. 1981. Internal migration in the Philippines: A review of research findings. Philippine Stud. 29:129–143.

Abejo, S. 1985. Migration to and from the National Capital Region: 1975–1980. J. Philippine Statist. 36:ix–xxii.

Agaloos, B. C. 1976. Aerial photography in forest surveys. Pp. 33–36 in Asian Forestry Industry Yearbook.

Agaloos, B. C. 1984. Silvicultural and logging systems in the Philippines. Pp. 210–233 in Proceedings of the First ASEAN Forestry Congress. Quezon City, Philippines: Bureau of Forest Development.

Aggarwal, P. K., and D. P. Garrity. 1989. Intercropping of legumes to contribute nitrogen in low-input upland rice-based cropping systems. Pp. 209–228 in Nutrient Management for Food Crop Production in Tropical Farming Systems. Haren, Netherlands: Institute of Soil Fertility.

Aggarwal, P. K., D. P. Garrity, S. P. Liboon, and R. A. Morris. 1992. Resource use and plant interactions in a rice-mungbean intercrop. Agron. J. 84:71–78.

Agricultural Policy and Strategy Team. 1986. Agenda for Action for the Philippine Rural Sector. Los Baños: University of the Philippines.

Alano, B. P. 1984. Import smuggling in the Philippines: An economic analysis. J. Philippine Devel. 11:157–190.

Alferez, A. C. 1980. Utilization of leucaena as organic fertilizer to food crops. Second SEARCA Professorial Chair Lecture, Agronomy Department, University of the Philippines, Los Baños, December 16, 1980.

Allied Geographic Section. 1944. Timber resources of the Philippine islands. Allied Geographic Section, Southwest Pacific Area. Unpublished document.

Aquino, B. A. 1987. Politics of Plunder. Quezon City: University of the Philippines.

Asian Development Bank. 1976. The Forest Economy of the Philippines. Manila: Asian Development Bank.

Atienza, R. N. 1989. Research status on management and utilization of acid soils in the Philippines. Paper presented at the Workshop on Management of Acid Soils in Humid Tropical Asia, Kuala Lumpur, Malaysia, January 30 to February 3, 1989.

Baliña, F. T., L. Tung, and A. P. Obusa. 1991. An indigenous soil and water conservation technique observed in Matalom, Leyte, Philippines. Pp. 1–7 in On-Farm Research Notes, Vol. 7. Baybay, Leyte: Visayas State College of Agriculture.

Barker, T. C. 1984. Shifting Cultivation Among the Ikalahans. UPLB-PESAM Working Series No. 1. Los Baños: University of the Philippines.

Barker, T. C. 1990. Agroforestry in the tropical highlands. Pp. 195–227 in Agroforestry: Classification and Management, K. C. MacKicken, and N. T. Vergara, eds. New York: Wiley.

Bartlett, H. H. 1956. Fire, primitive agriculture, and grazing in the tropics. Pp. 692–720 in Man's Role in Changing the Face of the Earth, W. L. Thomas, ed. Chicago: University of Chicago Press.

Basri, I., A. Mercado, and D. P. Garrity. 1990. Upland rice cultivation using leguminous tree hedgerows on strongly acid soils. Paper presented at the American Society of Agronomy, San Antonio, Texas, October 21–26, 1990.

Belsky, J. M., and S. Siebert. 1985. Social stratification, agricultural intensification and environmental degradation in Leyte, Philippines: Implications for sustainable development. Paper presented at Sustainable Development of Natural Resources in the Third World—An International Symposium, Ohio University, Athens, September 3–6, 1985.

Blanche, C. A. 1975. An overview of the effects and implications of Philippine selective logging on the forest ecosystem. Pp. 97–109 in Proceedings of the Symposium on the Long-Term Effects of Logging in Southeast Asia, R. S. Suparto, ed. Biotrop Special Publication No. 3. Bogor, Indonesia: Regional Center for Tropical Biology.

Boado, E. L. 1988. Incentive policies and forest use in the Philippines. Pp. 165–203 in Public Policies and the Misuse of Forest Resources, R. Repetto and M. Gillis, eds. Cambridge, U.K.: Cambridge University Press.

Bonita, M. L., and A. Revilla. 1977. The Philippines forest resources, 1976–2026. Pp. 3–8 in Project Reports and Technical Papers, Vol. 2. Manila: Development Academy of the Philippines.

Borja, L. J. 1929. The Philippine lumber industry. Econ. Geogr. 5:194–202.

Borlagdan, S. B. 1990. Social forestry in upland Cebu. Pp. 266–283 in Keepers of the Forest: Land Management Alternatives in Southeast Asia, M. Poffenburger, ed. Quezon City, Philippines: Ateneo de Manila University Press.

Briones, A. M., and P. R. Vicente. 1985. Fertilizer usage of indigenous phosphate deposits. I. Application of apatitic phosphate rock for corn and upland rice in a hydric dystrandept. Philippine Agron. 68:1–17.

Bureau of Forestry. 1902. Report of the Bureau of Forestry of the Philippine Islands. Report 7/1/1901–9/1/1902. Manila: Philippine Commission.

Bureau of Soils. 1977. Land Capability Classes. Manila, Philippines: Department of Environment and Natural Resources.

Burgess, P. F. 1971. The effect of logging on hill dipterocarp forests. Malayan Nature J. 24:231–237.

Burgess, P. F. 1973. The impact of commercial forestry on the hill forests of the Malay Peninsula. Pp. 35–38 in Proceedings of the Symposium on Biological Resources and National Development, E. Soepadmo and K. G. Singh, eds. Kuala Lumpur, Malaysia: Malayan Nature Society.

Byron, N., and G. Waugh. 1988. Forestry and fisheries in the Asian-Pacific Region: Issues in natural resource management. Asian-Pacific Econ. Lit. 2:46–80.

Carroll, J. J. 1983. Agrarian reform, productivity and equity: Two studies. Pp. 15–23 in Second View from the Paddy, A. J. Ledesma, P. Q. Makil, and V. A. Miralao, eds. Manila, Philippines: Ateneo de Manila University Press.

Celestino, A. F. 1984. Establishment of Ipil-Ipil Hedgerows for Soil Erosion and Degradation Control in Hilly Land. FSSRI/UPLB-CA Monograph. Los Baños: University of the Philippines.

Celestino, A. F. 1985. Farming systems approach to soil erosion control and management. Pp. 64–70 in Soil Erosion Management, E. T. Craswell, J. V. Remenyi, and L. G. Nallana, eds. ACIAR Proceedings Series 6. Canberra: Australian Center for International Agricultural Research.

Census Office of the Philippine Islands. 1920. Census of the Philippine Islands, 1918. Manila: Bureau of Printing.

Chambers, R. 1986. Normal Professionalism, New Paradigms, and Development. Discussion Paper 227. Brighton, U.K.: Institute of Development Studies, University of Sussex.

Concepcion, M. B., ed. 1983. Population of the Philippines: Current Perspectives and Future Prospects. Manila, Philippines: National Economic Development Authority.

Conklin, H. C. 1957. Hanunoo Agriculture in the Philippines. Forestry Development Paper No. 12. Rome, Italy: Food and Agriculture Organization of the United Nations.

Cook, M. G. 1989. Conservation districts: A model for conservation planning and implementation in developing countries. Paper presented at the International Workshop on Conservation Farming on Hillslopes, Taichung, Taiwan, March 20–29, 1989.

Consultative Group on International Agricultural Research (CGIAR), Technical Advisory Committee. 1991. An Ecoregional Approach to Research in the CGIAR. Rome, Italy: Consultative Group on International Agricultural Research.

Cornista, L. B., F. A. Javier, and E. F. Escueta. 1986. Land Tenure and Resource Use among Upland Farmers. Paper Series No. 2. Los Baños, Philippines: Agrarian Reform Institute.

Cornista, L. B., and E. F. Escueta. 1990. Communal forest leases as a tenurial option in the Philippines uplands. Pp. 134–144 in Keepers of the Forest: Land Management Alternatives in Southeast Asia, M. Poffenburger, ed. Quezon City, Philippines: Ateneo de Manila University Press.

Costello, M. A. 1984. Social change in Mindanao: A review of the research of a decade. Kinadman 6:1–41.

Cruz, C. A., and M. Segura-de los Angeles. 1984. Policy Issues on Commercial Forest Management. Working Paper 84-03. Manila: Philippine Institute for Development Studies.

Cruz, M. C., and I. Zosa-Feranil. 1988. Policy implications of population pressure in Philippine uplands. Paper prepared for the World Bank and Canadian International Development Agency Study on Forestry, Fisheries, and Agriculture Resource Management, University of the Philippines, Los Baños, Philippines.

Cruz, M. C., I. Zosa-Feranil, and C. L. Goce. 1986. Population Pressure and Migration: Implications for Upland Development in the Philippines. Working Paper 86-06. Los Baños, Philippines: Center for Policy and Development Studies.

Dacanay, P. 1943. The Forest Resources of the Philippines. Manila, Philippines: Bureau of Forestry and Fishery.

Dames and Moore International, Louis Berger International, and Institute for Development Anthropology. 1989. Sustainable Natural Resources Assessment—Philippines. Manila, Philippines: U.S. Agency for International Development.

Dar, W. D., and R. R. Bayaca. 1990. The Accelerated Agricultural Production Project—Research and Outreach Subproject (AARP-ROS): Institutionalizing a Community-Based and Participatory Approach in Farming Systems Development in the Philippines. Quezon City, Philippines: Department of Agriculture.

David, C. C. 1982. The impact of economic policies on agricultural incentives. Paper presented at the Development Academy of the Philippines, Manila, October 6, 1982.

David, C. C. 1983. Economic Policies and Philippine Agriculture. Working Paper 83-02. Manila: Philippine Institute for Development Studies.

Department of Environment and Natural Resources. 1990. Master Plan for Forestry Development. Manila, Philippines: Department of Environment and Natural Resources.

Domingo, I. 1983. Industrial pulpwood plantations. A paper presented during the First ASEAN Forestry Congress, Philippine International Convention Center, Manila, October 10–15, 1983.

Duckham, A. N., and G. B. Masefield. 1969. Farming Systems of the World. New York: Praeger.

Dugan, P. C. 1989. Returning the Forests to the People: Addressing Operational and Policy Constraints in Community-Based Forest Management. Manila, Philippines: U.S. Agency for International Development.

The Economist. 1989. A Brazilian tale. 310(7590):31–32.

Edgerton, R. K. 1983. Social disintegration on a contemporary Philippine frontier: The case of Bukidnon, Mindanao. J. Contemp. Asia 13:151–175.

Egerton, J. O. 1953. Notes on Logging in the Philippines. Malayan Forest. 16:146–156.

Evensen, C. L. I. 1989. Alley Cropping and Green Manuring for Upland Crop Production in West Sumatra. Ph.D. dissertation. University of Hawaii, Honolulu.

Fay, C., ed. 1989. Our Threatened Heritage. Manila, Philippines: Solidaridad.

Fearnside, P. M. 1986. Spatial concentration of deforestation in the Brazilian Amazon. Ambio 15:74–81.

Fernandes, E. C. M. 1990. Alley Cropping on Acid Soils. Ph.D. dissertation. North Carolina State University, Raleigh.

Food and Agriculture Organization (FAO). 1946. Forestry and Forest Products: World Situation, 1937–1946. Washington, D.C.: Food and Agriculture Organization of the United Nations.

FAO. 1948. Forest Resources of the World. Washington, D.C.: Food and Agriculture Organization of the United Nations.

Food and Agriculture Organization (FAO) and United Nations Environment Program (UNEP). 1981. Forest Resources of the World. Washington, D.C.: Food and Agriculture Organization of the United Nations.

FAO and UNEP. 1982. Tropical Forest Resources Assessment Project (4 vols.). Rome, Italy: Food and Agriculture Organization of the United Nations.

Forest Development Center. 1985. A 50-Year Development Program for the Philippines. Los Baños, Philippines: Forest Development Center.

Forest Management Bureau. 1988. Natural Forest Resources of the Philippines. Manila, Philippines: Philippine–German Forest Resources Inventory Project.

Fujisaka, S. 1986. Pioneer shifting cultivation, farmer knowledge, and an upland ecosystem: Co-evolution and systems sustainability in Calminoe, Philippines. Philippine Quart. Culture Soc. 14:137–164.

Fujisaka, S. 1989. The need to build upon farmer practice and knowledge: Reminders from selected upland conservation projects and policies. Agroforest. Syst. 9:141–153.

Fujisaka, S. 1990. Has Green Revolution Rice Research Paid Attention to Farmers' Technologies? Los Baños, Philippines: International Rice Research Institute.

Fujisaka, S., and D. P. Garrity. 1988. Developing sustainable food crop farming systems for the sloping acid uplands: A farmer-participatory approach. Pp. 1982–193 in Proceedings of the SUAN IV Regional Symposium on Agroecosystem Research. Khon Kaen, Thailand: Khon Kaen University.

Fujisaka, S., and E. Wollenburg. 1991. From forest to agroforest and logger to agroforester: A case study. Agroforest. Syst. 14:113–130.

Garcia-Padilla, V. 1990. Working Towards LEISA: A Register of Oganizations and Experiences in Low External Input and Sustainable Agriculture in the Philippines. Urdaneta, Philippines: Agtalon.

Garrity, D. P. 1989. Hedgerow systems for sustainable food crop production on sloping lands. Contour (Asia Soil Conservation Network Newsletter) 2:18–20.

Garrity, D. P., and P. Agustin. In press. Historical land use evolution in a tropical acid upland agroecosystem. Agric. Ecosyst. Environ.

Garrity, D. P., and P. E. Sajise. 1991. Sustainable land use systems in Southeast Asia: A regional assessment. Pp. 59–76 in Sustainable Land Use Systems Research, R. D. Hart and M. W. Sands, eds. Kutztown, Pa.: Rodale Institute.

Garrote, B. P., A. Mercado, and D. P. Garrity. 1986. Soil fertility management in acid upland environments. Philippine J. Crop Sci. 11(2):113–123.

Gibbs, C., E. Payauan, and R. del Castillo. 1990. The growth of the Philippine Social Forestry Program. Pp. 253–265 in Keepers of the Forest: Land Management Alternatives in Southeast Asia, M. Poffenburger, ed. Quezon City, Philippines: Ateneo de Manila University Press.

Gillis, M. 1988. The logging industry in tropical Asia. Pp. 177–184 in People of the Tropical Rain Forest, J. Denslow Sloan and C. Padoch, eds. Berkeley: University of California Press.

Granert, W. 1990. Final report of the agroforestry specialist. In Final Report: Technical Assistance for RRDP Natural Resources Component Cycle II. Quezon City, Philippines: Department of Environment and Natural Resources.

Granert, W. G., and T. Sabueto. 1987. Farmers' involvement and use of simple methods: Agroforestry strategies for watershed protection. In Agroforestry in the Humid Tropics, N. T. Vergara and N. Briones, eds. Los Baños, Philippines: East-West Center/Southeast Asian Regional College of Agriculture.

Guevara, A. B. 1976. Management of Leucaena leucocephala (Lam.) de Wet for Maximum Yield and Nitrogen Contribution to Intercropped Corn. Ph.D. dissertation. University of Hawaii, Honolulu.

Guiang, E. 1981. A Critical Analysis of Tree Plantation Ventures in the Philippines. Master's thesis. La Salle University, Manila, Philippines.

Guiang, E. 1991. Community Forestry Program: Assessment and Status of Implementation as of 31 December 1990. Quezon City, Philippines: Department of Environment and Natural Resources.

Guiang, E., and M. Gold. 1990. Use of "Pump-Priming" Strategy to Enhance the Employment-Regenerating Potential of Agroforestry Development: Experiences from the Philippines. East Lansing: Department of Forestry, Michigan State University. Mimeograph.

Gwyer, G. 1977. Agricultural Employment and Farm Incomes in Relation to Land Classes: A Regional Analysis. Technical Paper No. 6. Manila, Philippines: National Economic Development Authority–United Nations Development Program/World Bank Regional Planning Assistance Project.

Gwyer, G. 1978. Developing hillside farming systems for the humid tropics: The case of the Philippines. Oxford Agr. Stud. 7:1–37.

Hackenberg, R., and B. H. Hackenberg. 1971. Secondary development and anticipatory urbanization in Davao, Mindanao. Pacific Viewpoint 12:1–19.

Hainsworth, R. G., and R. T. Moyer. 1945. Agricultural Geography of the Philippines. Washington, D.C.: U.S. Department of Agriculture.

Harrington, L. W., M. D. Read, D. P. Garrity, J. Woolley, and R. Tripp. 1989. Approaches to on-farm client-oriented research: Similarities, differences, and future directions. Pp. 35–53 in Developments in Procedures for Farming Systems Research: Proceedings of an International Workshop. Grand Petit Mountain, Ark.: Winrock International.

Hart, R. D., and M. W. Sands. 1991. Sustainable land use systems research. Pp. 1–12 in Sustainable Land Use Systems Research, R. D. Hart, and M. W. Sands, eds. Kutztown, Pa.: Rodale Institute.

Herrin, A. N. 1985. Migration and agricultural development in the Philippines. Pp. 369–391 in Urbanization and Migration in ASEAN Development, P. M. Hauser, D. B. Suits, and N. Ogawa, eds. Tokyo: National Institute for Research Advancement.

Hicks, G. L., and G. McNicoll. 1971. Trade and Growth in the Philippines. Ithaca, N.Y.: Cornell University Press.

Hill, H., and S. Jayasuriya. 1984. Philippine economic performance in regional perspective. Contemporary Southeast Asia 6:135–158.

Hooley, R., and V. W. Ruttan. 1969. The Philippines. Pp. 215–250 in Agricultural Development in Asia, R. T. Shand, ed. Berkeley: University of California Press.

Institute of Population Studies. 1981. Migration in Relation to Rural Development: ASEAN Level Report. Bangkok, Thailand: Chulalongkorn University.

International Labour Office. 1974. Sharing in Development: A Programme of Employment, Equity and Growth for the Philippines. Geneva: International Labour Office.

International Rice Research Institute (IRRI). 1987. Annual Report for 1986. Los Baños, Philippines: International Rice Research Institute.

IRRI. 1991. Program Report for 1990. Los Baños, Philippines: International Rice Research Institute.

Johnson, N., and J. Alcorn. 1989. Ecological, Economic and Development Values of Biological Diversity in Asia and the Near East. Washington, D.C.: U.S. Agency for International Development.

Kang, B. T., L. Reynolds, and A. N. Atta-Krah. 1990. Alley farming. Advances Agron. 43:315–359.

Kerkvliet, B. J. 1974. Land reform in the Philippines since the Marcos coup. Pacific Affairs 47:286–304.

Kikuchi, M., and Y. Hayami. 1978. Agricultural growth against a land resource constraint: A comparative history of Japan, Taiwan, Korea and the Philippines. J. Econ. Hist. 38:839–864.

Krinks, P. 1974. Old wine in a new bottle: Land settlement and agrarian problems in the Philippines. J. Southeast Asian Stud. 5:1–17.

Kummer, D. 1990. Deforestation in the Post-War Philippines. Ph.D. dissertation. Boston University, Massachusetts.

Kummer, D. 1992. Deforestation in the Postwar Philippines. Chicago, Ill.: University of Chicago Press.

Lal, R. 1990. Soil Erosion in the Tropics. New York: McGraw-Hill.

Lopez-Gonzaga, V. 1987. Capital Expansion, Frontier Development, and the Rise of Monocrop Economy in Negros (1850–1898). Occasional Paper No. 1. Bacolod, Philippines: La Salle University.

Luning, H. A. 1981. The Need for Regionalized Agricultural Development Planning: Experiences from Western Visayas, Philippines. Los Baños, Philippines: Southeast Asian Regional College of Agriculture.

Ly, Tung. 1990. FARMI Newsletter. Baybay, Philippines: Farm and Resource Management Institute, Visayas State College of Agriculture.

Lynch, O. J., and K. Talbott. 1988. Legal responses to the Philippine deforestation crisis. J. Int. Law Politics 20:679–713.

MacDicken, K. G. 1990. Agroforestry management in the humid tropics. Pp. 99–149 in Agroforestry: Classification and Management, K. G. MacDicken and N. T. Vergara, eds. New York: Wiley.

Magbanua, R. D., R. O. Torres, and D. P. Garrity. 1988. Crop residue management to sustain productivity. Paper presented at the 4th Annual Scientific Meeting of the Federation of Crop Science Societies of the Philippines, Davao City, Philippines, April 27–30, 1988.

Magrath, W. B., and J. B. Doolette. 1990. Strategic issues in watershed development. Pp. 1–34 in Watershed Development in Asia. World Bank Technical Paper No. 127. Washington, D.C.: World Bank.

McIntosh, J. L., I. G. Ismail, S. Effendi, and M. Sudjadi. 1981. Cropping systems to preserve fertility of red-yellow podzolic soils in Indonesia. Pp. 409–429 in International Symposium on Distribution, Characterization, and Utilization of Problem Soils. Tsukuba, Japan: Tropical Agriculture Research Center.

Mercado, A. R., A. M. Tumacas, and D. P. Garrity. 1989. The establishment and performance of tree legume hedgerows in farmer's fields in a sloping acid upland environment. Paper presented at the 5th Annual Scientific Meeting of the Federation of Crop Science Societies of the Philippines, Iloilo City, Philippines, April 26–29, 1989.

Mercado, A., Jr., A. Montecalvo, D. P. Garrity, and I. H. Basri. 1992. Upland rice and maize response in a contour hedgerow system on a sloping acid upland soil. Paper presented at the 8th Annual Scientific Meeting of the Federation of Crop Science Societies of the Philippines, Zamboanga City, Philippines, May 24–28, 1992.

Myers, N. 1984. The Primary Source. New York: Norton.

National Census and Statistics Office. 1980. Population, Land Area, and Density: 1970, 1975, and 1980. Manila: National Census and Statistics Office.

National Census and Statistics Office. 1985. 1980 Census of Agriculture: National Summary. Manila, Philippines: National Census and Statistics Office.

National Economic Council. 1959. The Raw Materials Resources Survey: Series No. 1, General Tables. Manila, Philippines: Bureau of Printing.

National Economic Development Authority. 1981. Regional Development: Issues and Strategies on Agriculture. Regional Planning Studies Series No. 3. Manila, Philippines: National Economic Development Authority.

National Research Council. 1977. Leucaena: Promising Forage and Tree Crop for the Tropics. Washington, D.C.: National Academy of Sciences.

Ofreno, R. E. 1980. Capitalism in Philippine Agriculture. Quezon City, Philippines: Foundation for Nationalist Studies.

O'Sullivan, T. E. 1985. Farming systems and soil management: The Philippines/Australian development assistance program experience. Pp. 77–81 in Soil Erosion Management, E. T. Craswell, J. V. Remenyi, and L. G. Nallana, eds. ACIAR Proceedings Series 6. Canberra: Australian Center for International Agricultural Research.

Otsuka, K., V. G. Cordova, and C. C. David. 1990. Modern rice technology and regional wage differentials in the Philippines. Agric. Econ. 4:297–314.

Palmier, L. 1989. Corruption in the West Pacific. Pacific Rev. 2:11–23.

Paner, V. E. 1975. Multiple Cropping Research in the Philippines. Proceedings of the Cropping Systems Workshop. Los Baños, Philippines: International Rice Research Institute.

Pava, H. M., J. B. Arances, I. O. Mugot, J. M. Magallanes, J. M. Manubag, and I. S. Sealza. 1990. The Himaya MUSUAN Experience. Musuan, Philippines: Central Mindanao University.

Pelzer, K. J. 1941. An Economic Survey of the Pacific Area. Part I. Population and Land Utilization. New York: Institute of Pacific Relations.

Pernia, E. M. 1988. Urbanization and spatial development in the Asian and Pacific region: Trends and issues. Asian Devel. Rev. 6:86–105.

Philippine Council for Agriculture and Resources Research and Development. 1982. The Philippines Recommends for Dipterocarp Production. Los Baños: Philippine Council for Agriculture and Resources Research and Development.

Philippine Council for Agriculture and Resources Research and Development. 1986. Sloping agricultural land technology. Technology 8:5/86.

Poblacion, G. 1959. Logging in the Philippines. Filipino Forest. 11:89–106.

Population Reference Bureau. 1990. World Population Data Sheet 1990. Washington, D.C.: Population Reference Bureau.

Porter, G. D., and D. Ganapin. 1988. Resources, Population, and the Philippines' Future. Washington, D.C.: World Resources Institute.

Power, J. H., and T. D. Tumaneng. 1983. Comparative Advantage and Government Price Intervention Policies in Forestry. Working Paper 83-05. Manila: Philippine Institute for Development Studies.

Raintree, J. B., and K. Warner. 1986. Agroforestry pathways for the intensification of shifting cultivation. Agroforest. Syst. 4:39–54.

Repetto, R. 1988. The Forests for the Trees? Government Policies and the Misuse of Forest Resources. Washington, D.C.: World Resources Institute.

Revilla, A. V. 1988. The constraints to and prospects for forest development in the Philippines. Pp. 137–147 in Proceedings of the RP-German Forest Resources Inventory Application of Results to Forest Policy, R. Lennertz, and K. Uebelhor, eds. Quezon City, Philippines: Forest Management Bureau.

Reyes, M. R., and V. B. Mendoza. 1983. The Pantabangan watershed management and erosion control project. Pp. 485–555 in Forest and Watershed Development and Conservation in Asia and the Pacific, L. S. Hamilton, ed. Boulder, Colo.: Westview.

Rocamora, J. E. 1979. Rural development strategies: The Philippine case. Pp. 63–106 in Approaches to Rural Development: Some Asian Experiences, Inayatullah, ed. Kuala Lumpur, Malaysia: Asian and Pacific Development Administrative Center.

Rosenberg, J. G., and D. A. Rosenberg. 1980. Landless Peasants and Rural Poverty in Indonesia and the Philippines. Ithaca, N.Y.: Center for International Studies, Cornell University.

Roth, D. M. 1983. Philippine forests and forestry: 1565–1920. Pp. 30–49 in Global Deforestation and the Nineteenth Century World Economy, R. P. Tucher, and J. R. Richards, eds. Durham, N.C.: Duke University Press.

Sajise, P. E. 1987. Stable upland farming in the Philippines: Problems and prospects. Pp. 633–644 in Impact of Man's Activities on Tropical Upland Forest Ecosystems, Y. Hadi, K. Awang, N. M. Majid, and S. Mohamed, eds. Darul Ehsan, Malaysia: Faculty of Forestry, Universiti Pertanian.

Savonen, C. 1990. Ashes in the Amazon. J. Forest. 88(9):20–25.

Schade, J. 1988. Consequences of the FRI for forest policy. Pp. 95–102 in Proceedings of the RP-German Forest Resources Inventory Application of Results to Forest Policy, R. Lennertz and K. Uebelhor, eds. Quezon City, Philippines: Forest Management Bureau.

Segura-de los Angeles, M. 1985. Economic and social impact analysis of an upland development project in Nueva Ecija, Philippines. J. Philippine Devel. 12:324–394.

Serevo, T., F. Asiddao, and M. Reyes. 1962. Forest resources inventory in the Philippines. Philippine J. Forest. 18:1–19.

Smyle, J., W. Magrath, and R. G. Grimshaw. 1990. Vetiver grass—A hedge against erosion. Paper presented at the American Society of Agronomy, San Antonio, Texas, October 21–26, 1990.

Southgate, D., and D. Pearce. 1988. Agricultural Colonization and Environmental Degradation in Frontier Developing Economies. Working Paper No. 9. Washington, D.C.: World Bank.

Sulit, C. 1947. Forestry in the Philippines during the Japanese occupation. Philippine J. Forest. 5:22–47.

Sulit, C. 1963. Brief history of forestry and lumbering in the Philippines. J. Amer. Chamber Com. Philippines 39:16–24.

Swedish Space Corporation. 1988. Mapping of the Natural Conditions of the Philippines. Solna: Swedish Space Corporation.

Szott, L. T., C. A. Palm, and P. A. Sanchez. 1991. Agroforestry in acid soils of the humid tropics. Advances Agron. 45:275–300.

Talbot, L. M., and M. A. Talbot. 1964. Renewable Natural Resources in the Philippines—Status, Problems, and Recommendations. Manila, Philippines: International Union for the Conservation of Nature and Natural Resources.

Tamesis, F. 1937. General Information on Philippine Forests. Manila, Philippines: Bureau of Printing.

Tamesis, F. 1948. Philippine forests and forestry. Unasylva 6:316–325.

Thapa, B. B. 1991. Evaluation of Infiltration, Surface Runoff, and Soil Losses at Various Levels of Rainfall in Relation to Surface Cover, Tillage and Soil Management Practices. Master's thesis. University of the Philippines, Los Baños.

Thung, H. L. 1972. An Evaluation of the Impact of a Highway on a Rural Environment in Thailand by Aerial Photographic Methods. Ph.D. dissertation. Cornell University, Ithaca, New York.

Tiongzon, M. L., A. Regalado, and R. Pascual. 1986. Philippine Agriculture in the 70s and 80s: TNC's Boon, Peasants' Doom. Agricultural Policy Studies No. 2. Quezon City, Philippines: Philippine Peasant Institute.

Torres, R. O., and D. P. Garrity. 1990. Evaluation of batao (Lablab purpureus L. Sweet) as a dual purpose component of upland cropping patterns. Paper presented at the 6th Annual Scientific Meeting of the Federation of Crop Science Societies of the Philippines, Naga City, Philippines, May 16–18, 1990.

Torres, R. O., R. D. Magbanua, and D. P. Garrity. 1988. Evaluation of cowpeas as legume component in acid upland rice-based cropping systems. Philippine J. Crop Sci. 13(2):91–98.

Torres, R. O., D. P. Garrity, R. J. Buresh, R. K. Pandey, R. T. Bantilan, F. M. Tumacas, and A. Montecalvo. 1989. Production in rice-based rainfed upland cropping systems. Paper presented at the 5th Annual Scientific Meeting of the Federation of Crop Science Societies of the Philippines, Iloilo City, Philippines, April, 26–28, 1989.

Umali, R. M. 1981. Forestland assessment and management for sustainable uses in the Philippines. Pp. 289–301 in Assessing Tropical Forestlands: Their Suitability for Sustainable Uses, R. A. Carpenter, ed. Dublin: Tycooly.

U.S. Agency for International Development. 1980. Preliminary Analysis of Philippine Poverty as a Base for a U.S. Assistance Strategy. Manila, Philippines: U.S. Agency for International Development.

U.S. Bureau of the Census. 1905. Census of the Philippine Islands, 1903. Washington, D.C.: U.S. Government Printing Office.

Vandermeer, C. 1963. Corn cultivation on Cebu: An example of an advanced stage of migratory farming. J. Trop. Geogr. 17:172–177.

Vandermeer, C., and B. C. Agaloos. 1962. Twentieth century settlement of Mindanao. Papers Michigan Acad. Sci. Arts Lett. 47:537–548.

van Oosterhout, A. 1983. Spatial conflicts in rural Mindanao, the Philippines. Pacific Viewpoint 24:29–49.

Vergara, N. T. 1982. Sustained outputs from legume tree-based agroforestry systems. In New Directions in Agroforestry: The Potential of Tropical Legume Trees, N. T. Vergara, ed.

Watson, H. R., and W. L. Laquihon. 1987. Sloping agricultural land technology: An agroforestry model for soil conservation. In Agroforestry in the Humid Tropics, N. T. Vergara, and T. N. Briones, eds. College, Laguna, Philippines: East-West Center/Southeast Asian Regional College of Agriculture.

Weidelt, H. J., and V. S. Banaag. 1982. Aspects of Management and Silviculture of Philippine Dipterocarp Forests. Eschborn, Federal Republic of Germany: German Agency for Technical Cooperation.

Wernstedt, F. L., and P. D. Simkins. 1965. Migration and the settlement of Mindanao. J. Asian Stud. 25:83–103.

Wernstedt, F. L., and J. E. Spencer. 1967. The Philippine Island World. Berkeley: University of California Press.

Westoby, J. 1981. Who's Deforesting Whom? IUCN Bull. 14:124–125.

Westoby, J. 1989. Introduction to World Forestry. Oxford: Basil Blackwell.

Whitford, H. N. 1911. The Forests of the Philippines. Bulletin No. 10. Manila, Philippines: Bureau of Forestry.

Whitmore, T. C. 1984. Vegetation map of Malesia at scale of 1:5,000,000. J. Biogeogr. 11:461–471.

Williams, R. D., and E. D. Lavey. 1986. Selected Buffer Strip References. Durant, Okla.: Water Quality and Watershed Research Laboratory.

World Bank. 1989a. Philippines: Environment and Natural Resource Management Study. Washington, D.C.: World Bank.

World Bank. 1989b. Philippines: Toward Sustaining the Economic Recovery. Washington, D.C.: World Bank.

World Resources Institute. 1990. World Resources, 1990–91. Washington, D.C.: World Resources Institute.

Wurfel, D. 1983. The development of post-war Philippine land reform: Political and sociological explanations. Pp. 1–14 in Second View from the Paddy, A. Ledesma, P. Makil, and V. Miralao, eds. Manila, Philippines: Ateneo de Manila University Press.

Zon, R. 1910. The Forest Resources of the World. Washington, D.C.: U.S. Government Printing Office.

Zon, R., and W. N. Sparhawk. 1923. Forest Resources of the World. New York: McGraw-Hill.

Zosa-Feranil, I. 1987. Persisting and Changing Patterns of Population Redistribution in the Philippines. Quezon City: Population Institute, University of the Philippines.

Zaire

Mudiayi S. Ngandu and Stephen H. Kolison, Jr.

Zaire is located directly on the equator in the central part of the African continent. It is the third largest country in Africa, with an area of 2,344,885 km², three times the size of the state of Texas. Zaire has three distinct land areas: the tropical rain forests, located in the central and northern parts of the country; the savannahs, located in the northern and southern parts of the country; and the highlands, which consist of the plateaus, rolling meadows, and mountains found along the country's eastern border, all along the Great Rift valley. The highest point in this area is 5,809 m, on Ruwenzori Peak in Kivu Province.

Zaire's rivers and lakes are probably its most important natural resources. The most prominent is the Zaire River (formerly the Congo River). It is the fifth longest river in the world and is second only to the Amazon in the volume of water it carries. The Zaire River flows for about 4,667 km, but together with its tributaries, navigability of up to about 11,500 km is possible. In some parts of the country, however, the Zaire River is not navigable because of falls and rapids. The country also has several deep lakes, including Lake Tanganyika in the southeast.

Mudiayi S. Ngandu is an associate professor of agricultural economics and Stephen H. Kolison, Jr., is an assistant professor and coordinator of the forestry resources program at the School of Agriculture, The George Washington Carver Agricultural Experiment Station, Tuskegee University, Tuskegee, Alabama.

FOREST TYPES

Forest types range from dry semideciduous to swamps. Figure 1 shows the geographic distributions of forestland areas by the four distinguishable types: (1) evergreen rain forests and swamp forests in the central basin; (2) dry and moist semideciduous forests to the north and south of the evergreen forests; (3) montane forests in the eastern uplands on the borders with Tanzania, Rwanda, and Burundi; and (4) woodland and wooded savannahs in the far south. The variety of forest types is due to both soil types and a variety of climatic conditions.

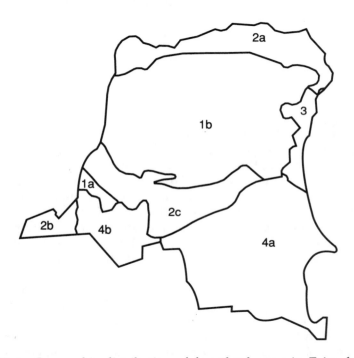

FIGURE 1 Geographic distribution of forestland areas in Zaire, by type. Forest types are as follows: 1. (a) Evergreen rain forests and swamp forests and (b) closed forests of the central basin. 2. (a) Dry semideciduous forests, substantially degraded, and (b and c) moist semideciduous forests of Mayumbe in the lower Zaire River region. 3. Montane forests of Kivu Province. 4. (a) Open forests, woodlands, and wooded savannahs, mainly in Shaba, and (b) part of Bandundu. Source: Government of Zaire and Canadian International Development Agency. 1990. Plan d'Action Forestier Tropical. Vol. I, Annex 2, Forestry Map. Kinshasa, Zaire, and Ottawa, Canada: Government of Zaire and Canadian International Development Agency.

CLIMATE

Climatic conditions vary with almost each region of the country. In the tropical rain forests, average annual rainfall reaches 220 cm, and the average daytime temperature is about 30°C. The equator runs through the center of this region, and the weather is hot and humid throughout the year. In the savannahs, the average annual rainfall is about 120–160 cm, and the average daytime temperature is 24°C. The climate of the highlands is characterized by an average daytime temperature of about 21°C and average annual rainfall of about 160–240 cm.

POPULATION

In 1988, Zaire had a population of about 35.4 million (Table 1) and an estimated annual population growth rate of 3 percent. The estimated population for 1991 was 39.2 million for an average population density of about 14 people per km². The population of Zaire is about 30 percent urban and 70 percent rural. Kinshasa, the capital and largest city, has a population of about 5 million. Matadi, in the Zaire delta (formerly the Congo), is the major port for exports. (For more information, see U.S. Department of State [1988].)

Society and Culture

There are about 700 local languages and dialects spoken in Zaire. Four of these—Lingala, Swahili, Tshiluba, and Kikongo—serve as official languages, in addition to French, which was introduced by the Belgians. All 700 languages belong to the Bantu group of languages. French is used in schools and in conducting official business and is used in particular by those with about 8 years or more of schooling. As regards religion, the U.S. Department of State (1988) noted that the population is about 80 percent Christian (Roman Catholics, Protestants, and indigenous Christians), and 10 percent syncretic and traditional religions.

LAND TENURE

Zaire has two recognized land tenure systems: the modern and the customary. Under the modern system, all land is owned by the government. The right to use land is therefore assigned or given by the government through the Department of Land Affairs, Environment, Nature Conservation, and Tourism (DLAENCT). In many parts of the country, however, the customary land tenure system is used.

TABLE 1 Selected Macroeconomic Indicators and Agricultural Statistics for Zaire

Indicator or Statistic	Units	1970	1975	1980	1985	1987	1988	1989
Macroeconomic indicators								
Population	Million	20.3	23.2	26.7	31.1	33.2	34.3	35.4
Rate of growth	Percent/year	2.8	2.7	2.8	3.1	3.3	3.3	3.3
Agricultural population	Percent	75.9	72.7	70.7	67.8	66.5	65.9	—
Pop/km² of total area	No.	87.0	99.0	114.0	133.0	141.0	146.0	151.0
Total labor force	Percent	42.8	40.9	39.1	37.9	37.3	37.0	—
Agricultural labor force	Percent	79.1	75.4	71.5	68.7	67.5	66.9	—
GNP total	Million current $	3,653	6,963	11,473	4,976	4,976	5,826	—
Rate of growth	Percent/year	—	13.8	10.5	-15.4	—	17.1	—
GNP per capita	$/person	80	300	430	160	150	170	—
GDP in agriculture	Percent	16.3	18.8	28.8	32.9	32.2	31.0	—
Consumer price index	1985 = 100	—	1	16	100	279	510	1,041
Energy conservation/coal equity	1,000 megatons	1,438	1,743	1,948	2,043	2,134	2,211	—
Energy conservation/person/year	kg of coal	?	75	73	66	64	65	—
Energy conservation/$1,000 of GNP	kg of coal	394	250	170	411	429	380	—
Exchange rate	Local currency/$	500	500	2,800	49,873	112,403	187,070	381,445
Total debt	Million $	—	2,152	4,462	5,117	6,497	7,055	—
Total debt service	Million $	—	208	393	365	302	400	—
Debt service ratio	Percent	—	25	24	38	31	36	—
Consumption indicators								
Per capita util-cereals	kg/year	34	43	42	41	42	42	40
Calorie intake/person/day	Calories	2,141	2,162	2,043	2,049	2,046	2,001	—
Protein intake/person/day	Grams	33	32	29	30	30	30	—
Factors of production								
Total land	1,000 ha	234,541	234,541	234,541	234,541	234,541	234,541	234,541
Arable and permanent cropland	1,000 ha	7,250	7,450	7,600	7,800	7,850	7,850	—
Permanent pastures	1,000 ha	15,000	15,000	15,000	15,000	15,000	15,000	—
Irrigated area	1,000 ha	—	7	7	9	9	—	—
Agricultural labor/1,000 ha	No.	947	961	981	1,037	1,065	—	—
Tractors/1,000 ha	No.	—	—	—	—	—	—	—

Tractors/1,000 agricultural workers	No.	—	—	—	—	—	—	—
Fertilizer use	kg/ha	—	—	—	—	—	—	—
Production values								
Total agriculture	Million $	1,920	2,131	2,207	2,567	2,675	2,740	—
Total food	Million $	1,739	1,946	2,046	2,363	2,454	2,515	—
Total crop	Million $	1,811	2,009	2,079	2,430	2,533	2,597	—
Total livestock	Million $	110	122	128	137	141	143	—
Crop production/ha	International $	250	270	274	312	323	—	—
Agricultural production/ agricultural worker	International $	280	298	296	317	320	—	—
Production indices (1979–1981 = 100)								
Total agriculture	Index	87.9	96.5	100.7	115.4	120.5	122.5	
Total agricultural per capita	Index	119.1	113.7	100.7	99.1	97.2	92.7	
Total food	Index	85.6	94.6	100.2	114.1	118.7	121.2	
Total food per capita	Index	115.9	111.4	100.2	98.0	95.8	91.7	
Total crop	Index	86.9	96.4	100.8	116.6	121.9	123.8	
Total livestock	Index	95.6	97.5	99.7	106.7	110.5	114.0	
Crop production/ha	Index	92.1	99.5	101.0	115.0	119.1	—	
Agricultural production/ agricultural worker	Index	95.5	101.6	100.9	108.1	109.1	—	
Production, quantities								
Cereals—total	1,000 Mtons	666	758	862	1,109	1,145	1,155	1,230
Rate of growth	Percent/year	9.2	2.6	2.6	5.2	1.6	.9	6.5
Wheat	1,000 Mtons	3	2	4	20	25	35	35
Rice, paddy	1,000 Mtons	180	208	234	297	300	300	315
Corn (maize)	1,000 Mtons	428	496	594	726	730	730	790
Barley	1,000 Mtons	2	—	—	—	—	—	—
Oats	1,000 Mtons	5	—	—	—	—	—	—
Sorghum	1,000 Mtons	25	27	14	32	60	60	60
Millet	1,000 Mtons	23	25	16	29	30	30	30
Soybeans	1,000 Mtons	2	2	10	17	18	18	18
Cottonseed	1,000 Mtons	46	36	19	50	50	50	50
Peanuts, in shell	1,000 Mtons	267	308	337	386	394	410	400

continued

TABLE 1 Continued

Indicator or Statistic	Units	1970	1975	1980	1985	1987	1988	1989
Palm oil	1,000 Mtons	232	181	168	160	165	178	178
Sugar, ?	1,000 Mtons	46	63	48	63	68	76	70
Roots and tubers	1,000 Mtons	10,861	12,424	13,664	16,232	16,995	17,007	17,051
Pulses	1,000 Mtons	122	145	139	121	129	140	140
Fruits	1,000 Mtons	2,135	2,429	2,421	2,540	2,601	2,669	2,666
Vegetables	1,000 Mtons	385	451	485	515	535	550	555
Coffee	1,000 Mtons	70	83	80	92	102	97	98
Tea	1,000 Mtons	7	7	3	5	6	6	6
Cocoa	1,000 Mtons	6	5	4	4	6	6	6
Meat	1,000 Mtons	170	174	174	186	193	197	200
Beef and veal	1,000 Mtons	18	21	22	24	26	26	26
Mutton and lamb	1,000 Mtons	2	2	2	3	3	3	3
Pork	1,000 Mtons	25	27	27	28	29	29	29
Poultry	1,000 Mtons	11	13	13	14	15	16	16
Eggs	1,000 Mtons	6	7	8	8	8	6	8
Cotton lint	1,000 Mtons	23	18	10	26	26	26	26
Hides and skins	1,000 Mtons	5	5	6	6	—	—	—
Area by selected crops								
Cereals—total	1,000 ha	916	1,014	1,094	1,292	1,356	1,356	1,357
Rate of growth	Percent/year	6.8	2.0	1.5	3.4	2.4	—	—
Wheat	1,000 ha	6	4	4	26	27	28	28
Rice	1,000 ha	244	268	293	334	344	345	345
Corn (maize)	1,000 ha	595	675	743	849	874	874	875
Barley	1,000 ha	3	—	—	—	—	—	—
Oats	1,000 ha	8	—	—	—	—	—	—
Sorghum	1,000 ha	29	32	28	41	66	66	66
Millet	1,000 ha	31	35	26	41	43	42	42
Soybean	1,000 ha	2	2	10	15	15	15	15
Peanuts	1,000 ha	384	432	482	553	567	567	570

Seed cotton	1,000 ha	175	175	175	175	100	151	179
Yields, selected crops								
Cereals—total	kg/ha	907	852	846	859	788	747	727
Wheat	kg/ha	1,250	1,250	926	769	925	492	530
Rice, paddy	kg/ha	913	870	873	889	800	776	736
Corn (maize)	kg/ha	903	835	835	855	800	734	719
Barley	kg/ha	600	600	600	600	600	1,199	790
Oats	kg/ha	—	—	—	—	—	—	625
Sorghum	kg/ha	909	913	915	900	516	841	867
Millet	kg/ha	714	714	700	700	605	725	735
Soybean	kg/ha	1,241	1,241	1,207	1,138	1,012	739	842
Peanuts	kg/ha	702	723	694	698	700	713	695
Seed cotton	kg/ha	440	440	440	440	292	358	390
Yields, livestock products								
Beef and veal	kg/animal	146	147	150	150	149	144	144
Pork	kg/animal	51	50	50	50	50	51	50
Mutton and lamb	kg/animal	10	10	10	11	11	11	11
Milk	kg/animal	883	883	883	855	838	783	754
Eggs	kg/animal	3	3	3	3	3	3	3
Trade								
Self-sufficiency ratio								
Cereals	Percent	70	72	72	75	75	68	74
Food	Percent	—	100	98	100	100	103	102
Agriculture	Percent	—	—	—	99	103	103	—
Value of total exports	Million $	—	1,120	983	949	1,632	826	775
Value of total imports	Million $	—	763	764	793	835	900	535
Trade balance	Million $	—	357	219	156	797	-74	240
Value of agricultural exports	Million $	—	164	212	222	235	192	112
Value of agricultural imports	Million $	—	201	221	237	167	153	63
Agricultural trade balance	Million $	—	-37	-9	-15	68	39	49
Imports from United States	Million $	—	122	103	102	153	187	62
Exports to United States	Million $	—	384	321	415	370	67	41
Agricultural imports from United States	Million $	—	34	35	25	44	18	3
Agricultural exports to United States	Million $	—	5	11	5	31	24	20

continued

TABLE 1 Continued

Indicator or Statistic	Units	1970	1975	1980	1985	1987	1988	1989
Agricultural exports								
Cereals	1,000 $	—	192	—	—	—	—	—
Wheat and flour	1,000 $	—	187	—	—	—	—	—
Rice	1,000 $	—	4	—	—	—	—	—
Corn (maize)	1,000 $	—	1	—	—	—	—	—
Oil crops and feed products	1,000 $	3,180	2,827	1,627	1,800	1,400	1,700	1,700
Vegetable oils	1,000 $	40,964	35,330	15,265	9,235	4,000	7,500	7,500
Fibers	1,000 $	4,602	615	3,500	—	230	370	370
Cotton lint	1,000 $	4,095	615	3,500	—	230	370	370
Other	1,000 $	507	—	—	—	—	—	—
Tobacco and products	1,000 $	22	412	—	—	—	—	—
Vegetables, fruits, and nuts	1,000 $	16	49	—	—	—	—	—
Sugar	1,000 $	75	7	—	—	—	—	—
Beverages	1,000 $	38,368	114,320	174,889	181,440	178,286	123,500	6,500
Coffee, green and roasted	1,000 $	33,900	103,910	166,440	169,640	168,186	116,000	—
Cocoa beans	1,000 $	2,576	6,575	6,941	8,200	8,100	6,500	6,500
Tea and mate	1,000 $	1,892	3,834	1,508	3,600	2,000	1,000	—
Wine and beer	1,000 $	—	1	—	—	—	—	—
Live animals	1,000 $	—	217	—	—	—	—	—
Meat products	1,000 $	—	—	—	—	—	—	—
Dairy products and eggs	1,000 $	—	6	—	—	—	—	—
Agricultural imports								
Cereals	1,000 $	17,644	56,989	84,900	65,200	80,800	80,500	—
Wheat and flour	1,000 $	8,928	23,714	41,400	33,200	39,700	39,500	—
Rice	1,000 $	4,235	3,246	6,500	9,000	18,100	21,000	—
Corn (maize)	1,000 $	4,456	30,000	37,000	23,000	23,000	20,000	20,000
Barley	1,000 $	22	—	—	—	—	—	—

	Unit							
Oil crops and feed products	1,000 $	—	82	—	—	—	—	—
Vegetable oils	1,000 $	133	1,095	45	7,856	6,500	6,000	6,000
Fibers	1,000 $	—	314	6,300	6,000	6,000	6,000	6,000
Cotton lint	1,000 $	—	—	5,400	1,856	500	6,000	6,000
Other	1,000 $	—	314	900	1,856	500	—	—
Tobacco and products	1,000 $	5,029	9,049	1,600	6,968	4,800	6,200	6,200
Vegetables, fruits, and nuts	1,000 $	1,863	1,552	1,330	3,330	3,225	2,580	—
Sugar	1,000 $	1,660	468	8,600	6,296	12,600	6,000	6,000
Beverages	1,000 $	3,182	1,980	1,800	3,304	3,500	3,400	1,800
Coffee, green and roasted	1,000 $	41	31	—	—	—	—	—
Tea and mate	1,000 $	2	11	—	—	—	—	—
Wine and beer	1,000 $	3,139	1,938	1,800	3,304	3,500	3,400	1,800
Live animals	1,000 $	934	1,589	400	260	310	310	—
Meat products	1,000 $	9,744	21,682	18,030	72,430	53,930	40,630	—
Dairy products and eggs	1,000 $	9,285	19,444	18,300	31,292	21,070	22,000	—
Farm inputs	1,000 $	8,853	14,349	17,730	21,500	24,020	—	—
Tractors	1,000 $	3,760	4,606	6,000	9,500	11,000	10,500	—
Fertilizers	1,000 $	1,554	5,652	3,280	2,000	2,200	—	—
Pesticides	1,000 $	1,994	2,606	4,500	5,200	5,700	3,600	—

NOTE: All dollar values are in U.S. dollars. International dollars, a value unit developed by the Food and Agriculture Organization of the United Nations to avoid the use of exchange rates (for example, 1 metric ton of wheat has the same price in whichever country it is produced); GNP, gross national product; GDP, gross domestic product.

SOURCE: U.S. Department of Agriculture, Economic Research Service. 1990. Pp. 541–544 in World Agriculture: Trends and Indicators, 1970–1989. Statistical Bulletin No. 815. Washington, D.C.: U.S. Department of Agriculture.

Under this system, which varies depending on the region and people, land ownership is collective—that is, land is held by groups or clans. The group, through its appointee, assigns land for use to its members. Land used by a family over a long period of time is recognized by the group or clan as belonging to that family, but the family may not sell the land because, in practice, land ownership rights belong, ultimately, to the national government. (This reflects the nature of the existing power relationship between the central government and the local communities.)

THE MACROECONOMIC SETTING

In the 1980s, management of Zaire's macroeconomy was constrained by the heavy external debt-servicing burden (by 1988, as much as 60 percent of exports of goods and services and 65 percent of the operating budget). This debt arose from the country's borrowings in the late 1960s and the 1970s when Zaire's export earnings were relatively higher and expected to grow and the country benefited from favorable terms of trade. With the deterioration in export earnings in the 1980s, a rising debt burden, and the accumulated effects of past economic mismanagement, Zaire, in cooperation with its major creditors, embarked on a series of economic adjustment programs. Unfortunately, these programs were unsuccessful and have resulted in drastic declines in the standard of living, public-sector employment, wages, and salaries. (See Table 1 for selected macroeconomic performance indicators.)

The related tight budgetary measures did not produce results because they were not accompanied by the institutional reforms necessary to strengthen policy formulation and implementation. The forestry sector has been adversely affected by the ongoing economic adjustment programs, and these constraints are likely to continue. Reforms mandated by the economic adjustment programs offer an opportunity to initiate a meaningful dialogue between the Zairian government and the international aid donor community regarding long-term forestry policy issues and deforestation. In this context, debt-for-nature swaps, as proposed for the heavily indebted Latin American countries, should also be applicable to African countries like Zaire (Government of Zaire and the Canadian International Development Agency, 1990; Hines, 1988).

It is, however, a tenuous hypothesis to link deforestation with foreign exchange to service external debt. In a study by Capistrano (1990), external foreign exchange earnings and the external debt-servicing burden were identified as significant macroeconomic factors contributing to accelerated deforestation in a number of countries, including Zaire, during 1967–1989. However, 98 percent of total wood production in Zaire is for domestic consumption and only 1 percent is exported. Deforestation is more appropriately linked to in-country uses of wood. In addition, because of extensive underinvoicing at the Matadi Port and inadequate export statistics related to other leakages, the reliability of Zaire's data on export earnings from logs and other wood products may also be questionable. To date, the forestry sector has not contributed significantly to the country's export strategy or to alleviation of its external debt. These findings encourage strong support for the establishment of a reliable data base as part of any long-term investigation of reforestation in Zaire.

FOREST RESOURCE DISTRIBUTION AND THE STATE OF FOREST MANAGEMENT

Zaire has 207 million of the 436 million ha of forests in central Africa or 47.56 percent of the total in the region that includes Angola, Cameroon, Central African Republic, Congo, Equatorial Guinea, Gabon, and Zaire (see Table 2). In addition, 75 percent of Zaire's national territory is covered by forests. In 1975, Persson (as cited in World Resources Institute [1988]) and the Government of Zaire and the Canadian International Development Agency (1990) estimated that Zaire's total forest cover in 1970 amounted to about 234 million ha, including lakes and rivers (Table 3).

About 101 million ha of closed forests is situated in the central basin and Mayumbe regions (Table 4), while the montane forests occupy about 300,000 ha (Table 3). The band of montane forests spreads from the Haut Zaire Province in the northeast through the Kivu and northern Shaba provinces. The savannah-type formations are found mainly in the northern- and southernmost parts of the country (see Figure 1).

Distributions

COMMERCIAL FOREST AREA

Commercial forestland is classified as that forestland capable of producing at least 20 ft^3 (0.56 m^3) of industrial roundwood per acre (0.4 ha) annually (Blyth et al., 1984). This means that 1 ha of forest should be capable of producing at least 1.4 m^3 of industrial roundwood annually. According to the World Bank (1986), about 139 million ha of forestland in Zaire is commercially exploitable. About

89.43 percent of this estimated area consists of closed forest in the central basin (Food and Agriculture Organization and United Nations Environment Program, 1981b). Furthermore, on the basis of the World Bank report, each hectare is capable of producing 5 m^3 of industrial roundwood annually; this is very different from the 1.4 m^3 estimated by Blyth (1984). (The average of 5 m^3/ha applies to forestland areas that have been logged several times. The figure for the first harvest is on the order of 25–35 m^3/ha. The difference between these two figures is an indication of inefficiencies in logging methods [Food and Agriculture Organization and United Nations Environment Program, 1981b:562–563].)

Although 89.43 percent of the commercial forestlands is situated in the central basin, it does not mean that these forest resources are accessible. In fact, some studies indicate that up to 30 percent of the entire central basin is on waterlogged or seasonally flooded soils, thus making them less attractive for commercial logging (World Resources Institute, 1988).

TABLE 2   Areas of Natural Woody Vegetation in Zaire, 1980 (in Thousands of Hectares)

Vegetation	Area
Tree formations	
Closed	105,750
Open	71,840
All	
Total	177,590
Percent of region	52.87
(Central Africa)	
Percent of country	59.83
Fallow of	
Closed formation	7,800
Open formation	10,600
Shrub formation	11,300
Woody formations and fallows	
Total	207,290
Percent of region	15.80
(Central Africa)	
Percent of country	88.36

SOURCE: Food and Agriculture Organization and United Nations Environment Program. 1981a. Pp. 41–44 in Tropical Forest Resources Assessment Project. Forest Resources of Tropical Africa, Part I. Rome, Italy: Food and Agriculture Organization of the United Nations.

TABLE 3   Types of Forests in Zaire, 1970
(in Thousands of Hectares)

Forest Type, Location	Area
Closed forest, central basin	101,000
Closed forest, Mayumbe	240
Montane forests	300
Subtotal, closed forests	101,540
Dry forest, Shaba region	20,000
Savannah woodland, Guinean type	85,000
Savannah woodland, Sudan-Zambian type	27,000
Gallery forests	760
Subtotal, open forests	132,760
Total, all forest types	234,300

NOTE: Values include the areas of lakes and rivers within the areas of the forest types; there is no allowance for urban or agricultural land use. Note the inconsistencies between these estimates of Persson (1975) and those of the Food and Agriculture Organization of the United Nations (Rome, Italy) in Table 2. Hence the need for reconciling them.

SOURCES: Persson (1975) cited in World Resources Institute. 1988. P. 86 in Zaire Forestry Policy Review and Related Studies. Draft Summary Report. Kinshasa, Zaire, and Washington, D.C.: World Resources Institute; Government of Zaire and Canadian International Development Agency. 1990. Unnumbered appendix table and Table 2.1, p. 19 in Plan d'Action Forestier Tropical. Vols. I and II. Kinshasa, Zaire, and Ottawa, Canada: Government of Zaire.

NATIONAL PARKS AND RESERVES

About 22 million ha of forestlands in Zaire are classified as national parks, wildlife and forest reserves, reforestation sites, and gardens. Of this, 60 percent has been allocated to wildlife, hunting, and nature reserves, while only 3 percent of the area has been set aside for forest reserves (Table 5).

Forest Management

There is limited documentation on forest management in Zaire, and there is no evidence that timber is managed on a sustainable yield basis. The Zairian government indicates (World Resources Institute, 1988) that industrial wood production and forest manage-

TABLE 4   Forest Cover in the Central Basin of Zaire

Cover Type	Inventoried Area (ha)	Percentage of Total Area Inventoried	Total Area if Extrapolated to Entire Basin (1,000 ha)
Evergreen forest	425,234	8.0	8,000
Semideciduous forest	2,287,981	43.2	43,632
Mature secondary forest	737,465	13.9	14,039
Immature secondary forest	110,898	2.1	2,121
Subtotal, upland forest	3,561,578	67.2	67,792
Seasonally flooded forest	1,563,475	29.5	29,795
Nonforested areas	173,176	3.3	3,333
Total area	5,298,229	100.0	101,000

SOURCE: Department of Land Affairs, Environment, Nature Conservation, and Tourism and International Institute for Environment and Development, World Resources Institute. 1990. Appendix table in Zaire Forest Policy Review. Draft Summary Report. Kinshasa, Zaire: Department of Land Affairs, Environment, Nature Conservation, and Tourism, and Washington, D.C.: World Resources Institute.

TABLE 5   Uses of Forestlands in Zaire (in Thousands of Hectares)

Forestland Use	Approximate Area
National parks	8,360
Wildlife, hunting, nature reserves	13,091
Classified forests and forest reserves	753
Reforestation sites	112
Botanical zoological gardens	3
Subtotal, classified land	22,319
Forestland allocated for wood production	21,500
Other nonclassified forestland	85,419
Total	129,310

NOTE: Note the inconsistencies in the estimates of total forestlands, especially compared with estimates in Tables 2 and 3. Also note that of all forestlands classified, only 22 million ha of a total of 129 million ha have been classified.

SOURCE: Department of Land Affairs, Environment, Nature Conservation, and Tourism and International Institute for Environment and Development, World Resources Institute. 1990. Appendix table in Zaire Forest Policy Review. Draft Summary Report. Kinshasha, Zaire: Department of Land Affairs, Environment, Nature Conservation, and Tourism, and Washington, D.C.: World Resources Institute.

ment consist of prescribed allowable cuts combined with guidelines on harvesting practices. It seems, however, that the vast majority of timber extractors do not adhere to the cut or harvesting guidelines. There is evidence that modest reforestation efforts took place some 40 years ago but that very little took place in the 1960s, 1970s, or 1980s (World Resources Institute, 1988). Other forest management plans are in the form of protection and conservation of areas designated as national parks and wildlife reserves (World Resources Institute, 1988).

FOREST-BASED INDUSTRY

Industrial Roundwood Production

In 1988, about 113,000 m^3 of logs worth US$15 million and 20,000 m^3 of sawn wood worth US$4 million were exported from Zaire (International Society of Tropical Foresters News, 1990). It has been estimated that Zaire produces about 2.6 million m^3 of industrial roundwood annually. About 81 percent is cut by small-scale operators and domestic pit-sawers. (Domestic pit-sawers are individuals who cut logs by using pits and hand-operated saws. Usually the process requires two persons, with one person in the pit holding one end of the saw and the other person standing over the log holding the other end of the saw.) The remaining 19 percent is cut by forest concessionaires (owners of companies that produce forest products on a large scale). These concessionaires control extensive tracts of forestland leased from the government. Most of this production is used to meet domestic demands, with less than 1 percent being exported (World Resources Institute, 1988). There are between 100 and 200 large- and medium-scale forestry-based companies in Zaire (Government of Zaire and the Canadian International Development Agency, 1990; World Resources Institute, 1988).

Tax Policies and Investment Procedures

There appears to be a discrepancy between the value of exported wood products and the government's estimated value on which the export tax is based. The government levies taxes on exported wood products on the basis of actual export market prices. Therefore, for the government to collect the full, prescribed amount of taxes on these products, it must be fully aware of the prevailing prices in the international markets so that it can make the necessary adjustments in the required taxes. Because the government does not keep track of price trends, however, exporters take advantage of the situation and report prices far below actual market prices. Thus, the value of certain species, which is based theoretically on the value on international markets, is, in effect, unrelated to the tax levied. It is estimated that the true market value and corresponding government revenue are reduced by about 50 percent (Government of Zaire and the Canadian International Development Agency, 1990:42). The government of Zaire loses an estimated 50 percent of its potential forest products tax revenue.

Bureaucratic red tape, extra taxes, and uneven collection of taxes place a particularly heavy burden on domestic pit-sawers and small-scale operators. At the same time, the relatively lower extra taxes assessed to higher valued primary tree species create an incentive for large operators to engage in selective logging (World Resources Institute, 1988). Thus, not only are taxes enforced unevenly between operators of forest concessions and small operators but also selective logging by larger concessionaires removes the more valuable tree species, leaving the lesser valued species for the small-scale and individual loggers, and in the process of removal damages what trees remain.

The cumbersome investment and export procedures have had an adverse effect on potential investors. It is equally true that the absence of policy and lack of enforcement of measures that have been enacted have created an environment in which existing companies familiar with the rules of the game benefit immensely. These unsustainable forest management practices and the underlying public policies reduce the long-term contribution of the forestry industry to the national income.

Domestic Loggers

The structure of the logging industry points to the important role of small-scale logging operators and domestic pit-sawers. Of the total industrial wood production of 500,000 m^3 per year in the late 1980s, domestic loggers accounted for about 70 percent of sawn wood for domestic processing and consumption. Not only have the local pit-sawers and small-scale operators successfully supplied domestic markets in many parts of the country with wood at a fraction of the cost charged by large logging companies but also their contribution to employment and income creation is significant. The estimated value of locally produced sawn wood is on the order of US$200 million, compared with an estimated US$37 million of exported forest products produced by large companies. The level of dynamism, resilience, and productivity of domestic loggers is remarkable, given the many adverse policy biases, particularly heavier export taxation than on large commercial loggers (Government of Zaire and the Canadian International Development Agency, 1990), facing local pit-sawers and small-scale logging operators.

Equally significant, however, is the impact of domestic loggers on deforestation, forestland degradation, and depletion. The depletion of the Mayumbe forests in Bas-Zaire attributable to domestic loggers, the severe degradation of the montane forests in the Kivu region, and the depletion of the woodlands and wooded savannahs in the Shaba and Bandundu regions cannot be dismissed. To promote long-term and sustainable forest management practices in commercial logging, public policies need to be reoriented so that they serve traditional domestic pit-sawers and small-scale operators better than they do at present (World Resources Institute, 1988). One of the policies that needs to be reoriented is the policy on land tenure. Because local communities cannot own forestlands or have the security of long-term tenure, they have no guarantee that they will be able to reap the benefit of any time or labor they might put toward sustainable practices and, therefore, have no incentive to replant trees they cut down. This lack of security progresses to depletion of fuelwood supplies and forest destruction. To prevent this destructive sequence, the government, in association with its major aid donors, needs to enter into a dialogue with local communities to resolve these issues.

DEFORESTATION AND ITS CAUSES

There are many causes of deforestation: the advancement of agricultural frontiers, demand for fuelwood, commercial logging, overgrazing of forested lands, and demand for land because of high population density. Another cause central to the problem is that the institutions responsible for formulating and implementing forestry policies are ineffective and inefficient in carrying out these functions. In Zaire, assessing the significance of the causes of deforestation is hampered by the lack of adequate and reliable data on such factors as estimates of forest cover, agricultural land use, and extent of forest regeneration. This is evidenced by the widely different forest area totals noted by various sources (see Tables 2–5). The lack of a national forestry policy has nurtured an environment that is not supportive of data collection. Fortunately, the seed of a policy has been planted through the remarkable efforts of a few dedicated national academics, civil servants, and a handful of foreign advisers, so that there is now a greater interest in the forestry sector than there was in previous years. However, this seed, in terms of policy formulation and implementation, has yet to take root.

The limited data available, mainly from the Food and Agriculture Organization (FAO) and United Nations Environment Program (1981a), indicate that during the period 1976–1980, the average annual deforestation rate of closed broadleaf forests was about 165,000 ha (Table 6). It is difficult to estimate the relative weights of the various factors responsible for deforestation, namely, agricultural crop conversion, perennial cash crops, small-scale farming, traditional subsistence agriculture, logging on commercial concessions, and tree-cutting for fuelwood. However, on the basis of 1976–1980 discussions with Zairian forestry experts, which were superseded by information from regional assessments, FAO projected that 180,000 ha would be deforested annually from 1981 to 1985.

Not all forestland clearing results in deforestation—some land

TABLE 6   Average 5-Year Deforestation
of Closed Broadleaf Forests in Zaire
(in Thousands of Hectares)

Type of Area Deforested	Area
Productive	
Undisturbed	
1976–1980	145
1981–1985	155
Logged	
1976–1980	20
1981–1985	25
Total	
1976–1980	165
1981–1985	180
Unproductive	
1976–1980	2
1981–1985	2
All areas	
1976–1980	167
1981–1985	182

SOURCE: Food and Agriculture Organization and United Nations Environment Program. 1981a. P. 86 in Tropical Forest Resources Assessment Project. Forest Resources of Tropical Africa. Part I. Rome, Italy: Food and Agriculture Organization of the United Nations.

TABLE 7   Area Logged for Industrial Hardwood, 1975–1981 (in Hectares)

Region	1975	1976	1977	1978	1979	1980	1981
Bas-Zaire	33,099	19,536	22,500	56,501	26,912	38,080	24,810
Bandundu	7,533	19,935	16,838	18,179	18,021	32,221	42,134
Equateur	8,297	13,380	13,184	1,220	5,423	5,232	12,558
Haut-Zaire	8,213	11,924	6,910	5,510	8,717	6,400	16,036
Kivu	3,045	2,976	1,905	1,500	4,739	—	6,797
Shaba	6	7,446	9,165	—	—	1,000	1,000
Kasai-Occidental	200	652	1,541	200	850	1,380	4,390
Kasai-Oriental	225	260	276	—	180	—	780
Total	65,814	76,109	72,319	87,313	64,842	84,313	108,505

NOTE: The reliability and accuracy of these data cannot be ascertained.

SOURCE: Department of Land Affairs, Environment, Nature Conservation, and Tourism and International Institute of Environment and Development, World Resources Institute. 1990. Appendix table in Zaire Forest Policy Review. Draft Summary Report. Kinshasa, Zaire: Department of Land Affairs, Environment, Nature Conservation, and Tourism, and Washington, D.C.: World Resources Institute.

reverts back to forests, at least temporarily. In many instances, however, degradation is irreversible, so forest regeneration is not possible. Of the 80,000–100,000 ha logged for industrial hardwood production for export and domestic consumption each year, an unknown portion is permanently deforested (World Resources Institute, 1988). Table 7 gives the amount of forestland logged for industrial wood production during 1975–1981 (Department of Land Affairs, Environment, Nature Conservation, and Tourism and International Institute for Environment and Development, World Resources Institute, 1990), when a total of 559,215 ha and an average of 80,000 ha/year were logged. This is not in agreement with FAO's estimate of 180,000 ha (Food and Agriculture Organization and United Nations Environment Program, 1981a); indeed, FAO's estimate is about 100 percent higher. Perhaps this inconsistency is an indication that there is much more logging than is reported.

Because of the abysmal record of commercial concessions regarding replanting, which has been required since 1982 but not enforced, one can infer that deforestation attributable to unsound logging practices is significant. A crude ordinal ranking based on the available data related to the major causes underlying deforestation and estimates of the area of forestland permanently removed in Zaire each year is given in Table 8 (in descending order). This ordinal ranking is based on (1) a review of the existing literature from the standpoint of

TABLE 8   Ordinal Ranking of Causes of Deforestation

Cause of Deforestation	Ordinal Ranking	Estimated Area of Forestland Removed Each Year (1,000 ha)
Fuelwood harvesting, including charcoal (individual and commercial)	1	5,500
Traditional farming practices	2	2,000
Commercial logging (large companies, traditional pit-sawers and small-scale loggers)	3	500
Perennial crops and livestock (mainly coffee and cattle)	4	400

NOTE: The ordinal ranking is from the authors and is based on data on hectares removed.

SOURCES: Government of Zaire and the Canadian International Development Agency. 1990. Plan d'Action Forestier Tropical. Vols. I and II. Kinshasa, Zaire, and Ottawa, Canada: Government of Zaire; Department of Land Affairs, Environment, Nature Conservation, and Tourism and International Institute for Environment and Development, World Resources Institute. 1990. Appendix table in Zaire Forest Policy Review. Draft Summary Report. Kinshasa, Zaire: Department of Land Affairs, Environment, Nature Conservation, and Tourism, and Washington, D.C.: World Resources Institute.

the relative weights assigned to the various causes of deforestation, (2) interviews with national experts, and (3) the authors' knowledge of the country. An important realistic assumption underlying the ranking is that virtually no significant replanting has taken place. In addition, there seems to be more logging by large- and small-scale operators than official statistics indicate.

There is a lack of adequate time-series data on permanent forest removal for various cropland uses. However, there are other indicators of forestland degradation, impoverishment, and depletion that, if combined with the lack of reforestation, point to an unsustainable rate of forest resource exploitation. Using the conservatively estimated rate of deforestation of closed broadleaf forests—165,000–180,000 ha/year (Capistrano, 1990; Food and Agriculture Organization and United Nations Environment Program, 1981a)—it can be inferred that, on average, about 1 percent of Zaire's total forestlands may have been permanently removed each year in the past decade. These estimates are extremely conservative, especially since they apply to the deforestation of broadleaf forests and not to savannahs. Since reforestation is insignificant and many forest areas are not under government control, the rate of forest destruction in Zaire is probably much higher than these numbers suggest.

Other indirect evidence, such as the shortened fallow period in traditional subsistence agricultural systems combined with the demographic pressures on land in many areas of Zaire, supports the thesis that permanent forest removal, along with forestland degradation and depletion, has worsened in the past 10 years. The magnitude of this increase is not known with certainty, however, a 1 percent permanent deforestation rate annually is considered to be detrimental to the environment, especially without reforestation.

Advancement of Agricultural Frontiers

For Zaire, there are at least three challenges to analyzing the long-term effects of traditional farming on forest areas. First, all farming does not necessarily take place on lands classified as commercial forests. Second, not all of the forestland converted to cropland remains in crop production. Usually the land is farmed for a number of years and then abandoned; depending on the soil's capabilities, some soil types easily allow regeneration over time, others do not. Third, adequate and reliable data are not available.

Fuelwood Demand and Harvesting

Fuelwood is an important source of energy for rural and urban households in Zaire, but more than 66 percent of the population lives in parts of the country where there is an increasing imbalance between fuelwood demand and supply. World Bank projections (World Bank and United Nations Development Program, 1983) to the year 2000 point to a growing demand for fuelwood, which is reflected in ever-increasing prices for charcoal along with pervasive shortages. According to these projections, each year about 5.5 million ha of forestlands would have to be depleted to meet the increasing fuelwood requirements. Without meaningful alternatives to fuelwood as a source of energy and given the dubious success of isolated and limited experiments with fuelwood plantations and more efficient wood-burning furnaces, the demand for fuelwood harvesting is likely to continue to put pressure on forests and increase the level of their destruction.

According to the World Resources Institute (1988), annual fuelwood demand is about 25 million m³, and annual production of industrial roundwood is 2.6 million m³. This means that about 27.6 million m³ of wood would be required annually to meet the estimated demand. Assuming annual growth of 700 million m³ of wood on commercial forestlands, it can be inferred that about 4 percent of the growth of commercial forests would need to be removed annually just to meet the demands for fuelwood and industrial roundwood. On the basis of forest productivity, which is estimated to be 5 m³ ha, this level of wood consumption will require logging about 6 million ha annually.

If reforestation is carried out and/or fuelwood plantations are established at a rate at least equivalent to the rate of removal, then the present rate of removal may not present a problem in the long run. Under the present circumstances, however, this would be an optimistic scenario because it is unlikely that such measures will be adopted in the near future. The worst-case scenario, one in which the area of forestland continues declining while the demand for wood (fuelwood and industrial roundwood) accelerates, appears to be the more likely for Zaire's future; and on the basis of current information, this appears to be the case. Unless this is reversed, not only will the rate of consumption or removal exceed the rate of growth, but also the growing stock itself will be threatened.

No systematic analysis of fuelwood plantations or the related issues of local community ownership and control has been undertaken to date. Also it is unclear whether the more efficient wood-burning furnaces have been thoroughly tested in various regions of Zaire or whether their rate of adoption by farmers and private charcoal-producing businesses justifies large-scale investments.

Unregulated Commercial Logging

Each hectare of commercial forestland in Zaire is capable of producing at least 5 m³ of industrial roundwood/year according to the Government of Zaire and the Canadian International Development Agency (1990). Given this estimate, one can infer that the 139 million ha of forestlands classified as commercial produces about 700 million m³ and can be considered the total annual growth for those areas classified as commercial forestlands (Food and Agriculture Organization and United Nations Environment Program, 1981b; World Bank, 1986).

The logging industry, despite prescribed management practices and regulations enacted since 1982, has been virtually unregulated because of weak administrative capabilities of key forest management institutions. These weak capabilities concern planning, organizing, and monitoring harvesting and management methods to achieve sustained yields. This lack of performance in managing existing forest resources allocated to industrial wood production (estimated at 100,000–150,000 ha per year in the 1980s) casts serious doubt on DLAENCT's capacity to manage the 600,000 ha of forest to be used to produce a target of 6 million m³ of industrial wood by the year 2000. This production level is 12 times the present production level of about 500,000 m³ (Government of Zaire and the Canadian International Development Agency, 1990:37–42).

Large-scale operators (mostly foreign), domestic pit-sawers, and small-scale operators use many methods that lead to unsustainable logging. High-value species most in demand in export markets are logged selectively; however, there is waste and destruction of the surrounding low-value species. A few logging companies (and special interest groups) control larger areas than is allowed by law, areas that are larger than can be sustainably exploited. Logging companies often exceed annual cut ceilings specified in concession agreements and cut immature trees whose diameters are below the limit. Loggers operate without forest-use permits and harvest forestlands that are not allocated to industrial wood production. Finally, there are inadequate reforestation efforts because of the lack of policy and penalties.

Added to these unsustainable logging practices are the government's flat-tax policies based on incorrectly quoted prices for higher value species, with the effect that high-value species are taxed at the same rate as low-value ones and are selectively harvested to the destruction of surrounding species. In terms of biodiversity, this policy encourages questionable tree-grading and does not help the promotion of lesser known species. It is estimated that Zaire has about 70 species of tropical woods, but only a dozen are known and marketed.

Population Density and Forest Removal

The relationship between population density and forest resource exploitation is not well known, but it is known that there is a high correlation between the two (Government of Zaire and the Canadian International Development Agency, 1990). The most densely forested areas, such as the central basin, tend to have below-average population densities. Zaire's fast-growing population (in excess of 3 percent annually [U.S. Department of State, 1988]) is concentrated in areas with fertile land and in economic enclaves.

Areas of greatest population concentration and urbanization are associated with permanent forest removal, as in the Mayumbe forests (Figure 1, area 2b); with forest degradation and impoverishment, as

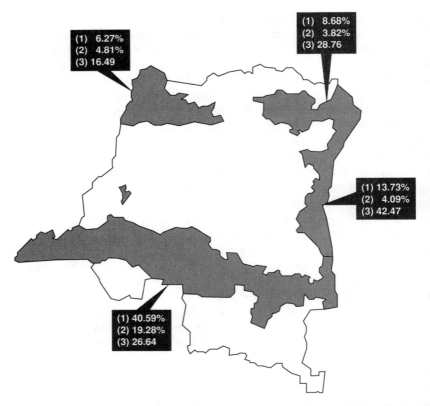

FIGURE 2   Areas of population concentration in Zaire.   (1) Percent total
population.   (2) Percent total land area (square kilometers).   (3) Population
density per square kilometer.   Source: Government of Zaire and the Canadian International Development Agency.   1990.   Plan d'Action Forestier Tropical.   Vol. I, Annex 2.   Forestry Map.   Kinshasa, Zaire, and Ottawa, Canada:
Government of Zaire.

in the northern rim along the border (Figure 1, area 2a) and in the
montane forests of Kivu (area 3); and with depleted woodlands and
wooded savannahs in the south, Bandundu, and southwestern, Shaba,
areas of the country (Figure 1, areas 4a and 4b).   These areas are
highly urbanized and heavily populated.   Figure 2 shows that about
70 percent of Zaire's estimated population of 35 million (1988 estimate) lives on less than 33 percent of the total land area.

The population concentration associated with large urban centers
is clustered along three major areas: the west-southeast band, which
extends from the lower Zaire River region (Bas-Zaire) through Kinshasa

to north Shaba; the mid-southeast/northeast band, which extends from north Shaba across the Kivu region to the higher Zaire River region (Haut-Zaire); and the densely populated centers of Gemena in the Equateur region and Kisangani and Isiro in the Haut-Zaire region. This pattern of population concentration is, to some extent, related to the relatively fertile soils (derived from volcanic materials and known to support agriculture), particularly in the Kivu region, and also to the uneven pattern of economic growth, urbanization, and administration established by the colonial government. This pattern has been reinforced in the postindependence era beyond the capacities of current physical and social infrastructures, especially in the major urban centers. The forestland areas in the three major areas with the highest population concentrations are the most impoverished and depleted.

The declining income of the rural population, because of the government's inadequate pricing and market policies and general neglect of agriculture, has caused farmers to stress cultivation practices beyond their technical limits. In addition, to compensate for declining yields and low prices, farmers have had to bring more forestland into cultivation to sustain their families, thus aggravating the permanent removal of natural forests (Government of Zaire and the Canadian International Development Agency, 1990).

INSTITUTIONAL ARRANGEMENTS AND POSSIBLE REFORMS

Responsibility for forest resources management in Zaire has shifted frequently in the past 20 years as ministries and agencies have been reshuffled, reorganized, or relabeled. One constant has been that the department responsible for forestry, DLAENCT, been the orphan child of the Ministry of Agriculture, Rural Development and Extension or the Ministry of Land Affairs or the Ministry of Tourism. DLAENCT was always several steps removed from the centers of financial, personnel, and political decision making, for example, the president's office and the National Executive Council. As a subordinate entity to a larger ministry, DLAENCT always fell prey to overriding national budget priorities within the agricultural sector. In fact, although DLAENCT has been separated from the Department of Agriculture for some time, its budgetary allocation is still often combined with that of the Department of Agriculture. It amounted to about 0.4 percent of the country's total operating budget in 1985 and rose slightly to 0.5 percent in 1987 (compared with an average of 1.1 percent for agriculture during 1985–1987 [World Bank, 1986:Table A, Annex A]). Moreover, these budgetary allocations fall far short of actual spending because of unusual central financial control practices; that is, the line ministries are not allowed to spend the amounts allocated to them.

Despite these meager budgetary allocations, DLAENCT's mandate has broadened over the years to include the goals of revenue generation, forest conservation and wildlife protection, and the development of local community-based forests. These objectives are in addition to the traditional goal of forest service management in spite of the fact that there is some agreement that DLAENCT sufficiently structured or empowered to effectively formulate appropriate policies or to implement relevant strategies for dealing with sustainability and resource management issues in the forestry sector.

In Zaire, austerity measures arising from the economic adjustment programs of recent years have lead to chronic problems for DLAENCT: inadequate funding and staffing, delinking of budget allocations from the amount of revenue generated from forestry activities, and a narrow and short-term view of forest resource exploitation. All of these are incompatible with long-term and sustainable resource management. Since civil service salaries today have declined dramatically in real terms to about 20 percent of their level 10 years ago, it is not surprising that the focus of DLAENCT is on the more lucrative and most visible aspects of its management activities: logging concessions and negotiations with large companies. These transactions can yield tangible individual recognition and monetary benefits to participants, often in the form of legal and illegal, but tolerated, payments.

The monetary returns on activities such as community-based forestry programs are low, however. DLAENCT lacks an active social, economic, and political constituency with vested interests in DLAENCT's objectives of forest management on a sustainable yield basis. There seems to be a lack of concern about the control and ownership of forestlands by local communities and the need to train a national cadre of technocrats to design suitable corrective policies and institutions to carry out these policies.

DLAENCT also has a responsibility to serve small-scale foresters. The government should adequately concentrate on their needs by establishing community-owned and -managed forests and providing agroforestry extension advice. Therefore, an urgent need in DLAENCT is a well-trained cadre of technocrats with the ability to inventory the forest and design corrective policies. There also needs to be appropriate administrative and financial support to implement those policies.

It follows that training of skilled (secondary school education level) as well as advanced-level technicians (post-secondary school education level) is also needed, as suggested by the Department of Land Affairs, Environment, Nature Conservation, and Tourism and International Institute for Environment and Development, World Resources Institute (1990). This could be accomplished by providing basic training in Zaire and specialized graduate training overseas. This would provide support for and serve to strengthen the forestry option at the Bengemisa College of Agronomy and the regular 5-year agronomy/forestry program proposed for the University of Kinshasa and the University of Kisangani at Yangambi.

Given the autonomy of each campus within the National University of Zaire and the agricultural and forestry development challenges facing each university's surrounding community, it makes sense for each campus to have a B.S.-level agronomy/forestry program. Although traditional training in forestry has focused on providing students with forest management skills (for example, forestry management regulations and measurement techniques), what is needed is a much stronger orientation in environmental and resource management, with a specific emphasis on problem-solving abilities related to research and policy.

Given the autonomy of each campus, however, it will be difficult to use all three separate campuses in the most economic manner. One possibility is a central core curriculum at each university, with optional curricula distributed among the three campuses. Ultimately, however, it will be necessary to send graduate students overseas for specialization, for example, to study environmental sciences and sustainable agriculture practices, including those related to forestry.

There is a critical need to understand how key forest management institutions such as DLAENCT function and the institutional reforms that are required to make them function better. Given the nature of critical or sectoral linkages among forestry institutions in Zaire, the second-class stature of DLAENCT within the power structure of the government erodes its coordination capacity with other key departments that address forestry and sustainable agriculture, such as energy, transport, rural development, and agriculture.

In sectoral matters such as access to and ownership of forestland, fuelwood harvesting, soil erosion, and reduced fertility caused by shifting cultivation practices, the opinions of the DLAENCT are the informed voice in the government, and these opinions must have great weight in decision making. The prerequisite for such intersectoral linkages is that DLAENCT be given a greater role in policy formation and implementation. It must also be given greater prominence in forestry matters vis-à-vis the central decision-making departments (for example, central planning, finance, and the president's office).

Cooperation and coordination between the institutions and the departments that address the forestry needs of Zaire are absolutely necessary.

SUGGESTIONS FOR SUSTAINABLE MANAGEMENT

Sustainable management will not be possible until the limited and unreliable data base on Zaire's forest cover, causes and extent of deforestation, and corrective measures is expanded with verifiable information.

Research Agenda

The following two-phase research agenda is proposed.

PHASE I: AGRICULTURAL AND FORESTRY AGENDA

Concurrently with field trials, there should be a detailed forestry survey of the following five regions: Yuki, Kisangani, Mayumbe (the Tshela site), Yamgambi, and Kaniameshi.

• Yuki is in the Bandundu region in the heart of the central basin rain forest where ebony trees are logged for export and low-value species are used for charcoal for Kinshasa.

• Kisangani is on the fringes of the central basin rain forest just north of the equator. Logging in this area is entirely for local consumption.

• Mayumbe, in the Mayumbe forests of lower Zaire, is an area where large logging companies cut timber for export.

• Yamgambi is the region where the corridor system was tried before the independence of Zaire in 1960. Yamgambi is located about 100 km from Kisangani and is the site of the National Agronomic Research Institute.

• Kaniameshi is on the fringes of open wooded savannah forests in southern Shaba near the Zambian border.

These savannah forests are subject to shifting-cultivator's seasonal brush fires, an important land-clearing practice. There is a need to field test low-input types of farming systems to alleviate the destructive effects of shifting cultivation even in areas with relatively low population density (see Table 9).

Ongoing research on various agroforestry systems implemented in other tropical countries should be tested in Zaire at the five selected sites to determine whether these systems are suitable to specific ecosystems within Zaire. (Table 9 supplies available data from some of these five sites.) This research will require staff at all levels and must be a long-term effort that leads to lasting results on the suitability of specific agroforestry systems. It must include those subjects pertaining to forestry that environmental scientists deem necessary, as defined by FAO (1981a). The objective is to assess problems of ecosystems and devise the means to correct these problems (see Table 9). The most promising and suitable farming systems will then be replicated and strictly observed at 12 other sites representative of different forest growth systems. Again, there will be a need for trained staff.

Some of the systems that should be tested at these sites are alley cropping, improved fallow, low-input cropping, livestock pasture, forest/ farming mosaic, continuous cropping, and the corridor system. The corridor system has been practiced in Yamgambi (Jurion and Henry, 1969) and was found to be technically and economically viable, but it was terminated because it restricted the movement of a population traditionally accustomed to the nomadic life of shifting cultivation (Ruthenberg, 1971). Local culture and practices must be considered and incorporated into any research practice implemented.

Because commercial logging is done on exclusive private concessions, it will be necessary to collect data from areas proximate to commercial production areas if access to private concessions is not possible.

PHASE II: EXTENSION OF DATA AND SERVICES TO POTENTIAL USERS

Large- and small-scale operators and, in particular, those who practice shifting cultivation must be made aware of the results of Phase I and all research and resources must be made available to them. Therefore, an efficient extension service with appropriately trained personnel will be required.

Human Resources Development

The pervasive shortage of well-trained staff in forestry and environmental management at all levels—technical, undergraduate, graduate, and specialized—must be rectified by extension workers, including those already employed, trained to carry out the requirements of the improved and restructured programs. There should be courses for in-house staff, training at Zaire's three universities mentioned above, and specialized training at overseas institutions.

TABLE 9 Proposed Agricultural and Forestry Research: Selected Commercial Logging Sites, Deforestation, and Charcoal Production

A. *Site-Specific Characteristics*

Logging Company (Year of Establishment)	Location	Soil Type	Altitude (m)
Siforzal (1950)	Kisangani (northern central basin)	Oxisol	485
Onatra (publicly owned) (1896)	Yuki (central basin)	Oxisol	500
Agrifor Forescom (publicly owned) (1883)	Tshela (south)	Ultisol	300

B. *Soil and Weather Data for Locations Close to the Three Sites in Part A*

Location	Horizon	Depth (cm)	pH	Cal-cium	Potas-sium	Cation Exchange Capacity	Clay	Silt	Sand
Yangambi[a]	A_{11}	0–22	4.1	0.3	0.07	5.3	14.3	3.2	82.5
	A_{12}	22–45	4.4	0.3	0.05	3.7	18.3	4.2	77.5
	A/B	45–60	4.3	0.4	0.04	3.8	19.8	3.8	76.4
	B_{21}	60–220	4.5	0.5	0.05	3.0	20.9	3.6	73.5
	B_{22}	220–250	4.3	0.3	0.03	2.5	19.4	3.2	77.4

Yuki, Lodja[b]								
A$_1$	0–8	4.1	0.85	0.06	4.25	11.8	1.5	86.7
A$_{31}$	8–23	4.1	0.30	0.04	2.95	11.1	1.4	87.5
A$_{32}$	23–38	4.2	0.80	0.03	2.1	17.1	2.1	81.8
B$_{21}$	38–85	4.3	0.50	0.06	1.6	15.5	2.0	82.5
B$_{22}$	85–200	4.5	0.55	0.06	1.5	17.6	1.7	81.7
Tshela[c]								
A$_1$	0–18	5.1	2.0	0.19	6.7	38.4	8.0	53.6
A$_3$	18–38	4.9	1.3	0.04	6.4	43.6	7.8	48.6
B$_2$t$_1$	38–63	4.7	1.0	0.02	5.3	54.6	6.5	38.9
B$_2$t$_2$	63–98	4.7	0.7	0.02	4.5	52.8	6.3	40.9
B$_{31}$	98–150	4.9	0.7	0.02	5.2	54.8	6.0	39.2
B$_{32}$	150–200	5.0			4.3	59.1	7.0	33.9

[a]The elevation of Yamgambi is ±500 m, rainfall is 1,875 mm/year, and the soil is typic haplorthox.

[b]For Yuki, Lodja (a research station site), data for forested Kasai are used. The elevation of Yuki is ±625 m, rainfall is 1,800 mm/year, and the soil is typic eutrorthox.

[c]For Tshela, data for Mayumbe are used. The elevation of Mayumbe is ±300 m, rainfall is 1,000 mm/year, and the soil is oxic paleustult.

SOURCES: (Part A) Kande, M. 1991. Draft doctoral dissertation. North Carolina State University, Raleigh; Sys, C. 1972. Caracterisation Morphologique et Physico-Chimique de Profils Types de l'Afrique Centrale. Serie Hors, Publications INEAC. Brussels: Institut National d'Etudes Agronomiques du Congo. (Part B) Jurion, J., and J. Henry. 1969. Can Primitive Farming be Modernised? Serie Hors, Publication INEAC. Brussels: Institut National d'Etudes Agronomiques du Congo; Smith, G. D., C. Sys, and A. Van Wamberke. 1975. Application of Soil Taxonomy to the Soils of Zaire (Central Africa). Bulletin de la Africa. Bulletin de la Societé Belge de Pedologie; N. Spec. 5. Brussels: Societeé Belge de Pedologie; Sys, C. 1972. Caracterisation Morphologique et Physico-Chimique de Profils Types de l'Afrique Centrale. Serie Hors, Publications INEAC. Brussels: Institut National d'Etudes Agronomique du Congo.

Land Tenure Policies

Community-owned and -managed forests with proper reforestation will not be possible in Zaire until the land tenure and land ownership rights of local communities are more secure. Whatever laws do exist, they appear to be applied in such a way as to favor large commercial operators. Therefore, the existing relevant laws must be modified. The new regulations must be structured to strengthen communal or local government and individual ownership rights and to ensure that enforcement of all forestry laws and regulations is uniform.

Strengthening the Forestry Department

Forestry policy formulation and the implementation of forestry projects involve the ministries of DLAENCT—agriculture, rural development, environmental—and the transportation ministry. These ministries have a significant impact on forestry management policies; therefore, coordination and consultation between these ministries on matters pertaining to forestry should be mandated in any government policy to minimize conflict. Furthermore, the budget should clearly state what funds are disbursed directly for forestry.

Funding

The government of Zaire has stated to its citizens and to international organizations that it wishes to sustain its environment. This statement should be translated into action. All funds allocated to sustaining forestry should be spent for that purpose, fair and responsible taxation policies should be augmented, and agencies that provide aid that supports sustainable agroforestry systems should commit to a long-term but strictly monitored environmental and resource management system in Zaire.

With some modifications and refinements, these suggestions will meet the objectives for formulating appropriate measures and policies to avoid the potentially disastrous effects of the destruction, depletion, and degradation of tropical forest cover in Zaire.

ACKNOWLEDGMENTS

The authors express their gratitude to the School of Agriculture and Home Economics of Tuskegee University and the George Washington Carver Agricultural Experiment Station for the valuable support they provided. Matungulu Kande of North Carolina State University at Raleigh and of the Faculty of Agronomy, University of Kisangani, Kisangani, Zaire, deserves special credit for sharing so generously of his private data base and collections on Zaire during his May 1991 visit to Tuskegee University. Finally, the authors are much indebted to a group of dedicated support staff in the School of Agriculture and Home Economics, Tuskegee University, especially Mary Cade, Judy Kinebrew, Sibyl Caldwell, and Marva Ballard.

REFERENCES

Blyth, J. E., Jr., J. Tibben, and W. B. Smith. 1984. Primary Forest Product, Industry and Timber Use, Iowa, 1980. USDA Research Bulletin NC-82. Washington, D.C.: U.S. Department of Agriculture.

Capistrano, A. D. N. 1990. Macroeconomic Influences on Tropical Forest Depletion: A Cross-Country Analysis. 1967–1989. Ph.D. dissertation. University of Florida, Gainesville.

Department of Land Affairs, Environment, Nature Conservation, and Tourism and International Institute for Environment and Development, World Resources Institute. 1990. Zaire Forest Policy Review. Draft Summary Report. Kinshasa, Zaire: Department of Land Affairs, Environment, Nature Conservation, and Tourism, and Washington D.C.: World Resources Institute.

Food and Agriculture Organization and United Nations Environment Program. 1981a. Tropical Forest Resources Assessment Project. Forest Resources of Tropical Africa. Part I. Rome, Italy: Food and Agriculture Organization of the United Nations.

Food and Agriculture Organization. 1981b. Tropical Forest Resources Assessment Project. Forest Resources of Tropical Africa. Part II. Rome, Italy: Food and Agriculture Organization of the United Nations.

Government of Zaire and the Canadian International Development Agency. 1990. Plan d'Action Forestier Tropical. Vols. I and II. Kinshasa, Zaire, and Ottawa, Canada: Government of Zaire.

Hines, D. 1988. Zaire Forestry Resources: Economic and Policy Perspectives. Working Paper. Washington, D.C.: World Resources Institute.

International Society of Tropical Foresters News. 1990. Log and sawnwood sources reported. Int. Soc. Trop. Foresters News 11(4):9.

Jurion, F., and J. Henry. 1969. Can Primitive Farming Be Modernised? Hors Serie, Publications INEAC. Brussels: Institut National d'Etudes Agronomiques du Congo.

Kande, M. 1991. Draft doctoral dissertation. North Carolina State University, Raleigh.

Ruthenberg, H. 1971. Farming Systems in the Tropics. London: Oxford University Press.

Smith, G. D., C. Sys, and A. Van Wamberke. 1975. Application of Soil

Taxonomy to the Soils of Zaire (Central Africa). Bulletin de la Societé Belge de Pedologie, N. Spec. 5. Brussels: Societé Belge de Pedologie.

Sys, C. 1972. Characterisation Morphologique et Physico-Chimique de Profils Types de l'Afrique Centrale. Serie Hors, Publications INEAC. Brussels: Institut National d'Etudes Agronomique du Congo.

U.S. Department of Agriculture, Economic Research Service. 1990. World Agriculture: Trends and Indicators, 1970–1989. Statistical Bulletin No. 815. Washington, D.C.: U.S. Department of Agriculture.

U.S. Department of State. 1988. Zaire Background Notes. Washington, D.C.: U.S. Department of State.

World Bank. 1986. Zaire: Toward Sustained Agricultural Development. Agriculture Sector Memorandum. Washington, D.C.: World Bank.

World Bank and United Nations Development Program. 1983. Zaire Energy Assessment Report. Washington, D.C.: World Bank.

World Resources Institute (WRI). 1988. Zaire Forestry Policy Review and Related Studies. Draft Summary Report. Kinshasa, Zaire, and Washington, D.C.: World Resources Institute.

WRI. 1990. World Resources 1990–91. New York: Basic Books.

Glossary

agricultural frontier Areas where agricultural expansion is resulting in forest conversion.

agrisilviculture An agrofroestry system that uses crops and trees, including shrubs or vines.

agrisilvopastoral system An agroforestry system that combines crops, pastures (with or without animals) and trees.

agroecological zones Geographic areas in which ecologic conditions (soil, water, climate) dictate the agricultural practices that are used.

agroecology The application of ecological concepts and principles to the study, design, and management of agricultural systems. By integrating cultural and environmental factors into its examination of food production systems, agroecology seeks to evaluate the full effect of system inputs and outputs and to use this knowledge to improve these systems, taking into account the needs of both the ecosystem as a whole and the people within it.

agroecosystem A model for the functioning of an agricultural system with all its inputs and outputs.

agroforestry A land use system in which woody perennials are deliberately used on the same land management unit as annual agricultural crops or animals, either sequentially or simultaneously, with the aim of obtaining greater outputs on a sustained basis.

agropastoral system Farming systems that combine animals and crop production.

Alfisols One of 10 soil orders. A mineral soil, usually formed under forest.

allelopathic compounds Various metabolic substances, such as terpenes, camphor, and cineole, released by plants that biochemically inhibit other plants or microorganisms.

allelopathic effects The results of the biochemical suppression of the growth of one plant species by another, thus reducing competition for resources. For example, in a field succession, the pioneer weed stage is replaced by annual grasses because the weeds produce substances that inhibit the growth of other weeds.

alley cropping An agroforestry system in which annual food crops are grown in alleys formed by hedgerows of nutrient-cycling trees or shrubs. The hedgerow plants are pruned throughout the cropping season to prevent competition for sunlight, water, and nutrients.

alluvial soils Soils made of materials deposited by running water (for example, clay, silt, sand, and gravel).

annual A plant that completes its life cycle (from seed to seed production and death) within a year or single season (for example, cultivated rice).

apiculture An agroforestry system that involves the selection of trees and their management for beekeeping.

aquasilviculture An agroforestry system that integrates fisheries and trees into a production system.

arable soils Soils that are fit for plowing or tillage to produce crops.

base saturation percentage The percentage of the cation exchange capacity occupied by cations other than hydrogen or aluminum.

biochannel The paths made by roots, animals, insects, and other soil biota that act as conduits for water and air through the soil.

biocontrol (biological control) Controlling crop pests by using living organisms harmless to the plants but which destroy or reduce the number of harmful pests.

biodiversity (biological diversity) The variety and variability among living organisms and the ecological complexes in which they occur.

biogeography The study of the origin, geography, and distribution of organisms.

biomass The total weight of organic material present per unit area.

biome A major ecological community type (for example, grassland);

a major biotic unit consisting of plant and animal communities having similarities in form and environmental conditions.

biosphere The largest, all-encompassing ecosystem that includes soil, water, and the atmosphere.

biosphere reserves A series of protected areas linked through a global network and established under the Man and the Biosphere Program of the United Nations Educational, Scientific, and Cultural Organization. They are intended to demonstrate the relationship between conservation and development.

biota The living organisms of a region.

biotic Of or relating to life; caused or produced by living things.

boundary planting The method of planting trees specifically to function as boundary markers, live fences, windbreaks, or firebreaks. Additional benefits include microclimate regulation and protection and the production of green manure, fodder, or fuelwood.

broadcast seeding The action of seeding by casting or scattering seed rather than transplanting seedlings.

broadleaf forests A type of closed forest where broadleaf species (dicotyledons or monocotyledons) predominate. The broadleaf trees (especially the dicotyledons) are often referred to as "hardwoods."

browse Tender shoots, twigs, and leaves of trees and shrubs used by animals for food.

buffer zones Areas on the edge of protected areas that have land use controls and allow only those activities (such as research, recreation, and tourism) that are compatible with protecting the core area.

bund An embankment used to control the flow of water.

canopy The more or less continuous cover of branches and foliage formed collectively by the crowns of adjacent trees and other woody vegetation. Layers of the canopy may be distinguished (that is, understory and overstory).

capoeira Secondary forest.

carbon fixation The conversion of atmospheric carbon dioxide into organic compounds by plants through the process of photosynthesis.

cash crop Crops produced for sale (such as cacao, rice, and wheat) as opposed to hay and other crops grown principally as feed for animals or as seed.

cassava A tropical plant grown for its fleshy edible rootstocks, which yield a nutritious starch. Also known as manioc and tapioca.

cation exchange Exchange between a cation in solution and one adsorbed on a soil colloid. The negative charge of soil colloids plays a key role in the way nutrients behave in the soil; the ability of a soil to hold nutrients is directly related to the number of cation exchange sites.

cereal crops Flowering plants of the family Poaceae (formerly Gramineae) that are grown to produce grain for human and animal consumption.

cerrado A savannah of the central Brazilian plateau that supports dwarf woody species.

closed forest Forest in which the stand density is greater than 20 percent of the area and tree crowns approach general contact with one another.

cocoyam A widely cultivated root crop of the tropics.

compaction, soil The squeezing together of soil particles by the weight of farm and construction equipment, vehicles, and animal and foot traffic. Compaction reduces average pore size and total air space in the soil.

continuous cropping One crop planting following soon after harvest, without seasonal fallowing.

contour cropping The use of tillage that follows the contours of a slope, rather than up and down a slope. It helps prevent erosion and runoff.

coppice A thicket, grove, or growth of small trees or a forest that has grown from shoots or root suckers rather than seed.

coppicing Cutting trees close to ground level so they will regrow from coppice shoots.

corridor system *See* alley cropping.

cover crop A crop grown for its value as ground cover to reduce soil erosion, retain soil moisture, provide nitrogen for subsequent crops, control pests, improve soil texture, increase organic matter, or control erosion; also known as living mulch and green manure. In the humid tropics, they can include some woody species and many legume or grass fodders.

crop residue The organic material that remains in the field following harvest.

crop rotation The successive planting of different crops in the same field over a period of years, usually to reduce the pest population or to prevent soil exhaustion.

cropping patterns The yearly sequence and spatial arrangement of crops or alternating crops and fallow within a given area. The fallow crop may be natural or planted.

crusting The formation of a surface layer on soils, ranging in thickness from a few millimeters to an inch, that, when dry, is much more compact, hard, and brittle than the material immediately beneath it.

cultivars (cultivated variety) A variety of a plant produced through selective breeding and improved specifically for agricultural or horticultural purposes.

cultivation To mechanically loosen or break up soil, uproot weeds, and aerate the soil between rows of growing crops. Soil around crops is generally cultivated one to three times per season, depending on soil type, weather, weed pressure, and herbicide use.

deforestation The conversion of forests to land uses that have a tree cover of less than 10 percent.

degradation Refers to changes within the biological, physical, and chemical processes of the forest that negatively affect the area or site and lower its productive capacity or potential (for example, soil erosion and loss of valuable or potentially valuable genetic types).

domestic species Plants or animals that have evolved either naturally or through artificial selection to forms more useful to people. These characteristics of domestication are frequently absent in wild types of the organism and may constitute a negative genetic load for survival in the wild state.

dooryard forest garden A garden around a dwelling with a tree overstory and animals below.

ecosystem The complex of an ecological community, together with the nonliving components of the environment, that function together as a stable system and in which exchange of material follows a circular path.

ecotourism Environmentally oriented recreational travel.

endemic Restricted or peculiar to a locality or region.

Entisols One of 10 soil orders. Soils of such recent development that they do not show a significant degree of horizon differentiation. This order includes Fluvents (well-drained young alluvial soils), Psamments (acid infertile, deep sands), and Lithosols (shallow soils of steep regions or near rock outcrops).

epiphytes A plant that derives its moisture and nutrients from air and rain. It usually grows on another plant.

erosion (soil) The removal or loss of rock or soil by water, wind, biotic factors, or human interference.

ethnobotany The study of the folklore and history of plant use.

evapotranspiration Loss of water, usually from the soil, both directly by changes into vapor or invisible minute particles and by transpiration from plants growing on the soils or in water.

even-aged forest   Forest that is managed to produce trees of the same age class for commercial use.

ex situ   In a place other than the original location.

extended-fallow swidden system   A food-crop production system that involves partial clearing of vegetation followed by flash burning and an extended fallow period sufficiently long (10 to 20 years) to allow for soil regeneration and weed suppression. *See also* fallow, shifting cultivation, swidden cultivation.

extension   Agricultural activities that involve dissemination of agricultural materials, technologies, and information (for example, varieties, chemical inputs, dates of farm operation, special training) to a relatively large number of farmers or associated agricultural workers or agents.

extensive agriculture   A method of farming using large areas and minimum inputs to raise livestock or crops.

extractive reserve   Forest areas for which use rights are granted by governments to residents whose livelihoods customarily depend on extracting forest products from the specified area.

fallow   The period during which land is left to recover its productivity (reduced by cropping) mainly through accumulation of water, nutrients, attrition of pathogens, or a combination of all three. During this period, the land may be bare or covered by natural or planted vegetation. The term may be applied to the land itself or to the crop growing on it.

fodder   Dried or cured plant material of crops, such as maize and sorghum, grown and processed for animal feed.

forage   Unharvested plant material available as food for domestic animals. It may be grazed or cut for hay, in which case it is termed feed.

forest conversion   The alteration of forest cover and forest conditions through human intervention, ranging from marginal modification to fundamental transformation.

forest regeneration   The process of a forest regrowing, without human intervention, as a result of both natural seed dispersal from adjacent undisturbed forest and stump sprouting.

forest reserve   An area of forest that is protected by laws against excessive tree cutting and burning, enabling protection of ecosystem functions, environmental services, cultural values, and biological diversity, and providing opportunities for research, education, recreation, and tourism.

fragmentation The breaking up of extensive landscape features into disjunct, isolated, or semi-isolated patches as a result of land use changes.

frugivorous Fruit-eating.

fuelwood Wood used as fuel for cooking, heating, or producing power; includes wood for charcoal, kilns, and ovens.

gallery forests A forest growing among a watercourse in a region otherwise devoid of trees.

germplasm The genetic material that forms the physical basis of heredity and is transmitted from one generation to the next by means of the germ cells. Also, an individual or clone representing a type, species, or culture that may be held in a repository for agronomic, historic, or other reasons.

girdling To cut the bark and cambium in a ring around a tree, which kills it by interrupting the circulation of water and nutrients.

green revolution A term coined following the success of the International Rice Research Institute with rice and Centro Internacional de Mejoramiento de Maíz y Trigo with wheat when newly developed high-yielding varieties greatly increased crop production and changes occurred in research principles, management techniques, pesticide use, and other agroeconomic and sociopolitical aspects of food crop agriculture.

greenhouse effect Warming of the earth's surface and the lower layers of atmosphere that tends to increase with greater atmospheric carbon dioxide concentration. Solar radiation is coverted into heat in a process involving selective transmission of short-wave solar radiation by the atmosphere, its absorption by the earth's surface, and reradiation as infrared that is absorbed and partly reradiated back to the surface by carbon dioxide and water vapor in the air.

greenhouse gases Gases, including water vapor, carbon dioxide, methane, nitrous oxide, chlorofluorocarbons, and ozone, that insulate the earth, letting sunlight through to the earth's surface while trapping outgoing radiation.

gross domestic product (GDP) Identical to gross national product (GNP), but, unlike GNP, GDP includes both nonresidents who contributed to the domestic economy and payment of foreign debt. *See also* gross national product.

gross national product (GNP) The total market value of the final goods and services produced during a specific period of time (usually 1 year) by the residents of a country. *See also* gross domestic product.

gully erosion The erosion process whereby water accumulates in narrow channels and, over short periods, removes the soil from this narrow area to considerable depths, ranging from 1 to 90 m (3 to 300 ft).

hectare (ha) One hectare equals 2.47 acres. One square kilometer equals 100 hectares. One square mile equals 259 hectares. Thus, 1.2 billion hectares of closed tropical forest is equal to 3 billion acres or 4.6 million square miles.

hedgerow A row of shrubs or trees enclosing or separating fields.

herbaceous Vegetation that has little or no woody tissue.

home garden A cultivated and managed area, adjacent to or surrounding a house, in which mixtures of plant species are grown and livestock is kept.

humid tropics Those areas of the earth's land surface where the mean annual biotemperature in the lowlands is greater than 24°C (75°F) and where annual rainfall exceeds or equals potential evaporative return on water to the atmosphere. In general, the humid tropics correspond to tropical areas that originally supported broadleaf evergreen forests and the humid component of vegetation above timberline. As for lowlands, this definition includes all areas receiving a total annual rainfall in excess of 1,500 mm (60 in). These areas are frost free and usually have no more than 2 dry months (precipitation <100 mm [4 in] per month) per year.

hydrological systems/processes The system by which moisture reaches the ground and percolates through the soil to a particular watercourse or body of water.

hydromorphic soils A suborder of intrazonal soils, all formed under conditions of poor drainage in marshes, swamps, seepage areas, or flats.

in situ In the original location.

Inceptisols One of 10 soil orders. Young soils of sufficient age to show horizon layers. Three major types are in the humid tropics: Aquepts (poorly drained), Andepts (well drained, volcanic origin), and Tropepts (well drained, nonvolcanic origin). They are of moderate to high fertility and support dense human populations.

indigenous Native to a specified area or region; not introduced.

infiltration rate The rate at which water enters the soil, or other porous material, in a given condition.

inputs Items purchased to carry out a farm's operation. Such items include fertilizers, pesticides, seed, fuel, and animal feeds and drugs.

integrated pest management An ecologically based strategy that relies on natural mortality factors, such as natural enemies, weather, and crop management, and seeks control tactics that disrupt these factors as little as possible while enhancing their effectiveness.

intensification The fuller use of land, water, and biotic resources to enhance agronomic performance.

intensive Use of multiple cropping techniques, usually with significant nutrient inputs, to achieve high levels of crop productivity and high use of available water and sunlight throughout the year.

intercropping The growing of more than one crop species on the same plot of ground, where the respective growing periods overlap for most of the crops' life cycles.

kitchen garden *See* home garden.

land tenure The right to exclusively occupy and use a specified area of land.

landrace An early, cultivated form of a crop species evolved from a wild population.

landscape The combination of soil type, slope, rivers, streams, ponds, and other topographical features and the extent of uniform areas that determine appropriate land use systems and their patterns. A landscape generally has no fixed size or boundary. It is used ecologically to designate an area of intensive biological interaction. It also can be synonymous with watershed, political township, or community.

landscape design The selection and use of agricultural and forestry options that protect and use a landscape in a manner compatible with the social and economic environment.

leaching The removal of useful chemicals or other materials in solution from the soil through water percolation.

leguminous Of or relating to, or consisting of, a large family (Leguminosae) of dicotyledonous herbs, shrubs, and trees having fruits that are legumes or loments (peas, bans, clovers), bearing nodules on the roots that contain nitrogen-fixing bacteria.

life zone Large portions of the earth's land area that have generally uniform climate and soil, and, consequently, a biota showing a high degree of uniformity in species composition and environmental adaptation; related terms are vegetational formation and biome. Holdridge defines life zones through the effects of three weighted climatic indexes: mean annual heat, precipitation, and atmospheric moisture.

littoral savannah A savannah situated near the sea.

lowlands Fertile low or level ground.

manioc *See* cassava.

marginal land Land that is relatively infertile or unproductive for agriculture without extraordinary capital inputs (such as irrigation, fertilizers).

microclimate The immediate environmental conditions surrounding an individual organism, as in a crop canopy, for example.

milpas A small field cleared from the jungle, cropped for a few seasons, and then abandoned for a fresh clearing.

mixed cropping system Two or more crops grown without distinct row divisions.

mixed tree plantation A plantation on which a mixture of perennial and annual tree crops are cultivated and harvested.

modified forest An ecosystem that has been managed in subtle but sophisticated ways to provide the human inhabitants with sustainable livelihoods.

monocropping (monoculture) The growing of a single plant species in one area, usually the same type of crop grown year after year.

montane In the context of this report, of, being, or related to the biogeographic zone made up of relatively moist cool upland slopes below the timberline and dominated by tropical evergreen trees and plants.

mulch Any material such as straw, sawdust, leaves, plastic film, and loose soil that is spread on the surface of the soil to protect the soil and plant roots from the effects of raindrops, soil crusting, freezing, evaporation, and other stresses.

multipurpose tree A tree that has several uses (food production, shade, erosion control) and from which a number of products can be gleaned (food, fuel, lumber).

mycorrhiza The symbiotic association of the mycelium of a fungus with the roots of a seed plant.

natural forest management system Controlled and regulated harvesting of forest trees, combined with silvicultural and protective measures, to sustain and increase the commercial value of subsequent stands; relies on natural regeneration of native species.

nitrogen fixation The conversion of atmospheric nitrogen gas to ammonia, nitrates, and other nitrogen-containing compounds, by nitrogen-fixing bacteria, photosynthetic bacteria, and blue-green algae. The nitrogen-fixing bacteria include clostridium and azotobacter (which are free-living and are believed to contribute minimally to soil nitrogen) and rhizobium (which lives symbiotically in root nodules). Atmospheric nitrogen fixation can be caused by lightning.

nitrogen-fixing trees Trees that are capable of converting free nitrogen into combined forms useful especially as starting materials for fertilizers.

nongovernmental organization A private organization that may be international or indigenous, community-based, or nationally associated, and that consists of rural farmers as well as technical and financial support intermediaries who network for information dissemination and for cross-cultural exchange.

nutrient cycling The process of retaining and efficiently recycling essential nutrients and micronutrients within the ecosystem.

nutrient depletion The detrimental removal of nutritional elements from the soil.

nutrient recycling *See* nutrient cycling.

off-farm resources External support systems or components that are not available on the farm including artificial fertilizers, pesticides, and irrigation sources or systems, as well as markets, labor, machinery, and funding.

on-farm resources Internal support systems or components that are available on the farm including sunlight, natural fertilizers, seeds, biological processes, irrigation sources or systems, labor, and knowledge.

organic matter Living biota present in the soil or the decaying or decayed remains of animals or plants. The living organic matter in the soil decomposes the dead organic matter. Organic matter in soil can reduce soil erosion and increase moisture and soluble nutrient retention, cation exchange, and water infiltration.

oxidation A chemical reaction that increases the oxygen content of a compound; a chemical reaction in which a compound or radical loses electrons, that is, in which the positive valence is increased.

Oxisols One of 10 soil orders. Generally deep, well-drained red or yellowish soils with excellent granular structure and little contrast between horizon layers. Due to poor chemical properties, however, these soils are low in available nutrients and acidic.

pastureland Land where grass or other plants are grown for use as food by grazing animals.

pathogen An organism (usually parasitic) capable of causing a disease in another organism (host).

perennial A plant that lives for more than 2 years, often for a number of years; many flower annually.

perverse policies National economic and land use policies that promote the inefficient and nonsustainable conversion of forests to other uses by measures such as tax incentives and credits, subsidized credit, timber pricing procedures, land subsidies and rents, concessions, tenure, and property rights.

pest Any form of plant or animal life or any pathogenic agent that is injurious or potentially injurious to plants, animals, or their products.

pH A quantitative expression of the degree of acidity or alkalinity of a solution.

photosynthesis The synthesis, by chlorophyll-containing plant or bacterial cells, of organic compounds (primarily carbohydrates) from carbon dioxide and a hydrogen source such as water. There is a simultaneous liberation of oxygen. The energy for the reaction is light energy in the form of photons.

physicochemical Physical and chemical in nature.

pioneer species A plant or animal capable of establishing itself in a bare or barren area and initiating an ecological cycle.

plantation A forest crop or stand established artificially either by sowing or planting. The term includes reforestation (reestablishment of a tree cover on deforested or degraded forestlands) and replacement of natural forest by a different tree crop. It does not include artificial regeneration (the application of postharvesting techniques to accelerate the regrowth of species that had been logged).

pollarding Cutting back of a tree to the trunk to promote the growth of a dense head of foliage.

polyculture The growing of more than one crop at once in the same field.

premontane A biogeographic zone that usually lies between 500 and 1,500 m (between 1,600 and 5,000 ft) in elevation and has a mean annual biotemperature between 18° and 24°C (between 34° and 75°F); this zone is excellent for coffee and tea culture as well as for other agricultural activities.

primary forest Relatively intact forest that has been essentially unmodified by human activity for the past 60 to 80 years.

primary productivity The accumulation of plant biomass as a direct result of photosynthesis and the fixing of atmospheric carbon.

private voluntary organization In the context of this report, a nongovernment organization, funded by private citizens and/or businesses, that sponsors projects and programs to study or enhance agricultural productivity.

privatization To alter the status of a business, industry, or land from public or government ownership or control to private ownership or control.

pruning To cut off or cut back parts of a plant for better shape or more fruitful growth.

pulse The edible seed of peas, beans, lentils, and related plants having pods.

reforestation The replacement of trees in cut-over forest areas.

resettlement policies Plans and programs that involve moving large numbers of people from heavily populated areas to less densely populated areas generally for the purposes of alleviating overcrowding and unemployment.

restoration The re-creation of entire communities of organisms closely modeled on communities that occur naturally. It is closely linked to reclamation.

rhizobium A genus of bacterium that is capable of forming a symbiotic relationship with plants of the Fabaceae (Leguminosae) and is found in root nodules, where the bacteria fix nitrogen in return for carbon from the host plant. *See* nitrogen fixation.

rill erosion An erosion process in which numerous small channels several inches deep are formed; occurs mainly on recently cultivated soils.

root crops A plant cultivated for its underground food-storing organ.

rotation The systematic growing of different kinds of crops in recurrent succession on the same piece of land.

ruminant Characterized by the act of regurgitation and rechewing of food. A mammal belonging to the suborder Ruminantia.

runoff The portion of precipitation that is discharged from an area through stream channels. That which is lost without entering the soil is called surface runoff, and that which enters the soil before reaching a stream is called groundwater runoff or seepage flow from groundwater.

salinization The process of accumulation of salts in soil.

savannah Tropical grassland containing scattered trees and drought-resistant undergrowth.

scale The relative size of an area. In this report, small scale usually refers to a farm under 1 to 2 ha in size.

secondary forest Natural forest growth after some major interference (for example, logging, serious fire, or insect attack).

sedimentation The action or process of forming or depositing sediment.

seed dispersers Natural means of sowing or distributing seeds.

sheet erosion The removal of a fairly uniform layer of soil from the land surface by runoff water.

shifting cultivation Any farming system where land is periodically cleared, cropped, and returned to fallow; synonymous with slash-and-burn or swidden agriculture.

siltation To choke, fill over, or obstruct with silt or mud.

silviculture The science and art of cultivating forest crops, based on a knowledge of forest tree characteristics.

silvopastoral system An agroforestry system that combines pastures (with or without animals) and trees.

sink Anything that can absorb and store carbon circulating in the atmosphere.

soil amendment Any substance such as lime, sulfur, gypsum, or sawdust used to alter the properties of a soil, generally to make it more productive. Fertilizers are soil amendments, but the term is used most commonly for materials other than fertilizers.

soil biota Organisms that live in the soil.

soil degradation Degeneration of the soil through erosion, nutrient depletion, and other degenerative processes.

spatial integration Interaction of land uses or agroecosystem components because of physical proximity as, for example, in strip cropping systems.

species A group of actually or potentially interbreeding natural populations that are reproductively isolated from other such groups. Species is the smallest of the commonly used units of classification and the easiest to recognize intuitively.

staple crop A crop that is used, enjoyed, or needed constantly by many people in a given area or country. It is provided or imported in large quantity into the area; examples are maize in Kenya and rice in Liberia.

subsistence farming Farming or a system of farming that provides all or almost all goods required by a farm family, usually without any significant surplus for sale.

subtropics The region bordering the tropical zone.

sustainable An agricultural production system in which the farmer increases or maintains productivity at levels that are economically viable, ecologically sound, and culturally acceptable, through the efficient management of resources with minimum damage to the environment or human health.

swidden A temporary agricultural plot produced by cutting back and burning off vegetative cover.

swidden cultivation A traditional food-crop production system that involves partial clearing of vegetation (forest or bush fallow) followed by flash burning and short-term mixed intercropping; synonymous with shifting cultivation or slash-and-burn agriculture. The fallow period must be sufficiently long to allow for soil regeneration and weed suppression. This system is based solely on the restorative properties of woody species.

symbionts Two dissimilar organisms that each benefit from the presence of the other, for example, rhizobium and the members of the Fabaceae (Leguminosae).

temperate zone The area or region between the tropic of Cancer and the arctic circle or between the tropic of Capricorn and the antarctic circle.

temporal integration Interaction of land uses or agroecosystem components over time, as, for example, in crop rotations where a previous crop affects those planted after it.

terracing The agricultural practice of using a raised strip of earth, more or less level or horizontal, usually constructed on or near a contour and designed to make the land suitable for tillage and to prevent accelerated erosion.

tillage The act of preparing the soil by mechanical manipulations for crop production.

trophic level One of the hierarchical strata of a food web characterized by organisms that are the same number of steps removed from the primary producers.

tropic Either of the two small circles of the celestial sphere on each side of and parallel to the equator at a distance of 23.5 degrees, which the sun reaches at its greatest declination north or south.

tropical moist forest Lowland, premontane, and montane tropical forest formations.

Ultisols One of 10 soil orders. Soils that are similar to Oxisols but exhibit a marked increase of clay content with depth. They are usually deep, well-drained red or yellowish soils somewhat higher in weatherable minerals than Oxisols but still acidic and low in fertility.

understory Vegetation growing in the shade of taller plants.

uplands Infertile sloping or hilly lands.

usufruct The legal right of using and enjoying the fruits or profits of something belonging to another.

volunteer species A plant that has grown from self-sown seed.

watershed A region or area draining ultimately to a particular watercourse or body of water.

weed An individual plant or species that grows where humans do not wish it to grow.

wetlands Land or areas (as tidal flats and swamps) containing much soil moisture.

wild species Species of flora or fauna that have not been domesticated or manipulated by humans.

woodlot A restricted area of woodland, usually privately maintained as a source of fuel, posts, and lumber.

woody shrub Shrubs rich in xylem and associated structures.

yield The weight or volume of the economic part of a plant harvested during plant growth or at maturity.

Authors

MARY E. CARTER is special assistant to the assistant secretary for science and education, U.S. Department of Agriculture (USDA). She received her Ph.D. degree in natural polymers and products from the University of Edinburg, Scotland. Her industry research was in natural and synthetic polymers and fibers and her USDA experience as associate administrator of the Agricultural Research Service encompassed management of a broad research program from soil and water, plants, and animals to human nutrition.

RODRIGO GÁMEZ is director general of the National Biodiversity Institute (Instituto Nacional de Biodiversidad) of Costa Rica. His research experience includes plant virology with an emphasis in virus identification and characterization and insect transmission of plant viruses. He received his Ph.D. degree in plant pathology from the University of Illinois in 1967.

STEPHEN GLIESSMAN Since 1982 Gliessman has been director of the agroecology program at the University of California at Santa Cruz, where he is also professor of environmental studies. He has a Ph.D. degree in plant ecology from the University of California at Santa Barbara, and has lived and worked extensively in Latin America. His primary areas of research are agroecology, the ecology of traditional agroecosystems, and the analysis of indicators of agricultural sustainability.

ARTURO GÓMEZ-POMPA is director of the University of California Consortium on Mexico and the United States and professor of botany at the University of California, Riverside. He received his D.Sc. degree in biology from the National Autonomous University of Mexico in 1966. He is adviser on tropical ecology to Mexican President Carlos Salinas de Gortari and consultant to the Mexican undersecretary of forestry for development of the Tropical Forest Action Plan for Mexico.

LOWELL HARDIN is emeritus professor of agricultural economics and assistant director for international programs at Purdue University. He has a Ph.D. degree in agricultural economics from Cornell University. His areas of expertise include agricultural and economic development.

RICHARD HARWOOD (*Chair*) holds the C.S. Mott Foundation Chair of Sustainable Agriculture at Michigan State University, where he received his Ph.D. degree in plant breeding. His areas of expertise include small farm agricultural systems in the humid tropics and integration methods for agronomic cropping systems.

WALTER A. HILL Since 1987 Hill has been Tuskegee University's dean of the School of Agriculture and Home Economics and director of the George Washington Carver agricultural experiment station. He received his Ph.D. degree in agronomy from the University of Illinois in 1978. His research has focused on plant and environment relationships.

RATTAN LAL In 1981 Lal left the International Institute of Tropical Agriculture, Ibadan, Nigeria, to become a professor at Ohio State University in the Department of Agronomy. In 1968 he received his Ph.D. degree in agronomy (soil physics) from the Ohio State University. His research interests include soil erosion and its control, soil structure and management, soil compaction and drainage, emission of radiatively active gases from soil-related processes, ecological impact of tropical deforestation, viable alternatives to shifting cultivation, and sustainable management of soil and water resources.

GILBERT LEVINE is emeritus professor of agricultural engineering at Cornell University, where he received his Ph.D. degree. He directed the university's Center for Environmental Research and currently is a senior associate of the International Irrigation Management Institute.

ARIEL E. LUGO In 1980 Lugo became director and project leader of the Institute of Tropical Forestry, Forest Service, U.S. Department of Agriculture, in Puerto Rico. He received his Ph.D. degree in ecology in 1969 from the University of North Carolina at Chapel Hill.

ALISON G. POWER is associate professor in the Section of Ecology and Systematics and in the Department of Science Technology Studies at Cornell University. She received her Ph.D. degree in ecology from the University of Washington. Her research interests include agricultural ecology, integrated pest management, environmental impacts of agricultural technology, and tropical agriculture.

VERNON W. RUTTAN is regents professor in the Department of Agricultural and Applied Economics at the University of Minnesota. He has a Ph.D. degree in economics from the University of Chicago. His major interests are within the areas of research policy and agricultural economics and development.

PEDRO A. SANCHEZ is director general of the International Center for Research in Agroforestry, Nairobi, Kenya. During most of the duration of this study, he was professor of soil science and coordinator of the Tropical Soils Program, North Carolina State University. He has a Ph.D. degree in soil science from Cornell University. His primary areas of interest are the mitigation of tropical deforestation, land depletion, and rural poverty through improved agroforestry systems.

E. ADILSON SERRÃO is an agronomist with the Center for Agroforestry Research of the Eastern Amazon (Centro de Pesquisa Agroflorestal da Amazônia Oriental) at the Brazilian Enterprise for Agricultural Research (Empresa Brasileira de Pesquisa Agropecuária), Belém, Brazil. He received his Ph.D. degree in agronomy from the University of Florida in 1976. His major area of research is pasture and animal production in the context of sustainable agricultural development in the humid tropics, with an emphasis on the Brazilian Amazon.

PATRICIA C. WRIGHT is an anthropology professor at the State University of New York at Stony Brook and is director of the Ranomafana National Park Project in Madagascar, where she has coordinated an integrated program of sustained rural development for long-term conservation of a rain forest. She received her Ph.D. degree at the City University of New York. She also devotes her efforts to studying lemurs in their natural habitat and to preserving biodiversity.

Index

RECENT PUBLICATIONS OF THE BOARD ON AGRICULTURE

Policy and Resources

Agriculture and the Undergraduate: Proceedings (1992), 296 pp., ISBN 0-309-04682-3.

Water Transfers in the West: Efficiency, Equity, and the Environment (1992), 320 pp., ISBN 0-309-04528-2.

Managing Global Genetic Resources: Forest Trees (1991), 244 pp., ISBN 0-309-04034-5.

Managing Global Genetic Resources: The U.S. National Plant Germplasm System (1991), 198 pp., ISBN 0-309-04390-5.

Sustainable Agriculture Research and Education in the Field: A Proceedings (1991), 448 pp., ISBN 0-309-04578-9.

Toward Sustainability: A Plan for Collaborative Research on Agriculture and Natural Resource Management (1991), 164 pp., ISBN 0-309-04540-1.

Investing in Research: A Proposal to Strengthen the Agricultural, Food, and Environmental System (1989), 156 pp., ISBN 0-309-04127-9.

Alternative Agriculture (1989), 464 pp., ISBN 0-309-03987-8; ISBN 0-309-03985-1 (pbk).

Understanding Agriculture: New Directions for Education (1988), 80 pp., ISBN 0-309-03936-3.

Designing Foods: Animal Product Options in the Marketplace (1988), 394 pp., ISBN 0-309-03798-0; ISBN 0-309-03795-6 (pbk).

Agricultural Biotechnology: Strategies for National Competitiveness (1987), 224 pp., ISBN 0-309-03745-X.

Regulating Pesticides in Food: The Delaney Paradox (1987), 288 pp., ISBN 0-309-03746-8.

Pesticide Resistance: Strategies and Tactics for Management (1986), 480 pp., ISBN 0-309-03627-5.

Pesticides and Groundwater Quality: Issues and Problems in Four States (1986), 136 pp., ISBN 0-309-03676-3.

Soil Conservation: Assessing the National Resources Inventory, Volume 1 (1986), 134 pp., ISBN 0-309-03649-9; Volume 2 (1986), 314 pp., ISBN 0-309-03675-5.

New Directions for Biosciences Research in Agriculture: High-Reward Opportunities (1985), 122 pp., ISBN 0-309-03542-2.

Genetic Engineering of Plants: Agricultural Research Opportunities and Policy Concerns (1984), 96 pp., ISBN 0-309-03434-5.

Nutrient Requirements of Domestic Animals Series and Related Titles

Nutrient Requirements of Horses, Fifth Revised Edition (1989), 128 pp., ISBN 0-309-03989-4; diskette included.

Nutrient Requirements of Dairy Cattle, Sixth Revised Edition, Update 1989 (1989), 168 pp., ISBN 0-309-03826-X; diskette included.

Nutrient Requirements of Swine, Ninth Revised Edition (1988), 96 pp., ISBN 0-309-03779-4.

Vitamin Tolerance of Animals (1987), 105 pp., ISBN 0-309-03728-X.

Predicting Feed Intake of Food-Producing Animals (1986), 95 pp., ISBN 0-309-03695-X.

Nutrient Requirements of Cats, Revised Edition (1986), 87 pp., ISBN 0-309-03682-8.

Nutrient Requirements of Dogs, Revised Edition (1985), 79 pp., ISBN 0-309-03496-5.

Nutrient Requirements of Sheep, Sixth Revised Edition (1985), 106 pp., ISBN 0-309-03596-1.

Nutrient Requirements of Beef Cattle, Sixth Revised Edition (1984), 90 pp., ISBN 0-309-03447-7.

Nutrient Requirements of Poultry, Eighth Revised Edition (1984), 71 pp., ISBN 0-309-03486-8.

More information, other titles (before 1984), and prices are available from the National Academy Press, 2101 Constitution Avenue NW, Washington, DC 20418 USA, 202/334-3313 (information only); 800/624-6242 (orders only); 202/334-2451 (fax).

RECENT PUBLICATIONS OF THE BOARD ON SCIENCE AND TECHNOLOGY FOR INTERNATIONAL DEVELOPMENT

Energy

Alcohol Fuels: Options for Developing Countries (1983), 128 pp., ISBN 0-309-04160-0.

Producer Gas: Another Fuel for Motor Transport (1983), 112 pp., ISBN 0-309-04161-9.

The Diffusion of Biomass Energy Technologies in Developing Countries (1984), 120 pp., ISBN 0-309-04253-4.

Technology Options

Priorities in Biotechnology Research for International Development: Proceedings of a Workshop (1982), 261 pp., ISBN 0-309-04256-9.

Fisheries Technologies for Developing Countries (1987), 167 pp., ISBN 0-309-04260-7.

Applications of Biotechnology to Traditional Fermented Foods (1992), 199 pp., ISBN 0-309-04685-8.

Plants

Amaranth: Modern Prospects for an Ancient Crop (1983), 81 pp., ISBN 0-309-04171-6.

Jojoba: New Crop for Arid Lands (1985), 102 pp., ISBN 0-309-04251-8.

Quality-Protein Maize (1988), 130 pp., ISBN 0-309-04262-3.

Triticale: A Promising Addition to the World's Cereal Grains (1988), 105 pp., ISBN 0-309-04263-1.

Lost Crops of the Incas (1989), 415 pp., ISBN 0-309-04264-X.

Saline Agriculture: Salt-Tolerant Plants for Developing Countries (1989), 150 pp., ISBN 0-309-04266-6.

Innovations in Tropical Forestry

Mangium and Other Fast-Growing Acacias for the Humid Tropics (1983), 63 pp., ISBN 0-309-04165-1.

Calliandra: A Versatile Small Tree for the Humid Tropics (1983), 56 pp., ISBN 0-309-04166-X.

Casuarinas: Nitrogen-Fixing Trees for Adverse Sites (1983), 118 pp., ISBN 0-309-04167-8.

Leucaena: Promising Forage and Tree Crop for the Tropics (1984), 2d ed., 100 pp., ISBN 0-309-04250-X.

Neem: A Tree that Could Help the World (1992), 149 pp., ISBN 0-309-04686-6.

Managing Tropical Animal Resources

Butterfly Farming in Papua New Guinea (1983), 36 pp., ISBN 0-309-04168-6.

Crocodiles as a Resource for the Tropics (1983), 60 pp., ISBN 0-309-04169-4.

Little-Known Asian Animals with a Promising Economic Future (1983), 133 pp., ISBN 0-309-04170-8.

Microlivestock: Little-Known Small Animals with a Promising Economic Future (1990), 449 pp., ISBN 0-309-04265-8.

Resource Management

Environmental Change in the West African Sahel (1984), 96 pp., ISBN 0-309-04173-2.

Agroforestry in the West African Sahel (1984), 86 pp., ISBN 0-309-04174-0.

Conserving Biodiversity: A Research Agenda for Development Agencies (1992), 127 pp., ISBN 0-309-04683-1.

Additional titles and ordering information are available from the Board on Science and Technology for International Development, Publications and Information Services (FO-2060), Office of International Affairs, National Research Council, 2101 Constitution Avenue NW, Washington, DC 20418 USA, 202/334-2688.